An Introduction to Ansys® Fluent® 2022

John E. Matsson, Ph.D., P.E.

SDC
PUBLICATIONS

SDC Publications
P.O. Box 1334
Mission, KS 66222
913-262-2664
www.SDCpublications.com
Publisher: Stephen Schroff

ISBN-13: 978-1-63057-569-4
ISBN-10: 1-63057-569-0

Printed and bound in the United States of America.

Acknowledgements

I would like to thank Stephen Schroff of SDC Publications for his help in preparing this book for publication.

About the Author

Dr. John Matsson is a Professor of Engineering and Chair of the Engineering Department at Oral Roberts University in Tulsa, Oklahoma. He earned M.S. and Ph.D. degrees from the Royal Institute of Technology in Stockholm, Sweden in 1988 and 1994, respectively and completed postdoctoral work at the Norwegian University of Science and Technology in Trondheim, Norway. His teaching areas include Finite Element Methods, Fluid Mechanics, Manufacturing Processes, and Aerodynamics. He is a member of the American Society of Mechanical Engineers ASME Mid-Continent Section. Please contact the author jmatsson@oru.edu with any comments, questions, or suggestions on this book.

Notes:

TABLE OF CONTENTS

TABLE OF CONTENTS

TABLE OF CONTENTS

TABLE OF CONTENTS

TABLE OF CONTENTS

TABLE OF CONTENTS

TABLE OF CONTENTS

TABLE OF CONTENTS

TABLE OF CONTENTS

TABLE OF CONTENTS

TABLE OF CONTENTS

Notes:

14

CHAPTER 1. INTRODUCTION

A. Ansys Workbench

Ansys Workbench is an integrated simulation platform that includes a wide range of systems in the toolbox including analysis systems, component systems, custom systems, and design exploration, see Table 1.1.

Analysis Systems	Component Systems	Custom Systems	Design Exploration
Design Assessment	ACP (Post)	FSI: Fluid Flow (CFX) > Static Structural	Direct Optimization
Eigenvalue Buckling	ACP (Pre)	FSI: Fluid Flow (FLUENT) > Static Structural	Parameters Correlation
Electric	Autodyn	Pre-Stress Modal	Response Surface
Explicit Dynamics	BladeGen	Random Vibration	Response Surface Optimization
Fluid Flow (CFX)	CFX	Response Spectrum	ROM Builder
Fluid Flow (Fluent)	Engineering Data	Thermal-Stress	Six Sigma Analysis
Harmonic Acoustics	External Data		
Harmonic Response	External Model		
IC Engine (Fluent)	Fluent		
IC Engine (Forte)	Fluent (with Fluent Meshing)		
Magnetostatic	Forte		
Modal	Geometry		
Modal Acoustics	Mechanical APDL		
Random Vibration	Mechanical Model		
Response Spectrum	Mesh		
Rigid Dynamics	Microsoft Office Excel		
Static Acoustics	Performance Map		
Static Structural	Results		
Steady-State Thermal	System Coupling		
Thermal-Electric	Turbo Setup		
Topology Optimization	Turbo Grid		
Transient Structural	Vista AFD		
Transient Thermal	Vista CCD		
TurbomachineryFluid Flow	Vista CPD		
	Vista RTD		

Table 1.1 Available systems in Ansys Workbench Toolbox

Ansys Workbench also has a Project Schematic where the details for the process of execution of the different analysis systems are available. These details of each analysis system in the project schematic include Engineering Data, Geometry, Model, Setup, Solution and Results. Furthermore, properties of schematics are available for user input.

B. Ansys DesignModeler

The geometry for the Fluid Flow (Fluent) flow study project can be created either in DesignModeler or alternatively using SpaceClaim. DesignModeler is a parametric modeler designed to draw 2D sketches and model 3D CAD parts. Ansys DesignModeler has nine different toolbars as listed in Table 1.2.

Toolbars
Active Plane/Sketch toolbar
Display toolbar
Feature toolbar
File toolbar
Graphics Options toolbar
Menus toolbar
Rotation Modes toolbar
Selection toolbar
Undo/Redo toolbar

Table 1.2 Toolbars in Ansys DesignModeler

Moreover, the Sketching Toolboxes in DesignModeler include Draw, Modify, Dimensions, Constraints and Settings as shown in Table 1.3.

Constraints	Dimensions	Draw	Modify	Settings
Auto Constraints	Angle	Arc by 3 Points	Chamfer	Grid
Coincident	Animate	Arc by Center	Copy	Major Grid Spacing
Concentric	Diameter	Arc by Tangent	Corner	Minor-Steps per Major
Equal Distance	Display	Circle	Cut	Snaps per Minor
Equal Length	Edit	Circle by 3 Tangents	Drag	
Equal Radius	General	Construction Point	Duplicate	
Fixed	Horizontal	Construction Point at Intersection	Extend	
Horizontal	Length/Distance	Ellipse	Fillet	
Midpoint	Move	Line	Move	
Parallel	Radius	Line by 2 Tangents	Offset	
Perpendicular	Semi-Automatic	Oval	Paste	
Symmetry	Vertical	Polygon	Replicate	
Tangent		Polyline	Spline Edit	
Vertical		Rectangle	Split	
		Rectangle by 3 Points	Trim	
		Spline		
		Tangent Line		

Table 1.3 Sketching Toolboxes in Ansys DesignModeler

C. Ansys Meshing

Ansys Meshing produces the mesh using Mesh Methods and Mesh Controls as shown in Tables 1.4 and Table 1.5.

Hexahedral Meshing	Tetrahedral Meshing	Surface Meshing
Cut cell Cartesian	ANSYS CFX-Mesh	Default quad, quad/tri or tri
General Sweep	Patch conforming	Uniform quad or quad tri
Hex-dominant	Patch independent	
MultiZone		
Thin Sweep		

Table 1.4 Mesh methods in Ansys Meshing

Global Mesh Control	Local Mesh Control
Curvature-based refinement settings	Automatic Contact Detection
Element midside node settings	Body, face, edge curvature-based refinement
Inflation settings	Body, face, edge, vertex sizings
Physics preference settings	Body, face, edge, vertex sphere of influence
Pinch (defeaturing) settings	Body mesh method controls
Proximity-based refinement settings	Body of Influence
Quality settings	Contact Sizing
Relevance settings	Gap tool
Rigid-body behavior settings	Inflation controls
Smoothing settings	Mapped-face meshing controls
Transition/growth settings	Match mesh controls
	Pinch controls
	Solver-based refinement controls
	Virtual topologies

Table 1.5 Mesh controls in Ansys Meshing

D. Ansys Fluent

Ansys Fluent is a computational fluid dynamics software package that is written in the C language. Fluent has numerous capabilities for simulations as shown in Table 1.6.

Fluent Capabilities
2D Planar flows, Axisymmetric flows, Axisymmetric flows with swirl, 3D Flows
Acoustics
Cavitation flows
Chemical species mixing and reaction
Compressible flows, Incompressible flows
Forced heat transfer, Mixed convection heat transfer, Natural heat transfer
Free surface flows
Ideal gases, Real gases
Inviscid flows, Laminar flows, Turbulent flows
Lumped Parameter Models
Melting and Solidification
Multiphase flows
Newtonian flows, Non-Newtonian flows
Porous media
Steady flows, Time-dependent flows

Table 1.6 Some of the capabilities for Ansys Fluent

There are eleven add on modules available in Ansys as shown in Table 0.7.

Fluent Addon Modules
Adjoint Solver
Dual-Potential MSMD Battery Model
Fiber Model
Fuel Cell and Electrolysis Model
Macroscopic Particle Model
MHD Model
PEM Fuel Cell Model
Population Balance Model
Single-Potential Battery Model
SOFC Model with Unresolved Electrolyte
Reduced Order Model

Table 1.7 Add on modules for Ansys Fluent

CHAPTER 1. INTRODUCTION

The user interface of Fluent includes the Outline View and the Task Page. The Outline View includes Setup, Solution, Results and Parameters & Customization, see Table 1.8. The different categories and variables available in setup, solution and results are shown in Tables 1.9-1.12.

Setup	Solution	Results	Parameters & Customization
Boundary Conditions	Calculation Activities	Animations	Custom Field Functions
Cell Zone Conditions	Cell Registers	Graphics	Parameters
Dynamic Mesh	Controls	Plots	User Defined Functions
General	Initialization	Reports	User Defined Memory
Materials	Methods	Scene	User Defined Scalars
Models	Monitors	Surfaces	
Named Expressions	Report Definitions		
Reference Frames	Run Calculation		
Reference Values			

Table 1.8 Outline view options in Ansys Fluent

Setup Models
Acoustics (Ffowcs-Williams & Hawkings, Wave Equation)
Discrete Phase (Numerics, Parallel, Physical Models, Tracking, UDF)
Electric Potential (Potential Equation)
Energy (Energy Equation)
Heat Exchanger (Dual Cell Model, Macro Model Group, Ungrouped Macro Model)
Multiphase (Eulerian, Mixture, Volume of Fluid, Wet Stream)
Radiation (Discrete Ordinates DO, Discrete Transfer DTRM, P1, Rosseland, Surface to Surface S2S)
Solidification and Melting (Back Diffusion, Solidification/Melting)
Species (Composition PDF Transport, Non-Premixed Combustion, Partially Premixed Combustion, Premixed Combustion, Species Transport)
Structure (Linear Elasticity)
Viscous (Detached Eddy Simulation DES, Inviscid, k-epsilon 2 eqn, k-omega 3 eqn, Laminar, Reynolds Stress 5 eqn, Scale-Adaptive Simulation SAS, Spalart-Allmaras 1 eqn, Transition k-kl-omega 3 eqn, Transition SST 4 eqn)

Table 1.9 Models options in Ansys Fluent

Setup Reference Values
Area (m2)
Density (kg/m3)
Enthalpy (j/kg)
Length (m)
Pressure (pascal)
Ratio of Specific Heats
Temperature (k)
Velocity (m/s)
Viscosity (kg/m-s)

Table 1.10 Reference values in Ansys Fluent

Solution	
Calculation Activities	Monitors
Autosave	Convergence Conditions
Execute Commands	Report Files
Cell Register Operations	Report Plots
Solution Animation	Residual

Table 1.11 Calculation activities and monitors options in Ansys Fluent

Results			
Animations	**Graphics**	**Plots**	**Reports**
Scene Animation	Contours	FFT	Discrete Phase
Solution Animation Playback	Mesh	File	Fluxes
Sweep Surface	Vectors	Histogram	Forces
	Particle Tracks	Interpolated Data	Surface Integrals
	Pathlines	Profile Data	Volume Integrals
	Vectors	XY Plot	

Table 1.12 Animation, graphics, plots and reports options in Ansys Fluent

The different options that are available for the General Setup are Gravity, Mesh and Solver with the details as shown in Table 1.13.

General		
Gravity	**Mesh**	**Solver**
X (m/s2)	Check	2D Space (Axisymmetric, Axisymmetric Swirl, Planar)
Y (m/s2)	Display…	Time (Steady, Transient)
Z (m/s2)	Report Quality	Type (Density-Based, Pressure-Based)
	Scale…	Velocity Formulation (Absolute, Relative)
	Units…	

Table 1.13 Gravity, mesh and solver options in Ansys Fluent

The different Solution Methods are shown in Table 1.14.

Solution Methods	
Pressure-Velocity Coupling	**Pressure-Based Solver , Spatial Discretization**
Coupled	Gradient (Green-Gauss Cell Based, Green-Gauss Node Based, Least Squares Cell Based)
PISO	Pressure (Body Force Weighted, Linear, PRESTO!, Second Order, Standard)
SIMPLE	Momentum (First Order Upwind, Power Law, QUICK, Second Order Upwind, Third-Order MUSCL)
SIMPLEC	Energy (First Order Upwind, Power Law, QUICK, Second Order Upwind, Third-Order MUSCL)
	Density-Based Solver , Spatial Discretization
	Gradient (Green-Gauss Cell Based, Green-Gauss Node Based, Least Squares Cell Based)
	Flow (First Order Upwind, Second Order Upwind, Third-Order MUSCL)

Table 1.14 Solution methods options in Ansys Fluent

E. References

1. ANSYS Fluent Getting Started Guide, Release 19.1, April 2018, ANSYS, Inc.
2. ANSYS FLUENT Theory Guide, Release 15.0, January 2013, ANSYS, Inc.
3. ANSYS Fluent Tutorial Guide, Release 18.0, January 2017, ANSYS, Inc.
4. ANSYS FLUENT User's Guide, Release 15.0, November 2013, ANSYS, Inc.
5. ANSYS Meshing User's Guide, Release 15.0, November 2013, ANSYS, Inc.
6. ANSYS Workbench User's Guide, Release 12.1, November 2009, ANSYS, Inc.
7. DesignModeler User Guide, Release 14.5, October 2012, ANSYS, Inc.

F. Notes

CHAPTER 2. FLAT PLATE BOUNDARY LAYER

A. Objectives

- Creating Geometry in Ansys Workbench for Ansys Fluent
- Setting up Ansys Fluent for Laminar Steady 2D Planar Flow
- Setting up Mesh
- Selecting Boundary Conditions
- Running Calculations
- Using Plots to Visualize Resulting Flow Field
- Compare with Theoretical Solution using Mathematica Code

B. Problem Description

In this chapter, we will use Ansys Fluent to study the two-dimensional laminar flow on a horizontal flat plate. The size of the plate is considered being infinite in the spanwise direction and therefore the flow is 2D instead of 3D. The inlet velocity for the 1 m long plate is 5 m/s and we will be using air as the fluid for laminar simulations. We will determine the velocity profiles and plot the profiles. We will start by creating the geometry needed for the simulation.

velocity
profile

flat plate

C. Launching ANSYS Workbench and Selecting Fluent

1. Start by launching ANSYS Workbench. Double click on Fluid Flow (Fluent) that is located under Analysis Systems in Toolbox.

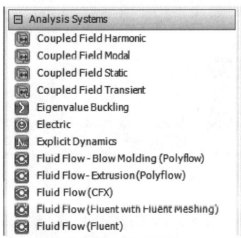

Figure 2.1 Selecting Fluid Flow

D. Launching ANSYS DesignModeler

2. Select Geometry under Project Schematic in ANSYS Workbench. Right-click on Geometry and select Properties. Select 2D Analysis Type under Advanced Geometry Options in Properties of Schematic A2: Geometry. Right-click on Geometry in Project Schematic and select to launch New DesignModeler Geometry. Select Units>>Millimeter as the length unit from the menu in DesignModeler.

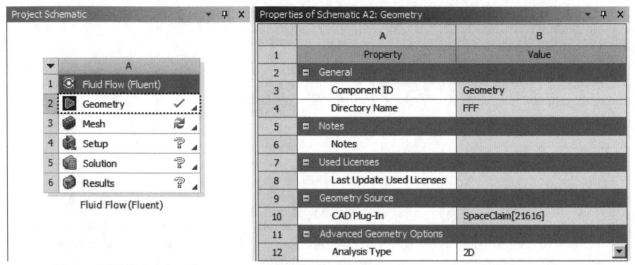

Figure 2.2a) Selecting Geometry and 2D Analysis Type

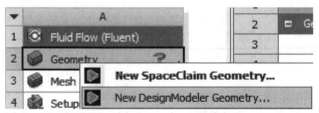

Figure 2.2b) Launching DesignModeler

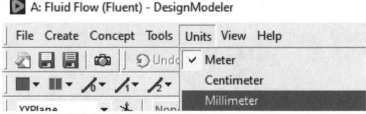

Figure 2.2c) Selecting the length unit

3. Next, we will be creating the geometry in DesignModeler. Select XYPlane from the Tree Outline on the left-hand side in DesignModeler. Select Look at Sketch. Click on the Sketching tab in the Tree Outline and select the Line sketch tool.

    Draw a horizontal line 1,000 mm long from the origin to the right. Make sure you have a P at the origin when you start drawing the line. Also, make sure you have an H along the line so that it is horizontal and a C at the end of the line.

Select Dimensions within the Sketching options. Click on the line and enter a length of 1,000 mm. Draw a vertical line upward 100 mm long starting at the end point of the first horizontal line. Make sure you have a *P* when starting the line and a *V* indicating a vertical line. Continue with a horizontal line 100 mm long to the left from the origin followed by another vertical line 100 mm long.

The next line will be horizontal with a length 100 mm starting at the endpoint of the former vertical line and directed to the right. Finally, close the rectangle with a 1,000 mm long horizontal line starting 100 mm above the origin and directed to the right.

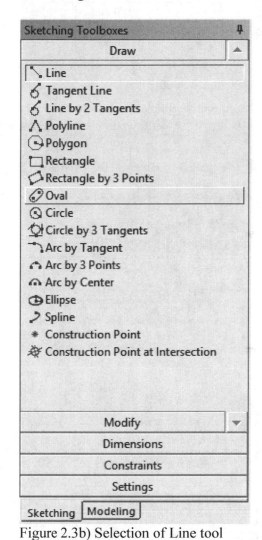

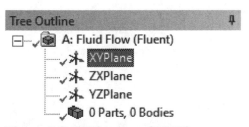

Figure 2.3a) Selection of XYPlane   Figure 2.3b) Selection of Line tool

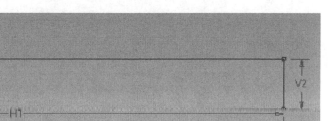

Figure 2.3c) Rectangle with dimensions

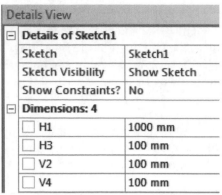

Figure 2.3d) Dimensions in Details View

4. Click on the Modeling tab under Sketching Toolboxes. Select Concept>>Surfaces from Sketches in the menu. Control select the six edges of the rectangle as Base Objects and select Apply in Details View. Click on Generate in the toolbar ⚡ Generate. The rectangle turns gray. Right click in the graphics window and select Zoom to Fit and close DesignModeler.

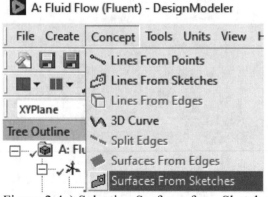

Figure 2.4a) Selecting Surfaces from Sketches

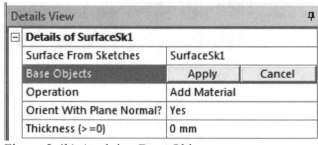

Figure 2.4b) Applying Base Objects

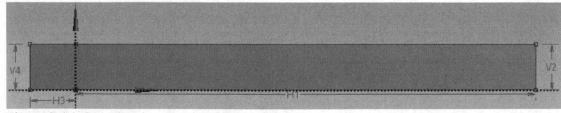

Figure 2.4c) Completed rectangle in DesignModeler

E. Launching ANSYS Meshing

5. We are now going to double click on Mesh under Project Schematic in ANSYS Workbench to open the Meshing window. Select Mesh in the Outline of the Meshing window. Right click and select Generate Mesh. A coarse mesh is created. Select Unit Systems>>Metric (mm, kg, N …) from the bottom of the graphics window.

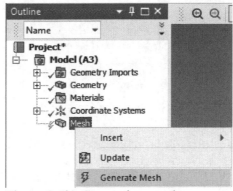

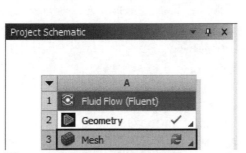

Figure 2.5a) Launching mesh

Figure 2.5b) Generating mesh

Figure 2.5c) Coarse mesh

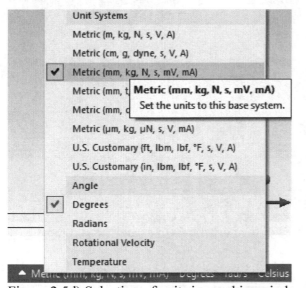

Figure 2.5d) Selection of units in graphics window

Select Mesh>> Controls>>Face Meshing from the menu. Click on the yellow region next to Geometry under Scope in Details of Face Meshing. Select the rectangle in the graphics window. Click on the Apply button for Geometry in Details of "Face Meshing". Select Mesh>>

Controls>>Sizing from the menu and select Edge above the graphics window.

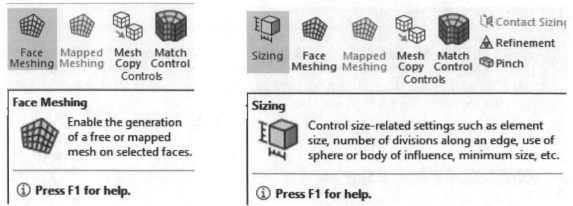

Figure 2.5e) Selection of face meshing   Figure 2.5f) Sizing

Select the upper longer horizontal edge of the rectangle, control select the lower horizontal edges and the vertical edges for a total of 5 edges. Click on Apply for the Geometry in "Details of Edge Sizing". Under Definition in "Details of Edge Sizing", select Element Size as Type, 1.0 mm for Element Size, and Hard as Behavior. Select the second Bias Type - - — —— and enter 12.0 as the Bias Factor.

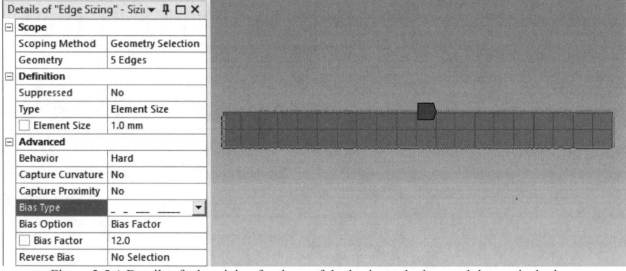

Figure 2.5g) Details of edge sizing for three of the horizontal edges and the vertical edges

Repeat the selection of Mesh>>Controls>>Sizing from the menu once again but this time select the upper horizontal edge to the left of the origin. Enter the same Element Size, Behavior, and Bias Factor but the first Bias Type —— — - - . Click on Home>>Generate Mesh in the menu and select Mesh in the Outline. The finished mesh is shown in the graphics window.

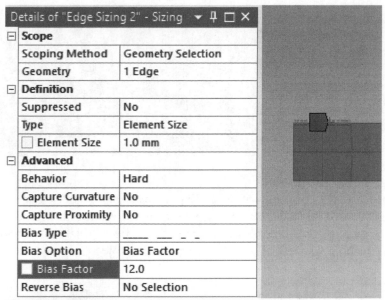

Details of "Edge Sizing 2" - Sizing	
Scope	
Scoping Method	Geometry Selection
Geometry	1 Edge
Definition	
Suppressed	No
Type	Element Size
Element Size	1.0 mm
Advanced	
Behavior	Hard
Capture Curvature	No
Capture Proximity	No
Bias Type	_____ ___ _ _
Bias Option	Bias Factor
Bias Factor	12.0
Reverse Bias	No Selection

Figure 2.5h) Details of edge sizing for the remaining horizontal edge

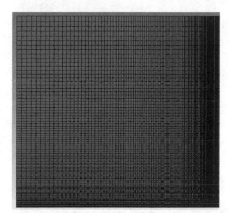

Figure 2.5i) Details of finished mesh

Why did we create a biased mesh? The reason for using a biased mesh is that we need a finer mesh close to the wall where we have velocity gradients in the flow. We also included a finer mesh where the boundary layer starts to develop on the flat plate.

We are now going to rename the edges for the rectangle. Select the left edge of the rectangle, right click and select Create Named Selection. Enter *inlet* as the name and click on the OK button.

Repeat this step for the right vertical edge of the rectangle and enter the name *outlet*. Create a named selection for the lower longer horizontal right edge and call it *wall*. Finally, control-select the remaining three horizontal edges and name them *ideal wall*. An ideal wall is an adiabatic and frictionless wall.

Figure 2.5j) Named selections

Select File>>Export...>>Mesh>>FLUENT Input File>>Export from the menu. Select Save as type: FLUENT Input Files (*.msh). Enter *boundary-layer-mesh* as file name and click on the Save button. Select File>>Save Project from the menu. Name the project *Flat Plate Boundary Layer*. Close the ANSYS Meshing window. Right click on Mesh in Project Schematic and select Update.

F. Launching ANSYS Fluent

6. We are now going to double click on Setup under Project Schematic in ANSYS Workbench to open Fluent. Launch the Dimension 2D and Double Precision solver of Fluent. Check Double Precision under Options. Set the number of Solver Processes equal to the number of computer cores. To check the number of physical cores, press the Ctrl + Shift + Esc keys simultaneously to open the Task Manager. Go to the Performance tab and select CPU from the left column. You'll see the number of physical cores on the bottom-right side. Close the Task Manager window. Click on Show More Options at the bottom of the Fluent Launcher. Write down the location of the Working Directory as you will use this information later. Click on the Start button to launch ANSYS Fluent. Click OK to close the Key Behavioral Changes window if it appears.

> **Why do we use double precision?** Double precision will give more accurate calculations than single precision.

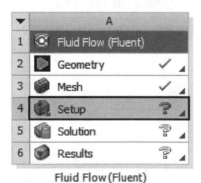

Fluid Flow (Fluent)

Figure 2.6a) Launching Setup

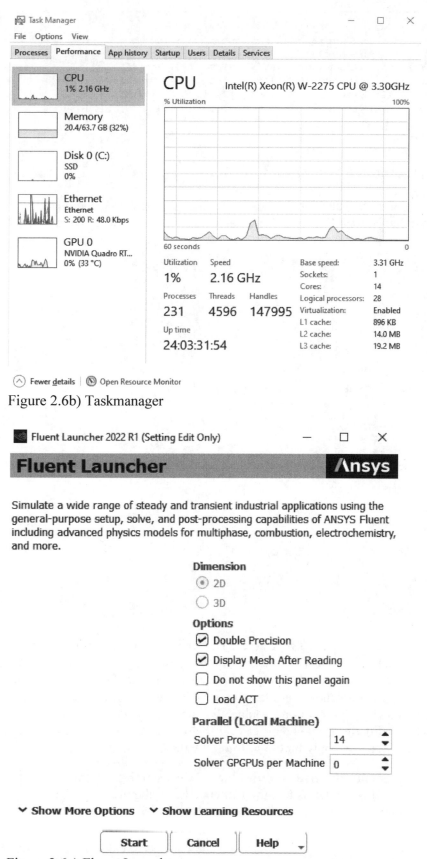

Figure 2.6b) Taskmanager

Figure 2.6c) Fluent Launcher

```
-------------------------------------------------------------------------
ID     Hostname    Core   O.S.         PID     Vendor
-------------------------------------------------------------------------
n3     ljmatsson5  4/12   Windows-x64  31460   Intel(R) Xeon(R) E-2186G
n2     ljmatsson5  3/12   Windows-x64  35540   Intel(R) Xeon(R) E-2186G
n1     ljmatsson5  2/12   Windows-x64  21792   Intel(R) Xeon(R) E-2186G
n0*    ljmatsson5  1/12   Windows-x64  33420   Intel(R) Xeon(R) E-2186G
host   ljmatsson5         Windows-x64  32060   Intel(R) Xeon(R) E-2186G

MPI Option Selected: intel
Selected system interconnect: default
```

Figure 2.6d) Example of console printout for four cores

7. Check the scale of the mesh by selecting the Scale button under Mesh in General on the Task Page. Make sure that the Domain Extent is correct and close the Scale Mesh window.

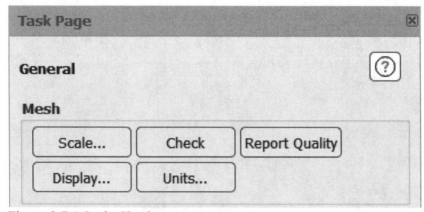

Figure 2.7a) Scale Check

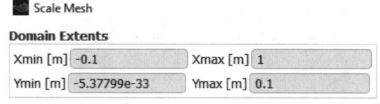

Figure 2.7b) Scale Mesh

8. Double click on Models and Viscous (SST k-omega) under Setup in the Outline View. Select Laminar as Viscous Model. Click OK to close the window. Double click on Boundary Conditions under Setup in the Outline View. Double click on *inlet* under Zone on the Task Page. Choose Components as Velocity Specification Method and set the X-Velocity [m/s] to 5. Click on the Apply button followed by the Close button. Double click on *ideal_wall* under Zones. Check Specified Shear as Shear Condition and keep zero values for specified shear stress since an ideal wall is frictionless. Click on the Apply button followed by the Close button.

> **Why did we select Laminar as Viscous Model?** For the chosen free stream velocity 5 m/s the Reynolds number is less than 500,000 along the plate and the flow is therefore laminar. Turbulent flow along a flat plate occurs at Reynolds numbers above 500,000.

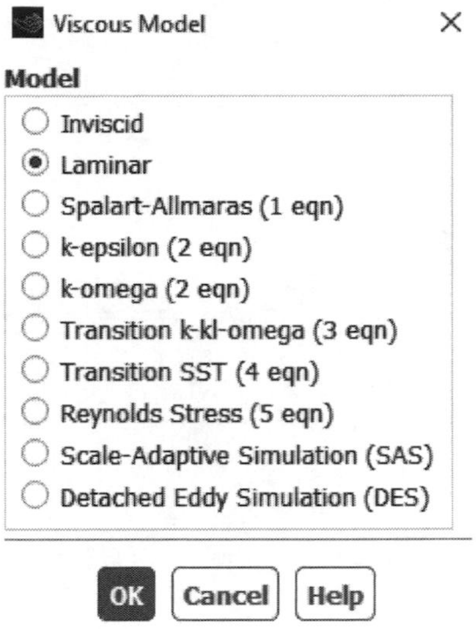

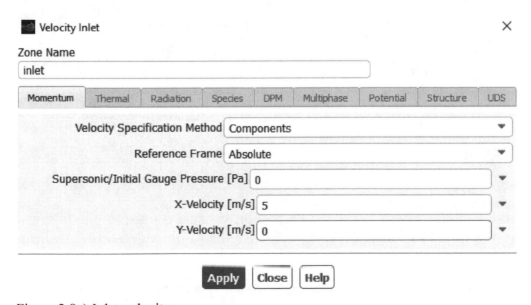

Figure 2.8a) Viscous model

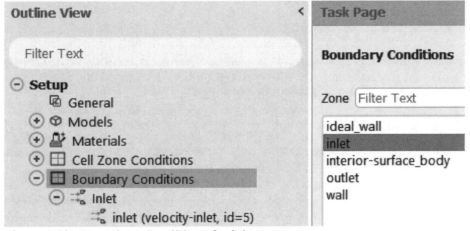

Figure 2.8b) Boundary Conditions for inlet

Figure 2.8c) Inlet velocity

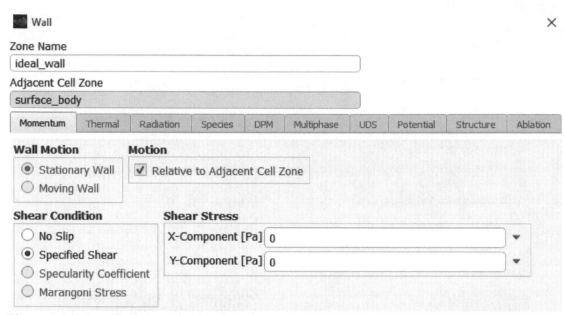

Figure 2.8d) Specified shear for ideal wall

9. Double click on Methods under Solution in the Outline View. Select *Standard* for Pressure and *First Order Upwind* for Momentum. Double click on Reference Values under Setup in the Outline View. Select *Compute from inlet* on the Task Page.

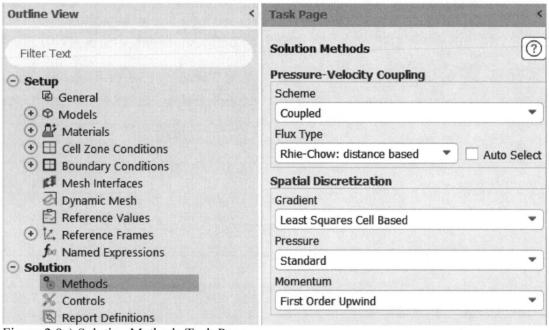

Figure 2.9a) Solution Methods Task Page

Why do we use First Order Upwind method for Spatial Discretization of Momentum?
The First Order Upwind method is generally less accurate but converges better than the Second Order Upwind method. It is common practice to start with the First Order Upwind method at the beginning of calculations and continue with the Second Order Upwind method.

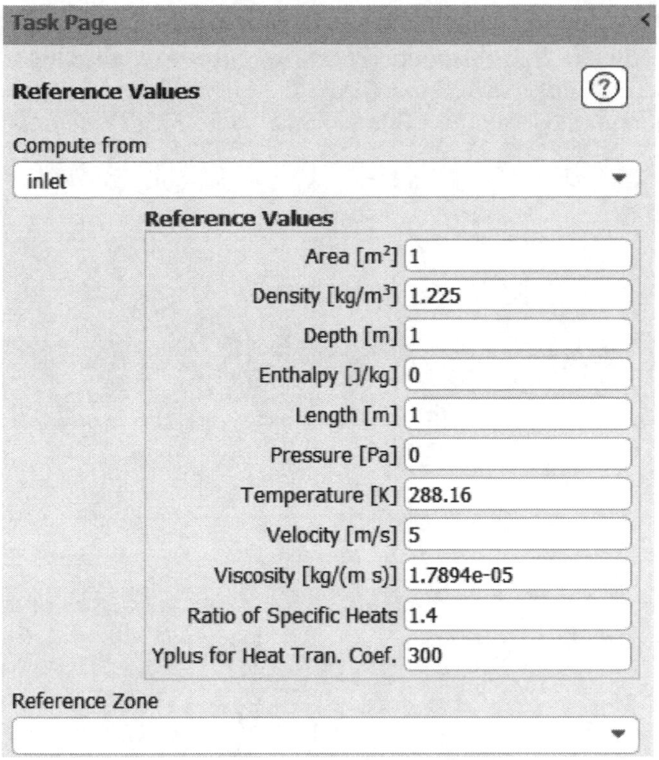

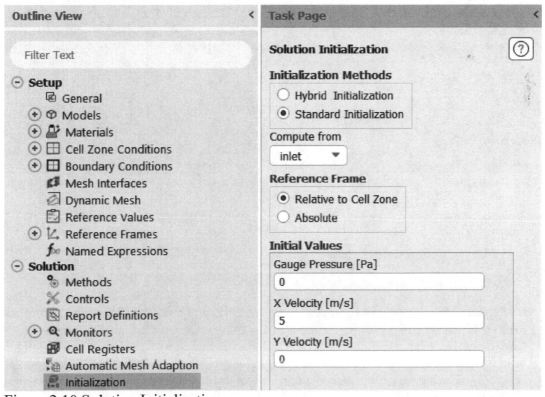

Figure 2.9b) Reference values

10. Double click on Initialization under Solution in the Outline View, select *Standard Initialization*, select *Compute from inlet*, and click on the *Initialize* button.

Figure 2.10 Solution Initialization

11. Double click on Monitors under Solution in the Outline View. Double click on Residual under Monitors in the Outline View and enter 1e-9 as *Absolute Criteria* for all Residuals. Click on the OK button to close the window. Select File>>Save Project from the menu. Select File>>Export>>Case… from the menu. Save the Case File with the name *Flat Plate Boundary Layer.cas.h5*

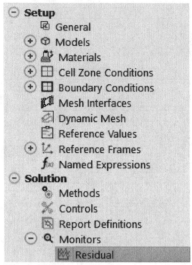

Figure 2.11a) Residual monitors

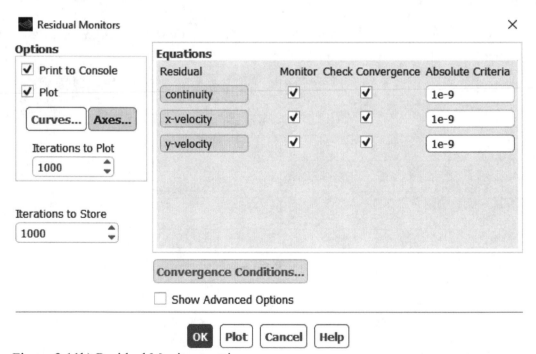

Figure 2.11b) Residual Monitors settings

Why did we set the Absolute Criteria to 1e-9? Generally, the lower the absolute criteria, the longer time the calculation will take and give a more exact solution. We see in Figure 2.12b) that the *x*-velocity and *y*-velocity equations have lower residuals than the continuity equation. The slopes of the residual curves for all three equations are about the same with a sharp downward trend.

12. Double click on Run Calculation under Solution and enter 5000 for *Number of Iterations*. Click on the *Calculate* button. The calculations will be complete after 196 iterations, see Figure 2.12b). Click on Copy Screenshot of Active Window to Clipboard, see Figure 2.12c). The Scaled Residuals can be pasted into a Word document.

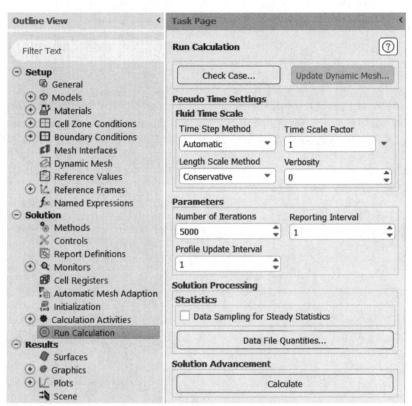

Figure 2.12a) Running the calculations

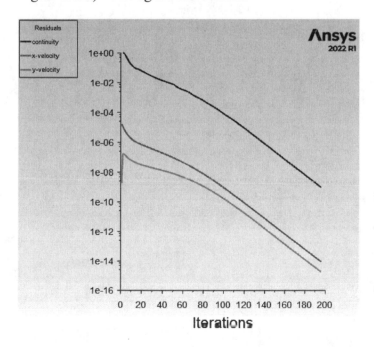

Figure 2.12b) Scaled Residuals

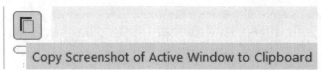

Figure 2.12c) Copying screenshot

G. Post-Processing

13. Select the Results tab in the menu and select Create>>Line/Rake under Surface. Enter 0.2 for x0 (m), 0.2 for x1 (m), 0 for y0 (m), and 0.02 m for y1 (m). Enter *x=0.2m* for the New Surface Name and click on Create. Repeat this step three more times and create vertical lines at x = 0.4m with length 0.04 m, x=0.6m with length 0.06 m, and x=0.8m with length 0.08 m. Close the window.

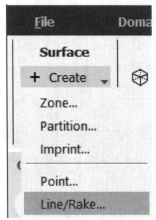

Figure 2.13a) Selecting Line/Rake from the Post-processing menu

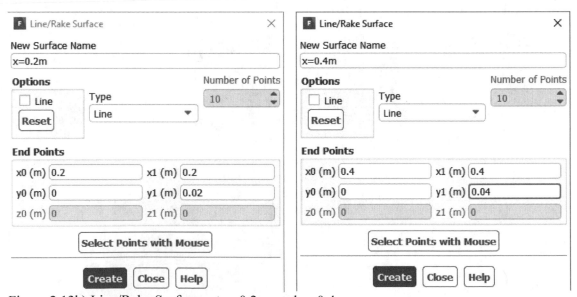

Figure 2.13b) Line/Rake Surfaces at x=0.2 m and x=0.4m

14. Double click on Plots and XY Plot under Results in the Outline View. Uncheck Position on X Axis under Options and check Position on Y Axis. Set Plot Direction for X to 0 and 1 for Y. Select Velocity… and X Velocity as X Axis Function. Select the four lines x=0.2m, x=0.4m, x=0.6m, and x=0.8m under Surfaces.

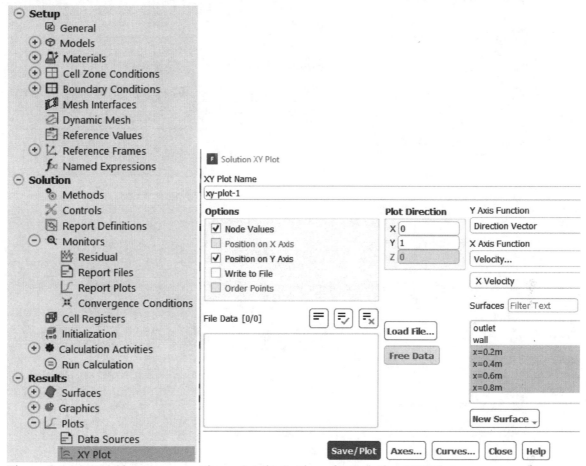

Figure 2.14a) XY Plot setup   Figure 2.14b) Settings for Solution XY Plot

15. Click on the *Axes*... button in the Solution XY Plot window. Select the X Axis, uncheck Auto Range under Options, enter 6 for Maximum Range, select general Type under Number Format and set Precision to 0. Click on the Apply button. Select the Y Axis, uncheck the Auto Range, enter 0.01 for Maximum Range, select general Type under Number Format, and click on the Apply button. Close the Axes window.

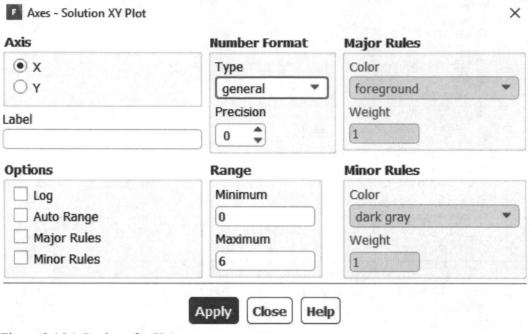

Figure 2.15a) Settings for X Axes

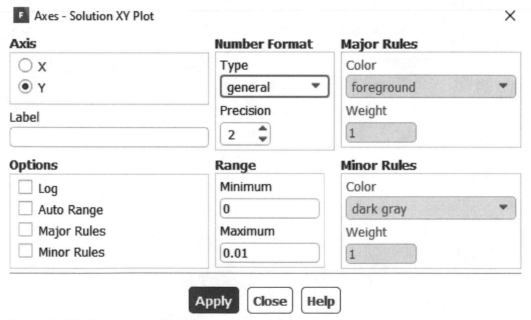

Figure 2.15b) Settings for Y Axes

16. Click on the Curves... button in the Solution XY Plot window. Select the first pattern under Line Style for Curve # 0. Select no Symbol for Marker Style and click on the Apply button. Next select Curve # 1, select the next available Pattern for Line Style, no Symbol for Marker Style, and click on the Apply button. Continue this pattern of selection with the next two curves # 2 and # 3. Close the Curves – Solution XY Plot window. Click on the Save/Plot button in the Solution XY Plot window and Close this window. Click on Copy Screenshot of Active Window to Clipboard, see Figure 2.16c). The XY Plot can be pasted into a Word document.

Figure 2.16a) Settings for curve # 0

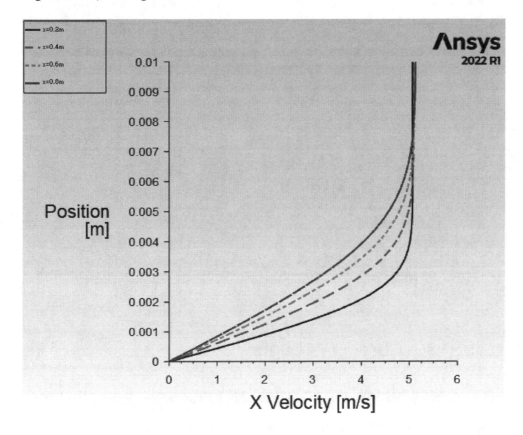

Figure 2.16b) Velocity profiles for a laminar boundary layer on a flat plate

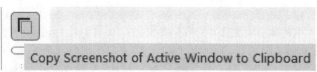

Figure 2.16c) Copying screenshot

Select the *User Defined* tab in the menu and *Custom*. Select a specific *Operand Field Functions* from the drop-down menu by selecting *Mesh...* and *Y-Coordinate*. Click on *Select* and enter the definition as shown in Figure 2.16f). You need to select *Mesh...* and *X-Coordinate* to include the x coordinate and complete the definition of the field function. Enter *eta* as *New Function Name*, click on *Define* and close the window.

Repeat this step to create another custom field function. This time, we select *Velocity...* and *X Velocity* as *Field Functions* and click on *Select*. Complete the *Definition* as shown in Figure 2.16g) and enter *u-divided-by-freestream-velocity* as *New Function Name*, click on *Define* and close the window.

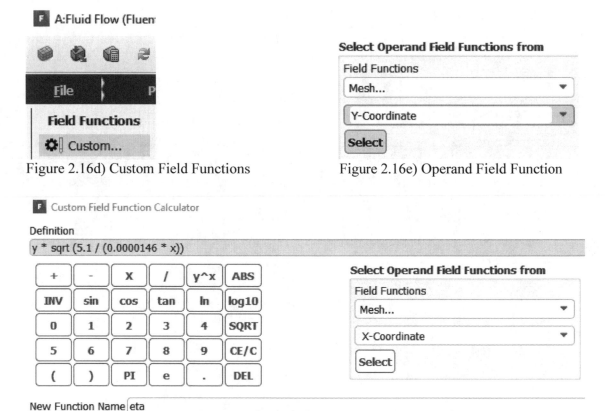

Figure 2.16d) Custom Field Functions   Figure 2.16e) Operand Field Function

Figure 2.16f) Custom field function for self-similar coordinate

> **Why did we create a self-similar coordinate?** It turns out that by using a self-similar coordinate, the velocity profiles at different streamwise positions will collapse on one self-similar velocity profile that is independent on the streamwise location.

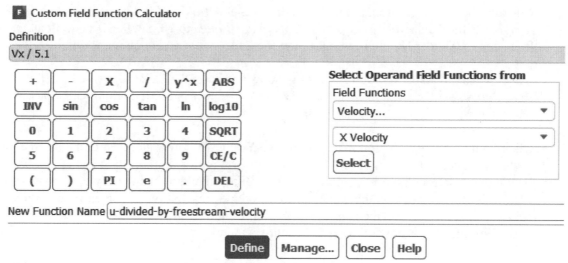

Figure 2.16g) Custom field function for non-dimensional velocity

17. Double click on Plots and XY Plot under Results in the Outline View. Set X to 0 and Y to 1 as Plot Direction. Uncheck *Position on X Axis* and uncheck *Position on Y Axis* under *Options*. Select *Custom Field Functions* and *eta* for *Y Axis Function* and select *Custom Field Functions* and *u-divided-by-freestream-velocity* for *X Axis Function*.

Place the file *blasius.dat* in your working directory. This file can be downloaded from *sdcpublications.com*. See Figure 2.19 for the Mathematica code that can be used to generate the theoretical Blasius velocity profile for laminar boundary layer flow over a flat plate. As an example, in this textbook the working directory is *C:\Users\jmatsson*. Click on *Load File*. Select Files of type: All Files (*) and select the file *blasius.dat* from your working directory. Select the four surfaces *x=0.2m, x=0.4m, x=0.6m, x=0.8m* and the loaded file *Theory*.

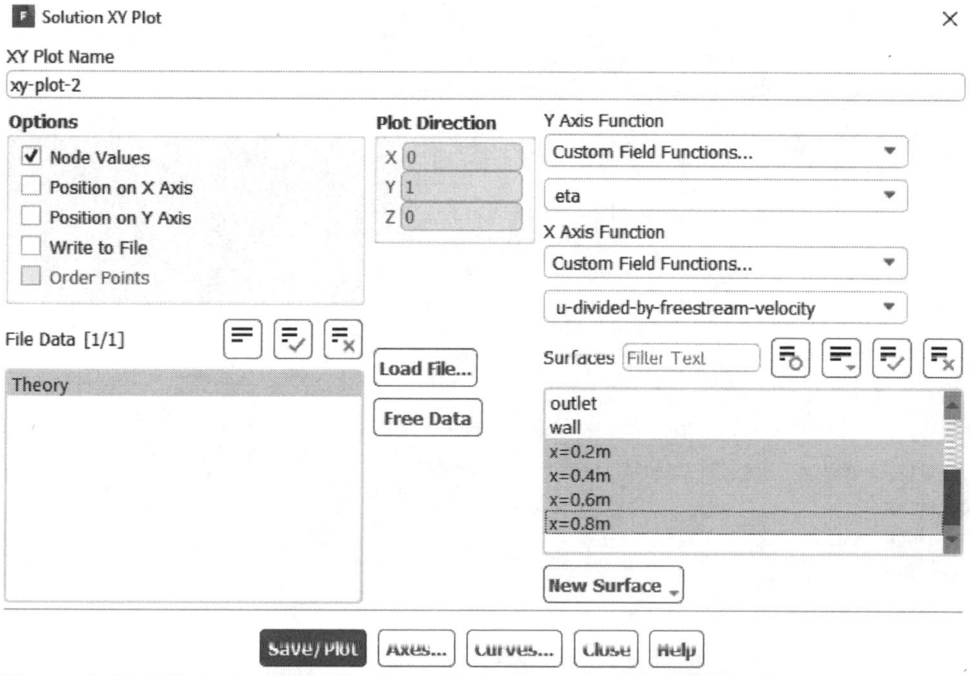

Figure 2.17a) Solution XY Plot for self-similar velocity profiles

Click on the Axes… button. Select *Y Axis* in Axes-Solution XY Plot window and uncheck *Auto Range*. Set the *Minimum Range* to 0 and *Maximum Range* to 10. Set the Type to float and Precision to 0 under Number Format. Enter the Label as *eta* and click on *Apply*.

Select *X Axis* in Axes-Solution XY Plot window and enter the Label to *u/U*. Check the box for *Auto Range*. Set the Type to float and Precision to 1 under Number Format. Click on Apply and Close the window.

Click on the Curves… button in the Solution XY Plot window. Select the first pattern under Line Style for Curve # 0, see Figure 2.16a). Select no Symbol for Marker Style and click on the Apply button. Next select Curve # 1, select the next available Pattern for Line Style, no Symbol for Marker Style, and click on the Apply button. Continue this pattern of selection with the next two curves # 2 and # 3. Close the Curves – Solution XY Plot window. Click on the Save/Plot button in the Solution XY Plot window and close this window.

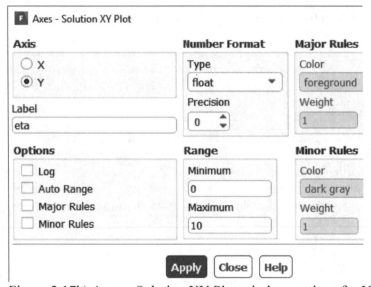

Figure 2.17b) Axes – Solution XY Plot window settings for Y Axis

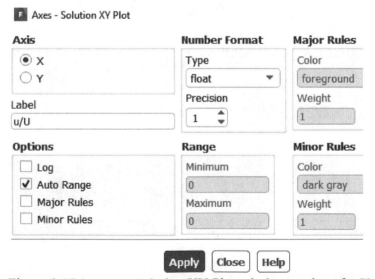

Figure 2.17c) Axes – Solution XY Plot window settings for X Axis

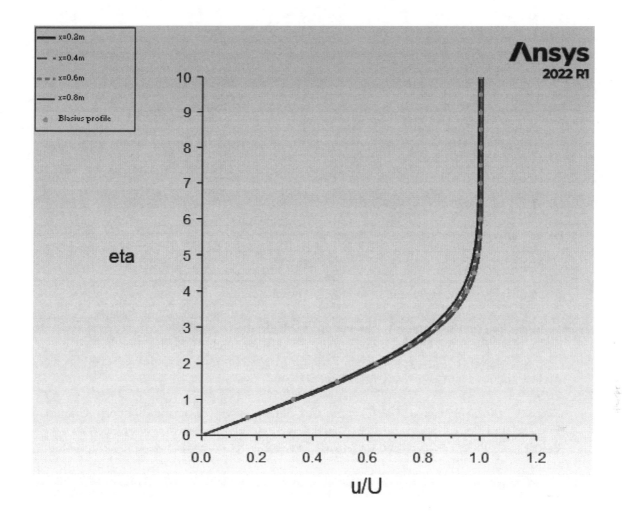

Figure 2.17d) Self-similar velocity profiles for a laminar boundary layer on a flat plate

Click on Copy Screenshot of Active Window to Clipboard, see Figure 2.16c). The XY Plot can be pasted into a Word document.

Select the User Defined tab in the menu and Custom. Select a specific Operand Field Functions from the drop-down menu by selecting Mesh… and X-Coordinate. Click on Select and enter the definition as shown in Figure 2.17e). Enter *re-x* as New Function Name, click on Define and Close the window.

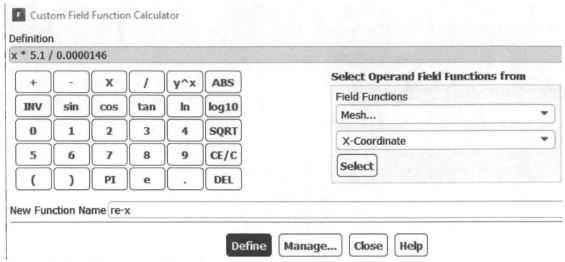

Figure 2.17e) Custom field function for Reynolds number

18. Double click on Plots and XY Plot under Results in the Outline View. Set X to 0 and Y to 1 under Plot Direction. Uncheck Position on X Axis and uncheck Position on Y Axis under Options. Select *Wall Fluxes* and *Skin Friction Coefficient* for Y Axis Function and select Custom Field Functions and *re-x* for X Axis Function.

Place the file "*Theoretical Skin Friction Coefficient*" in your working directory. Click on Load File. Select Files of type: All Files (*) and select the file "*Theoretical Skin Friction Coefficient*". Select *wall* under Surfaces and the loaded file Skin Friction under File Data. Click on the Axes button.

Figure 2.18a) Solution XY Plot for skin friction coefficient

Check the X Axis, check the box for Log under Options, enter Re-x as Label, uncheck Auto Range under Options, set Minimum to 100 and Maximum to 1000000. Set Type to float and Precision to 0 under Number Format and click on Apply.

Check the Y Axis, check the box for Log under Options, enter Cf-x as Label, uncheck Auto Range, set Minimum to 0.001 and Maximum to 0.1, set Type to float, Precision to 3 and click on Apply. Close the window. Click on Save/Plot in the Solution XY Plot window.

Click on the Curves… button in the Solution XY Plot window. Select the first pattern under Line Style for Curve # 0. Select no Symbol for Marker Style and click on the Apply button.

Next select Curve # 1, select the next available Pattern for Line Style, no Symbol for Marker Style, and click on the Apply button. Close the Curves – Solution XY Plot window. Click on the Save/Plot button in the Solution XY Plot window and close this window.

Figure 2.18b) Axes – Solution XY Plot window settings for X Axis

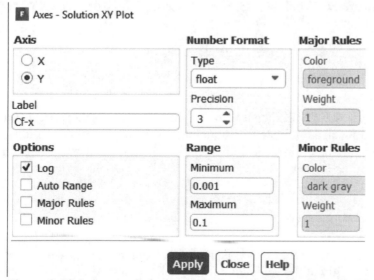

Figure 2.18c) Axes – Solution XY Plot window settings for Y Axis

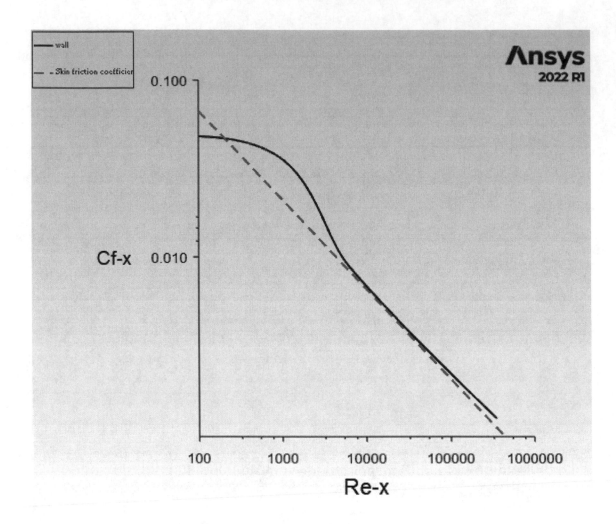

Figure 2.18d) Comparison between ANSYS Fluent and theoretical skin friction coefficient (dashed line) for laminar boundary layer flow on a flat plate

Click on Copy Screenshot of Active Window to Clipboard, see Figure 2.16c). The XY Plot can be pasted into a Word document.

H. Theory

19. In this chapter we have compared ANSYS Fluent velocity profiles with the theoretical Blasius velocity profile for laminar flow on a flat plate. We transformed the wall normal coordinate to a similarity coordinate for comparison of profiles at different streamwise locations. The similarity coordinate is defined by

$$\eta = y\sqrt{\frac{U}{\nu x}} \qquad (2.1)$$

where y (m) is the wall normal coordinate, U (m/s) is the free stream velocity, x (m) is the distance from the streamwise origin of the wall and ν (m^2/s) is the kinematic viscosity of the fluid.

We also used the non-dimensional streamwise velocity u/U where u is the dimensional velocity profile. u/U was plotted versus η for ANSYS Fluent velocity profiles in comparison with the Blasius theoretical profile and they all collapsed on the same curve as per definition of self-similarity. The Blasius boundary layer equation is given by

$$f'''(\eta) + \tfrac{1}{2}f(\eta)f''(\eta) = 0 \tag{2.2}$$

with the following boundary conditions

$$f(0) = f'(0) = 0, f'(\infty) = 0 \tag{2.3}$$

The Reynolds number for the flow on a flat plat is defined as

$$Re_x = \frac{Ux}{\nu} \tag{2.4}$$

The boundary layer thickness δ is defined as the distance from the wall to the location where the velocity in the boundary layer has reached 99% of the free stream value. For a laminar boundary-layer we have the following theoretical expression for the variation of the boundary layer thickness with streamwise distance x and Reynolds number Re_x.

$$\delta = \frac{4.91x}{\sqrt{Re_x}} \tag{2.5}$$

The corresponding expression for the boundary layer thickness in a turbulent boundary layer is given by

$$\delta = \frac{0.16x}{Re_x^{1/7}} \tag{2.6}$$

The local skin friction coefficient is defined as the local wall shear stress divided by dynamic pressure.

$$C_{f,x} = \frac{\tau_w}{\tfrac{1}{2}\rho U^2} \tag{2.7}$$

The theoretical local friction coefficient for laminar flow is determined by

$$C_{f,x} = \frac{0.664}{\sqrt{Re_x}} \qquad Re_x < 5 \cdot 10^5 \tag{2.8}$$

and for turbulent flow we have the following relation

$$C_{f,x} = \frac{0.027}{Re_x^{1/7}} \qquad 5 \cdot 10^5 < Re_x < 10^7 \tag{2.9}$$

```
sol = NDSolve[{f'''[η] + 0.5 f[η] f''[η] == 0, f[0] == f'[0] == 0, f'[10] == 1}, f, η];
myplot = Plot[f'[η] /. First[sol], {η, 0, 10}, AxesLabel → {"u/U", "η"}];
axisFlip ≠ /. {x_Line | x_GraphicsComplex :→ MapAt[≠~Reverse~2 &, x, 1],
    x : (PlotRange → _) :→ x~Reverse~2} &;
myplot // axisFlip

SetDirectory["C:\\Users\\jmatsson"];
mylist = Table[{f'[η] /. First[sol], η}, {η, 0, 10, 0.5}];
TableOfValues1 = Prepend[mylist, {""}];
TableOfValues1 = Prepend[TableOfValues1, {"((xy/key/label \"Blasius profile\")"}];
TableOfValues1 = Prepend[TableOfValues1, {""}];
TableOfValues1 = Prepend[TableOfValues1, {"(labels \"X Velocity\" \"η-Coordinate\")"}];
TableOfValues1 = Prepend[TableOfValues1, {"(title \"Theory\")"}];
TableOfValues1 = Append[TableOfValues1, {")"}];
Grid[TableOfValues1]
Export["blasius.dat", TableOfValues1]
```

Figure 2.19 Mathematica code for theoretical Blasius laminar boundary layer

I. References

1. Çengel, Y. A., and Cimbala J.M., Fluid Mechanics Fundamentals and Applications, 1st Edition, McGraw-Hill, 2006.
2. Richards, S., Cimbala, J.M., Martin, K., ANSYS Workbench Tutorial – Boundary Layer on a Flat Plate, Penn State University, 18 May 2010 Revision.
3. Schlichting, H., and Gersten, K., Boundary Layer Theory, 8th Revised and Enlarged Edition, Springer, 2001.
4. White, F. M., Fluid Mechanics, 4th Edition, McGraw-Hill, 1999.

J. Exercises

2.1 Use the results from the ANSYS Fluent simulation in this chapter to determine the boundary layer thickness at the streamwise positions as shown in the table below. Fill in the missing information in the table. U_δ is the velocity of the boundary layer at the distance from the wall equal to the boundary layer thickness and U is the free stream velocity.

x (m)	δ (mm) Fluent	δ (mm) Theory	Percent Difference	U_δ (m/s)	U (m/s)	ν (m²/s)	Re_x
0.2						.0000146	
0.4						.0000146	
0.6						.0000146	
0.8						.0000146	

Table 2.1 Comparison between Fluent and theory for boundary layer thickness

2.2 Change the element size to 2 mm for the mesh and compare the results in XY Plots of the skin friction coefficient versus Reynolds number with the element size 1 mm that was used in this chapter. Compare your results with theory.

2.3 Change the free stream velocity to 3 m/s and create an XY Plot including velocity profiles at $x = 0.1, 0.3, 0.5, 0.7$ and 0.9 m. Create another XY Plot with self-similar velocity profiles for this lower free stream velocity and create an XY Plot for the skin friction coefficient versus Reynolds number.

2.4 Use the results from the ANSYS Fluent simulation in Exercise 2.3 to determine the boundary layer thickness at the streamwise positions as shown in the table below. Fill in the missing information in the table. U_δ is the velocity of the boundary layer at the distance from the wall equal to the boundary layer thickness and U is the free stream velocity.

x (m)	δ (mm) Fluent	δ (mm) Theory	Percent Difference	U_δ (m/s)	U (m/s)	ν (m²/s)	Re_x
0.1						.0000146	
0.2						.0000146	
0.5						.0000146	
0.7						.0000146	
0.9						.0000146	

Table 2.2 Comparison between Fluent and theory for boundary layer thickness

K. Notes

CHAPTER 3. FLOW PAST A CYLINDER

A. Objectives

- Using ANSYS Workbench to Model the Cylinder and Mesh
- Inserting Boundary Conditions and Free Stream Velocity
- Running ANSYS Fluent Simulation for Laminar Transient 2D Planar Flow
- Using FFT Analysis, Velocity, Pressure and Vorticity Plots for Visualizations

B. Problem Description

We will study the flow of air past a cylinder and we will analyze the problem using ANSYS Fluent. The diameter of the cylinder is 50 mm and the free stream velocity is 0.06 m/s.

C. Launching ANSYS Workbench and Selecting Fluent

1. Start by launching ANSYS Workbench. Double click on Fluid Flow (Fluent) that is located under Analysis Systems in the Toolbox.

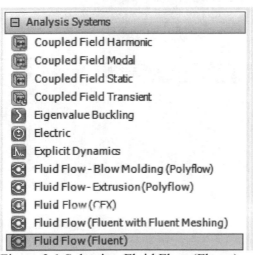

Figure 3.1 Selecting Fluid Flow (Fluent)

D. Launching ANSYS DesignModeler

2. Select Geometry under Project Schematic in ANSYS Workbench. Right-click on Geometry and select Properties. Select 2D Analysis Type under Advanced Geometry Options in Properties of Schematic A2: Geometry. Right-click on Geometry under Project Schematic and select to launch New DesignModeler Geometry. Select Units>>Millimeter as the length unit from the menu in DesignModeler.

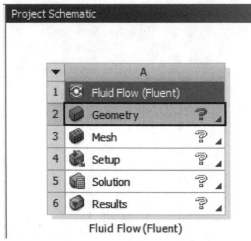

Figure 3.2a) Selecting Geometry

Properties of Schematic A2: Geometry

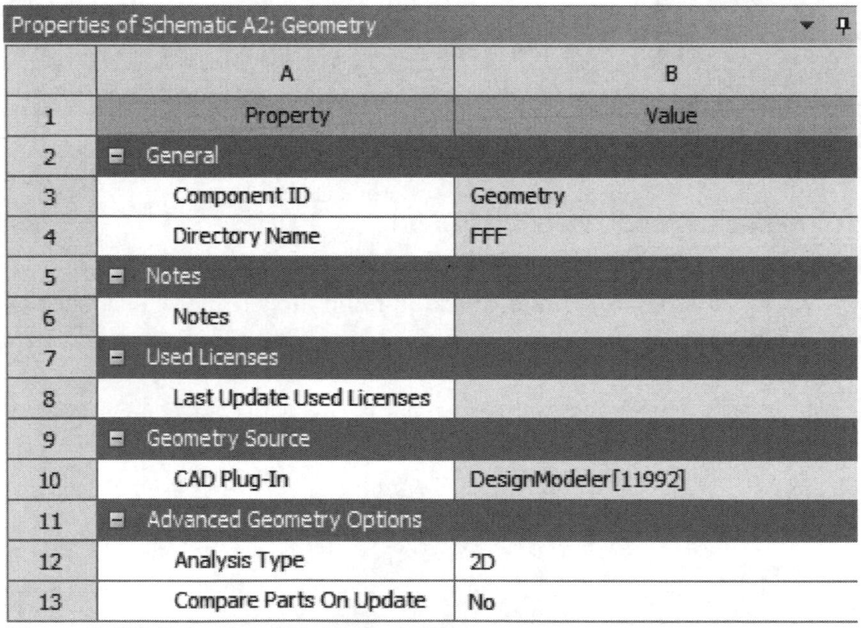

	A	B
1	Property	Value
2	General	
3	Component ID	Geometry
4	Directory Name	FFF
5	Notes	
6	Notes	
7	Used Licenses	
8	Last Update Used Licenses	
9	Geometry Source	
10	CAD Plug-In	DesignModeler[11992]
11	Advanced Geometry Options	
12	Analysis Type	2D
13	Compare Parts On Update	No

Figure 3.2b) Selecting 2D Analysis Type

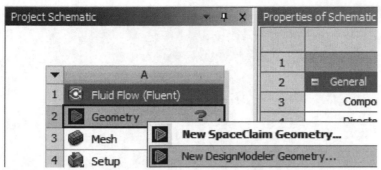

Figure 3.2c) Launching DesignModeler

Figure 3.2d) Selecting the length unit

3.   Next, we will be creating the geometry for the simulation. Select XYPlane from the Tree Outline on the left-hand side. Select Look at Sketch ![]. Click on the Sketching tab in the Tree Outline, select the Draw tab within the Sketching options, and select the Circle sketch tool.

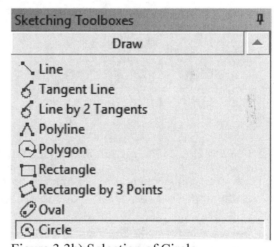

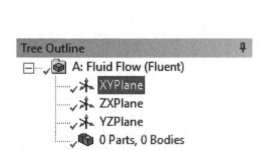

Figure 3.3a) Selection of XYPlane   Figure 3.3b) Selection of Circle

Draw a circle centered at the origin of the coordinate system in the graphics window. Make sure that you move the cursor so that you get the letter *P* in the graphics window indicating that your circle will be centered at the origin. Select the Dimensions tab within the Sketching options and select Diameter. Click on the circle in the graphics window and enter 50 mm for the diameter D1 under Dimensions:1 in Details View.

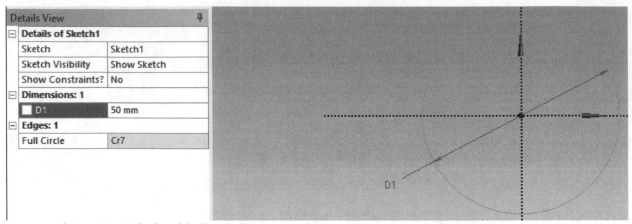

Figure 3.3c) Circle with dimension

Click on the Modeling tab in the Tree Outline and select Concepts>>Surfaces from Sketches from the menu. Click on the circle in the graphics window and select Apply as Base Objects in Details View. Click on Generate. The circle turns gray.

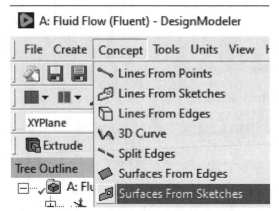

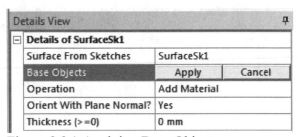

Figure 3.3d) Surfaces from Sketches   Figure 3.3e) Applying Base Objects

Click on the Modeling tab in the Tree Outline and select the XYPlane. Click on New Sketch. Select Rectangle by 3 Points from the Sketching tools. Click anywhere outside of the circle in the upper left corner (2nd quadrant) of the graphics window, followed by clicking anywhere outside of the circle in the lower left corner (3rd quadrant) of the graphics window (make sure you have a *V* before clicking), and finally click in the right-hand plane outside of the circle (4th quadrant).

Select the Dimensions tab under Sketching Toolboxes, select Horizontal and click on the vertical edges of the rectangle. Enter 1575 mm as the horizontal length of the rectangle. Select the vertical axis and the vertical edge of the rectangle on the left-hand side. Enter 575 mm as the length. Select the Vertical dimensioning tool and click on the horizontal edges of the rectangle. Enter 1250 mm as the vertical length of the rectangle. Click on the horizontal axis and the lower horizontal edge of the rectangle. Enter 625 mm as the length. Right-click in the graphics window and select Zoom to Fit.

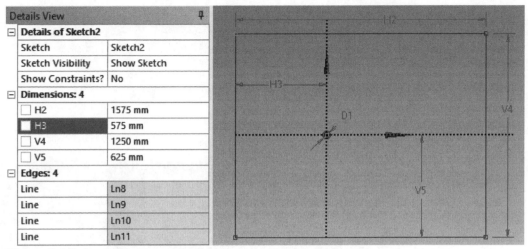

Details View		
Details of Sketch2		
Sketch	Sketch2	
Sketch Visibility	Show Sketch	
Show Constraints?	No	
Dimensions: 4		
☐ H2	1575 mm	
☐ H3	575 mm	
☐ V4	1250 mm	
☐ V5	625 mm	
Edges: 4		
Line	Ln8	
Line	Ln9	
Line	Ln10	
Line	Ln11	

Figure 3.3f) Applying Base Objects

Select Concept>>Surface from Sketches from the menu. Click on Sketch2 in the Tree Outline and select Apply for Base Objects in Details View. Select Operation>>Add Frozen in Details View followed by Generate.

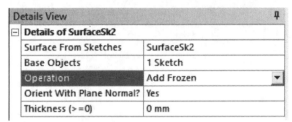

Details View		
Details of SurfaceSk2		
Surface From Sketches	SurfaceSk2	
Base Objects	1 Sketch	
Operation	Add Frozen	▼
Orient With Plane Normal?	Yes	
Thickness (>=0)	0 mm	

Figure 3.3g) Add frozen operation

Select Create>>Boolean from the menu. Select Subtract Operation from the Details View. Click on the mesh region around the cylinder in the graphics window and click on Apply as Target Bodies. Select the Cylinder as the Tool Body, see Figure 3.3j), and click on Generate.

Figure 3.3h) Creating a Boolean

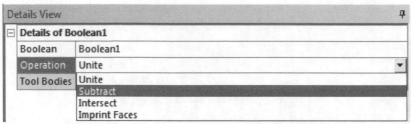

Figure 3.3i) Selecting a subtracting operation

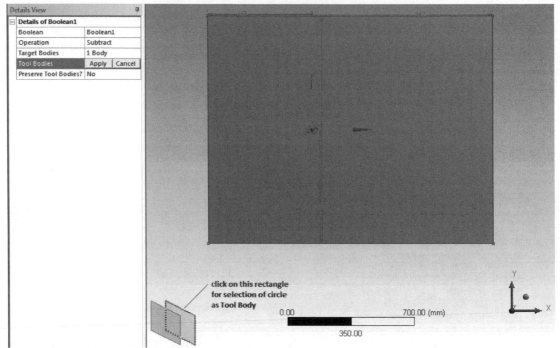

Figure 3.3j) Selecting tool body

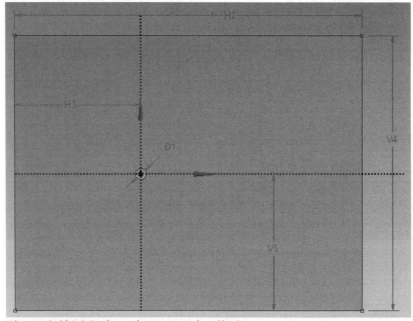

Figure 3.3k) Mesh region around cylinder

4. Select *XYPlane* in the Tree Outline and create a New Sketch 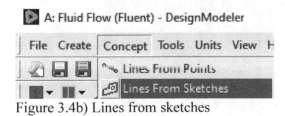. Select the Sketching tab and select Line under the Draw tab. Draw a vertical line in the mesh region from top to bottom to the left of the cylinder, see Figure 3.4a). Make sure you have a *C* when you start the line, a *V* indicating that the line is vertical and another *C* at the end of the line. The vertical line will intersect the entire mesh region. Position the line 75 mm from the vertical axis using the Horizontal dimension.

    Select Concepts>>Lines from Sketches from the menu. Select from the graphics window the newly created vertical line and Apply it as a Base Object in Details View. Click on Generate. Create another sketch in the *XYPlane* with a vertical line 75 mm to the right of the vertical axis and repeat all the different steps.

    Finally, repeat these steps two more times with one line from left to right intersecting the mesh region and positioned 75 mm above the cylinder and another line positioned 75 mm below the cylinder.

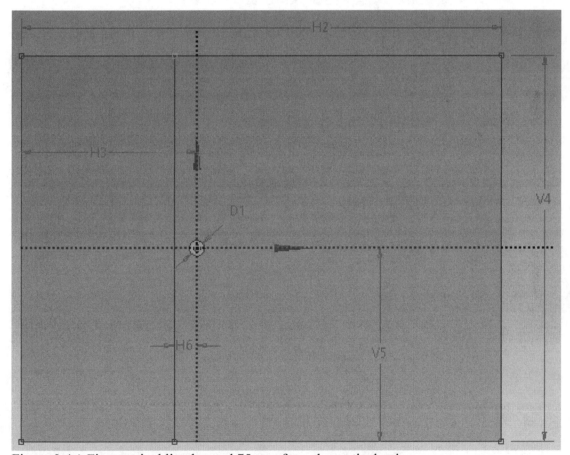

Figure 3.4a) First vertical line located 75 mm from the vertical axis

A: Fluid Flow (Fluent) - DesignModeler

File   Create   Concept   Tools   Units   View   H

Lines From Points

Lines From Sketches

Figure 3.4b) Lines from sketches

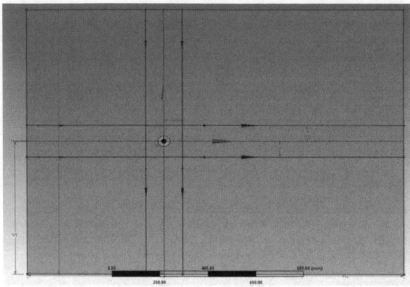

Figure 3.4c) Lines added to the mesh region

Select Tools>>Projection from the menu. Select Edges on Face as type in Details View. Control-select all 12 line-segments for the 4 lines that you have created and Apply as the Edges in the Details View. Select the mesh region and Apply it as the Target under Details View. Click Generate.

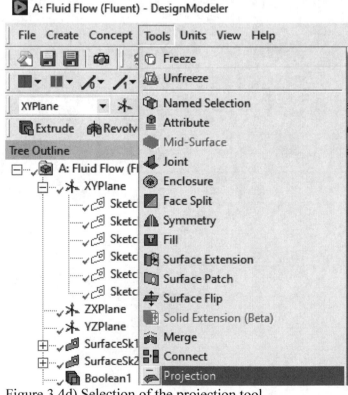

Figure 3.4d) Selection of the projection tool

You will now have 9 different regions for meshing. Select File>>Save Project from the menu and save the project with the name "Cylinder Flow Study". Close DesignModeler.

E. Launching ANSYS Meshing

5. Next, we will be creating the mesh around the cylinder. We are now going to double click on Mesh under Project Schematic in ANSYS Workbench to open the Meshing window.

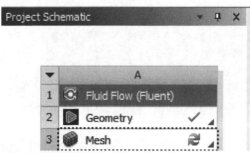

Figure 3.5a) Launching mesh

Select Mesh under Model (A3) in Project under Outline on the left-hand side. Select Mesh>> Controls>>Face Meshing from the menu. Control-select all 9 faces of the mesh. Apply the Geometry under Details of "Face Meshing".

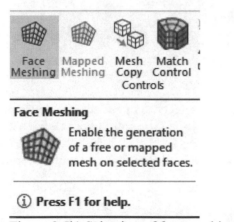

Figure 3.5b) Selection of face meshing

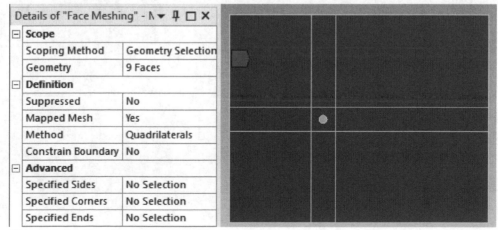

Figure 3.5c) Details of face meshing

Select Mesh>>Controls>>Sizing from the menu, right-click in the graphics window and select Cursor Mode>>Edge. Control-select the three edges to the left and bottom of the mesh region as shown in Figure 3.5d). Apply the Geometry under Details of "Edge Sizing". Select Number of Divisions as Type under Details of "Edge Sizing". Set the Number of Divisions to 100. Set the Behavior to Hard. Select the Bias Type at the top from the drop-down menu, see Figure 3.5d). Enter a Bias Factor of 10.

Details of "Edge Sizing" - Sizing	▾ 무 □ ×
⊟ **Scope**	
Scoping Method	Geometry Selection
Geometry	3 Edges
⊟ **Definition**	
Suppressed	No
Type	Number of Divisions
☐ Number of Divisions	100
⊟ **Advanced**	
Behavior	Hard
Capture Curvature	No
Capture Proximity	No
Bias Type	____ __ _ _
Bias Option	Bias Factor
☐ Bias Factor	10.0
Reverse Bias	No Selection

Figure 3.5d) Details for first mesh sizing

Select Mesh>>Controls>>Sizing from the menu. Control-select the three edges to the right and bottom of the mesh region as shown in Figure 3.5e). Apply the Geometry under Details of "Edge Sizing 2". Select Number of Divisions as Type under Details of "Edge Sizing 2". Set the Number of Divisions to 150. Set the Behavior to Hard. Select the second Bias Type from the drop-down menu, see Figure 3.5e). Enter a Bias Factor of 10.

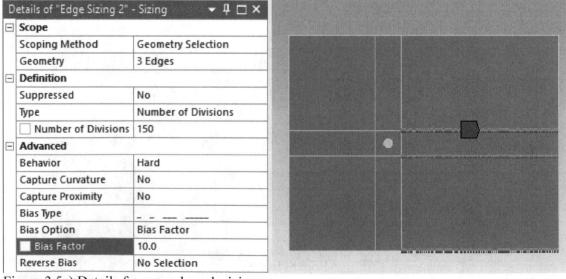

Details of "Edge Sizing 2" - Sizing	▾ 무 □ ×
⊟ **Scope**	
Scoping Method	Geometry Selection
Geometry	3 Edges
⊟ **Definition**	
Suppressed	No
Type	Number of Divisions
☐ Number of Divisions	150
⊟ **Advanced**	
Behavior	Hard
Capture Curvature	No
Capture Proximity	No
Bias Type	_ _ __ ____
Bias Option	Bias Factor
☐ Bias Factor	10.0
Reverse Bias	No Selection

Figure 3.5e) Details for second mesh sizing

Select Mesh>>Controls>>Sizing from the menu. Select the horizontal upper left edge. Apply the Geometry under Details of "Edge Sizing 3". Select Number of Divisions as Type under Details of "Edge Sizing 3". Set the Number of Divisions to 100. Set the Behavior to Hard. Select the second Bias Type from the drop-down menu. Enter a Bias Factor of 10, see Figure 3.5f).

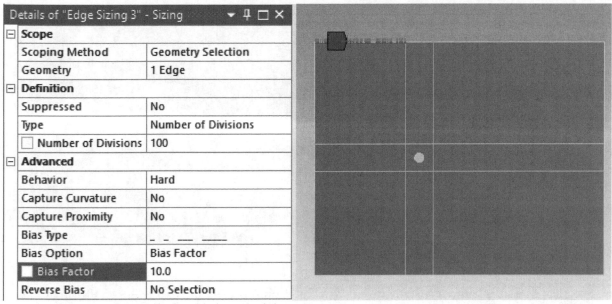

Figure 3.5f) Details for third mesh sizing

Select Mesh>>Controls>>Sizing from the menu. Select the upper horizontal right edge. Apply the Geometry under Details of "Edge Sizing 4". Select Number of Divisions as Type under Details of "Edge Sizing 4". Set the Number of Divisions to 150. Set the Behavior to Hard. Select the first Bias Type from the drop-down menu. Enter a Bias Factor of 10, see Figure 3.5g).

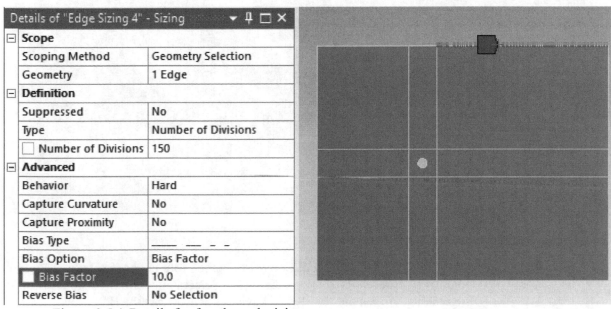

Figure 3.5g) Details for fourth mesh sizing

Select Mesh>>Controls>>Sizing from the menu. Control-select the 8 short edges around the cylinder and on the edges as shown in Figure 3.5h). Apply the Geometry under Details of "Edge Sizing 5". Select Number of Divisions as Type under Details of "Edge Sizing 5". Set the number of Divisions to 50. Set the Behavior to Hard.

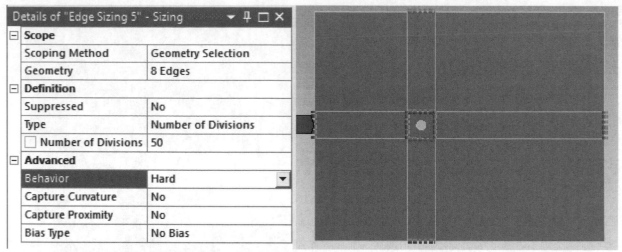

Figure 3.5h) Details for fifth mesh sizing

Select Mesh>>Controls>>Sizing from the menu. Control-select the four edges as shown in Figure 3.5i). Apply the Geometry under Details of "Edge Sizing 6". Select Number of Divisions as Type under Details of "Edge Sizing 6". Set the Number of Divisions to 100. Set the Behavior to Hard. Select the first Bias Type from the drop-down menu. Enter a Bias Factor of 20.

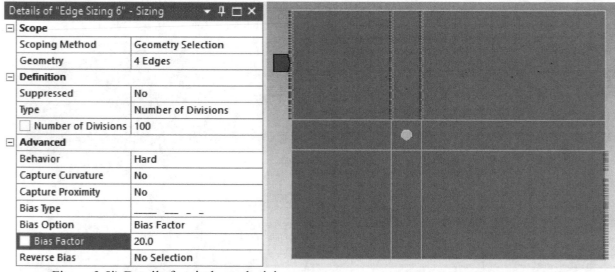

Figure 3.5i) Details for sixth mesh sizing

Select Mesh>>Control>>Sizing from the menu. Control-select the four edges as shown in Figure 3.5j). Apply the Geometry under Details of "Edge Sizing 7". Select Number of Divisions as Type under Details of "Edge Sizing 7". Set the Number of Divisions to 100. Set the Behavior to Hard. Select the second Bias Type from the drop-down menu. Enter a Bias Factor of 20.
Right click on Mesh in the tree outline and select Generate Mesh.

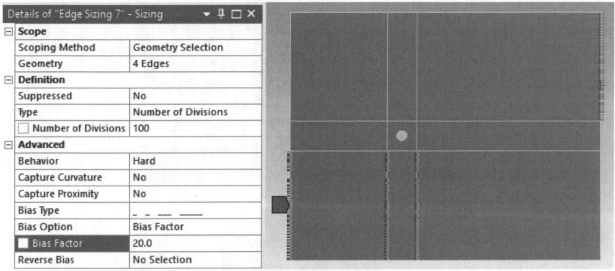

Details of "Edge Sizing 7" - Sizing	▼ ⊓ □ ×
Scope	
Scoping Method	Geometry Selection
Geometry	4 Edges
Definition	
Suppressed	No
Type	Number of Divisions
Number of Divisions	100
Advanced	
Behavior	Hard
Capture Curvature	No
Capture Proximity	No
Bias Type	_ _ __ ____
Bias Option	Bias Factor
Bias Factor	20.0
Reverse Bias	No Selection

Figure 3.5j) Details for seventh mesh sizing

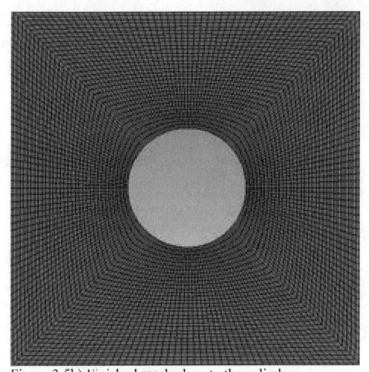

Figure 3.5k) Finished mesh close to the cylinder

Why do we use a biased mesh? A biased mesh enables us to create a finer mesh closer to the cylinder. In this way we create a fine mesh where it is needed the most and the calculation time will be much shorter compared with a finer mesh in the entire mesh region.

6. Select Geometry in the outline. Select the Edge Selection filter ⬚. Control-select the 3 vertical edges on the right-hand side, right click and select Create Named Selection. Name the edges "outlet" and click OK. Name the 3 vertical edges to the left "inlet" and the edge of the cylinder with the name "cylinder". Name the 6 horizontal edges (3 at the top and 3 at the bottom) as "symmetry".

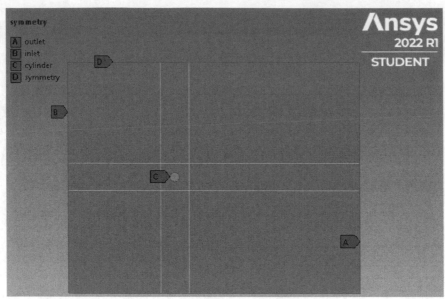

Figure 3.6a) Named selections

Open Geometry in the outline, select Line Body, and select Thermal Fluid as Model Type in Details of "Line Body". In Geometry, right click the Line Body and select Suppress Body.

Details of "Line Body"

Graphics Properties	
Definition	
Suppressed	No
Model Type	Thermal Fluid
Coordinate System	Default Coordinate System
Offset Mode	Refresh on Update
Offset Type	Centroid
Treatment	None
Fluid Cross Area	0. m^2
Fluid Discretization	Upwind/Linear
Material	
Assignment	
Fluid/Solid	Defined By Geometry (Solid)
Bounding Box	
Properties	
Statistics	

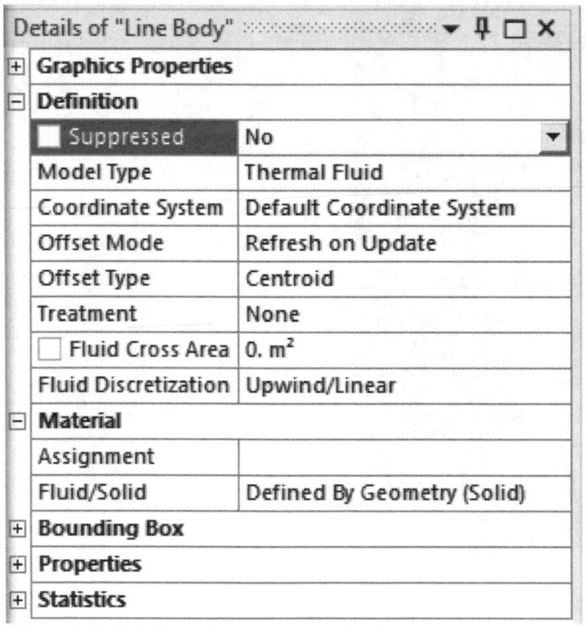

Figure 3.6b) Details of Line Body

Select File>>Export>>Mesh >>FLUENT Input File>>Export from the menu. Enter the name *cylinder-flow-mesh* and select FLUENT Input Files (*,msh) as Save as Type. Select File>>Save Project… from the menu. Close the meshing window. Right click on Mesh under Project Schematic in ANSYS Workbench and select Update.

Why do we export the mesh? When we save the mesh file, we can read the mesh using standalone mode directly in ANSYS Fluent without having to design the mesh all over again each time that we run the simulation for various flow variables.

F. Launching ANSYS Fluent

7. Double click Setup under Project Schematic in ANSYS Workbench. Check the Options box for Double Precision. Set the number of Solver Processes equal to the number of computer cores. To check the number of physical cores, press the Ctrl + Shift + Esc keys simultaneously to open the Task Manager. Select the Performance tab and select CPU from the left column.
You'll see the number of physical cores on the bottom-right side. Click on Show More Options. You can see the location of your Working Directory. Write down this location as you will need it later in this chapter. Click OK to launch Fluent. Click OK when you get the Key Behavioral Changes window. Select Transient Time under General on the Task Page.

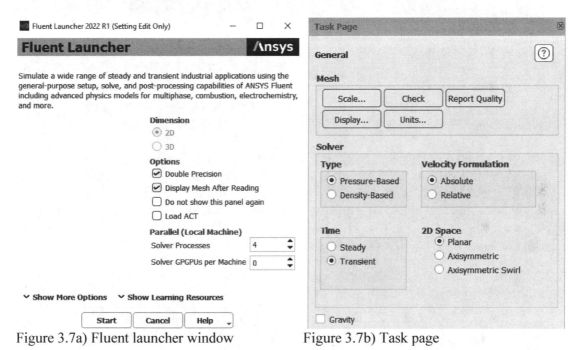

Figure 3.7a) Fluent launcher window   Figure 3.7b) Task page

> **Why do we use the Pressure-Based Solver?** The pressure-based solver is usually being chosen for incompressible flows such as the flow past a cylinder at low Reynolds numbers.

> **Why do we use a planar 2D Space study?** The plane 2D Space study is completed under the assumption that the cylinder is infinitely long such that there is no variation of the flow along the length of the cylinder.

8. Double click on Models and Viscous (SST k-omega) under Setup in the Outline View. Select Laminar as Viscous Model. Click OK to close the window. Double click on Boundary Conditions under Setup in the Outline View. Double click on *inlet* under Zone on the Task Page.

> **Why do we use a Transient study and the Laminar model?** The flow past a cylinder is time dependent at the Reynolds number that we are using in this chapter. Vortex shedding will occur behind the cylinder and we are interested in determining the frequency and Strouhal number for the vortex shedding process. The Reynolds number is not sufficiently high to have turbulent flow and therefore we choose Laminar as Viscous model in this case.

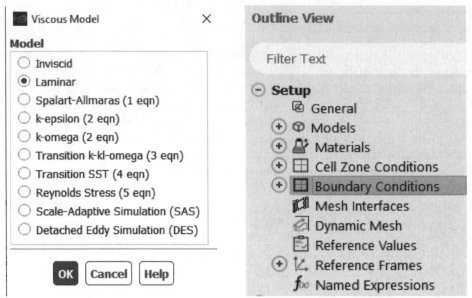

Figure 3.8a) Viscous model   Figure 3.8b) Defining boundary conditions

Select Components as Velocity Specification Method. Enter 0.06 m/s as X-Velocity. Click Apply and Close the Velocity Inlet window. Double-click on Methods in the Outline View under Solution. Set the Pressure-Velocity Coupling Scheme to PISO. Uncheck the box for Skewness-Neighbor Coupling. Select Green-Gauss Cell Based Gradient under Spatial Discretization.

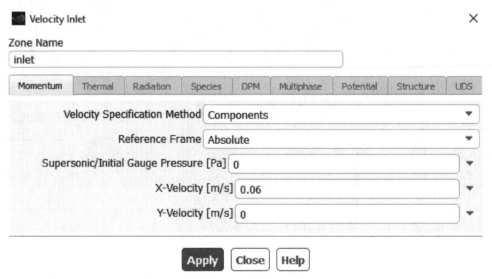

Figure 3.8c) Velocity inlet boundary condition

Why do we choose PISO as Pressure-Velocity Coupling Scheme? PISO is used for transient solutions.

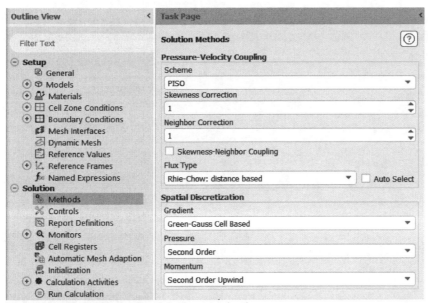

Figure 3.8d) Settings for solution methods

9. Double-click on Initialization in the Outline View under Solution. Select Standard Initialization under Solution Initialization on the Task Page and enter 0.06 for X-Velocity (m/s). Click on Initialize.

Double-click on Reference Values under Setup in the Outline View. Select Compute from inlet from the drop-down menu. Set the value for Area (m2) to 0.05 and Length (m) to the same value. Select *solid-surface_body* as Reference Zone.

Double-click on Report Definitions under Solution in the Outline View. Select New>>Force Report>>Lift… from the drop-down menu. Select Lift Coefficient as Report Output Type and set the Force Vector to X 0, Y 1 and Z 1. Select cylinder under Zones. Check the boxes for Report File, Report Plot and Print to Console under Create. Click on the OK button to close the Lift Report Definition window. Close the Report Definitions window.

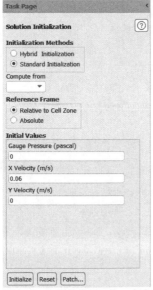

Figure 3.9a) Solution initialization

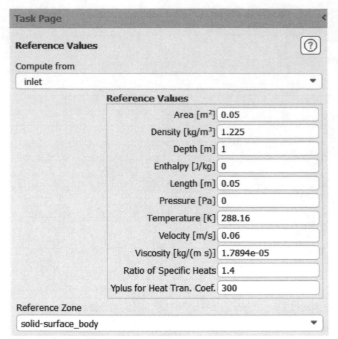

Figure 3.9b) Reference values

Figure 3.9c) New Force Report

Figure 3.9d) Lift report definition

Select File>>Save Project from the menu. Select File>>Export>>Case & Data… from the menu. Save the Case/Data File with the name Flow Past a Cylinder.cas.h5

Double-click on Run Calculation under Solution on the left-hand side in the Outline View. Enter 0.2 for the Time Step Size (s) and 600 for number of Time Steps. Enter 30 for Max Iterations/Time Step and click on Calculate.

> **Why do we choose 0.2 s as Time Step Size?** The Time Step Size must be sufficiently small to resolve the wave form of the time signal as shown in Figure 3.9g).

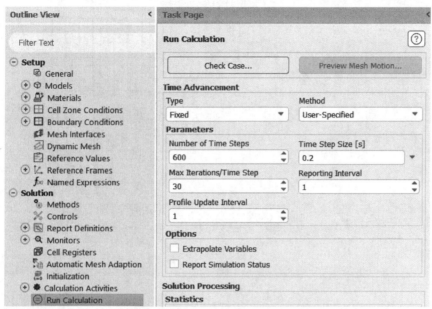

Figure 3.9e) Calculation settings

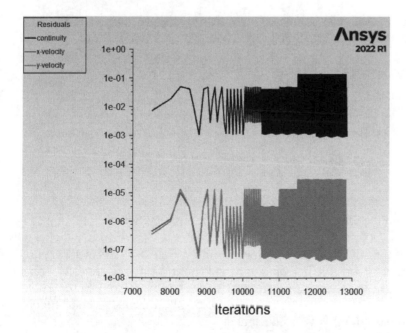

Figure 3.9f) Scaled residuals

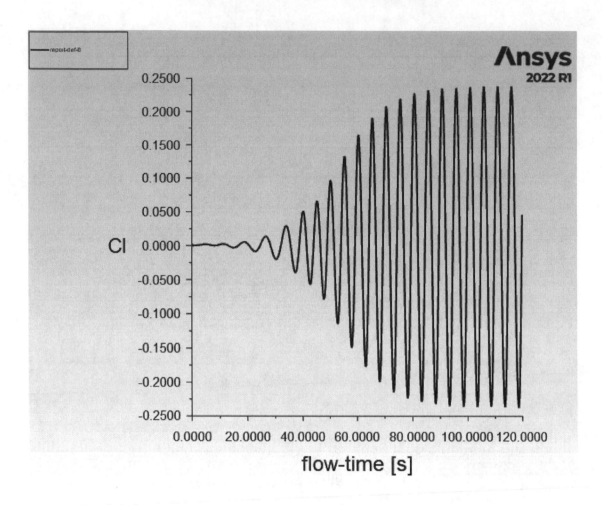

Figure 3.9g) Lift coefficient versus time

Click on Copy Screenshot of Active Window to Clipboard, see Figure 3.9h). The Scaled Residuals and lift coefficient versus flow-time graphs can be pasted into a Word document.

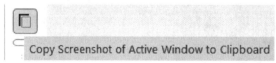

Figure 3.9h) Copying screenshot

G. Post-Processing

10. Open Plots under Results in the Outline View. Double-click on FFT. Select Load Input File in the Fourier Transform window. Select the file with the name *report-def-0-rfile.out* and click OK. Select Axes… at the bottom of the Fourier Transform window. Select Y axis, set Type to float and Precision to 1 as Number Format. Click on Apply and Close the Axes – Fourier Transform window. Click on Plot FFT in the Fourier Transform window. Click on Copy Screenshot of Active Window to Clipboard, see Figure 3.9h). The Power Spectral Density versus Frequency graph can be selected and pasted into a Word document.

Check the box for Write FFT to File under Options in the Fourier Transform window and click on the Write FFT button. Save the file with All Files (*) as file type and the name *fft-data* in the working directory. This file can be opened in Notepad, see Figure 3.10d). The power spectral density maximum is at 0.2 Hz. The power spectral density shows the variation of power in the different frequencies that makes up the time signal and is based on the mathematical theory for Fourier analysis.

Figure 3.10a) Fourier transform window

Figure 3.10b) Axes settings

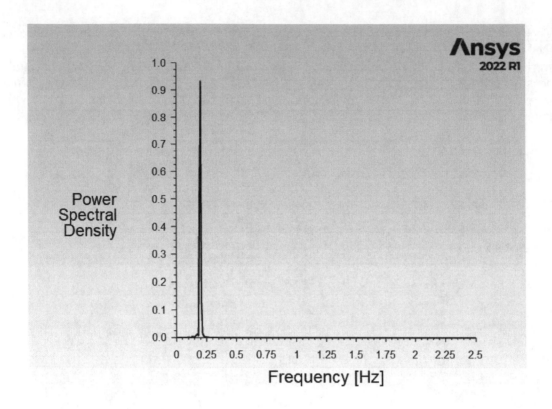

Figure 3.10c) Plot of power spectral density versus frequency

fft-data - Notepad

File   Edit   Format   View   Help

```
(title "Spectral Analysis of report-def-0-rfile")
(labels "Frequency [Hz]" "Power Spectral Density")

((xy/key/label "report-def-0")
8.3194673e-03    5.4002152e-04
1.6638935e-02    5.1596289e-04
2.4958402e-02    5.2659522e-04
3.3277869e-02    5.8455981e-04
4.1597337e-02    6.1115180e-04
4.9916804e-02    6.5809663e-04
5.8236271e-02    6.7306327e-04
6.6555738e-02    7.2979985e-04
7.4875206e-02    8.2850771e-04
8.3194673e-02    7.9917011e-04
9.1514140e-02    8.4782008e-04
9.9833608e-02    1.3149853e-03
1.0815307e-01    1.3446107e-03
1.1647254e-01    5.0098909e-04
1.2479201e-01    1.9232444e-03
1.3311148e-01    4.9760230e-03
1.4143094e-01    3.0031309e-03
1.4975041e-01    9.9321001e-04
1.5806988e-01    5.6467545e-03
1.6638935e-01    9.7872121e-03
1.7470881e-01    1.3283495e-02
1.8302828e-01    6.3943965e-03
1.9134775e-01    4.6670908e-01
1.9966722e-01    9.3324542e-01
2.0798668e-01    1.3551804e-01
2.1630615e-01    1.5349879e-02
2.2462562e-01    7.9118554e-03
2.3294508e-01    3.8762654e-03
```

Figure 3.10d) Data for plot in figure 3.10c)

Double-click on Contours in Graphics under Results in the Outline View. Enter *static-pressure* as Contour Name in the Contours window. Select Contours of Pressure… and Static Pressure. De-select all Surfaces and click on Save/Display.

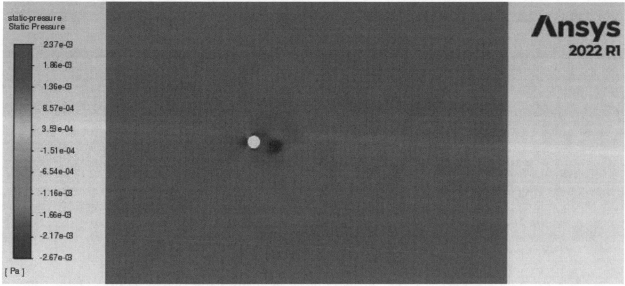

Figure 3.10e) Contours plot of pressure

Select Contours of Velocity and Vorticity Magnitude. Enter *vorticity-magnitude* as Contour Name in the Contours window. Uncheck the box for Filled under Options. De-select all surfaces and click on Save/Display.

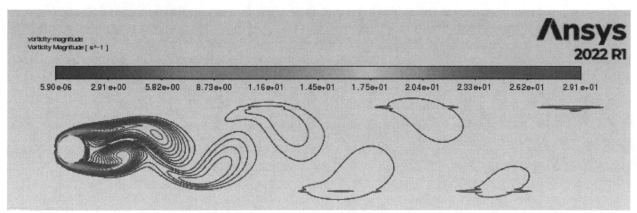

Figure 3.10f) Contour plot of vorticity

Click on Copy Screenshot of Active Window to Clipboard, see Figure 3.9h). The contour plots can be pasted into a Word document.

Find the coefficient of drag by double-clicking on Reports under Results in the Outline View. Double-click on Forces under Reports. Select the cylinder under Wall Zones and select Print and write down the value for the total drag coefficient that you have calculated: 1.1514.

```
Forces - Direction Vector (1 0 0)
                         Forces [N]                              Coefficients
Zone                     Pressure         Viscous        Total         Pressure         Viscous        Total
cylinder                 0.00010292177    2.4022755e-05  0.00012694452 0.93353077       0.21789347     1.1514242
----------------------   ---------------  -------------- ------------- ---------------  -------------- ----------
Net                      0.00010292177    2.4022755e-05  0.00012694452 0.93353077       0.21789347     1.1514242
```

Figure 3.10g) Forces and coefficients

H. Theory

11. We determine the Reynolds number for the flow around the cylinder as

$$Re = Ud/v = 0.06 * 0.050 / 1.5 * 10^{-5} = 200 \tag{3.1}$$

where U (m/s) is the magnitude of the free stream velocity, d (m) is the diameter of the cylinder, and v (m²/s) is the kinematic viscosity of air at room temperature. The drag coefficient from experiments for a smooth cylinder is found using the following curve-fit formula:

$$C_{D,Exp} = 1 + \frac{10}{Re^{2/3}} = 1.29 \qquad 0 \leq Re \leq 250,000 \tag{3.2}$$

The difference between ANSYS Fluent results from Figure 3.10g) and experiments for the drag coefficient is 11%.

The Strouhal number S is one of the many non-dimensional numbers used in fluid mechanics and based on the oscillation frequency of fluid flows. In this case the Strouhal number is based on the vortex shedding frequency f as shown in Figures 3.10c), 3.10d) and can be defined as

$$S = \frac{fd}{U} = \frac{0.2Hz*0.050m}{0.06 \ m/s} = 0.167 \tag{3.3}$$

This value of the Strouhal number from ANSYS Fluent can be compared with DNS (Direct Numerical Simulation) results by Henderson[3], see also Williamson and Brown[6]. The difference is 15%.

$$S_{DNS} = 0.2698 - \frac{1.0272}{\sqrt{Re}} = 0.1968 \tag{3.4}$$

I. References

1. Cimbala, J.M., Fluent-Laminar Flow over a cylinder, Penn State University, 29 January 2008.
2. Flow Over a Cylinder, Fluent Inc., FlowLab 1.2, April 16, 2007.
3. Henderson, R., Nonlinear dynamics and pattern formation in turbulent wake transition, J. of Fluid Mech., 352, (65-112), 1997.
4. Tutorial 6. Flow Past a Circular Cylinder. Fluent Inc. January 17, 2007.
5. Validation 3. Laminar Flow Around a Circular Cylinder, Fluent Inc. September 13, 2002.
6. Williamson, C. H. K., and Brown G.L., A series in 1/√Re to represent the Strouhal-Reynolds number relationship of the cylinder wake, Journal of Fluids and Structures, 12, (1073 – 1085), 1998.

J. Exercises

3.1 Rerun the simulation with the same free stream velocity and Reynolds number and the same mesh but use a Time Step Size of 0.25 s and 500 as the Number of Time Steps. Set Max Iterations /Time Step to 30. Determine the drag coefficient and compare with equation (3.1). Determine the Strouhal number and compare with equation (3.4). Include plots of the lift coefficient over time and power spectral density versus frequency.

3.2 Rerun the simulation shown in this chapter for a lower free stream 0.05 m/s but with the same mesh and for the same diameter 0.05 m of the cylinder. Use the same Time Step Size, Number of Time Steps and Max Iterations/Time Step as in this chapter. Determine the Reynolds number. Determine the drag coefficient and compare with equation (3.1). Determine the Strouhal number and compare with equation (3.4). Include plots of the lift coefficient over time and power spectral density versus frequency. Also, include contour plots of static pressure, vorticity and velocity magnitude.

3.3 Rerun the simulation shown in this chapter for the free stream velocities as shown in Table 3.1 with the same mesh and for the same diameter 0.05 m of the cylinder. Use the same Time Step Size, Number of Time Steps and Max Iterations/Time Step as in this chapter. Determine the drag coefficients and compare with equation (3.1). Determine the Strouhal numbers and compare with equation (3.4).

U (m/s)	f (Hz)	Re	$C_{D,Exp}$	$C_{D,Fluent}$	Percent Diff.	S_{DNS}	S_{Fluent}	Percent Diff.
0.00006		0.2						
0.0006		2						
0.006		20						
0.06	0.2	200	1.29	1.1513	11	0.1968	0.167	15
0.6		2000						
6		20000						
60		200000						

Table 3.1 Comparison between Fluent and theory for boundary layer thickness

3.4 Rerun the simulation shown in this chapter for the flow around a square with a size of 50 mm x 50 mm. Use the same Time Step Size, Number of Time Steps and Max Iterations/Time Step as in this chapter. Determine the Reynolds number. Determine the drag coefficient and the Strouhal number. Include plots of the lift coefficient over time and power spectral density versus frequency. Also, include contour plots of static pressure, vorticity and velocity magnitude.

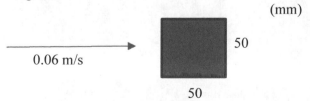

(mm)

50

0.06 m/s

50

3.5 Rerun the simulation shown in this chapter for the flow around an equilateral triangle with a side length of 50 mm. Use the same Time Step Size, Number of Time Steps and Max Iterations/Time Step as in this chapter. Determine the Reynolds number. Determine the drag coefficient and the Strouhal number. Include plots of the lift coefficient over time and power spectral density versus frequency. Also, include contour plots of static pressure, vorticity and velocity magnitude.

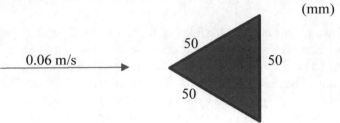

CHAPTER 4. FLOW PAST AN AIRFOIL

A. Objectives

- Using ANSYS Workbench to Model the NACA 2412 Airfoil and Mesh
- Inserting Boundary Conditions and Free Stream Velocity
- Running Turbulent Steady 2D Planar ANSYS Fluent Simulations
- Using Velocity and Pressure Plots for Visualizations
- Determine Lift and Drag Coefficients for Zero Angle of Attack

B. Problem Description

The purpose of this simulation is to become familiar with a few basic aerodynamic concepts and terms such as Reynold's number, coefficients of lift, drag, and pressure. We will therefore study the flow of air past the NACA 2412 airfoil and we will analyze the problem using ANSYS Fluent. The chord length of the airfoil is 230 mm, the angle of attack is 0° and the free stream velocity is 12.7 m/s.

Another airfoil shape that has been studied extensively is the Clark-Y airfoil. One of the first measurements of pressure distributions on this airfoil was completed by Jacobs et al.[1], followed by Marchman and Werme[2] and Stern et al.[3] that made low Reynolds number measurements. Warner[4] and Matsson[5] have also determined results from tests of the Clark-Y airfoil. The NACA 2412 airfoil has been tested by Matsson et al.[6].

For the four digit series[7] of NACA airfoils, the first digit stands for the camber which is the distance between chord line and mean camber line. The second digit stands for the location of maximum camber along the chord line from the leading edge. The last two digits stands for the maximum thickness of the airfoil. For the NACA 2412 studied in this chapter, there is a 2% max camber at 40% chord length with 12% thickness.

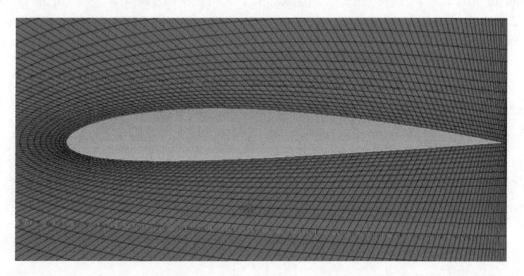

C. Launching ANSYS Workbench and Selecting Fluent

1. Start by launching ANSYS Workbench. Double click on Fluid Flow (Fluent) that is located under Analysis Systems in Toolbox.

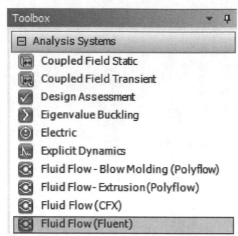

Figure 4.1 Selecting Fluid Flow in ANSYS Workbench

D. Launching ANSYS DesignModeler

2. Right click Geometry and select Properties. In Properties of Schematic A2: Geometry, select Analysis Type 2D under Advanced Geometry Options. Right-click on Geometry in the Project Schematic window and select New DesignModeler Geometry. Select **Units>>Millimeter** from the menu as the desired length unit in the new DesignModeler window.

    Select Create>>Point from the menu. We will use the coordinates file *NACA_2412_Airfoil.txt* that is available for download at *sdcpublications.com*. Open the Coordinates File from Details View and click on Generate. Select Look At Face 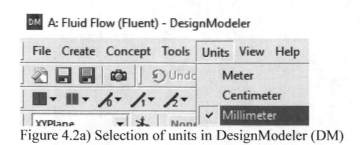. Select Concept>>Lines From Points in the menu. Right click in the graphics window and select Point Chain. Right click in the graphics window and select Select All. Apply the Point Segments in Details View. Click on Generate. Select Concept>>Surfaces From Edges from the menu. Right click in the graphics window and Select All. Apply Edges in Details View. Click on Generate.

Figure 4.2a) Selection of units in DesignModeler (DM)

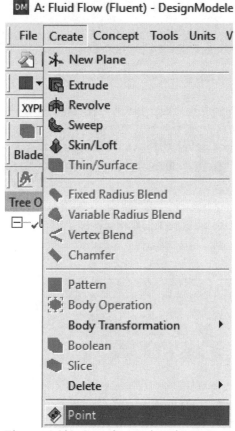

Figure 4.2b) Creating points in DM

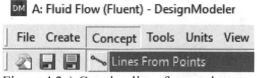

Figure 4.2c) Creating lines from points

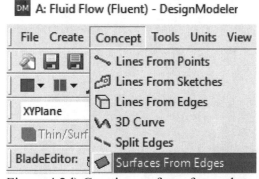

Figure 4.2d) Creating surfaces from edges

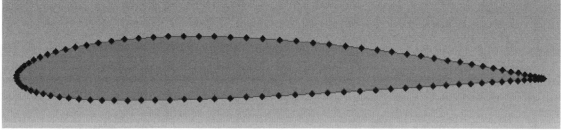

Figure 4.2e) NACA 2412 Airfoil in DesignModeler

3. Next, we will be sketching the mesh region around the airfoil. Create a new coordinate system by first clicking on New Plane ✴. Select Type>>From Coordinates in the Details View. Enter 230 mm for FD11, Point X. Click on Generate. Select the new Plane4 in the tree outline and click on ⬚ to generate a new sketch. Click on the Sketching tab and select Arc by Center from the Draw tab. Click on Look At Face/Plane/Sketch ⬚. Click on the origin of the coordinate system at the trailing edge of the airfoil. Make sure that you see the letter *P* when you move the cursor over the origin of the coordinate system. Next, click on the vertical axis above the origin (make sure you have a *C*) and finally on the vertical axis below the origin (make sure you have a *C* once again), see Figure 4.3b).

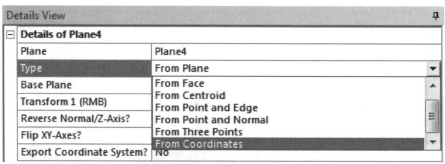

Figure 4.3a) A new plane from coordinates

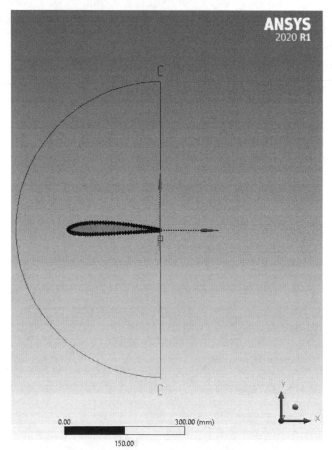

Figure 4.3b) Half-circle around the NACA 2412 airfoil

4. Select Rectangle by 3 Points from the Draw tab. Click at the intersection of the arc and the positive vertical axis, at the intersection between the arc and the negative vertical axis, and finally click in the right-hand side plane. Select Modify and Trim from Sketching Toolboxes. Click on the lines of the rectangle that are aligned with and on top of the vertical axis.

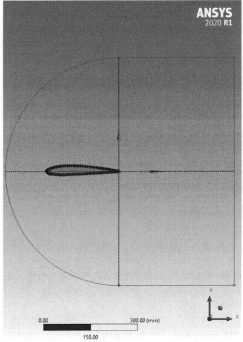

Figure 4.4a) Half-circle and rectangular mesh region around the NACA 2412 airfoil

Select Dimensions under Sketching Toolboxes, select Radius under Dimensions and select the arc in the graphics window. Set the radius of the arc to 2875 mm which gives a ratio of the radius of the arc to the chord length of 12.5. Next, select Horizontal under Dimensions from the Sketching Toolboxes and click on the vertical axis and the vertical edge of the rectangle on the right-hand side. Enter 5750 mm as the length.

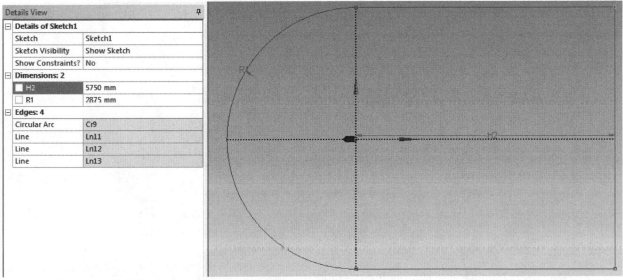

Figure 4.4b) Dimensions for mesh region

Select Concept>>Surface from Sketches. Select Sketch1 under Plane4 in the Tree Outline and Apply as Base Objects in Details View. Select Operation>>Add Frozen in Details View followed by Generate.

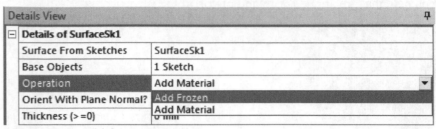

Figure 4.4c) Add frozen operation

5. Select Create>>Boolean from the menu. Select Subtract Operation from the Details View. Click on the mesh region in the graphics window and click on Apply as Target Bodies in Details View. Zoom in on the airfoil, select the Airfoil as Tool Bodies in Details View and click on Generate, see Figure 4.5c) and Figure 4.5d).

Figure 4.5a) Creating a Boolean

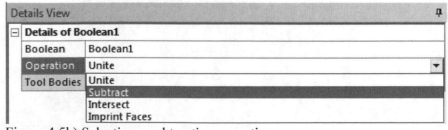

Figure 4.5b) Selecting a subtracting operation

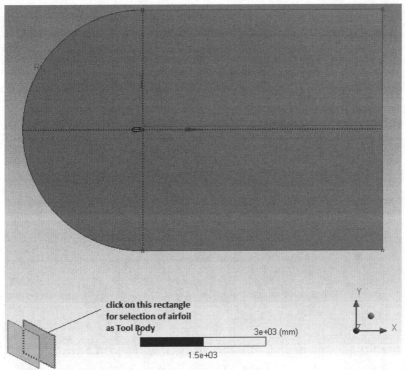

Figure 4.5c) Selection of Tool Body

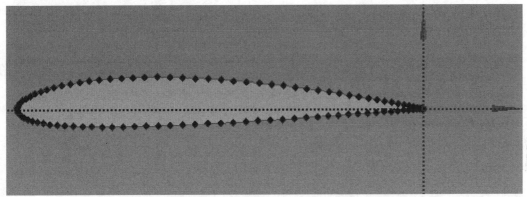

Figure 4.5d) Completed subtraction operation

6. Select Plane 4 in the tree outline and create a new sketch [ ]. Select the Sketching tab and the Line tool under Draw. Draw a line on the vertical axis from top to bottom of the mesh that intersects the entire mesh region. Select Concepts>>Lines from Sketches from the menu. Select the newly created vertical line and Apply it as a Base Object in the Details View. Click on Generate.

Repeat this step as described above and create a new sketch with a line on the horizontal axis in Plane4 that is drawn from left to right and goes through the whole mesh region including the leading and trailing edges of the airfoil. You will need to create a Coincident Constraint using the Sketching Toolbox and insert the constraint between the new line and the horizontal axis. This is completed by selecting the Coincident Constraint and clicking on both the new line and the horizontal axis. It will be helpful to draw the horizontal line a little bit below the horizontal axis and use the constraint to fix the line with the horizontal axis.

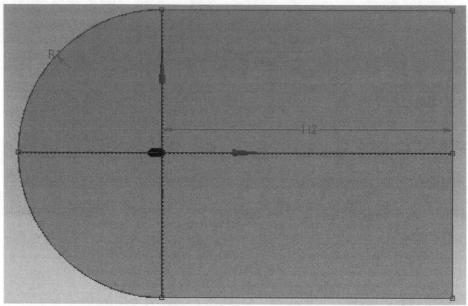

Figure 4.6 Completed vertical and horizontal lines through the mesh region

7. Select Tools>>Projection from the menu. Select Edges on Face as Type in Details View. Control-select both parts of the newly created vertical line and the two parts of the newly created horizontal line and Apply the four Edges in the Details View. Click in the yellow region next to Target in Details View. Select the mesh region around the airfoil and Apply it as the Target under Details View. Click Generate. You should now have four different mesh regions.

Figure 4.7a) Selection of the projection tool

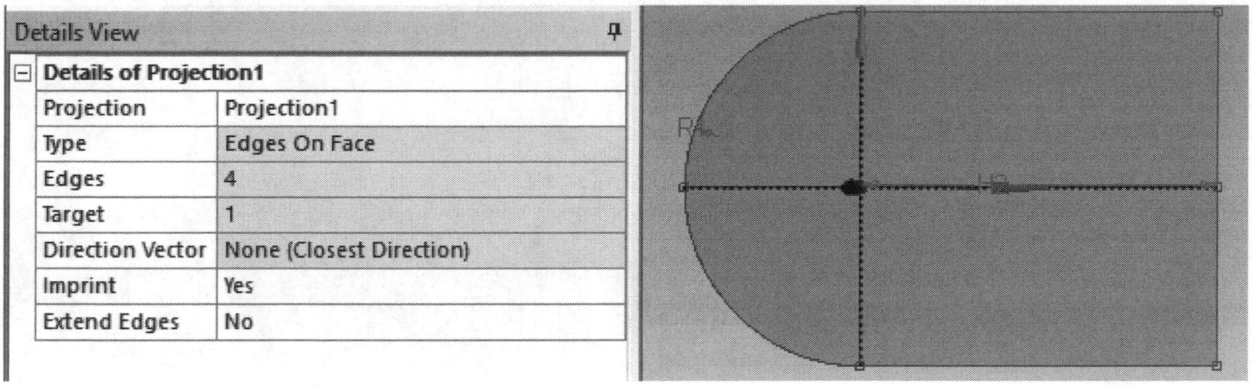

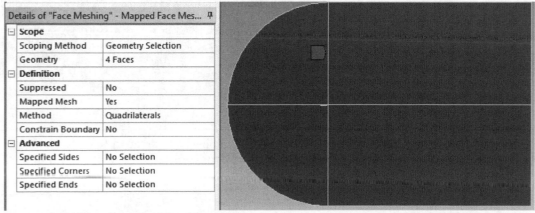

Figure 4.7b) Details view for projection

Click on the plus sign next to 2 Parts, 2 Bodies in the Tree Outline. Right click the Line Body and select Suppress Body. Select File>>Save Project from the menu and save the project with the name *NACA 2412 Airfoil Flow Study*. Close DesignModeler.

E. Launching ANSYS Meshing

8. Double click on Mesh under Project Schematic in ANSYS Workbench. Select Mesh in the Outline of the Meshing window and select Mesh>>Controls>>Face Meshing from the menu. Control-select all four faces of the mesh. Apply the Geometry under Details of "Face Meshing".

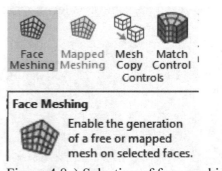

Figure 4.8a) Selection of face meshing

Details of "Face Meshing" - Mapped Face Mes...	
Scope	
Scoping Method	Geometry Selection
Geometry	4 Faces
Definition	
Suppressed	No
Mapped Mesh	Yes
Method	Quadrilaterals
Constrain Boundary	No
Advanced	
Specified Sides	No Selection
Specified Corners	No Selection
Specified Ends	No Selection

Figure 4.8b) Detail of face meshing

Select Mesh Control>>Sizing from the menu and click on the Edge Selection Filter . Control-select the four edges as shown in Figure 4.8c). Apply the Geometry under Details of "Edge Sizing". Select Number of Divisions as Type under Details of "Edge Sizing". Set the number of Divisions to 50. Set the Behavior as Hard. Select the first Bias Type from the drop-down menu. Enter a Bias Factor of 150.

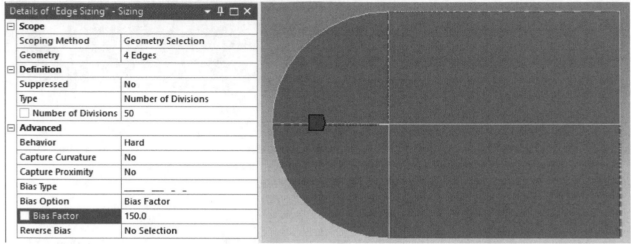

Figure 4.8c) Details for first edge sizing

Create another identical edge sizing but this time select the four remaining straight edges, see Figure 4.8d). Select the same number of divisions 50 and behavior Hard. Select the second Bias Type ‒ ‒ — ═══ from the top from the drop-down menu and the same bias factor of 150.

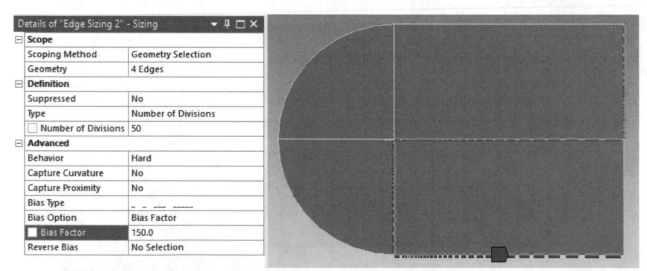

Figure 4.8d) Details for second edge sizing

> **Why do we use a biased mesh?** A biased mesh enables us to create a finer mesh closer to the airfoil. We create a fine mesh where it is needed the most and the calculation time will be much shorter compared with a finer mesh over the entire mesh region.

The final edge sizing will be for the two round edges. Set the Number of Divisions to 100 and Hard behavior. No bias will be used for the round edges. Right click on Mesh in the tree outline and select Generate Mesh.

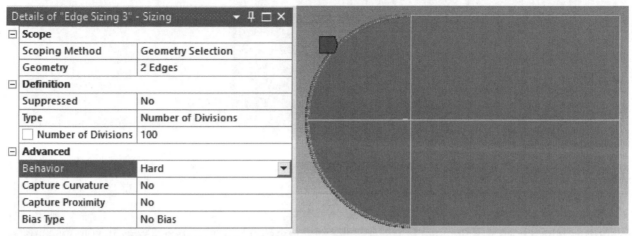

Figure 4.8e) Details of edge sizing for the two round edges

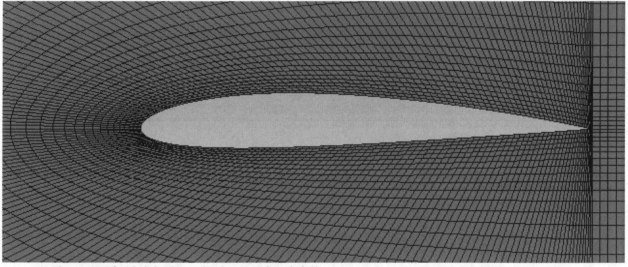

Figure 4.8f) Finished mesh close to the airfoil

9. Select Geometry in the Outline. Select the Edge Selection filter [icon]. Control-select the two vertical edges on the right-hand side of the mesh region, right click and select Create Named Selection. Name the edges "outlet" and click OK. Name the two round edges "inlet". Name the horizontal upper and lower edges "symmetry". Save the project. Select File>>Export>>Mesh>>FLUENT Input File>>Export from the menu. Enter *airfoil-flow-mesh* as file name and Save as type: FLUENT Input Files (*.msh). Close the meshing window. Right click on Mesh and select Update in ANSYS Workbench.

> **Why do we export the mesh?** When we save the mesh file, we can read the mesh using standalone mode directly in ANSYS Fluent without having to design the mesh all over again each time that we run the simulation for various flow variables.

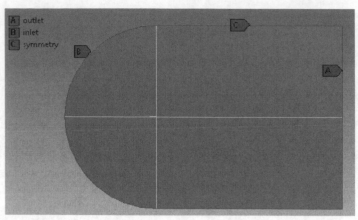

Figure 4.9 Named selections for mesh

F. Launching ANSYS Fluent

10. Double click on Setup under Project Schematic in ANSYS Workbench. Check the Options box for Double Precision. Select the number of Solver Processes under Parallel (Local Machine) to the same number as the number of cores for your computer. To check the number of physical cores, press the Ctrl + Shift + Esc keys simultaneously to open the Task Manager. Go to the Performance tab and select CPU from the left column. You'll see the number of physical cores on the bottom-right side. Close the Task Manager window. Click on Show More Options and take a note of the location of your working directory. Click Start to launch Fluent. Click OK when you get the Key Behavioral Changes window.

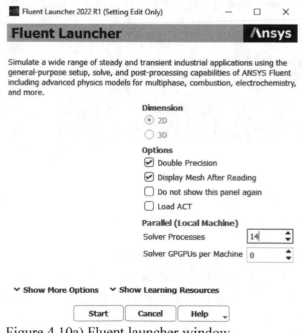

Figure 4.10a) Fluent launcher window

Select the Density-Based Solver under General on the Task Page. Double-click on Models under Setup in Outline View and double click on the Viscous (SST k-omega) on the Task Page. Select the Spalart-Allmaras (1 eqn) turbulence model. Select OK to close the Viscous Model window.

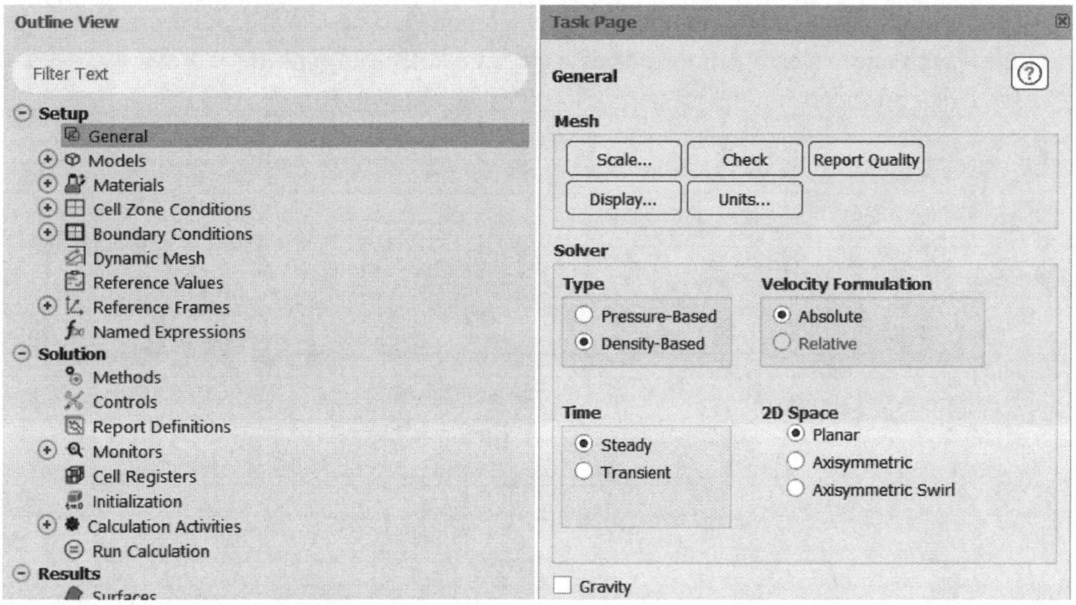

Figure 4.10b) General solution setup

Why did we select the Density-Based Solver? Both solver type options can be used for a variety of flow cases. Traditionally, the density-based solver has been used for compressible flow. For the flow past an airfoil that is studied in this chapter, by using the density-based solver we can choose to include the study of flows where compressibility of air is important.
Why do we use the Spalart-Allmaras Viscous Model? This viscous model was designed for boundary layers in aerospace applications.

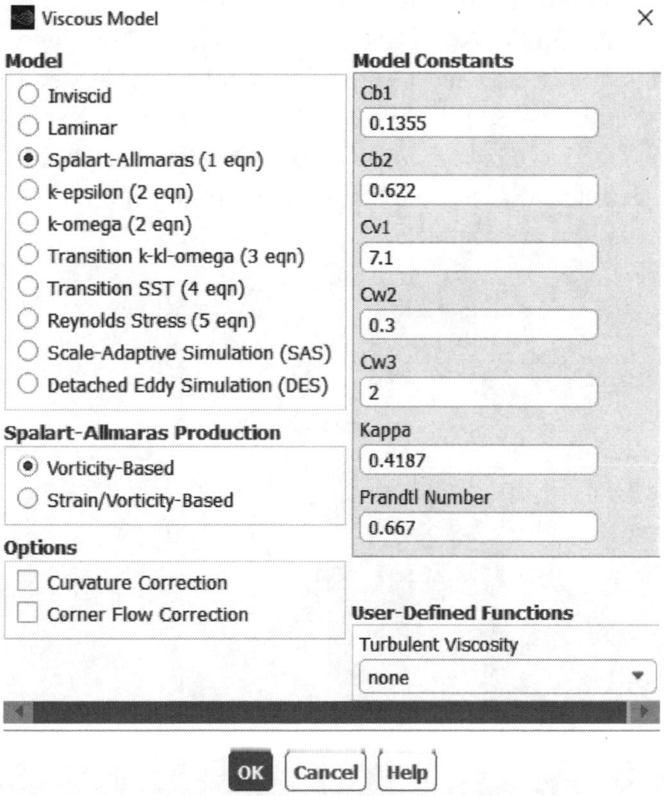

Figure 4.10c) Details for the viscous model

11. Double click on Boundary Conditions under Setup in Outline View. Double click the inlet Zone on the Task Page. Select Components as Velocity Specification Method. Enter 12.7 as X-Velocity [m/s]. Click on Apply and Close the Velocity Inlet window.

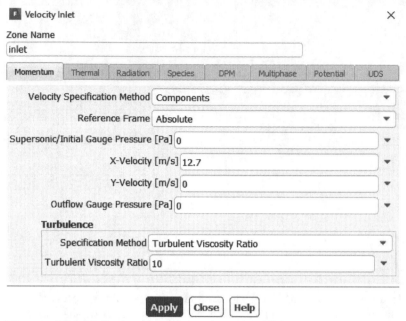

Figure 4.11 Velocity inlet boundary condition

12. Double click on Monitors under Solution in the Tree. Double click on Residual under Monitor. Change the Convergence Criterion for all four residuals to 1e-8. Click on OK to exit the Residual Monitors window. Double click on Initialization under Solution under the Outline View. Select Standard Initialization as Initialization Method, select Compute from inlet and click on Initialize. Select File>>Save Project from the menu. Select File>>Export>>Case & Data... from the menu. Save the Case/Data File with the name *Flow Past an Airfoil.cas.h5*. Double click on Run Calculation under Solution in the Outline View. Set number of Iterations to 10000. Click on Calculate. Click OK when the calculations are complete.

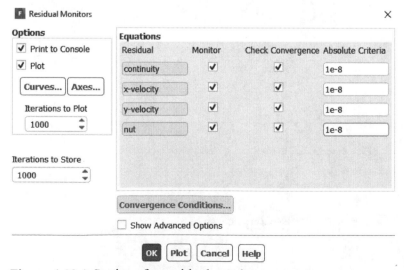

Figure 4.12a) Settings for residual monitors

Why did we select 1e-8 as the Absolute Criteria? We want to choose a sufficiently small value for the absolute criteria such that the drag and lift coefficients are not depending on the value of the absolute criteria. As an example, using 1e-12 instead of 1e-8 will not change the results for lift and drag coefficients. Therefore, the higher value was chosen as calculation times will be shorter.

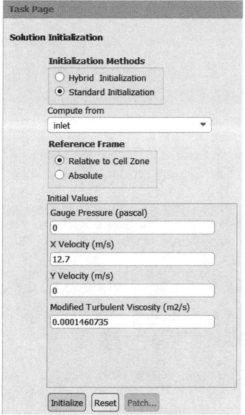

Figure 4.12b) Solution initialization

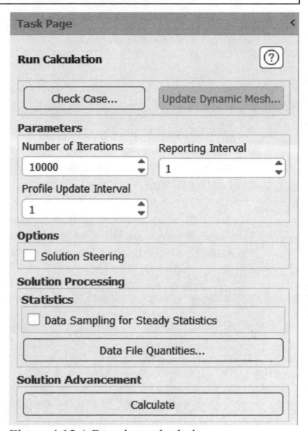

Figure 4.12c) Running calculations

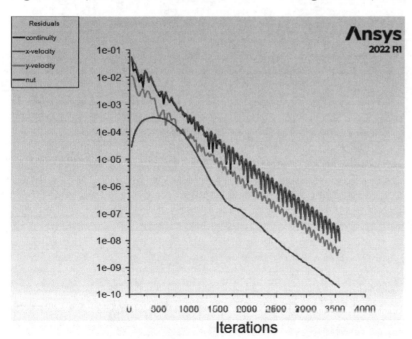

Figure 4.12d) Residuals window

Click on Copy Screenshot of Active Window to Clipboard, see Figure 4.12e). The Scaled Residuals can be pasted into a Word document.

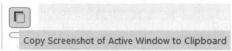

Figure 4.12e) Copying screenshot

G. Post-Processing

13. Double click on Graphics under Results in the Outline View. Display a plot of pressure contours by double clicking on Contours under Graphics, check the box for Filled under Options, select all Surfaces and select Pressure and Static Pressure under Contours of. Click on Colormap Options and set Colormap Alignment to Top. Set the Font Behavior to Fixed and Font Size under Font to 16. Set Type to general and Precision to 2 under Number Format. Select Apply followed by Close for the Colormap window. Click on Save/Display in the Contours window. Click on Copy Screenshot of Active Window to Clipboard, see Figure 4.12e). The Scaled Residuals can be pasted into a Word document. Create another contour plot for Velocity and Velocity Magnitude.

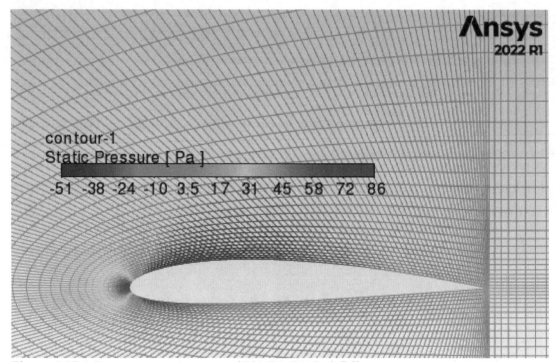

Figure 4.13a) Pressure contours around NACA 2412 airfoil

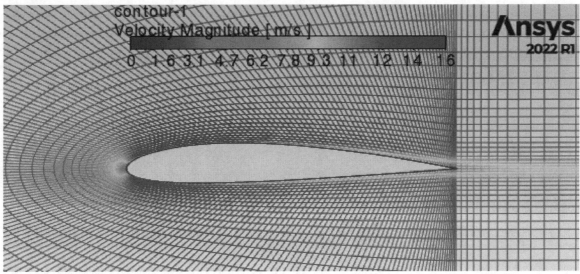

Figure 4.13b) Velocity contours around the NACA 2412 airfoil

> **Why did we select all surface when we created the pressure contour plots?** When we select all surfaces, the mesh will be shown on top of the contour plot. Deselecting all surfaces will cause the mesh to disappear.

14. Double click on Reference Values under Setup in the Outline View. Select Compute from inlet. Enter the value 0.23 for Length [m] which is the chord length and enter the value 12.7 for Velocity [m/s]. Enter the value 0.3048 for Depth [m] and 0.070104 for Area [m²]. The Area is equal to the Depth times the Length.

Task Page

Reference Values ⓘ

Compute from

inlet ▼

Reference Values

Area [m²]	0.070104
Density [kg/m³]	1.225
Depth [m]	0.3048
Enthalpy [J/kg]	0
Length [m]	0.23
Pressure [Pa]	0
Temperature [K]	288.16
Velocity [m/s]	12.7
Viscosity [kg/(m s)]	1.7894e-05
Ratio of Specific Heats	1.4
Yplus for Heat Tran. Coef.	300

Reference Zone

surface_body ▼

Figure 4.14a) Reference values

Find the coefficient of drag by double clicking on Reports under Results in the Outline View. Double click on Forces under Reports. Set the Direction Vector in the Force Reports window to 1 for X and 0 for Y. Select wall-surface_body under Wall Zones and click on Print. Write down the value for the total drag coefficient that you have calculated C_d = 0.017688834. Repeat this step to find the value for the lift coefficient using 0 for X and 1 for Y for the Direction Vector in the Force Reports window. The value for the lift coefficient is C_l = 0.19927453.

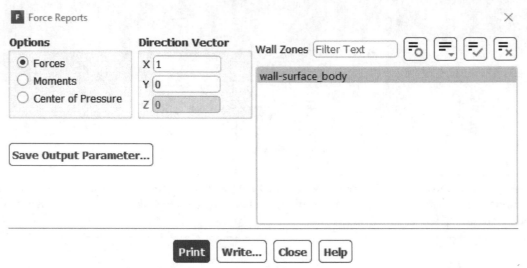

Figure 4.14b) Force reports

Select the Domain tab in the menu and Adapt>>Manual.... Select Cell Registers>>New>>Region in the Manual Mesh Adaption window. Define the region with X Min [m] -0.1, X Max [m] 0.5, Y Min [m] -0.1 and Y Max [m] 0.1. Click on Save and Close the Region Register window. Set the Maximum Refinement Level to 9 and the Minimum Edge Length [m] to 0 in the General Adaption Controls. Click on OK in the General Adaption Controls to close the window. Select *region_0* as Refinement Criterion in the Manual Mesh Adaption window. Select Adapt in the Manual Mesh Adaption window. Select OK in the Information window.

Reinitialize the solution and run the calculations. Enter the reference values once again. Recalculate the drag and lift coefficients, see Table 4.1. Repeat this step, refine the mesh once again, initialize the solution and rerun the calculations.

	Original Mesh	*Refinement Level 1*	*Refinement Level 1*	*Refinement Level 3*	*Refinement Level 4*
C_d	0.01769	0.01553	0.01560	0.01876	0.00629
C_l	0.19927	0.19885	0.20348	0.19557	0.18444
No. Cells	15000	29586	87765	319983	1244982
No. Faces	30300	59722	176582	642018	2494017
No. Nodes	15300	30136	88817	322035	1249035

Table 4.1 Drag and lift coefficient for different mesh sizes, *Re* = 200,000, *Angle of Attack* = 0

Why did we do mesh refinement? Any CFD study should be accompanied by a mesh independence study where you refine the mesh until the results converge. The ANSYS Student version has a maximum allowed number of elements and it can therefore sometimes be difficult to achieve mesh independence.

H. Theory

15. The Reynolds number for the flow over an airfoil is determined by

$$Re = \frac{Uc}{v} = 200{,}000 \qquad (4.1)$$

where U (m/s) is the magnitude of the free stream velocity, c (m) is the chord length of the airfoil, and v (m²/s) is the kinematic viscosity of air at room temperature. The angle of attack AoA is defined as the angle between the free stream and the chord line (the line between the leading edge and the trailing edge of the airfoil). For the flow case that we studied in this chapter the angle of attack was zero degrees.

The pressure distribution over the airfoil is expressed in non-dimensional form by the pressure coefficient

$$C_p = \frac{p_i - P}{\frac{1}{2}\rho U^2} \qquad (4.2)$$

where ρ (kg/m³) is the free stream density and where p_i (Pa) is the surface pressure at location i and p (Pa) is the pressure in the free stream. The pressure coefficient can alternatively be related to the velocity distribution u (m/s) around the airfoil through

$$C_p = 1 - \left(\frac{u}{U}\right)^2 \qquad (4.3)$$

The lift and drag coefficients can be determined from

$$C_l = \frac{F_l}{\frac{1}{2}\rho U^2 bc} \qquad (4.4)$$

$$C_d = \frac{F_d}{\frac{1}{2}\rho U^2 bc} \qquad (4.5)$$

, where F_l and F_d (N) are lift and drag forces and b (m) is the wingspan.

I. References

1. Jacobs E.N., Stack J. and Pinkerton R.M. Airfoil Pressure Distribution Investigation in the Variable Density Wind Tunnel, NACA Report No. 353, 1930.
2. Marchman III J.F and Werme T.D., Clark-Y Airfoil Performance at Low Reynolds Numbers, AIAA-84-0052, 1984.
3. Stern F., Muste M., Houser D., Wilson M. and Ghosh S., Measurement of Pressure Distribution and Forces acting on an Airfoil, Laboratory Experiment #3, 57:020 Mechanics of Fluids and Transfer Processes (http://css.engineering/uiowa.edu/fluidslab/pdfs/57-020/airfoil.doc)
4. Warner E.P., Airplane Design: Performance, McGraw-Hill, New York, 1936
5. Matsson J., A Student Project on Airfoil Performance, Annual Conference & Exposition, Honolulu, Hawaii, 2007.

6. Matsson J., Voth J., McCain C. and McGraw C., Aerodynamic Performance of the NACA 2412 Airfoil at Low Reynolds Number, ASEE 123rd Annual Conference & Exposition, New Orleans, LA, June 26-29, 2016.
7. Anderson, J., Fundamentals of Aerodynamics, 6th Edition, McGraw-Hill, 2017.
8. Abbott, I.H., Von Doenhoff, A.E., Theory of Wing Sections: Including a Summary of Airfoil Data, Dover Publications, 1959.
9. Saha, P. and Kaberwal, H.S., Numerical Analysis of Aerodynamic Performance Characteristics of NACA 2312 and NACA 2412, International Journal of Innovative Science and Research Technology, Volume 5, Issue 3, 2020.

J. Exercises

4.1 Determine contours of pressure and velocity for the flow field around an airfoil with a different NACA number than the one covered in this chapter. Include the corresponding mesh refinement level, number of iterations, refine threshold, number of cells refined, drag coefficient, lift coefficient, Reynolds number and the angle of attack as shown in Table 4.1 for the NACA 2412 airfoil. Include the residuals window after the solution has converged and include a printout of the mesh around the airfoil.

4.2 Continue with Exercise 4.1 using different angles of attack AoA = 2, 4, 6, 8, 10, 12 degrees for the same Reynolds number. For each AoA, refine the mesh until you get a converged solution. Plot the drag and lift coefficients versus AoA and compare with experimental results for the chosen airfoil using data from Abbott and Von Doenhoff[8].

4.3 Create an O-mesh mesh around the airfoil instead of the C-mesh that was modeled in this chapter. Run the simulations for the O-mesh and compare results with the mesh used in this chapter. Use the same radius for the O-mesh as for the C-mesh in this chapter.

CHAPTER 5. RAYLEIGH-BENARD CONVECTION

A. Objectives

- Using ANSYS Workbench to Create the Mesh for 3D Rayleigh-Bénard Convection
- Inserting Wall Boundary Conditions
- Running Laminar Steady 3D ANSYS Fluent Simulations
- Using Contour Plots for Visualizations
- Compare Results with Neutral Stability Theory

B. Problem Description

Bénard[1] was the first scientist to experimentally study convection in a fluid layer with a free surface heated from below. Rayleigh[2] used linear stability analysis to theoretically explain and study the stability of the fluid motion between horizontal parallel plates with the hotter plate at the bottom. Chandrasekhar[3] completed the linear stability analysis for Rayleigh-Bénard convection and Koschmieder[4] showed the development of the research in this area during the following couple of decades.

There are only a few experiments including Koschmieder and Pallas[5], Hoard et al.[6] and to some extent Matsson[6] that have produced the concentric ring cells in Rayleigh-Bénard convection that according to stability theory should appear with a rigid upper boundary condition. This instability with an upper rigid wall in experiments is very sensitive to the surface roughness and flatness of the upper wall. Torres et al.[7] completed 3D numerical simulations of Rayleigh-Benard convection in a cylindrical container and verified the existence of roll cells in the form of concentric circles.

In this chapter, we will study Rayleigh-Bénard convection and we will analyze the problem using ANSYS Fluent in 3D. The enclosure is filled with water and the temperature of the cold wall is 296.5 K while the hot wall is at 298 K. The diameter of the enclosure is 100 mm and the depth 4.8 mm.

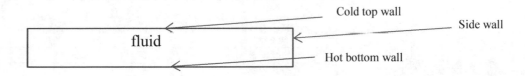

C. Launching ANSYS Workbench and Selecting Fluent

1. Start by launching ANSYS Workbench. Double click on Fluid Flow (Fluent).

Figure 5.1 Selecting Fluid Flow

D. Launching ANSYS DesignModeler

2. Right-click on Geometry in Project Schematic and select Properties. Select Geometry and 3D Analysis Type under Advanced Geometry Options in Properties of Schematic A2: Geometry. Right click on Geometry and open a New DesignModeler Geometry. Select **Millimeter as the Unit** from the menu in DesignModeler.

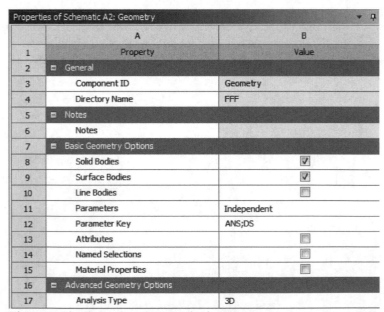

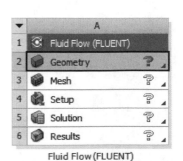

Figure 5.2a) Geometry   Figure 5.2b) Selecting 3D Analysis Type

3. Next, we will be creating the mesh region for the simulation. Select XYPlane from the Tree Outline. Select Look at Sketch 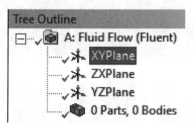. Select the Sketching tab and draw a circle from the origin. Select the Dimensions tab in Sketching Toolboxes. Click on the circle and enter a diameter of 100 mm. Select Create>>Extrude from the menu. Select Sketch1 under XYPlane in the Tree Outline and Apply the sketch as Geometry in Details View. Enter 4.8 mm as FD1, Depth (>0). Click on Generate ⫴ ⫽ Generate and Close DesignModeler.

Figure 5.3a) Selection of XYPlane

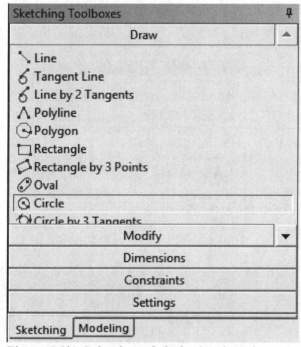

Figure 5.3b) Selection of circle sketch tool

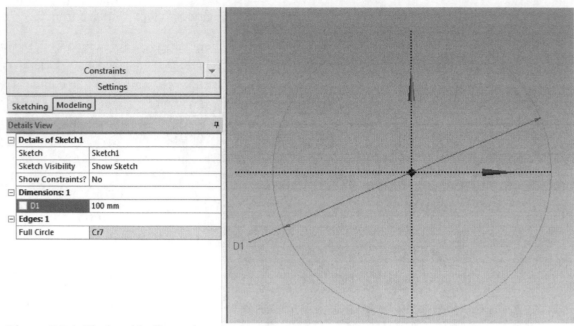

Figure 5.3c) Circle with dimensions in details view

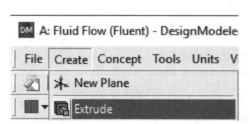

Figure 5.3d) Selecting extrusion

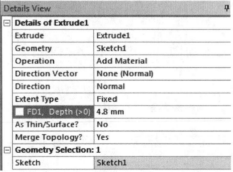

Figure 5.3e) Details of extrusion

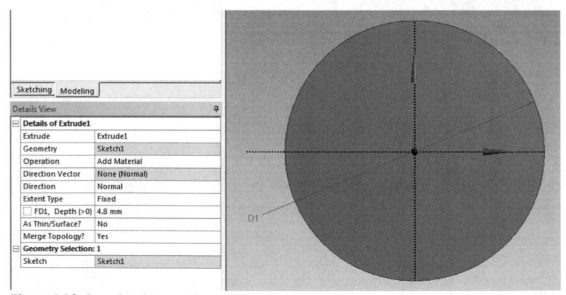

Figure 5.3f) Completed generation of 3D geometry

E. Launching ANSYS Meshing

4. We are now going to double click on Mesh under Project Schematic in ANSYS Workbench to open the meshing window. Select Mesh in the Outline. Select Unit Systems>>Metric (mm, kg, N …) from the bottom of the graphics window. Right click on Mesh in the Outline and select Update. A coarse mesh is created.

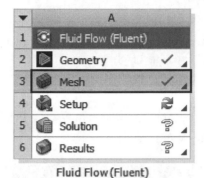

Figure 5.4a) Starting Mesh

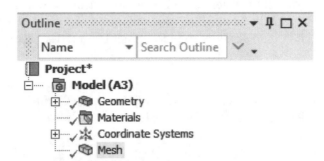

Figure 5.4b) Selection of Mesh in Outline

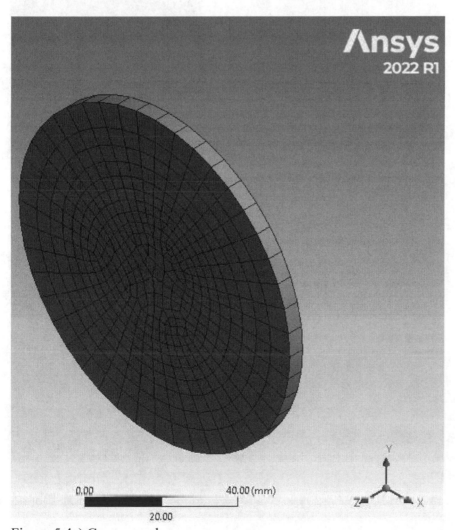

Figure 5.4c) Coarse mesh

5. Select Mesh>>Controls>>Face Meshing from the menu. Click on the cylindrical face of the mesh region in the graphics window. The face turns green. Click on the Apply button for Geometry in Details of "Face Meshing".

Select Mesh>>Controls>>Sizing from the menu and select Edge 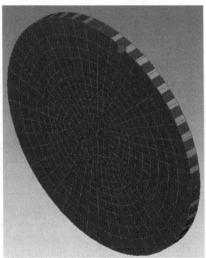. Control click on the two circular edges of the mesh region. Click on Apply for the Geometry in "Details of Edge Sizing". Under Definition in "Details of Edge Sizing", select Number of Divisions for Type, 500 for Number of Divisions, and Hard as Behavior.

Repeat the selection of Mesh>>Controls>>Sizing from the menu and select Face . Control select the two plane faces and the cylindrical face of the mesh region. You will need to rotate the model to select both faces. Enter 0.5 mm for Element Size and Hard Behavior.

Repeat the selection of Mesh>>Controls>>Sizing from the menu and select Body . Select and Apply the body of the mesh region. Enter 5 mm for Element Size and Hard Behavior. Click on Update and select Mesh in the Outline. Right click in the graphics window and select Isometric View. The finished mesh is shown in the graphics window.

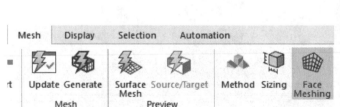

Figure 5.5a) Generation of Face Meshing

Figure 5.5b) Face for Face Meshing

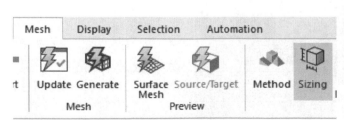

Figure 5.5c) Mesh Control Sizing

Figure 5.5d) Round edges for sizing

Details of "Edge Sizing" - Sizing	▾ ⊓ □ ✕
Scope	
Scoping Method	Geometry Selection
Geometry	2 Edges
Definition	
Suppressed	No
Type	Number of Divisions
☐ Number of Divisions	500
Advanced	
Behavior	Hard
Capture Curvature	No
Capture Proximity	No
Bias Type	No Bias

Figure 5.5e) Details of edge sizing

Figure 5.5f) Faces for face sizing

Details of "Face Sizing" - Sizing	▾ ⊓ □ ✕
Scope	
Scoping Method	Geometry Selection
Geometry	3 Faces
Definition	
Suppressed	No
Type	Element Size
☐ Element Size	0.5 mm
Advanced	
☐ Defeature Size	Default (3.5376e-002 mm)
Influence Volume	No
Behavior	Hard
Capture Curvature	No
Capture Proximity	No

Figure 5.5g) Details of face sizing

Details of "Body Sizing" - Sizing	▾ ⊓ □ ✕
Scope	
Scoping Method	Geometry Selection
Geometry	1 Body
Definition	
Suppressed	No
Type	Element Size
☐ Element Size	5.0 mm
Advanced	
☐ Defeature Size	Default (3.5376e-002 mm)
Behavior	Hard
Capture Curvature	No
Capture Proximity	No

Figure 5.5h) Details of body sizing

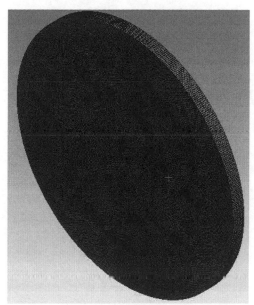

Figure 5.5i) Finished mesh

Details of "Mesh"	▾ ⊓ □ ✕
Display	
Display Style	Use Geometry Setting
Defaults	
Physics Preference	CFD
Solver Preference	Fluent
Element Order	Linear
☐ Element Size	Default (7.0751 mm)
Export Format	Standard
Export Preview Surface Mesh	No
⊞ **Sizing**	
⊞ **Quality**	
⊞ **Inflation**	
⊞ **Advanced**	
Statistics	
☐ Nodes	303057
☐ Elements	267720

Figure 5.5j) Details of mesh

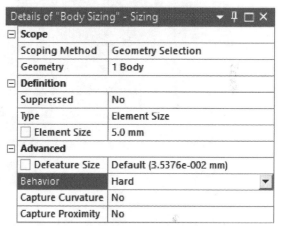
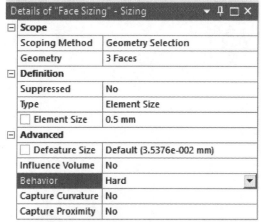

6. Select Face 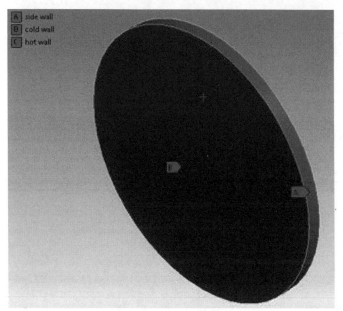 and select the cylindrical face of the mesh region, right click and select Create Named Selection. Enter *side wall* as the name and click on the OK button. Repeat this step for the upper and lower faces in the Z-direction of the mesh region and name these *cold wall* and *hot wall*, respectively.

Select File>>Export>>Mesh>>FLUENT Input File>>Export from the menu and save the mesh with the name *rayleigh-benard-3D.msh*. Select File>>Save Project from the menu and name the project "3D Rayleigh-Benard Convection". Select File>>Close Meshing to close the meshing window. Right click on Mesh in ANSYS Workbench and select Update.

Figure 5.6a) Named selections for the mesh

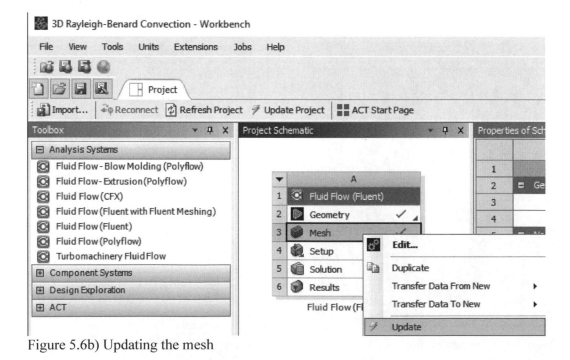

Figure 5.6b) Updating the mesh

F. Launching ANSYS Fluent

7. Double click on Setup under Project Schematic in ANSYS Workbench. Check the box for Double Precision. Select the number of Solver Processes under Parallel (Local Machine) to the same number as the number of cores for your computer. To check the number of physical cores, press the Ctrl + Shift + Esc keys simultaneously to open the Task Manager. Go to the Performance tab and select CPU from the left column. You'll see the number of physical cores on the bottom-right side. Close the Task Manager window. Click on Show More Options and take a note of the location of your Working Directory. Click Start to launch Fluent. Click Start to launch Fluent. Click OK when you get the Key Behavioral Changes window.

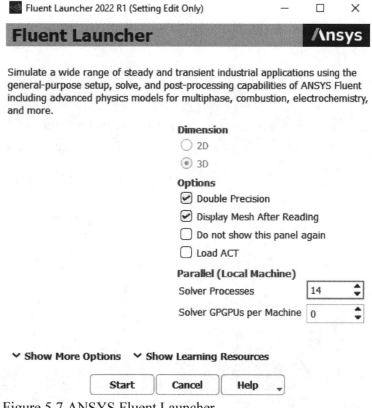

Figure 5.7 ANSYS Fluent Launcher

8. Check the mesh in ANSYS Fluent by selecting the Check button under Mesh in General Setup. Check the scale of the mesh by selecting the Scale button. Make sure that the Domain Extent is correct and Close the Scale Mesh window.

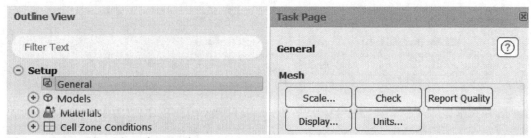

Figure 5.8a) Checking the mesh

```
Domain Extents:
  x-coordinate: min (m) = -5.000000e-02, max (m) = 5.000000e-02
  y-coordinate: min (m) = -5.000000e-02, max (m) = 5.000000e-02
  z-coordinate: min (m) = 0.000000e+00, max (m) = 4.800000e-03
Volume statistics:
  minimum volume (m3): 6.348532e-12
  maximum volume (m3): 3.542288e-10
    total volume (m3): 3.769812e-05
Face area statistics:
  minimum face area (m2): 1.058089e-08
  maximum face area (m2): 5.903813e-07
Checking mesh.....................................
Done.
```

Figure 5.8b) Console output generated by mesh check

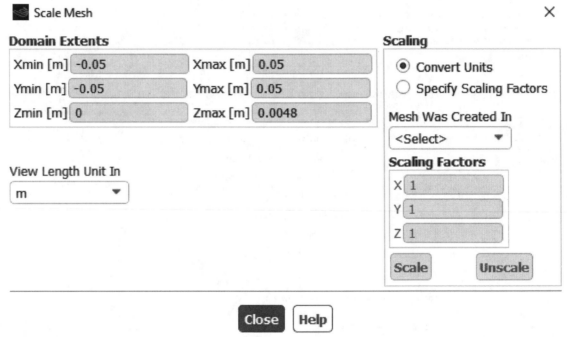

Figure 5.8c) Scale Mesh window

9. Check the Gravity box and enter -9.81 as the Gravitational Acceleration in the Z direction. Double click on Models under Setup in Outline View. Double click on the Energy model and check the box for the Energy equation. Click OK to exit the Energy Model window.

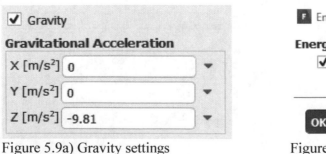

Figure 5.9a) Gravity settings

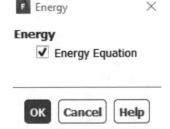

Figure 5.9b) Energy equation

10. Double click on Materials under Setup in the Outline View. Click on the Create/Edit button for Fluid on the Task Page to open the Create/Edit Materials window. Select Fluent Database…. Scroll down in the Fluent Fluid Materials window and select *water-liquid (h2o<l>)*. Click on the Copy button and Close the Fluent Database Materials window. Select *boussinesq* using the drop-down menu next to Density under Properties. Enter the value 997 for the Density [kg/m3]. Enter the value 4180 for Cp (Specific Heat) [J/(kg K)] , 0.607 for Thermal Conductivity [W/(m K)], and 0.000891for Viscosity [kg/(m s)]. Scroll down and enter the value 0.000247 for the Thermal Expansion Coefficient [K^{-1}]. Click on the Change/Create button and Close the window.

Why did we choose the boussinesq setting? The Boussinesq approximation is used for buoyancy driven flows. The reason for using boussinesq for density is that this will give faster convergence. Density is treated as a constant in all equations except for the buoyancy term in the momentum equation.

Double click on Boundary Conditions under Setup in the Outline View. Click on Operating Conditions on the Task Page. Set the value for the Operating Temperature [K] to 297.395 and click on OK to close the Operating Conditions box.

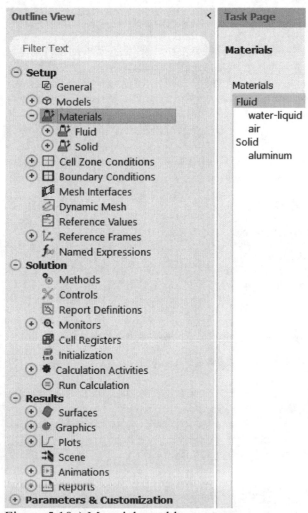

Figure 5.10a) Materials problem setup

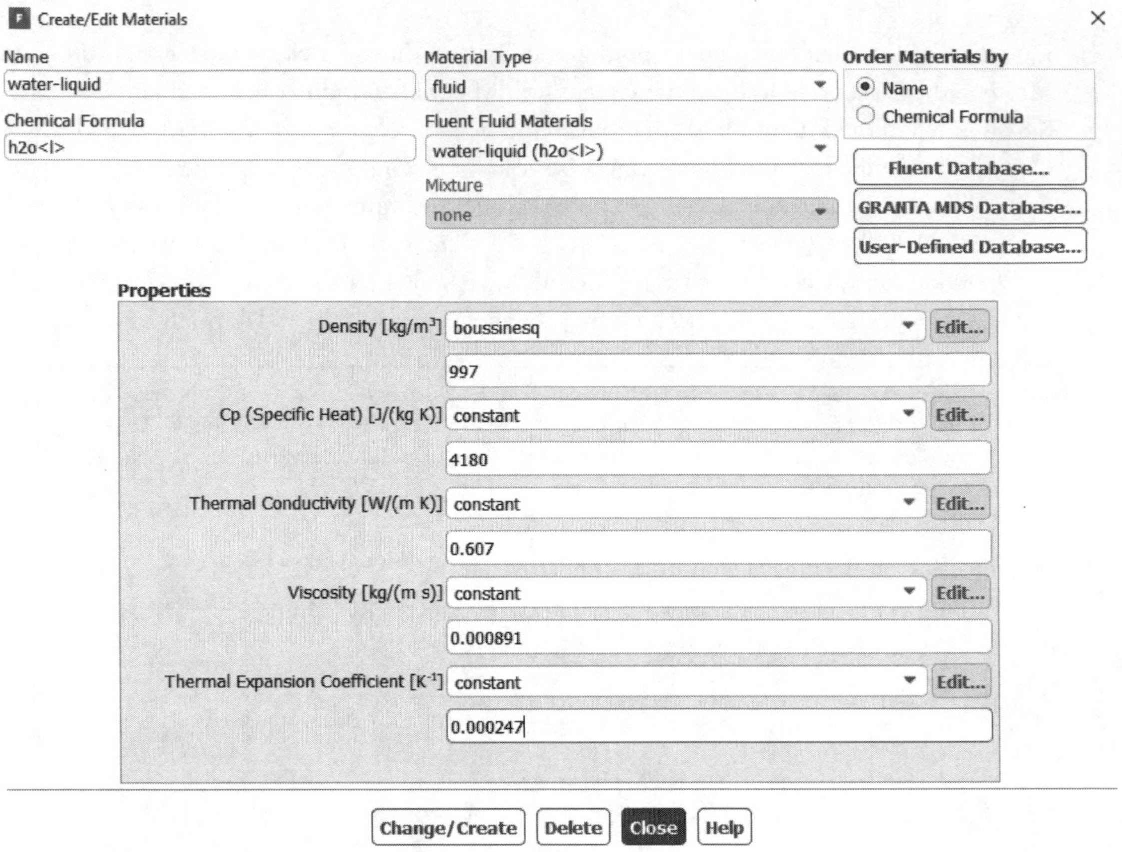

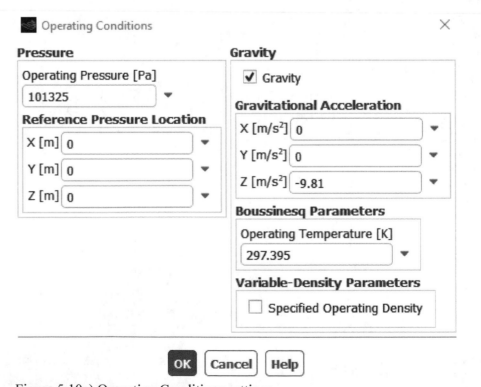

Figure 5.10b) Create/Edit Materials window

Figure 5.10c) Operating Conditions settings

11. Open Boundary Conditions under Setup in the Outline View. Double click on *cold_wall* under Boundary Conditions on the Task Page. Select the Thermal tab and select Temperature for Thermal Conditions. Enter the value 296.79 as Temperature [K]. Click on Apply and Close the window. Double click *hot_wall* under Boundary Conditions on the Task Page and set the Temperature [K] boundary condition to 298. Double click on *side_wall* under Boundary Conditions on the Task Page and set the Temperature [K] boundary condition to 296.79. Double click on Cell Zone Conditions under Setup in the Outline View. Double click on solid under Cell Zone Conditions on the Task Page. Select *water-liquid* from the Material Name drop-down menu in the Fluid window. Click on Apply and Close the window.

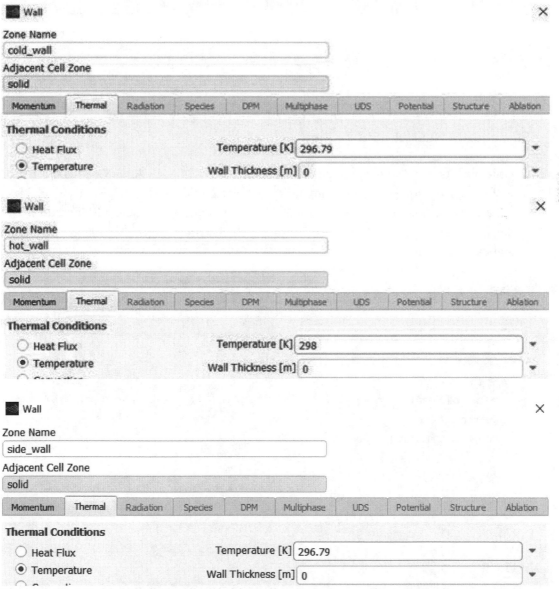

Figure 5.11a) Cold wall, hot wall and side wall temperature settings

Figure 5.11b) Fluid selection

12. Double click on Models under Setup in Outline View. Double click on Viscous (SST k-omega) model and check the box for the Laminar model. Click OK to exit the Viscous Model window. Double click on Initialization under Solution in the Outline View. Select Standard Initialization under Initialization Methods. Enter 0.01 for Z Velocity [m/s], enter 297.395 as Temperature [K] and click on Initialize.

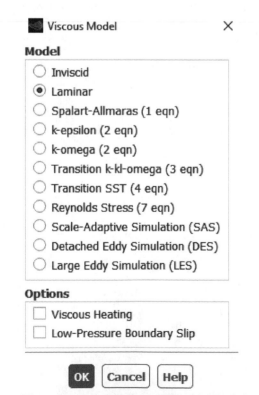

Figure 5.12a) Laminar Viscous Model

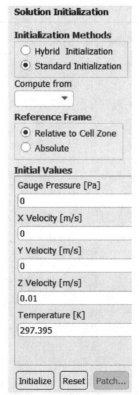

Figure 5.12b) Solution Initialization settings

110

13. Double click on Monitors under Solution in the Outline View and double click on Residual under Monitors. Set the Absolute Criteria to 1e-6 for all equations and click OK to close the window. Double click on Methods under Solution in the Outline View and select Body Force Weighted from the Pressure drop down menu under Spatial Discretization. Select SIMPLE as Pressure-Velocity Coupling Scheme.

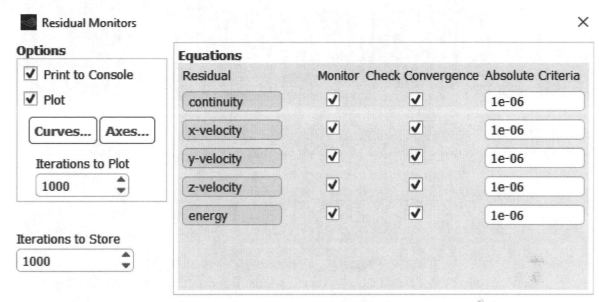

Figure 5.13a) Residual Monitors Settings

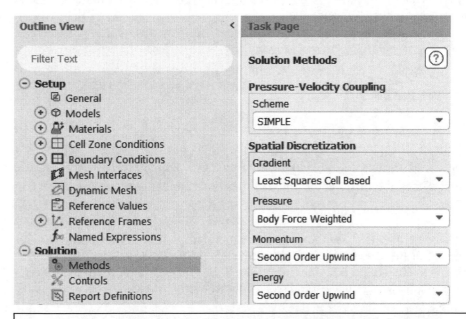

Why did we choose Body Force Weighted and SIMPLE Solution Methods? The reason for using SIMPLE as Pressure-Velocity Coupling algorithm is that we are studying steady laminar flow. SIMPLEC may alternatively be used as it can provide a faster solution. PISO is used for transient solutions. Coupled can also give an efficient steady state solution for single phase flows. The Body Force Weighted pressure scheme is recommended when we have large body forces such as gravity in Rayleigh-Benard convection.

Figure 5.13b) Solution Methods Settings

14. Select File>>Save Project from the menu. Select File>>Export>>Case… from the menu. Save the Case File with the name *Rayleigh-Benard Convection.cas.h5*. Double click Run Calculation under Solution in the Outline View. Set the number of iterations to 6000. Click on the Calculate button. Click on the OK button in the Information window that appears when the calculations are complete.

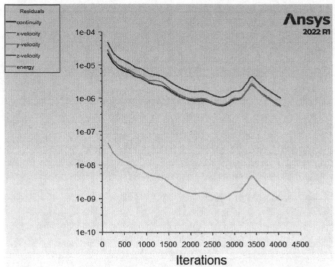

Figure 5.14a) Scaled Residuals for Rayleigh-Bénard convection after 4062 iterations

Click on Copy Screenshot of Active Window to Clipboard, see Figure 5.14b). The Scaled Residuals can be pasted into a Word document.

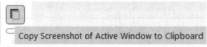

Figure 5.14b) Copying screenshot

G. Post-Processing

15. Double click on Graphics under Results in the Outline View. Double click on Contours in the Graphics section. Check the Filled and Contour Lines boxes under Options and select Contours of Velocity and Z Velocity.

Select New Surface>>Plane… from the drop-down menu in the Contours window. Select Three Points as Method. For Point 1, set (X (m), Y (m), Z (m)) to (0, -0.05, 0.0024), for Point 2 set (X (m), Y (m), Z (m)) to (0, 0.05, 0.0024) and for Point 3 set (X (m), Y(m), Z (m)) to (0.05, 0, 0.0024). Enter the name *mid-plane-z=0.0024m* as the New Surface Name. Click on Create and Close the window.

Select the new plane under Surfaces in the Contours window. Click on the Save/Display button and Close the window. Select the x-y plane view. Repeat this step but select Contours of Temperature and Static Temperature. Click on Colormap Options. Select float as Type for Number Format and set Precision to 2. Click on Apply and Close the Colormap window. Click on the Save/Display button. Select the x-y plane view.

Figure 5.15a) Creation of a new x-y plane   Figure 5.15b) Contours settings

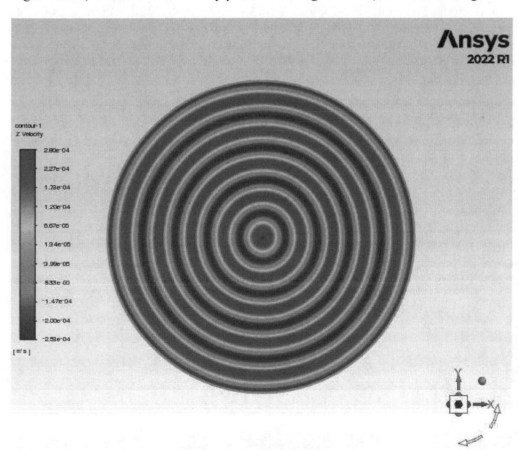

Figure 5.15c) Contours of Z Velocity in x-y plane, $Ra = 2491$

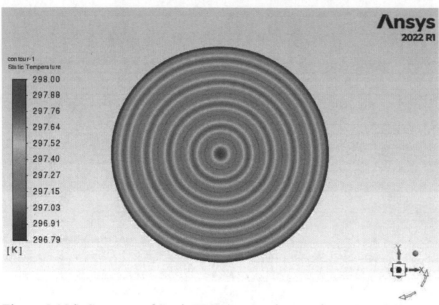

Figure 5.15d) Contours of Static Temperature in x-y plane, *Ra* = 2491

Click on Copy Screenshot of Active Window to Clipboard, see Figure 5.14b). The Contours can be pasted into a Word document.

16. Select New Surface>>Plane… from the drop-down menu in the Contours window. Select Three Points as Method. For Point 1, set (X (m), Y (m), Z (m)) to (0, 0, 0), for Point 2 set (X (m), Y (m), Z (m)) to (0, 0, 0.0048) and for Point 3 set (X (m), Y (m), Z (m)) to (0.05, 0, 0). Enter the name *x-z-plane-y=0* as the New Surface Name. Click on Create and Close the window. Select the new plane under Surfaces in the Contours window. Click on Colormap Options. Select float as Type for Number Format and set Precision to 2. Set Colormap Alignment to Top. Click on Apply and close the Colormap window. Click on the Save/Display button to display the Temperature field. Select the x-z plane view. Finally, display the Z Velocity field in the *x-z* plane, see Figure 5.16c). You will need to change Colormap Options. Select float as Type for Number Format and set Precision to 5.

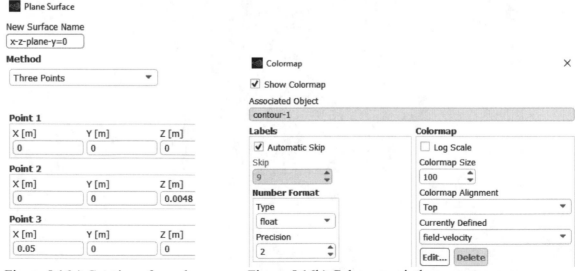

Figure 5.16a) Creation of *x-z* plane   Figure 5.16b) Colormap window

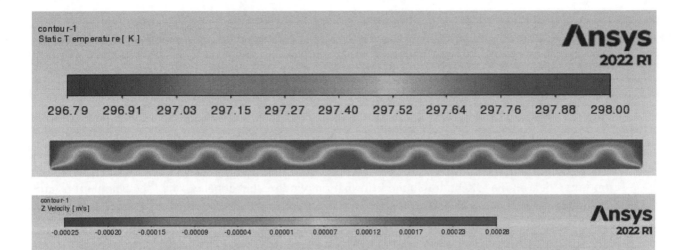

Figure 5.16c) Contours of static temperature and Z velocity in x-z plane, $Ra = 2491$

Click on Copy Screenshot of Active Window to Clipboard, see Figure 5.14b). The Contours can be pasted into a Word document.

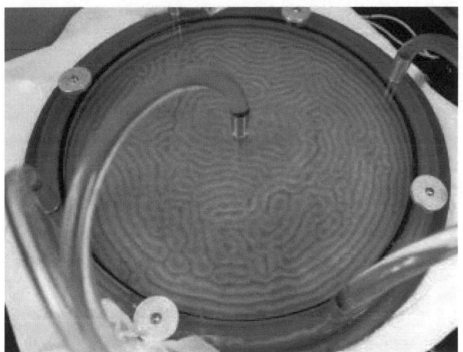

Figure 5.16d) Photo of experimental Rayleigh-Benard convection at $Ra = 2490$ with a rigid upper wall, Matsson[7].

Why do the experiments have a different appearance than the concentric roll cells from ANSYS Fluent simulations? In the experiments with a rigid upper wall, it is hard to get the perfectly concentric circular ring pattern as shown from simulations. The reason for this is that the experiments are sensitive to the non-flatness and imperfections of the bottom and top surfaces. The upper lid in experiments is an acrylic plate that is not perfectly flat.

17. Double click on Boundary Conditions under Setup in the Outline View. Double click on *cold_wall* under Boundary Conditions on the Task Page. Select the Momentum tab and select Marangoni Stress as Shear Condition. Enter the value -2e-5 as Surface Tension Gradient [N/(m K)] for Marangoni Stress. Click Apply and Close to exit the window. Uncheck the Gravity box under General in the Setup.

Double click on Initialization under Solution in the Outline View. Select Standard Initialization under Initialization Methods. Enter 0.01 for Z Velocity [m/s], enter 297.395 as Temperature [K] and click on Initialize. Click OK when you get a Question box. Double click Run Calculation under Solution in the Outline View. Set the number of iterations to 5000. Click on the Calculate button. Click on the OK button in the Information window that appears when the calculations are complete. Repeat steps **15 – 16**.

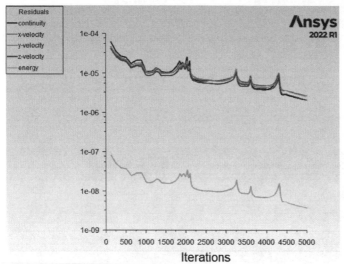

Figure 5.17a) Scaled Residuals for Marangoni convection

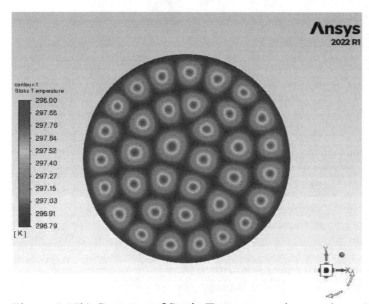

Figure 5.17b) Contours of Static Temperature in x-y plane, *M* = 895

Click on Copy Screenshot of Active Window to Clipboard, see Figure 5.14b). The Contours can be pasted into a Word document.

116

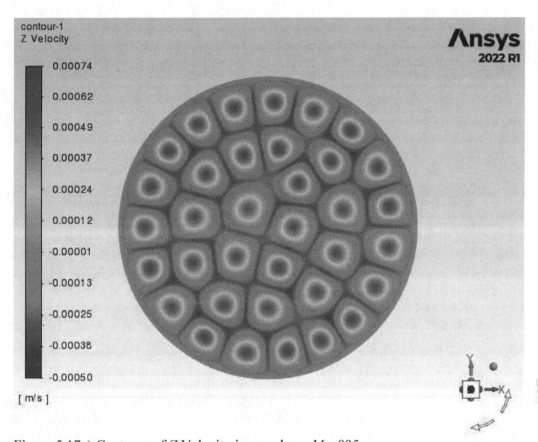

Figure 5.17c) Contours of Z Velocity in x-y plane, $M = 895$

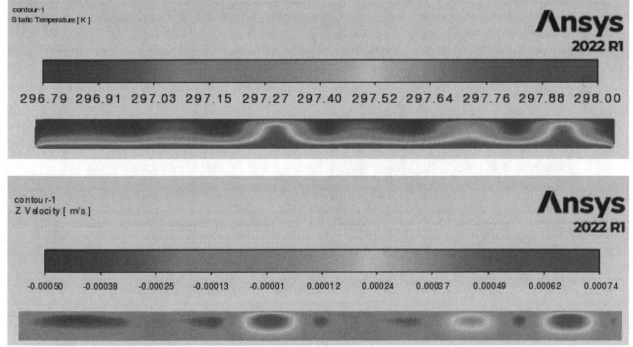

Figure 5.17d) Contours of static temperature and Z velocity in x-z plane, $M = 895$

Click on Copy Screenshot of Active Window to Clipboard, see Figure 5.14b). The Contours can be pasted into a Word document.

H. Theory

Comparison with Neutral Stability Theory for Rayleigh-Bénard Convection

18. The instability of the fluid layer between two parallel plates heated from below is governed by the Rayleigh number Ra.

$$Ra = \frac{g\beta\rho^2 C_p(T_1-T_2)L_c^3}{\mu k} \tag{5.1}$$

where g is acceleration due to gravity, β is the coefficient of volume expansion, ρ is density of the fluid, C_p is the specific heat, T_1 and T_2 are the temperatures of the hot and cold surfaces respectively, L_c is the distance between the surfaces (fluid layer thickness), k is thermal conductivity of the fluid, and μ is the dynamic viscosity of the fluid. Below the critical Ra_{crit} = 1708 for a rigid upper surface, the flow is stable but convective currents will develop above this Rayleigh number. For the case of a free upper surface, the theory predicts a lower critical Rayleigh number at 1101 and for two free boundaries the critical Rayleigh number is 657. The non-dimensional wave number α of this instability is determined by

$$\alpha = \frac{2\pi L_c}{\lambda} \tag{5.2}$$

where λ is the wave length of the instability. The determination of the average wave-length is determined from Figure 5.15d) and shown in Figure 5.18a). The diameter of the computational domain is 100 mm. The length for eight wave-lengths can be measured in Figure 5.18a) as a percentage of the 100 mm diameter using a ruler. In this case $8\lambda = 88.3\ mm$ which gives an average wave-length $\lambda = 11.0375\ mm$.

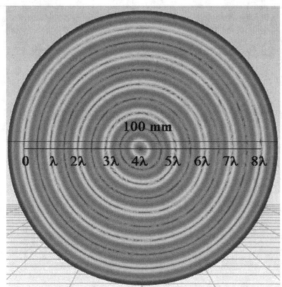

Figure 5.18a) Determination of average wave-length

The critical wave numbers are α_{crit} = 3.12, 2.68 for rigid and free surface boundary conditions, respectively. The wave number can be determined to be

$$\alpha = \frac{2\pi \cdot 0.0048}{0.0110375} = 2.73 \tag{5.3}$$

CHAPTER 5. RAYLEIGH-BENARD CONVECTION

The Rayleigh number in the flow simulation for rigid-rigid boundaries is

$$Ra = \frac{g\beta\rho^2 C_p(T_1-T_2)L_c^3}{\mu\,k} = \frac{9.81*0.000247*997^2*4180*1.5*0.0048^{\wedge}3}{0.000891*0.607} = 2491 \tag{5.4}$$

The neutral stability curve for Rayleigh-Benard convection with rigid-rigid boundaries can be given to the first approximation, see Figure 5.18b).

$$Ra = \frac{(\pi^2+\alpha^2)^3}{\alpha^2\left\{1-\dfrac{16\alpha\pi^2\cosh^2\left(\frac{\alpha}{2}\right)}{\left[(\pi^2+\alpha^2)^2(\sinh\alpha+\alpha)\right]}\right\}} \tag{5.5}$$

The corresponding expression the neutral stability curve to the first approximation for rigid-free boundary conditions is

$$Ra = \frac{(\pi^2+\alpha^2)^3}{\alpha^2\left\{1-\dfrac{8\alpha\pi^2\sinh^2(\alpha)}{\left[(\pi^2+\alpha^2)^2(\sinh 2\alpha-2\alpha)\right]}\right\}} \tag{5.6}$$

The neutral stability curve for Rayleigh-Benard convection with free-free boundaries is

$$Ra = \frac{(\pi^2+\alpha^2)^3}{\alpha^2} \tag{5.7}$$

Boundaries	$R_{c,\ exact}$	$R_{c,\ 1st\ approx.}$	α_c	α_{Fluent}	% diff.
Free-free	657.5	----------	2.22		
Rigid-free	1100.7	1112.7	2.68		
Rigid-rigid	1707.8	1715.1	3.12	2.73	12.5

Table 5.1 Critical parameters for Rayleigh-Benard convection

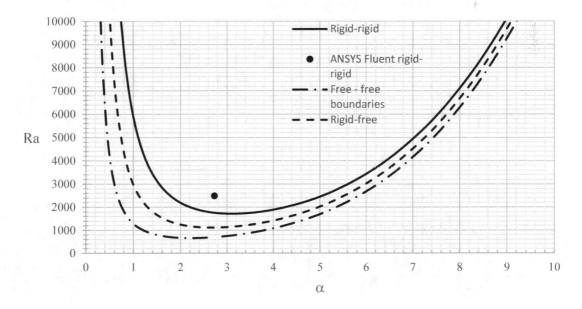

Figure 5.18b) Neutral stability curve for Rayleigh-Bénard convection. The filled circle represents result from ANSYS Fluent simulation.

Comparison with Neutral Stability Theory for Surface-Tension-Driven Benard Convection (Marangoni Convection)

The instability of the fluid layer heated from below and affected by surface tension gradients is governed by the Marangoni number M.

$$M = \frac{\sigma(T_1 - T_2)L_c}{\mu \kappa} \tag{5.8}$$

where $\sigma = -\frac{\partial S}{\partial T}$ is the surface tension S rate of change with temperature T, T_1 and T_2 are the hot and cold temperatures respectively, L_c is the fluid layer thickness, κ is thermal diffusivity, and μ is the dynamic viscosity of the fluid.

Below the critical $M_{crit} = 79.6$, the flow is stable. The non-dimensional wave number α of this instability is determined by

$$\alpha = \frac{2\pi L_c}{\lambda} \tag{5.9}$$

where λ is the wave length of the instability. The determination of the average wave-length is determined from Figure 5.17d) and shown in Figure 5.18c). The diameter of the computational domain is 100 mm. The length for five wave-lengths can be measured in Figure 5.18c) as a percentage of the 100 mm diameter using a ruler. In this case $5\lambda = 84.2 \; mm$ which gives an average wave-length $\lambda = 16.84 \; mm$.

100 mm

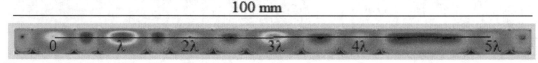

Figure 5.18c) Determination of average wave-length

The critical wave number is $\alpha_{crit} = 1.99$. The wave number can be determined to be

$$\alpha = \frac{2\pi \cdot 0.0048}{0.01684} = 1.79 \tag{5.10}$$

The Marangoni number in the flow simulation is

$$M = \frac{\sigma(T_1 - T_2)L_c}{\mu \kappa} = \frac{0.00002 * 1.5 * 0.0048}{0.000891 * 1.457E-7} = 895 \tag{5.11}$$

The neutral stability curve for Marangoni convection can be given by

$$M = \frac{8a(a \cosh\alpha + B \sinh\alpha)(\alpha - \sinh\alpha \cosh\alpha)}{a^3 \cosh\alpha - \sinh^3\alpha} \tag{5.12}$$

where $B = qL_c/k$ is the Biot number and $q = \frac{\partial Q}{\partial T}$ is the heat flux Q rate of change with temperature T.

B	M_c	α_c	α_{Fluent}	% diff.
0	79.6	1.99	1.79	10
5	250.6	2.60		
10	413.4	2.74		
15	575.0	2.81		
20	736.0	2.85		
25	896.8	2.88		

Table 5.2 Critical parameters for Marangoni convection

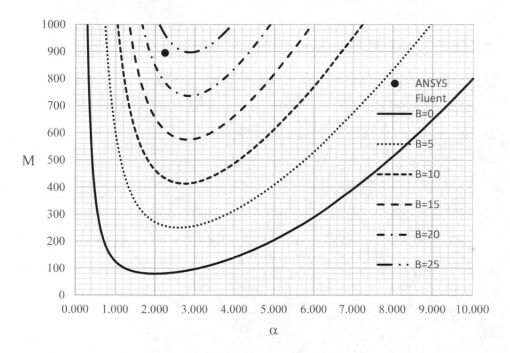

Figure 5.18d) Neutral stability curve for Marangoni convection.

I. References

1. Bénard, H. "Les tourbillons cellulaires dans une nappe liquide", *Rev. Gen. Sciences Pure Appl.* **11**, 1261-1271, 1309-1328, 1900
2. Rayleigh, L. "On convection currents in a horizontal layer of fluid when the higher temperature is on the underside.", Phil. Mag. **32**, 529-546, 1916.
3. Chandrasekhar, S. "Hydrodynamic and Hydromagnetic Stability", Dover, 1981.
4. Koschmieder, E.L. "Bénard cells and Taylor vortices", Cambridge University Press, 1993.
5. Koschmieder, E.L. and Pallas, S.G. "Heat transfer through a shallow, horizontal convecting fluid layer.", Int. J. Heat Mass Transfer **17**, 991-1002, 1974.
6. Hoard, C.Q., Robertson, C.R., and Acrivos, A. "Experiments on the cellular structure in Bénard convection", Int. J. Heat Mass Transfer **13**, 849-856, 1970.
7. Matsson J. "A Student Project on Rayleigh-Benard Convection", Annual ASEE Conference & Exposition, June 22-25, Pittsburg, Pa, 2008.
8. Sánchez Torres N.Y., López Sánches E.J., Hernández Zapata S., and Ruiz Chavarría G., 3D Numerical Simulation of Rayleigh-Bénard Convection in a Cylindrical Container. *Selected Topics of Computational and Experimental Fluid Mechanics,* Environmental Science and Engineering, J. Klapp et al.(eds.), Spring International Publishing, Switzerland, 2015.

J. Exercises

5.1 Run ANSYS Fluent simulations with a rigid upper boundary condition for Rayleigh number $Ra = 3088$ and compare the results with Figure 5.15d) by plotting contours of Z velocity and contours of temperature. Determine the wave number α in the simulations and compare with the neutral stability curve as shown in Figure 5.18b).

5.2 Run ANSYS Fluent simulations with a free upper boundary condition for Rayleigh number $Ra = 2120$ and compare the results with Figure 5.18e) by plotting contours of Z velocity and contours of temperature.

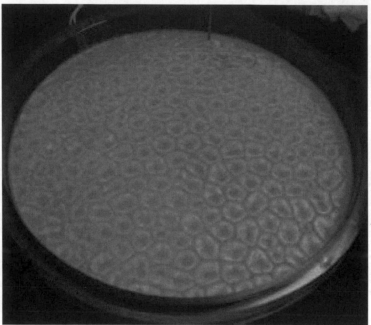

Figure 5.18e) Photo of experimental Rayleigh-Benard convection at $Ra = 2120$ with a free surface upper boundary condition (copied from Matsson[7]).

CHAPTER 6. CHANNEL FLOW

A. Objectives

- Using ANSYS Workbench to Model Geometry and Mesh
- Creating a User Defined Temperature Profile at Inlet
- Inserting Boundary Conditions
- Running Laminar Steady 2D Planar ANSYS Fluent Simulations
- Using Contour Plots for Visualizations of Pressure, Temperature and Velocity Fields
- Using XY Plots for Comparison of Inlet and Outlet Velocity and Temperature Profiles
- Comparing with Theoretical Solution Using Mathematica Code

B. Problem Description

We will study the development of the laminar flow of aqueous glycerin solution through a channel. The channel has a height of 90 mm and a length of 3,000 mm. The inlet temperature profile is linear where the bottom wall has a temperature of 0 °C and the upper wall is at 100 °C. The inlet velocity profile is laminar with a parabolic velocity profile in between the walls.

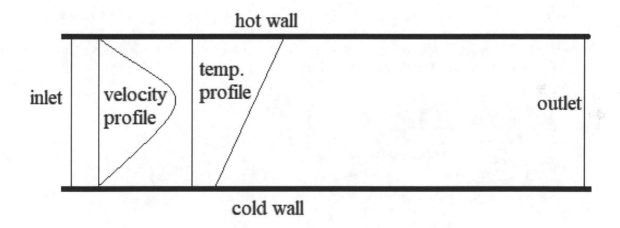

C. Launching ANSYS Workbench and Selecting Fluent

1. Start by launching the ANSYS Workbench. Double click on Fluid Flow (Fluent).

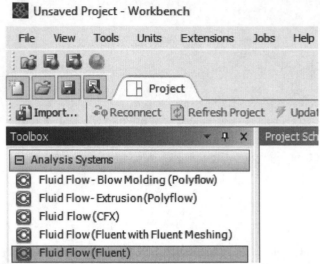

Figure 6.1 Selecting fluid flow Fluent

D. Launching ANSYS DesignModeler

2. Right click Geometry and select Properties. In Properties of Schematic A2: Geometry, select Analysis Type 2D under Advanced Geometry Options. Right-click on Geometry in the Project Schematic window and select New DesignModeler Geometry. Select **Units>>Meter** meter as the desired length unit from the menu in DesignModeler.

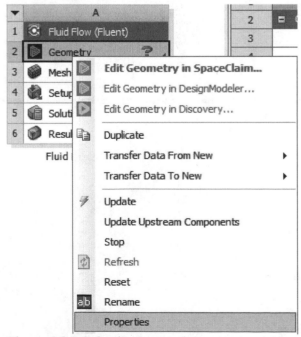

Figure 6.2a) Selecting properties

16	⊟ Advanced Geometry Options	
17	Analysis Type	2D
18	Use Associativity	☑

Figure 6.2b) Advanced geometry options

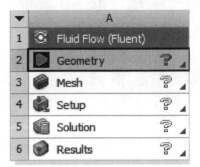

	A
1	⟳ Fluid Flow (Fluent)
2	▢ Geometry ? ◢
3	◈ Mesh ? ◢
4	◈ Setup ? ◢
5	◈ Solution ? ◢
6	◈ Results ? ◢

Fluid Flow (Fluent)

Figure 6.2c) Selecting the geometry

3. Select XY plane in the Tree Outline and select the Sketching tab. Select Look at Sketch .
Draw a rectangle from the origin and set the horizontal length of the rectangle in Details View to
3 m and the vertical height of the rectangle to 0.09 m, see Figure 6.3a). Click on the Modeling tab
in the Tree Outline and highlight Sketch1 under XYPlane. Select Concept>>Surfaces from
Sketches from the menu. Apply Sketch1 as the Base Object in Details View and click on
Generate. Close DesignModeler.

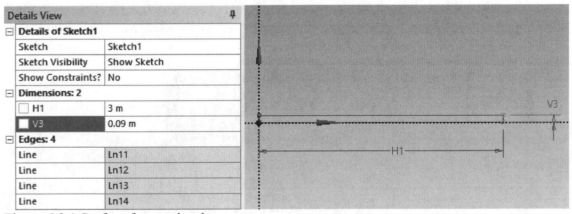

Details View	𝗽
⊟ **Details of Sketch1**	
Sketch	Sketch1
Sketch Visibility	Show Sketch
Show Constraints?	No
⊟ **Dimensions: 2**	
☐ H1	3 m
☐ V3	0.09 m
⊟ **Edges: 4**	
Line	Ln11
Line	Ln12
Line	Ln13
Line	Ln14

Figure 6.3a) Surface from a sketch

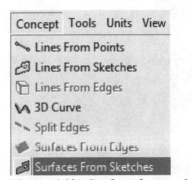

Concept	Tools	Units	View
✎ Lines From Points			
⌨ Lines From Sketches			
◳ Lines From Edges			
∿ 3D Curve			
✂ Split Edges			
◤ Surfaces From Edges			
⌨ Surfaces From Sketches			

Details View	
⊟ **Details of SurfaceSk1**	
Surface From Sketches	SurfaceSk1
Base Objects	1 Sketch
Operation	Add Material
Orient With Plane Normal?	Yes
Thickness (>=0)	0 m

Figure 6.3b) Surface from a sketch   Figure 6.3c) Details view

E. Launching ANSYS Meshing

4. Next, we will be creating the mesh for the flow field. Double click on Mesh under Project Schematic in ANSYS Workbench. Right click on Mesh under Outline in the Meshing window and select Generate Mesh. Select Unit Systems>>Metric (mm, kg, N …) from the bottom of the graphics window.

Figure 6.4a) Coarse mesh

Select Mesh>>Controls>>Face Meshing from the menu. Select the mesh region in the graphics window and apply it as Geometry under Scope in Details of "Face Meshing".

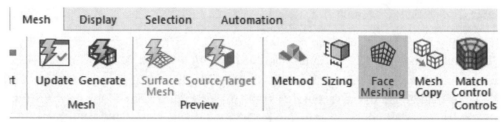

Figure 6.4b) Face meshing

Select Mesh>>Controls>>Sizing from the menu. Select Edge ⬚ , control select the four edges of the rectangle and apply them as the Geometry under Scope in Details of "Edge Sizing". Set the element size to 5 mm and the Behavior to Hard. Click on Generate Mesh and select Mesh in the Outline.

Figure 6.4c) Refined mesh

Select the left vertical edge of the mesh region, right click and select Create Named Selection. Name this edge "inlet". Select the right vertical edge, right click and select Create Named Selection. Name this edge "outlet". Select the lower horizontal edge, right click and select Create Named Selection. Name this edge "cold wall". Finally, select the upper horizontal edge, right click and select Create Named Selection. Name this edge "hot wall".

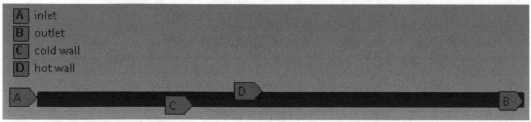

Figure 6.4d) Named selections

Select File>>Save Project from the menu and name the project *Channel Flow*. Select File>>Export...>>Mesh>>FLUENT Input File>>Export from the menu in the Meshing window and name the mesh *channel-flow-mesh.msh*. Close the meshing window. Right click on Mesh under the Project Schematic in ANSYS Workbench and select Update.

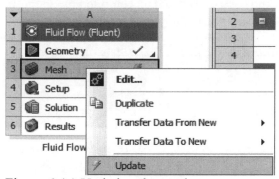

Figure 6.4e) Updating the mesh

F. Creating a UDF

5. We are now going to create a Used Defined Function for the inlet temperature profile. Start Notepad located in the Windows Accessories folder and include the following text as listed below in Figure 6.5. Save the file as a text document with the name *udf-inlet-temp-velocity-profiles.txt*.

```
#include "udf.h"
DEFINE_PROFILE(inlet_t_temperature, thread, position)
{
real x[ND_ND]; /* this will hold the position vector */
real y;
face_t f;
begin_f_loop(f, thread)
{
F_CENTROID(x,f,thread);
y = x[1];
F_PROFILE(f, thread, position) =273.15+(100*y/0.09);
}
end_f_loop(f, thread)
}
DEFINE_PROFILE(inlet_u_velocity, thread, position)
{
real x[ND_ND]; /* this will hold the position vector */
real y;
face_t f;
begin_f_loop(f, thread)
{
F_CENTROID(x,f,thread);
y = x[1];
F_PROFILE(f, thread, position) =-6*0.03*(y*y-y*0.09)/(0.09*0.09*0.09);
}
end_f_loop(f, thread)
}
```

Figure 6.5 Code for UDF

What is a UDF? A UDF is a user defined function that can be used to define your own boundary conditions, initial conditions or material properties. The C programming language is used for user defined functions and the UDF is saved with the extension .c. All user defined functions have the first line #include "udf.h". UDFs are defined using DEFINE macros and can be executed as compiled or interpreted functions. In this chapter we are using the UDF to define the temperature profile at the inlet.

G. Launching ANSYS Fluent

6. Double click Setup under Project Schematic in ANSYS Workbench. Check the Options box for Double Precision. Set the number of Solver Processes to the number of cores for your computer processor. To check the number of physical cores, press the Ctrl + Shift + Esc keys simultaneously to open the Task Manager. Go to the Performance tab and select CPU from the left column. You'll see the number of physical cores on the bottom-right side. Close the Task Manager window. Click on Show More Options, notice the location of the Working Directory and copy the file *udf-inlet-temp-velocity-profiles.txt* to this location. Click on Start to close the Fluent Launcher window. Click OK when you get the Key Behavioral Changes window.

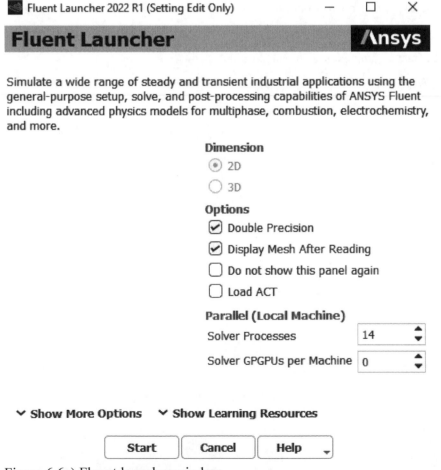

Figure 6.6a) Fluent launcher window

Select the Pressure-Based Solver under General on the Task Page and click on Units… under General, see Figure 6.6b). Scroll down the Quantities and select temperature. Set the Units to *C* for temperature and do the same for temperature-difference. Close the Set Units window. Open Models under Setup in the Outline View. Double click Energy-Off under Models and check the box for Energy Equation. Click OK to close the Energy window.

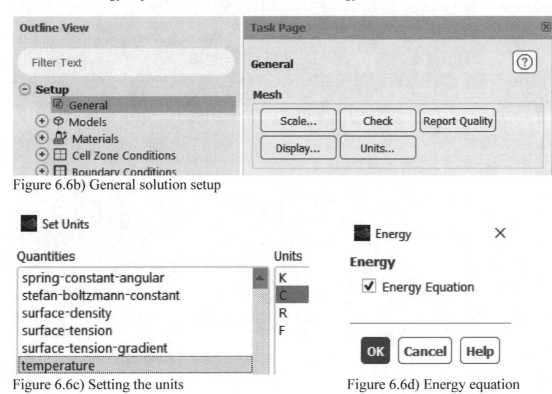

Figure 6.6b) General solution setup

Figure 6.6c) Setting the units   Figure 6.6d) Energy equation

7. Double click Materials under Setup in the Outline View. Click on Create/Edit… for Fluid on the Task Page and select Fluent Database in the new window. Scroll down in the Fluent Fluid Materials and select glycerin (c3h8o3). Click on Copy and Close the window.

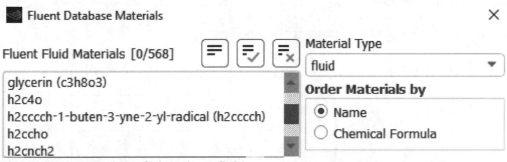

Figure 6.7a) Selection of glycerin as fluid

Select piecewise-linear for Density [kg/m³] under Properties in the Create/Edit Materials window. Increase the number of points to 11 in the Piecewise-Linear Profile window. Set the Temperature [C] for Point 1 to 0 and Value (kg/m3) to 1268.1. Set the values for the remaining points as shown in Table 6.1. Click on the OK button to close the window.

Select piecewise-linear for Viscosity (kg/m-s) under Properties. Increase the number of points to 11. Set the Temperature [C] for Point 1 to 0 and Value (kg/m-s) to 7.2162. Set the values for the remaining points as shown in Table 6.1. Click on the OK button to close the piecewise-linear profile window. Click on Change/Create and Close the Create/Edit Materials window.

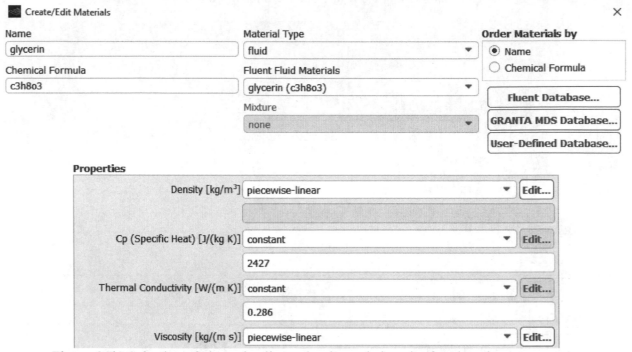

Figure 6.7b) Selection of piecewise-linear density and viscosity for glycerin

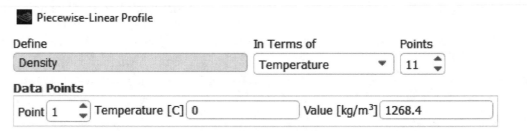

Figure 6.7c) Temperature and density for data points

Point	Temperature (c)	Density (kg/m3)	Viscosity (kg/m-s)
1	0	1268.4	7.2162
2	10	1262.2	2.3857
3	20	1256.0	0.92181
4	30	1249.8	0.40505
5	40	1243.7	0.19801
6	50	1237.6	0.10591
7	60	1231.5	0.06113
8	70	1225.5	0.03767
9	80	1219.5	0.024552
10	90	1213.5	0.016809
11	100	1207.6	0.012009

Table 6.1 Properties of glycerin (98 percent weight) at different temperatures

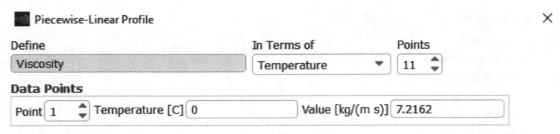

Figure 6.7d) Temperature and viscosity for data points

8. Open Cell Zone Conditions and Fluid under Setup in the Outline View. Double click *surface_body*. Select glycerin as the Material Name. Click on Apply and Close the window.

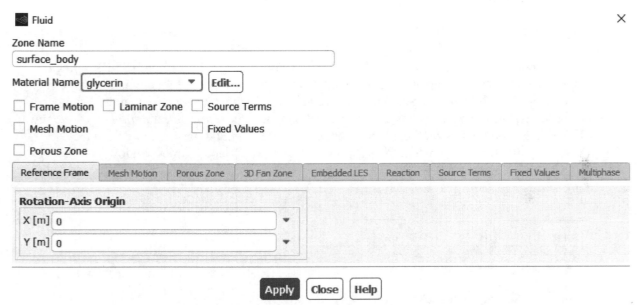

Figure 6.8a) Selection of glycerin as fluid

Select User-Defined>>Functions>>Interpreted… from the menu. Browse for the source file in your Working Directory and Select Files of type: All Files (*). Find the file named *udf-inlet-temp-velocity-profiles.txt* in the working directory and click OK. Click on Interpret and Close the Interpreted UDFs window.

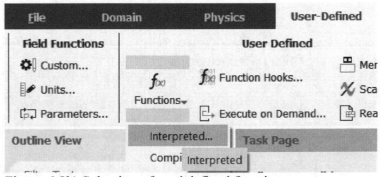

Figure 6.8b) Selection of used defined function

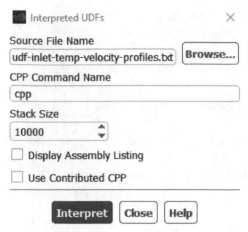

Figure 6.8c) Interpreted UDFs

9. Open Models under Setup in the Outline View. Double click on the Viscous (SST k-omega) model. Select the Laminar Model and click OK to close the Viscous Model window. Open Boundary Conditions under Setup in the Outline View. Double click on *inlet* under Boundary Conditions. Set the Velocity Specification Method to Components. Select *udf inlet_u_velocity* as X-Velocity [m/s]. Click on the Thermal tab and select *udf inlet_t_temperature* from the Temperature drop-down menu. Click Apply and Close the Velocity Inlet window.

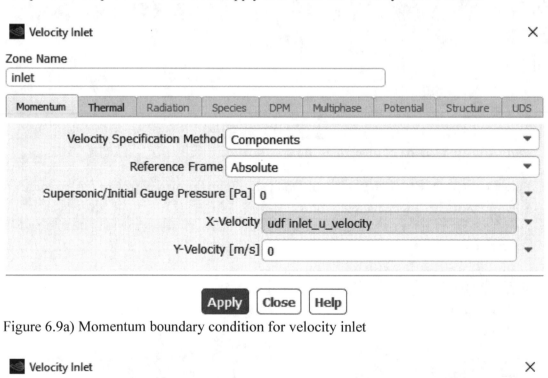

Figure 6.9a) Momentum boundary condition for velocity inlet

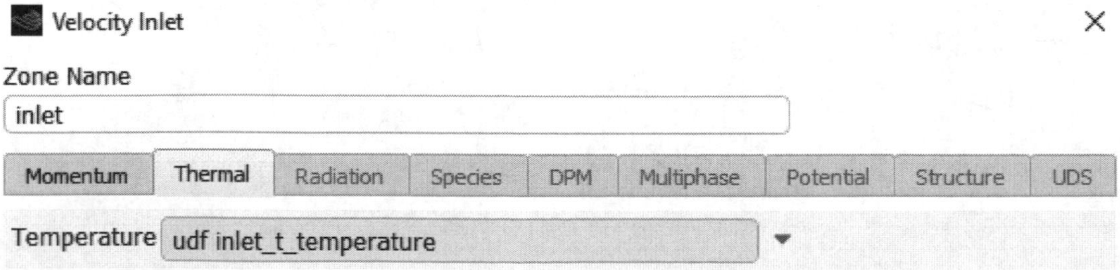

Figure 6.9b) Selection of *udf* for velocity inlet as thermal boundary condition

Double click on *cold_wall* under Boundary Conditions and Wall in the Outline View. Select the Thermal tab in the Wall window and select Temperature as Thermal Condition. Set the Temperature [C] to 0. Click Apply and Close the Wall window. Repeat this step for the *hot_wall* and set the Temperature [C] to 100.

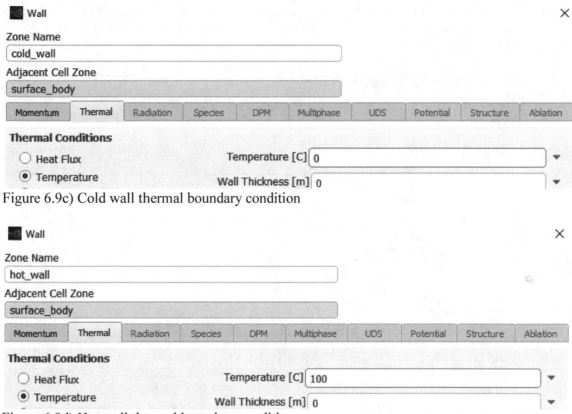

Figure 6.9c) Cold wall thermal boundary condition

Figure 6.9d) Hot wall thermal boundary condition

10. Open Monitors under Solution in the Outline View and double click on Residual under Monitors. Set Absolute Criteria to 1e-06 for all four Residuals. Click OK to close the Residual Monitors window.

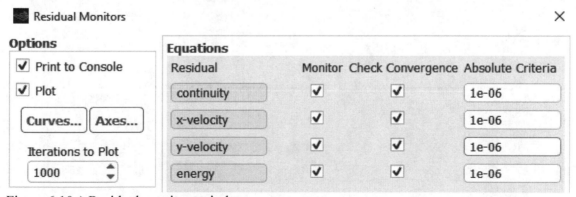

Figure 6.10a) Residual monitors window

Double click on Reference Values under Setup in the Outline View. Select *Compute from inlet* and select *surface_body* as Reference Zone under Reference Values on the Task Page. Double click on Initialization under Solution in the Outline View. Select Standard Initialization on the Task Page as Initialization Method and *Compute from inlet*. Click on *Initialize*. Select File>>Save Project from the menu. Select File>>Export>>Case and Data… from the menu. Save the Case and Data File with the name *Channel Flow.cas.h5*. Double click on Run Calculation under Solution in the Outline View. Enter 10000 as *Number of Iterations*. Click on *Calculate*. Click OK when calculation is complete.

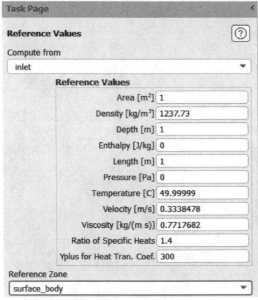

Figure 6.10b) Reference values

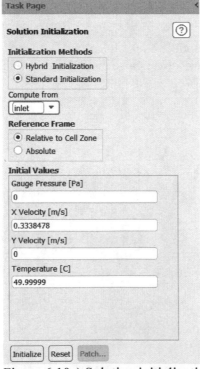

Figure 6.10c) Solution initialization

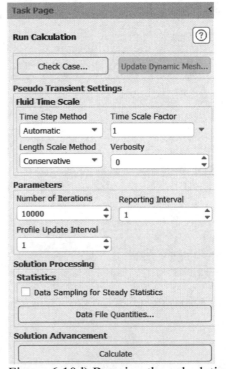

Figure 6.10d) Running the calculations

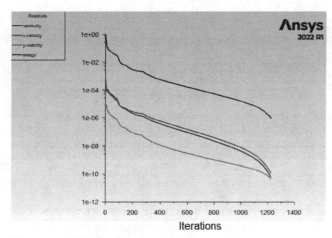

Figure 6.10e) Residuals window

Click on Copy Screenshot of Active Window to Clipboard, see Figure 6.10f). The Scaled Residuals can be pasted into a Word document.

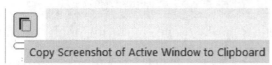

Figure 6.10f) Copying screenshot

H. Post-Processing

11. Open Graphics under Results in the Outline View. Double click on Contours under Graphics. Select all Surfaces and click on Colormap Options. Select float as Type and 0 for Precision under Number Format. Select Fixed Font Behavior and set Font Size to 12 under Font. Select Top as Colormap Alignment, click on Apply and Close the Colormap window. Click on Save/Display. Display another plot of temperature contours by selecting Contours of Temperature and Static Temperature and click on Save/Display. Finally, select Contours of Velocity and X Velocity and set the Precision to 2 in the Colormap window. Click on Save/Display.

Figure 6.11a) Contours window   Figure 6.11b) Colormap window

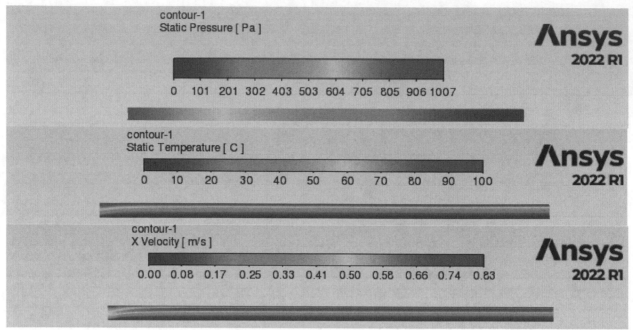

Figure 6.11c) Contours of static pressure, static temperature and X velocity for glycerin flow

Click on Copy Screenshot of Active Window to Clipboard, see Figure 6.10f). The contours can be pasted into a Word document.

12. Double click on XY Plot under Results and Plots in the Outline View. Uncheck Position on X Axis under Options. Select Mesh and Y-Coordinate as Y Axis Function. Select Velocity and X Velocity as X Axis Function. Click on Load File and load the file with the name *theory channel flow.dat* from the working directory. This file can be downloaded from *sdcpublications.com*. Select *inlet* and *outlet* as the Surfaces and click on Curves. Select the first Pattern for Curve # 0 and no Symbol. Click on Apply. Select the next available Pattern for Curve # 1 and no Symbol. Click on Apply. Select the next available Pattern for Curve # 2 with no Symbol and blue color. Click on Apply and Close the window.

Click on Axes… and uncheck the box for Auto Range under Options for the X Axis. Set Minimum to 0 and Maximum to 1 under Range. Set Type as float and set Precision to 1 under Number Format. Click on Apply. Check the Y Axis and set the Precision to 2 under Number Format. Click on Apply. Close the window and click on Save/Plot.

Repeat this step and create another XY Plot. Select Temperature and Static Temperature as X Axis Function. Select inlet and outlet as the Surfaces and unselect Theory under File Data. Click on Axes… and uncheck the box for Auto Range under Options for the X Axis. Set Minimum to 0 and Maximum to 100 under Range. Set Type as float and set Precision to 0 under Number Format. Click on Apply. Check the Y Axis and set the Precision to 2 under Number Format. Click on Apply. Click on Save/Plot.

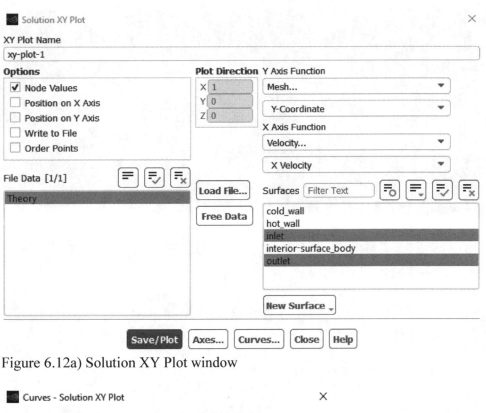

Figure 6.12a) Solution XY Plot window

Figure 6.12b) Solution XY Plot window

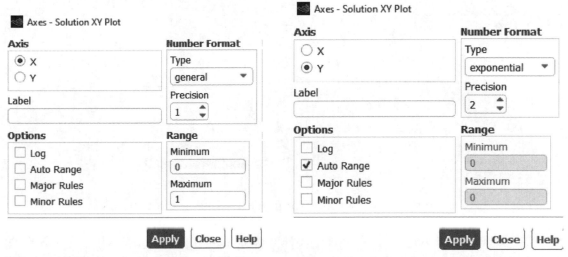

Figure 6.12c) Settings for X and Y Axes

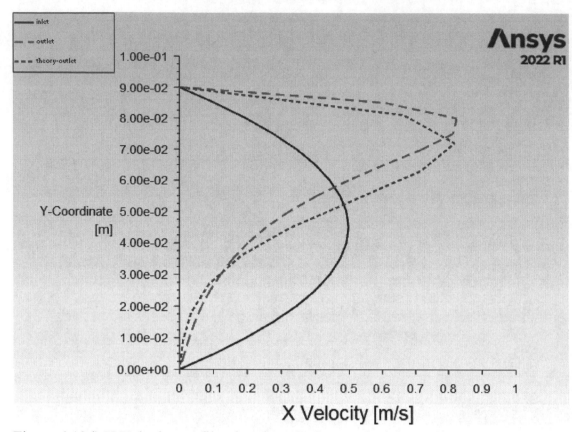

Figure 6.12d) X Velocity profiles for glycerin flow at the inlet and outlet

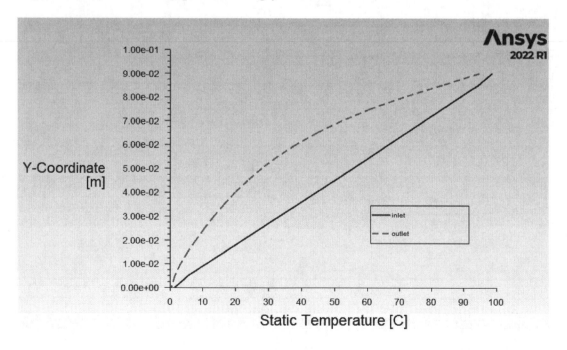

Figure 6.12e) Temperature profiles for glycerin flow at inlet and outlet

Click on Copy Screenshot of Active Window to Clipboard, see Figure 6.10f). The XY Plots can be pasted into a Word document.

I. Theory

13. The equation of motion for the fluid for fully developed flow is given by the following equation

$$\mu \frac{d^2u}{dy^2} = \frac{dp}{dx} \tag{6.1}$$

where u (m/s) is streamwise velocity, x (m) is streamwise coordinate, y (m) is the wall-normal coordinate, μ (kg/ms) is dynamic viscosity, and p (Pa) is the pressure. The boundary conditions are $u(0) = u(h) = 0$ where h is the height of the channel.

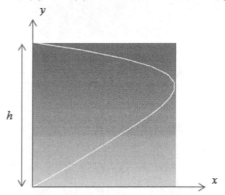

Figure 6.13a) Geometry for laminar flow of glycerin

Element i	Density (kg/m3)	Viscosity μ_i (kg/ms)	Pressure gradient $\left(\frac{dp}{dx}\right)_i$ (Pa/m)
1	1265.3	4.80095	-78.9249
2	1259.1	1.653755	-78.9249
3	1252.9	0.66343	-78.9249
4	1246.75	0.30153	-78.9249
5	1240.65	0.15196	-78.9249
6	1234.55	0.08352	-78.9249
7	1228.5	0.0494	-78.9249
8	1222.5	0.031111	-78.9249
9	1216.5	0.020681	-78.9249
10	1210.55	0.014409	-78.9249

Table 6.2 Average properties of glycerin (98 percent weight) and pressure gradient for each element

We can now compute the resistance or stiffness matrix for each element, see Moaveni[1]

$$K_i = \frac{10\mu_i}{h}\begin{pmatrix} 1 & -1 \\ -1 & 1 \end{pmatrix} \tag{6.2}$$

, and the forcing vector will be the following for each element

$$F_i = \frac{h}{20}\left(\frac{dp}{dx}\right)_i\begin{pmatrix} 1 \\ 1 \end{pmatrix} \tag{6.3}$$

Next, we assemble the stiffness matrices into the following matrix

$$
K = \frac{10}{h}
\begin{pmatrix}
\mu_1 & -\mu_1 & 0 & 0 & 0 & 0 & 0 & 0 & 0 & 0 & 0 \\
-\mu_1 & \mu_1+\mu_2 & -\mu_2 & 0 & 0 & 0 & 0 & 0 & 0 & 0 & 0 \\
0 & -\mu_2 & \mu_2+\mu_3 & -\mu_3 & 0 & 0 & 0 & 0 & 0 & 0 & 0 \\
0 & 0 & -\mu_3 & \mu_3+\mu_4 & -\mu_4 & 0 & 0 & 0 & 0 & 0 & 0 \\
0 & 0 & 0 & -\mu_4 & \mu_4+\mu_5 & -\mu_5 & 0 & 0 & 0 & 0 & 0 \\
0 & 0 & 0 & 0 & -\mu_5 & \mu_5+\mu_6 & -\mu_6 & 0 & 0 & 0 & 0 \\
0 & 0 & 0 & 0 & 0 & -\mu_6 & \mu_6+\mu_7 & -\mu_7 & 0 & 0 & 0 \\
0 & 0 & 0 & 0 & 0 & 0 & -\mu_7 & \mu_7+\mu_8 & -\mu_8 & 0 & 0 \\
0 & 0 & 0 & 0 & 0 & 0 & 0 & -\mu_8 & \mu_8+\mu_9 & -\mu_9 & 0 \\
0 & 0 & 0 & 0 & 0 & 0 & 0 & 0 & -\mu_9 & \mu_9+\mu_{10} & -\mu_{10} \\
0 & 0 & 0 & 0 & 0 & 0 & 0 & 0 & 0 & -\mu_{10} & -\mu_{10}
\end{pmatrix}
\tag{6.4}
$$

, and we assemble the forcing vector into the following

$$
F = 6h
\begin{pmatrix}
1 \\ 2 \\ 2 \\ 2 \\ 2 \\ 2 \\ 2 \\ 2 \\ 2 \\ 2 \\ 1
\end{pmatrix}
\tag{6.5}
$$

After inclusion of boundary conditions, we get system of equations (6.6) with the solution (6.7).

$$
\begin{pmatrix}
1 & 0 & 0 & 0 & 0 & 0 & 0 & 0 & 0 & 0 & 0 \\
-\mu_1 & \mu_1+\mu_2 & -\mu_2 & 0 & 0 & 0 & 0 & 0 & 0 & 0 & 0 \\
0 & -\mu_2 & \mu_2+\mu_3 & -\mu_3 & 0 & 0 & 0 & 0 & 0 & 0 & 0 \\
0 & 0 & -\mu_3 & \mu_3+\mu_4 & -\mu_4 & 0 & 0 & 0 & 0 & 0 & 0 \\
0 & 0 & 0 & -\mu_4 & \mu_4+\mu_5 & -\mu_5 & 0 & 0 & 0 & 0 & 0 \\
0 & 0 & 0 & 0 & -\mu_5 & \mu_5+\mu_6 & -\mu_6 & 0 & 0 & 0 & 0 \\
0 & 0 & 0 & 0 & 0 & -\mu_6 & \mu_6+\mu_7 & -\mu_7 & 0 & 0 & 0 \\
0 & 0 & 0 & 0 & 0 & 0 & -\mu_7 & \mu_7+\mu_8 & -\mu_8 & 0 & 0 \\
0 & 0 & 0 & 0 & 0 & 0 & 0 & -\mu_8 & \mu_8+\mu_9 & -\mu_9 & 0 \\
0 & 0 & 0 & 0 & 0 & 0 & 0 & 0 & -\mu_9 & \mu_9+\mu_{10} & -\mu_{10} \\
0 & 0 & 0 & 0 & 0 & 0 & 0 & 0 & 0 & 0 & 1
\end{pmatrix}
\begin{pmatrix}
u_1 \\ u_2 \\ u_3 \\ u_4 \\ u_5 \\ u_6 \\ u_7 \\ u_8 \\ u_9 \\ u_{10} \\ u_{11}
\end{pmatrix}
= \frac{6h^2}{10}
\begin{pmatrix}
0 \\ 2 \\ 2 \\ 2 \\ 2 \\ 2 \\ 2 \\ 2 \\ 2 \\ 2 \\ 0
\end{pmatrix}
\tag{6.6}
$$

$$
u =
\begin{pmatrix}
0 \\
0.0051 \\
0.0178 \\
0.0447 \\
0.0930 \\
0.1677 \\
0.2647 \\
0.3631 \\
0.4153 \\
0.3371 \\
0
\end{pmatrix}
\tag{6.7}
$$

```
h = 0.09; n = 10; w = 1; dpdx = -78.9249; a = -dpdx * h / n;
μ = {4.80095, 1.653755, 0.66343, 0.30153, 0.15196, 0.08352, 0.0494, 0.031111, 0.020681, 0.014409};
ρ = {1265.3, 1259.1, 1252.9, 1246.75, 1240.65, 1234.55, 1228.5, 1222.5, 1216.5, 1210.55};

{a1, b1, c1} = {Array[-μ[[#]] &, n-1], Array[μ[[#-2]]+μ[[#-1]] &, n, {2, 11}], Array[-μ[[#-1]] &, n, {2, 11}]};
K = (n / h) * SparseArray[{Band[{1, 2}] → c1, Band[{1, 1}] → b1, Band[{2, 1}] → a1}, 11];
K[[1, 1]] = h / n;
K[[1, 2]] = 0;
K[[n + 1, n + 1]] = h / n;
MatrixForm[Normal[K]];

s = LinearSolve[K, {0, a, a, a, a, a, a, a, a, a, 0}];
```

$$\dot{m} = \sum_{i=1}^{n} \rho[[i]] * w * (h/n) * (s[[i]] + s[[i+1]]) / 2;$$

```
ListLinePlot[Array[{s[[#]], (# - 1) * h / n} &, n + 1, {1, n + 1}], AxesLabel → {"u (m/s)", "y (m)"}, PlotLabels → {"Theory"}]

mylist = Array[{s[[#]], (# - 1) * h / n} &, n + 1, {1, n + 1}];
TableOfValues1 = Prepend[mylist, {"((xy/key/label \"theory-outlet\")"}];
TableOfValues1 = Prepend[TableOfValues1, {""}];
TableOfValues1 = Prepend[TableOfValues1, {"(labels \"X Velocity\" \"Y-Coordinate\")"}];
TableOfValues1 = Prepend[TableOfValues1, {"(title \"Theory\")"}];
TableOfValues1 = Append[TableOfValues1, {")"}];
Grid[TableOfValues1]
Export["theory.dat", TableOfValues1]
```

Figure 6.13b) Mathematica code for theoretical solution

J. References

1. S. Moaveni, Finite Element Analysis: Theory and Applications with ANSYS, 3rd Ed., Prentice Hall, 2007.

K. Exercises

6.1 Replace the fluid with engine oil from the Fluent Database and enter the piecewise-linear values for the four points as listed in the table below. Increase the Gage Total Pressure at the inlet to 500 Pa. Include residuals window, contour plots of Static Pressure, Static Temperature, X Velocity, and the X Velocity profile.

Point	Temperature (c)	Density (kg/m3)	Viscosity (kg/m-s)
1	20	881.5	0.23939
2	30	875.4	0.12842
3	40	869.3	0.07455
4	50	863	0.04643

Table 6.3 Properties for engine oil (SAE 30)

6.2 Change the size of the computational domain to 6 m in length and 0.2 m in height and change the mesh size to 0.015 m in horizontal direction and 0.01 m in the vertical direction. Modify the temperature of the cold lower wall to 0 (c) and 100 (c) for the upper hot wall. Rewrite and reinterpret the *udf* file for the new temperatures of the cold and hot wall and modify the *udf* file for the change in height of the computational domain. Replace the fluid with engine oil from the Fluent Database and enter the piecewise-linear values for the 11 points as listed in the table below. Increase the Gage Total Pressure at the inlet to 1000 pascal. Include *udf* code, zoomed in mesh that shows details, residuals window, contour plots of Static Pressure, Static Temperature, X Velocity, and the X Velocity profile.

Point	Temperature (c)	Density (kg/m3)	Viscosity (kg/m-s)
1	0	867.4	0.75352
2	10	860.9	0.37865
3	20	854.5	0.20689
4	30	848.3	0.12190
5	40	842.1	0.076551
6	50	835.8	0.050861
7	60	829.5	0.035409
8	70	823.2	0.025631
9	80	817	0.019181
10	90	810.6	0.014742
11	100	804.5	0.011619

Table 6.4 Properties for engine oil (SAE 5W-40)

CHAPTER 7. ROTATING FLOW IN A CAVITY

A. Objectives

- Using ANSYS Fluent to Study Rotating Flow in a Cavity
- Inserting Boundary Conditions and Rotation
- Running Laminar 2D Axi-Symmetric ANSYS Fluent Simulations with Swirl
- Using XY Plots for Profiles of Velocity Components
- Comparing with Results Using Mathematica Code
- Using Contour Plots for Visualizations of Streamlines and Velocity Components
- Using Pathlines to Visualize the Flow Patterns and Compare with Experiments

B. Problem Description

We will study the rotating flow in a cavity and we will analyze the problem using ANSYS Fluent. The rotating cavity filled with water has a cylindrical geometry with a diameter of 95 mm and a height with the same dimension. The top lid is rotating with a rotational speed of 0.445 rad/s while the remaining walls of the cavity are stationary.

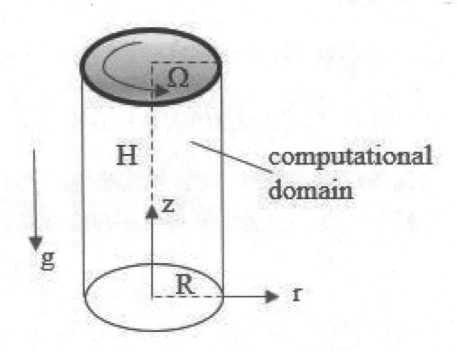

C. Launching ANSYS Workbench and Selecting Fluent

1. Start by launching ANSYS Workbench. Double click on Fluid Flow (Fluent) under Analysis Systems in Toolbox. Right click on Geometry in the Project Schematic and select Properties. Select 2D Analysis Type in Advanced Geometry Options under Properties of Schematic A2: Geometry.

Figure 7.1a) Selecting Fluid Flow

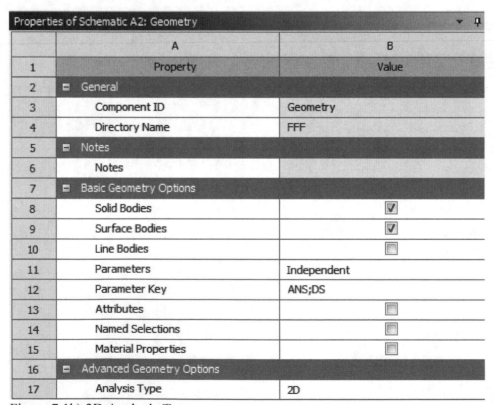

	A	B
	Property	Value
2	⊟ General	
3	Component ID	Geometry
4	Directory Name	FFF
5	⊟ Notes	
6	Notes	
7	⊟ Basic Geometry Options	
8	Solid Bodies	☑
9	Surface Bodies	☑
10	Line Bodies	☐
11	Parameters	Independent
12	Parameter Key	ANS;DS
13	Attributes	☐
14	Named Selections	☐
15	Material Properties	☐
16	⊟ Advanced Geometry Options	
17	Analysis Type	2D

Figure 7.1b) 2D Analysis Type

D. Launching ANSYS DesignModeler

2. Right click on Geometry under Project Schematic in ANSYS Workbench and select New DesignModeler Geometry. Select **Units Millimeter** from the menu in DesignModeler.

   Select the XYPlane in Tree Outline. Look At Face/Plane/Sketch . Select the Sketching tab and Rectangle under the Draw tab. Draw a rectangle from the origin in the first quadrant. Select the Dimensions tab. Click on the left vertical edge of the rectangle and the lower horizontal edge. Enter 47.5 mm as the vertical dimension and 95 mm as the horizontal dimension. Select Concept>>Surfaces from Sketches from the menu. Select Sketch 1 under XY Plane in the Tree Outline. Apply the Sketch as Base Object in the Details View. Click on Generate and close DesignModeler.

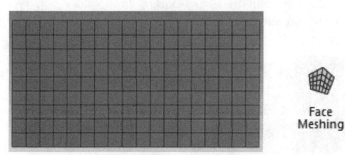

Figure 7.2 Surface sketch of the rectangle

E. Launching ANSYS Meshing

3. Double click on Mesh under Project Schematic in ANSYS Workbench. In the Meshing window, right-click on Mesh under Project and select Update. Select Mesh>>Controls>>Face Meshing from the menu. Select the rectangle in the graphics window and Apply it as Geometry in Details of Face Meshing.

Figure 7.3a) Coarse mesh   Figure 7.3b) Face meshing

Select Mesh>>Controls>>Sizing from the menu. Select the Edge tool , control click on the two vertical edges of the rectangle and Apply them as Geometry in Details of Sizing. Select Number of Divisions as Type and enter 20. Select Hard as Behavior. Select the third Bias type from the drop-down menu and enter 5.0 as Bias Factor. Repeat this step but select the two horizontal edges and 40 as the Number of Divisions. Select Hard as Behavior. Use the same Bias type and the same Bias Factor 5.0. Right-click on Mesh and select Update.

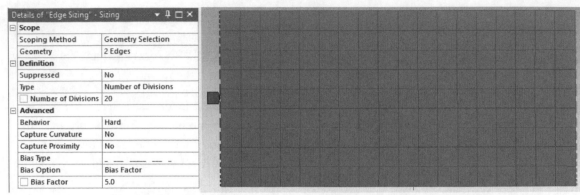

Figure 7.3c) Details of edge sizing for vertical edges

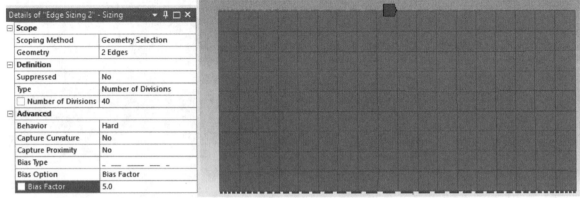

Figure 7.3d) Details of edge sizing for horizontal edges

4. Control-select the upper horizontal edge and left vertical edge, right-click and select Create Named Selection. Enter the name *wall* and click OK to close the window. Select the right vertical edge, right click and select Create Named Selection. Enter the name *lid*. Finally, name the lower horizontal edge as *symmetry*. Select File>>Save Project from the menu and enter the name *Rotating Lid Flow in Cylinder*. Select File>>Export…>>Mesh>>FLUENT Input File>>Export from the menu in ANSYS Meshing window and save the mesh with the name *rotating-lid-cylinder-flow-mesh.msh*. Close the Meshing window. Right-click on Mesh in the Project Schematic of ANSYS Workbench and select Update.

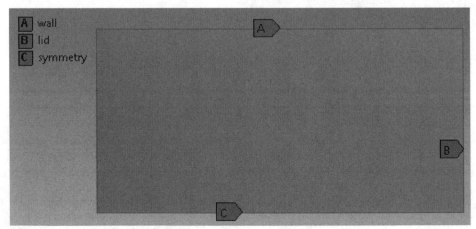

Figure 7.4 Named selections

F. Launching ANSYS Fluent

5. Double-click on Setup under Project Schematic in ANSYS Workbench. Select Double Precision. Set the number of Processes equal to the number of processor cores on your computer. To check the number of physical cores, press the Ctrl + Shift + Esc keys simultaneously to open the Task Manager. Go to the Performance tab and select CPU from the left column. You'll see the number of physical cores on the bottom-right side. Close the Task Manager window. Click on Show More Options and write down the location of the Working Directory as you will need it later. Click on the Start button in the Fluent Launcher window.

Figure 7.5a) Fluent launcher window

Click on Check and Scale… under Mesh in General on the Task Page to verify the Domain Extents. Close the Scale Mesh window. Select Axisymmetric Swirl as 2D Space Solver. Select Relative Velocity Formulation. Check the box for Gravity. Enter -9.81 as Gravitational Acceleration in the X direction.

> **Why did we select Axisymmetric Swirl and Relative Velocity Formulation?** This problem is axisymmetric and has a rotating lid boundary condition that will cause a swirling flow in the container. We choose Relative Velocity Formulation since we are using a small container where the lid is rotating and this will cause most of the flow in the container to be rotating.

Figure 7.5b) Scale mesh window

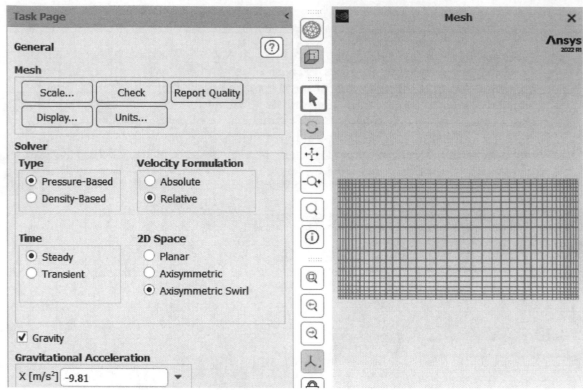

Figure 7.5c) Selecting axisymmetric swirl and gravitational acceleration

6. Next, we double click on Materials under Setup in the Outline View. Select Fluid under Materials on the Task Page and click on Create/Edit…. Click on Fluent Database… in the Create/Edit Materials window. Scroll down in the Fluent Fluid Material section and select water-liquid (h2o<l>). Click on Copy at the bottom of the Fluent Database Materials window. Close the two windows.

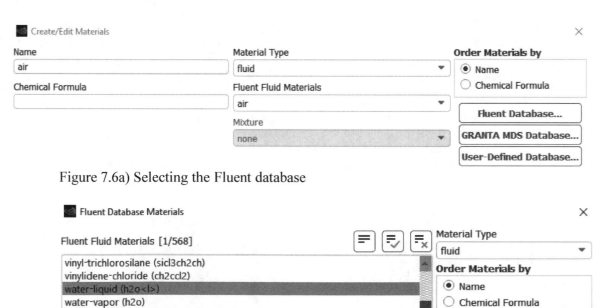

Figure 7.6a) Selecting the Fluent database

Figure 7.6b) Selection of water-liquid as fluid material

7. Open Cell Zone Conditions and Fluid under Setup in the Outline View. Double click on *surface_body* under Fluid. Select *water-liquid* as Material Name. Check the box for Frame Motion and enter Speed (rad/s) -0.4449 as Rotational Velocity. Click Apply and Close the Fluid window.

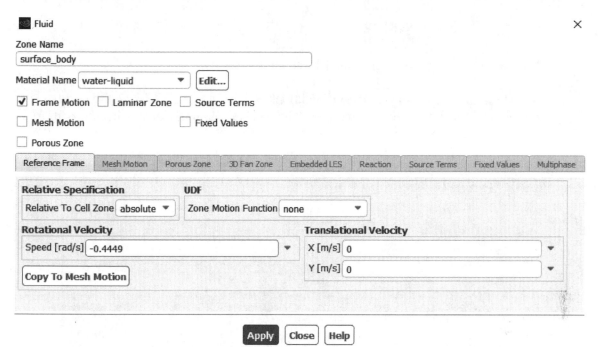

Figure 7.7 Selection of water-liquid and frame motion for the surface body

8. Double click on Boundary Conditions under Setup in the Outline View. Select *symmetry* under Zone in Boundary Conditions on the Task Page. Choose axis from the Type drop-down menu, rename the Zone Name to *axis*, click Apply and Close the Axis window.

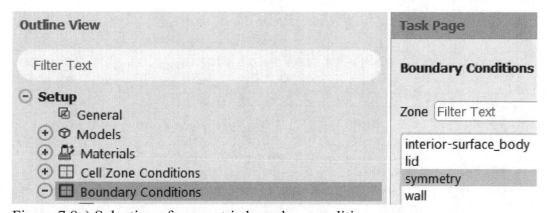

Figure 7.8a) Selection of symmetric boundary condition

Figure 7.8b) Selection of axis boundary condition

Select *lid* under Zone in Boundary Conditions on the Task Page. Click on the Edit… button for *wall* Type. Select *Moving Wall* under Wall Motion. Check Absolute and Rotational under Motion and enter 0.4449 as Speed [rad/s]. Click Apply and Close the window. Repeat this step for the *wall* Zone, select *Moving Wall* under Wall Motion, select Absolute and Rotational Motion and enter 0 as Speed [rad/s], see Figure 7.8d). Click Apply and Close the window.

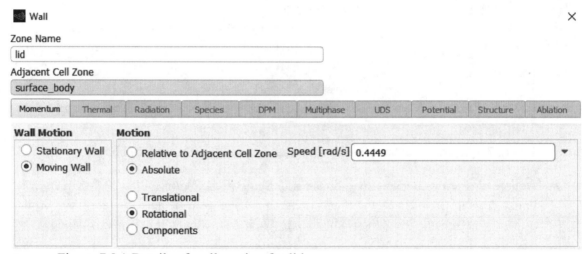

Figure 7.8c) Details of wall motion for lid

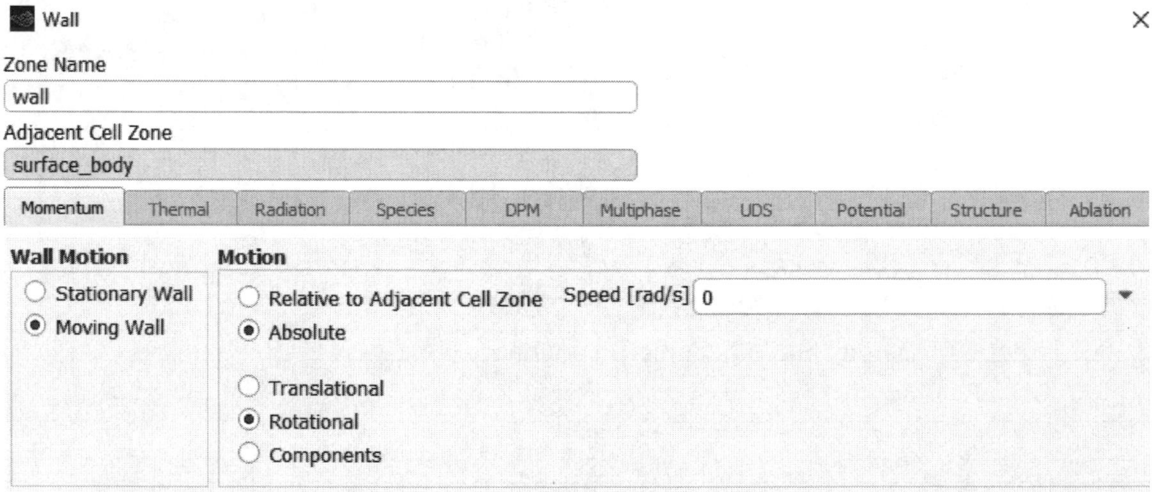

Figure 7.8d) Details of wall motion for wall

9. Double click on Models and Viscous (SST k-omega) under Setup in the Outline View. Choose Laminar as the Viscous Model and click OK to close the window. Double click on Methods under Solution in the Outline View and choose PRESTO! for Pressure Spatial Discretization on the Task Page. Select SIMPLEC as the Pressure-Velocity Coupling Scheme. Select Green-Gauss Node Based Gradient under Spatial Discretization. Select QUICK for both Momentum and Swirl Velocity under Spatial Discretization. Double click on Controls under Solution in the Tree. Set all Under-Relaxation Factors to 1 except Momentum that is set to 0.7.

> **Why did we choose Green-Gauss Node Based, QUICK, PRESTO! and SIMPLEC Solution Methods?** The reason for using SIMPLEC as Pressure-Velocity Coupling algorithm is that we are studying steady laminar flow and SIMPLEC is used instead of SIMPLE as it provides a faster solution and a higher under-relaxation. PISO is used for transient solutions. Coupled can alternatively give an efficient steady state solution for single phase flows. The PRESTO! pressure scheme is recommended when we have rotating, swirling flows with curvature. QUICK is used with quadrilateral meshes to provide better accuracy for rotating and swirling flows. The Green-Gauss Node Based gradient is more accurate than the cell based gradient.

Figure 7.9a) Solution method

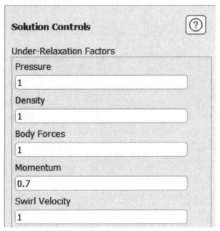
Figure 7.9b) Under-relaxation factors

10. Double click on Monitors and Residual under Solution in the Outline View. Make sure that the box for Plot under Options is checked. Set the Absolute Criteria to 1e-12 for all four residuals and click OK to close the window.

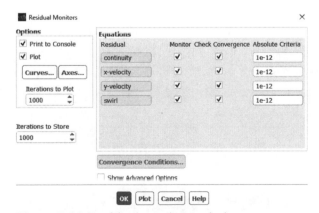
Figure 7.10 Residual monitors window

11. Double click on Initialization under Solution in the Outline View and select Standard Initialization as Initialization Method on the Task Page. Set Swirl Velocity [m/s] to 0.02. This value is approximately the speed of rotation 0.4449 rad/s times the radius 0.0475m. Click on Initialize. Select File>>Export>>Case and Data… from the menu. Save the Case and Data File with the name *Rotating Flow in a Cavity.cas.h5*. Double click on Run Calculation under Solution in the Outline View and set the Number of Iterations to 2000. Click on the *Calculate* button on the Task Page. Click OK in the Information window when the calculation is complete.

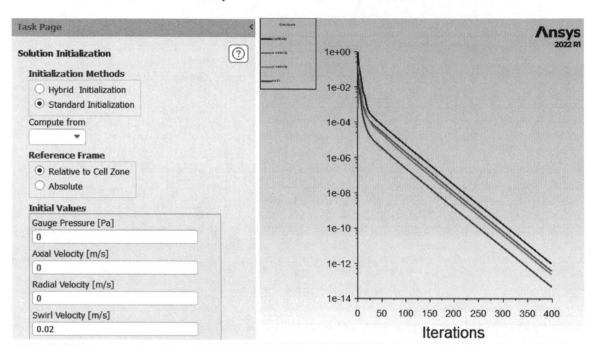

Figure 7.11a) Initialization   Figure 7.11b) Scaled residuals

Click on Copy Screenshot of Active Window to Clipboard, see Figure 7.11c). The Scaled Residuals can be pasted into a Word document.

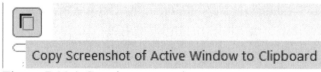

Figure 7.11c) Copying screenshot

G. Post-Processing

12. Open Graphics under Results in Outline View and double click on Contours under Graphics. Select Contours of Velocity and Stream Function. Deselect all surfaces including *interior-surface_body*. Select Colormap Options… in the Contours window. Set Colormap Alignment to Top, set Font Behavior to Fixed and set Font Size to 12 under Font. Set Type to exponential and Precision to 1 under Number Format. Set Length to 0.8 under Colormap Dimensions. Select Apply and Close the Colormap window. Click on Save/Display and Close the Contours window.

Figure 7.12a) Contours window

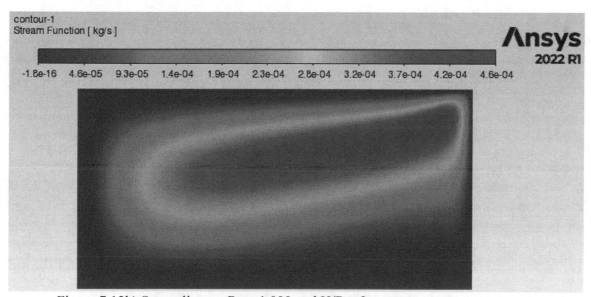

Figure 7.12b) Streamlines at Re = 1,000 and H/R = 2

Click on Copy Screenshot of Active Window to Clipboard, see Figure 7.11c). The Contours can be pasted into a Word document.

153

13. Open Plots and double click on XY Plot under Results in the Outline View. Uncheck Position on X Axis and Position on Y Axis under Options. Select Velocity… and Swirl Velocity as Y Axis Function. Select Mesh… and X-Coordinate as X Axis Function. Select New Surface>>Line/Rake…. Enter x0 (m) 0 and x1 (m) 0.095. Enter y0 (m) 0.0285 and y1 (m) 0.0285. Enter *y=0.0285m* as the New Surface Name and click on Create. Close the Line/Rake Surface window.

Select *y=0.0285m* under Surfaces in the Solution XY Plot window and click on Axes…. Set Type as *float* and set Precision to 2 under Number Format for X Axis. Click on Apply. Set Type as *float* and set Precision to 3 under Number Format for Y Axis. Click on Apply. Close the Axes window. Click on Curves…. Select the first available Line Style Pattern for Curve # 0 and use no symbol. Click on Apply. Select the next available Line Style Pattern for Curve # 1 and use no symbol. Click on Apply and Close the Curves – Solution XY Plot window.

Copy the file *swirl-velocity-y = 0.0285m.dat* to the Working Directory. This file and other files can be downloaded from *sdcpublications.com*. Click on Load File… and read the file *swirl-velocity-y = 0.0285m.dat* from the working directory. You will need to set the Files of Type: to All Files (*) to find the file in the Working Directory. Click on Save/Plot. Repeat this step twice but instead load *radial-velocity-y=0.0285m.dat* and *axial-velocity-y=0.0285m.dat*. Plot the Radial Velocity and the Axial Velocity components, respectively.

Create another Line/Rake… with x0 (m) 0.0475 and x1 (m) 0.0475. Enter y0 (m) 0 and y1 (m) 0.0475. Enter *x=0.0475m* as the New Surface Name. Plot the three velocity components versus Y-Coordinate and compare with loaded files. Close the Solution XY Plot window.

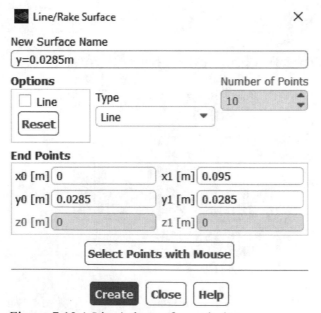

Figure 7.13a) Line/rake surface window

Figure 7.13b) X Axes settings   Figure 7.13c) Y Axes settings

Figure 7.13d) Solution XY Plot window

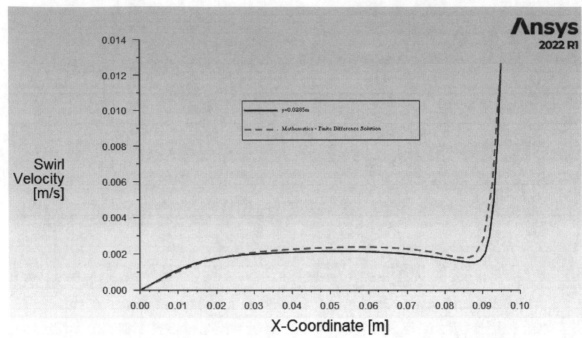

Figure 7.13e) Swirl velocity versus *x* (Re = 1,000 and H/R = 2) at *y* = 0.0285 m

Click on Copy Screenshot of Active Window to Clipboard, see Figure 7.11c). The XY plot can be pasted into a Word document.

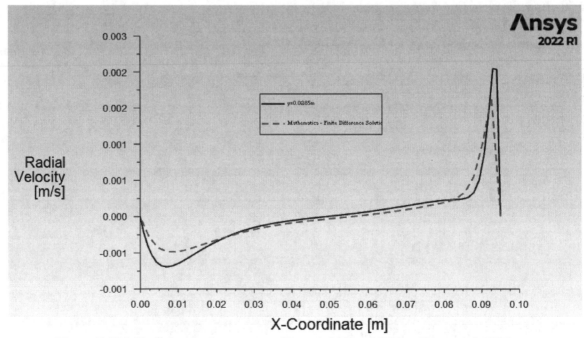

Figure 7.13f) Radial velocity versus *x* (Re = 1,000 and H/R = 2) at *y* = 0.0285 m

Click on Copy Screenshot of Active Window to Clipboard, see Figure 7.11c). The XY plot can be pasted into a Word document.

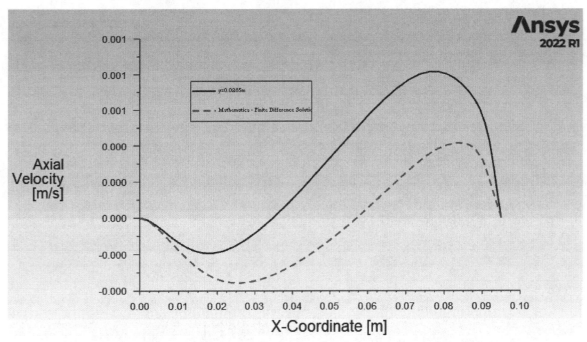

Figure 7.13g) Axial velocity versus *x* (Re = 1,000 and H/R = 2) at *y* = 0.0285m

Click on Copy Screenshot of Active Window to Clipboard, see Figure 7.11c). The XY plot can be pasted into a Word document.

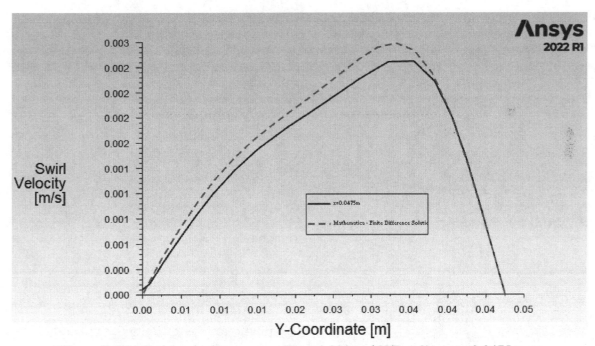

Figure 7.13h) Swirl velocity versus *y* (Re = 1,000 and H/R = 2) at *x* = 0.0475m

Click on Copy Screenshot of Active Window to Clipboard, see Figure 7.11c). The XY plot can be pasted into a Word document.

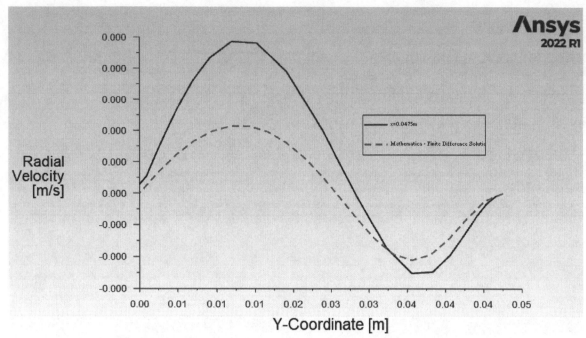

Figure 7.13i) Radial velocity versus y (Re = 1,000 and H/R = 2) at x = 0.0475m

Click on Copy Screenshot of Active Window to Clipboard, see Figure 7.11c). The XY plot can be pasted into a Word document.

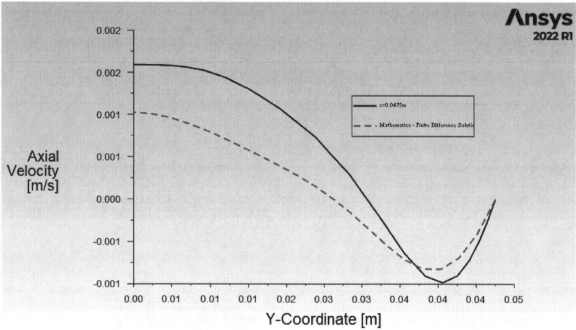

Figure 7.13j) Axial velocity versus y (Re = 1,000 and H/R = 2) at x = 0.0475m

Click on Copy Screenshot of Active Window to Clipboard, see Figure 7.11c). The XY plot can be pasted into a Word document.

14. Double click on Graphics and Contours under Results in the Outline View. Select Contours of Velocity and Stream Function. Deselect all Surfaces in the Contours window. Click on Save/Display.

Select the View tab in the menu and click on Views…. Select *axis* as Mirror Plane and click on Apply. Select the Camera… button in the Views window. Use your left mouse button to rotate the dial counter-clockwise until the cavity appears upright. Close the Camera Parameters window. Click on the Save button under Actions in the Views window and close the windows. Click on Copy Screenshot of Active Window to Clipboard, see Figure 7.11c). The Contour Plot can be pasted into a Word document.

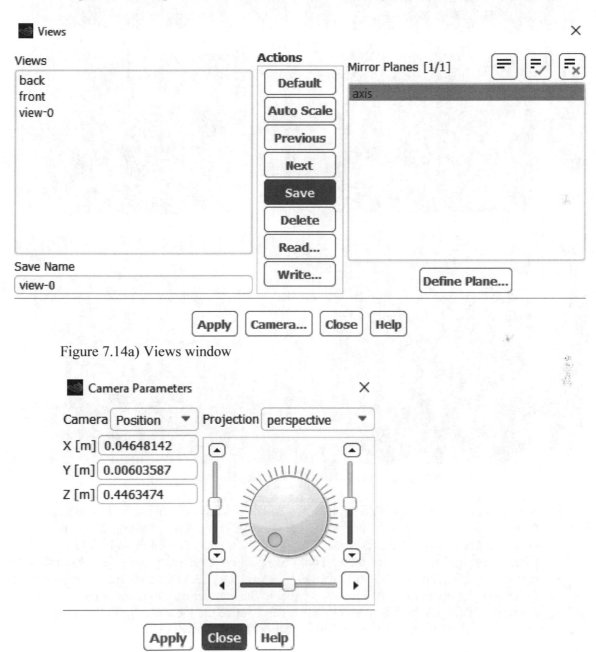

Figure 7.14a) Views window

Figure 7.14b) Camera parameters

Double click on Graphics and Contours under Results in the Outline View. Select Contours of Velocity… and Swirl Velocity in the Contours window. Deselect all surfaces. Click on Save/Display. Select the View tab in the menu and click on Views…. Select *view-0* under Views, select axis under Mirror Planes and click on Apply. Close the Views window. Repeat this for Radial Velocity and Axial Velocity.

Click on Copy Screenshot of Active Window to Clipboard, see Figure 7.11c). The Contour Plot can be pasted into a Word document.

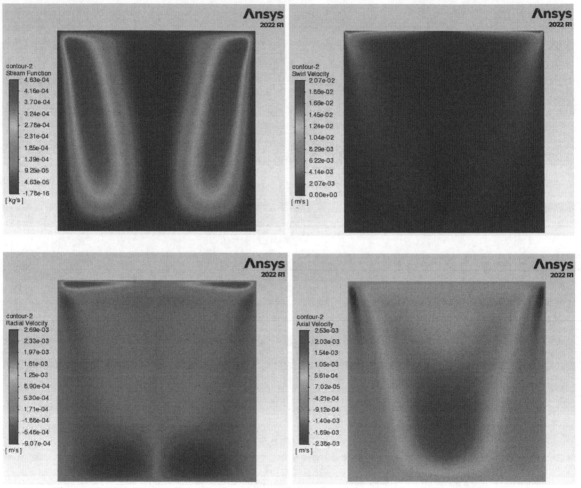

Figure 7.14c) Streamlines and swirl, radial and axial velocities at Re=1000

15. Start another separate ANSYS Workbench 2022 R1 Session from the Windows Start menu by clicking Start in the bottom-left corner of the Windows desktop screen and choosing Workbench 2022 R1 under ANSYS 2022 R1. Double click on Fluid Flow (Fluent) in Analysis Systems under Toolbox. Right click on Setup in Project Schematic and select Edit…. In the Fluent Launcher window, select 2D Dimension and Double Precision. Set the number of Processes equal to the number of processor cores on your computer. Click on the Show More Options and notice the location of the Working Directory. Click on the Start button in the Fluent Launcher window.

Copy the file *rotating-flow-mesh-lambda=2.5.msh* to the Working Directory. This file is available for download at *sdcpublications.com*. Select File>>Import>>Mesh from the Fluent menu. Answer *Yes* to the question that you get on the screen. Select the file *rotating-flow-mesh-lambda=2.5.msh*. Double click on General under Setup in Outline View. Click on Display… under Mesh and General on the Task Page. Select all Surfaces and click on Display in the Mesh Display window. Close the window. Select Axisymmetric Swirl as 2D Space Solver. Select Relative Velocity Formulation. Check the box for Gravity. Enter -9.81 as Gravitational Acceleration in the X direction. Repeat Steps *6* to *11* in this chapter but enter -1 rad/s as Rotational Velocity in Step *7*, set the Speed of the Lid in Step *8* to 1 rad/s corresponding to *Re* = 2,245 and set the Swirl Velocity [m/s] to 0.0475 in Step *11*. Furthermore, in Step **11**, save the Case and Data File with the name *Rotating Flow in a Cavity Lambda = 2.5.cas.h5*. Click on Copy Screenshot of Active Window to Clipboard, see Figure 7.11c). The Contour Plot can be pasted into a Word document.

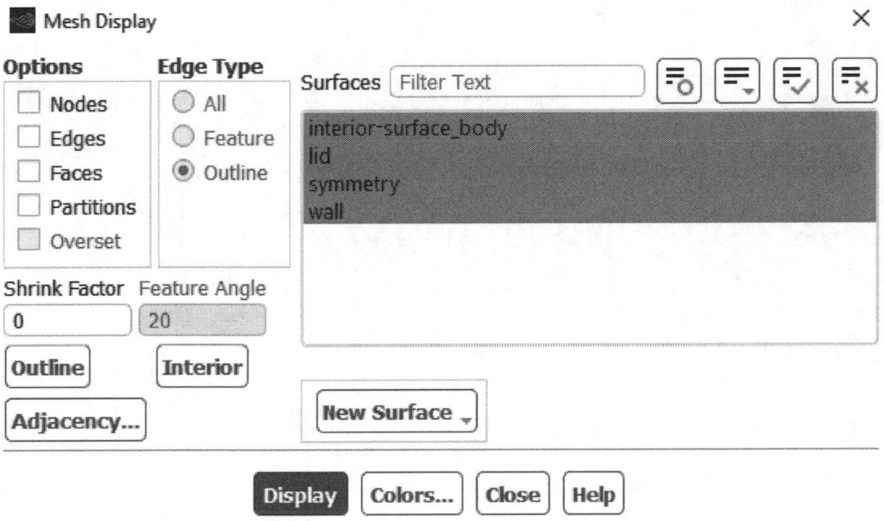

Figure 7.15a) Mesh display window

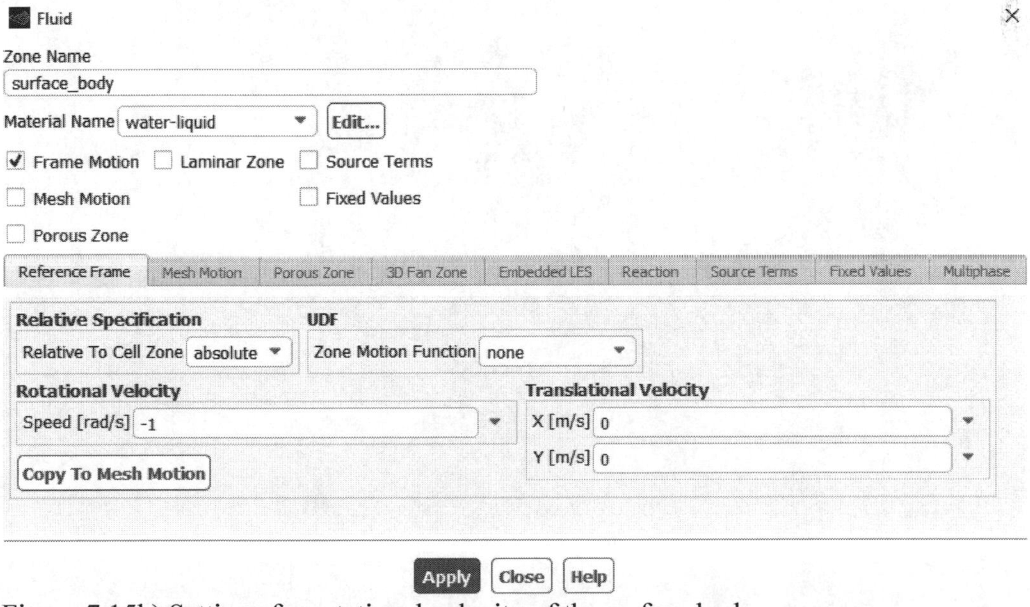

Figure 7.15b) Settings for rotational velocity of the surface body

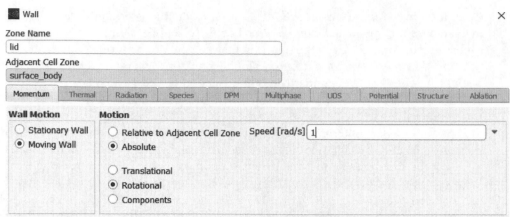

Figure 7.15c) Settings for speed of the lid corresponding to $Re = 2,245$

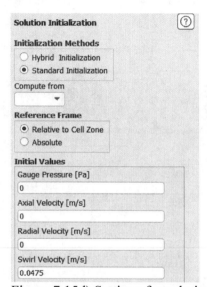

Figure 7.15d) Settings for solution initialization

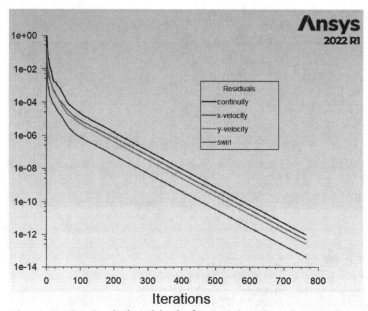

Figure 7.15e) Scaled residuals for rotating flow in a cavity at $Re = 2245$

16. Double click on Graphics and Contours under Results in the Outline View. Select Contours of Velocity and Stream Function. Select Colormap Options… in the Contours window. Set Font Behavior to Fixed and Font Size to 12 under Font in the Colormap window. Select Apply and Close the window. Click on Save/Display. Select the View tab in the menu and click on Views…. Select *axis* as Mirror Plane and click on Apply. Click on the Camera… button in the Views window. Use your left mouse button to rotate the dial counter-clockwise until the cavity appears upright. Close the Camera Parameters window. Select the Save button under Actions in the Views window and close the windows. Click on Copy Screenshot of Active Window to Clipboard, see Figure 7.11c).

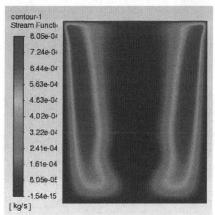

Figure 7.16a) Mirrored and rotated streamlines for Re = 2245 and H/R = 2.5

Double click on Graphics and Pathlines under Results in the Outline View. Select Color by Velocity and Velocity Magnitude. Select *interior-surface_body* under Release from Surfaces. Uncheck Relative Pathlines under Options, set the Tolerance to 1e-06, set the Number of Steps to 20000, set Path Skip to 100 and Path Coarsen to 2. Click on Save/Display. Click on Copy Screenshot of Active Window to Clipboard, see Fig. 7.11c).

Figure 7.16b) Settings for pathlines

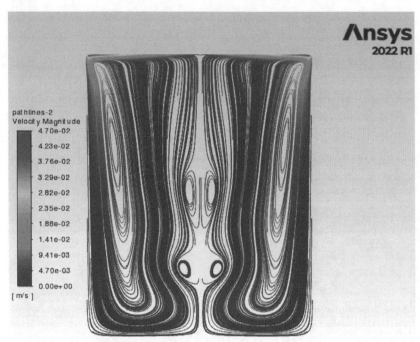

Figure 7.16c) Pathlines at Re = 2245 and H/R = 2.5

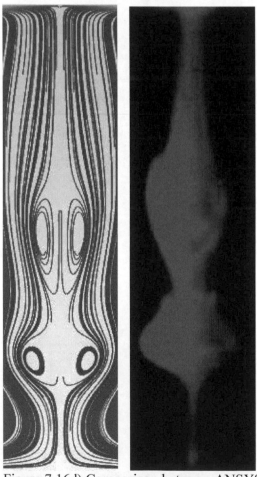

Figure 7.16d) Comparison between ANSYS Fluent simulations and experimental visualizations at Re = 2245, H/R = 2.5

H. Theory

17. We define the Reynolds number as

$$Re = \Omega R^2 / \nu \tag{7.1}$$

, where Ω (rad/s) is the angular velocity of the lid, R (m) is the radius of the cylindrical container and ν (m²/s) is kinematic viscosity. The aspect ratio of the cylindrical container is defined as

$$\lambda = H/R \tag{7.2}$$

, where H (m) is the height of the cylindrical container.

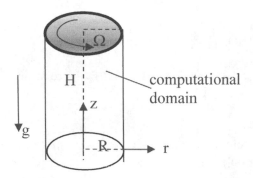

Figure 7.17a) Geometry for cylinder with rotating lid

The incompressible Navier-Stokes equations in cylindrical coordinates (r, θ, z) with velocity components $\big(u_r(r, \theta, z, t), u_\theta(r, \theta, z, t), u_z(r, \theta, z, t)\big)$ are given by the following equations:

Continuity equation: $\quad \dfrac{1}{r}\dfrac{\partial(ru_r)}{\partial r} + \dfrac{1}{r}\dfrac{\partial u_\theta}{\partial \theta} + \dfrac{\partial u_z}{\partial z} = 0 \tag{7.3}$

r-component: $\quad \rho\left(\dfrac{\partial u_r}{\partial t} + u_r\dfrac{\partial u_r}{\partial r} + \dfrac{u_\theta}{r}\dfrac{\partial u_r}{\partial \theta} - \dfrac{u_\theta^2}{r} + u_z\dfrac{\partial u_r}{\partial z}\right) =$

$$-\dfrac{\partial p}{\partial r} + \mu\left[\dfrac{1}{r}\dfrac{\partial}{\partial r}\left(r\dfrac{\partial u_r}{\partial r}\right) - \dfrac{u_r}{r^2} + \dfrac{1}{r^2}\dfrac{\partial^2 u_r}{\partial \theta^2} - \dfrac{2}{r^2}\dfrac{\partial u_\theta}{\partial \theta} + \dfrac{\partial^2 u_r}{\partial z^2}\right] \tag{7.4}$$

θ-component: $\quad \rho\left(\dfrac{\partial u_\theta}{\partial t} + u_r\dfrac{\partial u_\theta}{\partial r} + \dfrac{u_\theta}{r}\dfrac{\partial u_\theta}{\partial \theta} + \dfrac{u_r u_\theta}{r} + u_z\dfrac{\partial u_\theta}{\partial z}\right) =$

$$-\dfrac{1}{r}\dfrac{\partial p}{\partial \theta} + \mu\left[\dfrac{1}{r}\dfrac{\partial}{\partial r}\left(r\dfrac{\partial u_\theta}{\partial r}\right) - \dfrac{u_\theta}{r^2} + \dfrac{1}{r^2}\dfrac{\partial^2 u_\theta}{\partial \theta^2} + \dfrac{2}{r^2}\dfrac{\partial u_r}{\partial \theta} + \dfrac{\partial^2 u_\theta}{\partial z^2}\right] \tag{7.5}$$

z-component: $\quad \rho\left(\dfrac{\partial u_z}{\partial t} + u_r\dfrac{\partial u_z}{\partial r} + \dfrac{u_\theta}{r}\dfrac{\partial u_z}{\partial \theta} + u_z\dfrac{\partial u_z}{\partial z}\right) =$

$$-\dfrac{\partial p}{\partial z} + \mu\left[\dfrac{1}{r}\dfrac{\partial}{\partial r}\left(r\dfrac{\partial u_z}{\partial r}\right) + \dfrac{1}{r^2}\dfrac{\partial^2 u_z}{\partial \theta^2} + \dfrac{\partial^2 u_z}{\partial z^2}\right] \tag{7.6}$$

where ρ (kg/m³) is density, g (m/s²) is acceleration due to gravity and μ (kg/m-s) is dynamic viscosity. For axisymmetric flow there is no θ dependence for the velocity components so they can be expressed as $\left(u_r(r,z), u_\theta(r,z), u_z(r,z)\right)$. This together with a steady flow will reduce the equations above to the following equations:

Continuity equation: $\quad \dfrac{1}{r}\dfrac{\partial(ru_r)}{\partial r} + \dfrac{\partial u_z}{\partial z} = 0$ $\qquad$ (7.7)

r-component: $\quad \rho\left(u_r\dfrac{\partial u_r}{\partial r} - \dfrac{u_\theta^2}{r} + u_z\dfrac{\partial u_r}{\partial z}\right) = -\dfrac{\partial p}{\partial r} + \mu\left[\dfrac{1}{r}\dfrac{\partial}{\partial r}\left(r\dfrac{\partial u_r}{\partial r}\right) - \dfrac{u_r}{r^2} + \dfrac{\partial^2 u_r}{\partial z^2}\right]$ $\qquad$ (7.8)

θ-component: $\quad \rho\left(u_r\dfrac{\partial u_\theta}{\partial r} + \dfrac{u_r u_\theta}{r} + u_z\dfrac{\partial u_\theta}{\partial z}\right) = \mu\left[\dfrac{1}{r}\dfrac{\partial}{\partial r}\left(r\dfrac{\partial u_\theta}{\partial r}\right) - \dfrac{u_\theta}{r^2} + \dfrac{\partial^2 u_\theta}{\partial z^2}\right]$ $\qquad$ (7.9)

z-component: $\quad \rho\left(u_r\dfrac{\partial u_z}{\partial r} + u_z\dfrac{\partial u_z}{\partial z}\right) = -\dfrac{\partial p}{\partial z} + \mu\left[\dfrac{1}{r}\dfrac{\partial}{\partial r}\left(r\dfrac{\partial u_z}{\partial r}\right) + \dfrac{\partial^2 u_z}{\partial z^2}\right]$ $\qquad$ (7.10)

We make the velocity components and pressure non-dimensional using the following relations:

$$u_r^* = u_r/(\Omega R), u_\theta^* = u_\theta/(\Omega R), u_z^* = u_z/\Omega R, p^* = p/(\rho\Omega^2 R^2) \qquad (7.11)$$

and we make the coordinates non-dimensional using the following relations:

$$r^* = r/R, \theta^* = \theta, z^* = z/R \qquad (7.12)$$

The continuity equation and the different component equations in non-dimensional form after skipping the $*$ superscript symbol:

$$\dfrac{\partial u_r}{\partial r} + \dfrac{u_r}{r} + \dfrac{\partial u_z}{\partial z} = 0 \qquad (7.13)$$

$$u_r\dfrac{\partial u_r}{\partial r} - \dfrac{u_\theta^2}{r} + u_z\dfrac{\partial u_r}{\partial z} = -\dfrac{\partial p}{\partial r} + \dfrac{1}{Re}\left[\dfrac{\partial^2 u_r}{\partial r^2} + \dfrac{1}{r}\dfrac{\partial u_r}{\partial r} - \dfrac{u_r}{r^2} + \dfrac{\partial^2 u_r}{\partial z^2}\right] \qquad (7.14)$$

$$u_r\dfrac{\partial u_\theta}{\partial r} + \dfrac{u_r u_\theta}{r} + u_z\dfrac{\partial u_\theta}{\partial z} = \dfrac{1}{Re}\left[\dfrac{\partial^2 u_\theta}{\partial r^2} + \dfrac{1}{r}\dfrac{\partial u_\theta}{\partial r} - \dfrac{u_\theta}{r^2} + \dfrac{\partial^2 u_\theta}{\partial z^2}\right] \qquad (7.15)$$

$$u_r\dfrac{\partial u_z}{\partial r} + u_z\dfrac{\partial u_z}{\partial z} = -\dfrac{\partial p}{\partial z} + \dfrac{1}{Re}\left[\dfrac{\partial^2 u_z}{\partial r^2} + \dfrac{1}{r}\dfrac{\partial u_z}{\partial r} + \dfrac{\partial^2 u_z}{\partial z^2}\right] \qquad (7.16)$$

The boundary conditions are the following:

$$u_r(0,z) = u_r(1,z) = u_r(r,0) = u_r(r,2) = 0 \qquad (7.17)$$

$$u_\theta(0,z) = u_\theta(1,z) = u_\theta(r,0) = 0, \, u_\theta(r,2) = \Omega r \qquad (7.18)$$

$$\dfrac{\partial u_z(0,z)}{\partial r} = u_z(1,z) = u_z(r,0) = u_z(r,2) = 0 \qquad (7.19)$$

CHAPTER 7. ROTATING FLOW IN A CAVITY

```
Remove[XandYGrid,MakeVariables,FDMatrices];
XandYGrid[domain_List,pts_List]:=MapThread[N@Range[Sequence@@#1,Abs[Subtract@@#1]/#2]&,{domain,pts-1}];
BoundaryIndex[rgridlen_,zgridlen_]:=Module[{tmp,left,right,bot,top},tmp=Table[(n-1)zgridlen+Range[1,zgridlen],{n,1,rgridlen}];
  {left,right}=tmp[[{1,-1}]];{bot,top}=Transpose[{First[#],Last[#]}&/@tmp];{top,right[[2;;-2]],bot,left[[2;;-2]]}];
Attributes[MakeVariables]={Listable};MakeVariables[var_,n_]:=Table[Unique[var],{n}];
FDMatrices[deriv_,rzgrid_,difforder_]:=Map[NDSolve`FiniteDifferenceDerivative[#,rzgrid,"DifferenceOrder"->difforder]["Differen
tiationMatrix"]&,deriv];

Options[DrivenCylinderCavitySolver]={"InitialGuess"->1};
DrivenCylinderCavitySolver[Rey_,domain_List,pts_List,difforder_,OptionsPattern[]]:=Module[{rzgrid,rrgrid,rgrid,r2grid,nr,nz,top,r
ight,bot,left,ur,uz,uθ,p,urvar,uzvar,uθvar,pvar,dr,dz,dr2,dz2,eqnur,eqnuz,eqnuθ,eqncont,bcindx,sol,boundaries,deqns,dvars,grid}
,(*Get the grid and differentiation matrices*)rzgrid=XandYGrid[domain,pts];rrgrid=rzgrid[[1]];
  grid=Flatten[Outer[List,Sequence@@rzgrid],1];rgrid=grid[[All,1]];r2grid=rgrid*rgrid;
  {nr,nz}=Map[Length,rzgrid];{top,right,bot,left}=BoundaryIndex[nr,nz];
  {dr,dz,dr2,dz2}=FDMatrices[{{1,0},{0,1},{2,0},{0,2}},rzgrid,difforder];
  (*Get the discretized axisymmetric Navier Stokes equations*){urvar,uzvar,uθvar,pvar}=MakeVariables[{ur,uz,uθ,p},nr*nz];
  eqnur=r2grid urvar (dr.urvar)-rgrid uθvar uθvar+r2grid uzvar (dz.urvar)+r2grid(dr.pvar)-r2grid (1/Rey) (dr2+dz2).urvar-
rgrid(1/Rey) (dr.urvar)+(1/Rey) urvar;
  eqnuz=rgrid urvar(dr.uzvar)+rgrid uzvar (dz.uzvar)+rgrid(dz.pvar)-rgrid (1/Rey) (dr2+dz2).uzvar-(1/Rey) (dr.uzvar);
  eqnuθ=r2grid urvar (dr.uθvar)+rgrid urvar uθvar+r2grid uzvar (dz.uθvar)-r2grid (1/Rey) (dr2+dz2).uθvar-rgrid(1/Rey)
(dr.uθvar)+(1/Rey) uθvar;
  eqncont=rgrid (dr.urvar)+urvar+rgrid(dz.uzvar);
  (*Apply the boundary conditions and solve the
system*)boundaries=Join[top,right,bot,left];eqnur[[boundaries]]=urvar[[boundaries]];
  eqnuz[[boundaries]]=uzvar[[boundaries]];eqnuz[[left]]=(dr[[left]].uzvar);
  eqnuθ[[boundaries]]=uθvar[[boundaries]];eqnuθ[[top]]=uθvar[[top]]-(rrgrid-1);
  eqncont[[top[[1]]]]=pvar[[top[[1]]]];
  {deqns,dvars}={Join[eqnur,eqnuz,eqnuθ,eqncont],Join[urvar,uzvar,uθvar,pvar]};
  sol=dvars/.Quiet@FindRoot[deqns,Thread[{dvars,OptionValue["InitialGuess"]}]];
  (*Get Interpolating
functions*)grid=Flatten[Outer[List,Sequence@@rzgrid],1];Map[Interpolation@Join[grid,Transpose@List@#,2]&,Partition[sol,Len
gth[grid]]]];

(* Calculate Solution *)
Timing[res={ur,uz,uθ,p}=DrivenCylinderCavitySolver[1000,{{1,2},{0,2}},{30,60},4]]
StreamDensityPlot[{ur[r,z],uz[r,z]},{r,1,2},{z,0,2},AspectRatio->2,PlotLabel->Style[Text["Streamfunction at Re = 1000"]]]

(* Plot Results and Save Velocity Profiles for uθ *)
SetDirectory["C:\\Users\\johne"];r1=1;omega=0.4449;radius=0.0475;
Plot[uθ[0.6+r1,z]*omega*radius,{z,0,2},PlotRange->{{0,2},{0,0.014}},AxesLabel->{z,Uθ[m/s]},PlotLabel->Style[Text["r=1.6,
Re=1000"]]]
mylist=Table[{uθ[0.6+r1,z]*omega*radius,z*radius},{z,0,2,0.05}];
Plotex[mylist];Export["swirl-velocity-y=0.0285m.dat",TableOfValues1];

Plot[uθ[r,1]*omega*radius,{r,0+r1,1+r1},AxesLabel->{r+1,Uθ[m/s]},PlotLabel->Style[Text["z=1, Re=1000"]]]
mylist=Table[{uθ[r,1]*omega*radius,(r-r1)*radius},{r,0+r1,1+r1,0.05}];Plotex[mylist];Export["swirl-velocity-
x=0.0475m.dat",TableOfValues1];

Plotex[n_]:=(mylist[[All,{1,2}]]=n[[All,{2,1}]];TableOfValues1=Prepend[mylist,{""}];
 TableOfValues1=Prepend[TableOfValues1,{"((xy/key/label \"Mathematica - Finite Difference Solution\")"}];
 TableOfValues1=Prepend[TableOfValues1,{""}];
 TableOfValues1=Prepend[TableOfValues1,{"(labels \"Swirl Velocity (m/s)\" \"Position (m)\")"}];
 TableOfValues1=Prepend[TableOfValues1,{"(title
\"Mathematica\")"}];TableOfValues1=Append[TableOfValues1,{")"}];Grid[TableOfValues1];)

(* Plot Results and Save Velocity Profiles for ur *)
SetDirectory["C:\\Users\\johne"];r1=1;omega=0.4449;radius=0.0475;
Plot[ur[0.6+r1,z]*omega*radius,{z,0,2},PlotRange->{{0,2},{-0.001,0.0025}},AxesLabel->{z,Ur[m/s]},PlotLabel->Style[Text["r=1.6,
Re=1000"]]]
mylist=Table[{ur[0.6+r1,z]*omega*radius,z*radius},{z,0,2,0.05}];
Plotex[mylist];Export["radial-velocity-y=0.0285m.dat",TableOfValues1];

Plot[ur[r,1]*omega*radius,{r,0+r1,1+r1},AxesLabel->{r+1,Ur[m/s]},PlotLabel->Style[Text["z=1, Re=1000"]]]
mylist=Table[{ur[r,1]*omega*radius,(r-r1)*radius},{r,0+r1,1+r1,0.05}];Plotex[mylist];Export["radial-velocity-
x=0.0475m.dat",TableOfValues1];
```

```
Plotex[n_]:=(mylist[[All,{1,2}]]=n[[All,{2,1}]]];TableOfValues1=Prepend[mylist,{""}];
 TableOfValues1=Prepend[TableOfValues1,{"((xy/key/label \"Mathematica - Finite Difference Solution\")"}];
 TableOfValues1=Prepend[TableOfValues1,{""}];
 TableOfValues1=Prepend[TableOfValues1,{"(labels \"Radial Velocity (m/s)\" \"Position (m)\")"}];
 TableOfValues1=Prepend[TableOfValues1,{"(title
\"Mathematica\")"}];TableOfValues1=Append[TableOfValues1,{")"}];Grid[TableOfValues1];)


(* Plot Results and Save Velocity Profiles for uz *)
SetDirectory["C:\\Users\\johne"];r1=1;omega=0.4449;radius=0.0475;
Plot[uz[0.6+r1,z]*omega*radius,{z,0,2},PlotRange→{{0,2},{-0.0004,0.0010}},AxesLabel→{z,Uz[m/s]},PlotLabel→Style[Text["r=1.6,
Re=1000"]]]
mylist=Table[{uz[0.6+r1,z]*omega*radius,z*radius},{z,0,2,0.05}];
Plotex[mylist];Export["axial-velocity-y=0.0285m.dat",TableOfValues1];


Plot[uz[r,1]*omega*radius,{r,0+r1,1+r1},AxesLabel→{r+1,Uz[m/s]},PlotLabel→Style[Text["z=1, Re=1000"]]]
mylist=Table[{uz[r,1]*omega*radius,(r-r1)*radius},{r,0+r1,1+r1,0.05}];Plotex[mylist];Export["axial-velocity-
x=0.0475m.dat",TableOfValues1];


Plotex[n_]:=(mylist[[All,{1,2}]]=n[[All,{2,1}]]];TableOfValues1=Prepend[mylist,{""}];
 TableOfValues1=Prepend[TableOfValues1,{"((xy/key/label \"Mathematica - Finite Difference Solution\")"}];
 TableOfValues1=Prepend[TableOfValues1,{""}];
 TableOfValues1=Prepend[TableOfValues1,{"(labels \"Axial Velocity (m/s)\" \"Position (m)\")"}];
 TableOfValues1=Prepend[TableOfValues1,{"(title
\"Mathematica\")"}];TableOfValues1=Append[TableOfValues1,{")"}];Grid[TableOfValues1];)
```

Figure 7.17b) Mathematica 8.0 code for cylinder with rotating lid

I. References

1. ANSYS Fluid Dynamics Verification Manual, Release 15.0, November 2013.
2. Granger, R.A., Experiments in Fluid Mechanics, Dryden Press, 1988.
3. Michelsen, J.A., Modeling of Laminar Incompressible Rotating Fluid Flow, AFM 86-05, Ph.D. thesis, Department of Fluid Mechanics, Technical University of Denmark, 1986.
4. Mokhasi, Paritosh, Using Mathematica to Simulate and Visualize Fluid Flow in a Box, 2013.
5. Sorensen, J.N. and Loc,T.P., Higher-Order Axisymmetric Navier-Stokes Code: Description and Evolution of Boundary Conditions, *International Journal For Numerical Methods in Fluids*, 9:1517-1537, 1989.
6. Sorensen, J.N. and Christensen, E.A., Direct Numerical Simulation of Rotating Fluid Flow in a Closed Cylinder, *Physics of Fluids* 7:764, 1995.

J. Exercises

7.1 Use ANSYS Fluent to study the rotating cavity filled with water that has a cylindrical geometry with a radius of 100 mm and a height 300 mm. The top lid is rotating with a Reynolds number $Re = \frac{\Omega R^2}{\nu} = 2800$ while the remaining walls of the cavity are stationary. Visualize pathlines, streamlines, swirl velocity, radial velocity and axial velocity for this flow case.

CHAPTER 7. ROTATING FLOW IN A CAVITY

7.2 Use ANSYS Fluent to study the rotating cavity filled with water that has a cylindrical geometry with a radius of 100 mm and a height 150 mm. The top lid is rotating with a Reynolds number $Re = \frac{\Omega R^2}{v} = 1400$ while the remaining walls of the cavity are stationary. Visualize pathlines, streamlines, swirl velocity, radial velocity and axial velocity for this flow case.

7.3 Use ANSYS Fluent to study a rotating cavity filled with water that has a frustum geometry with a bottom radius of 50 mm, top radius of 100 mm and a height 150 mm. The top lid is rotating with a Reynolds number $Re = \frac{\Omega R t^2}{v} = 2000$ while the remaining walls of the cavity are stationary. Visualize pathlines, streamlines, swirl velocity, radial velocity and axial velocity for this flow case.

7.4 Use ANSYS Fluent to study a rotating cavity filled with water that has a frustum geometry with a bottom radius of 50 mm, top radius of 100 mm and a height 150 mm. The top lid is rotating with a Reynolds number $Ret = \frac{\Omega t R t^2}{v} = 2000$ and the bottom is rotating with the same Reynolds number $Reb = \frac{\Omega b R b^2}{v} = 2000$ while the remaining walls of the cavity are stationary. Visualize pathlines, streamlines, swirl velocity, radial velocity and axial velocity for this flow case.

169

K. Notes

CHAPTER 8. SPINNING CYLINDER

A. Objectives

- Using ANSYS Fluent to Study the Fluid in an Open Cylinder with a Free Surface and Solid Body Rotation
- Inserting Boundary Conditions and System Rotation
- Using Volume of Fluid Model for Multiphase Flow with Surface Tension
- Running Laminar 2D Axisymmetric ANSYS Fluent Simulations with Swirl
- Using Contour Plots for Visualizations of Volume Fraction and Swirl Velocity
- Using Excel for Free Surface Plots

B. Problem Description

We will study the startup flow in a partially filled spinning open cylindrical container where the fluid has a free surface. We will analyze the problem using ANSYS Fluent. The cylindrical container has a diameter of 304.8 mm and the same height. The rotational speed of the cylinder is 12 rad/s.

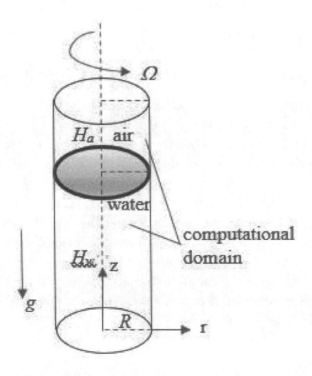

C. Launching ANSYS Workbench and Selecting Fluent

1. Start by launching ANSYS Workbench. Launch Fluid Flow (Fluent) that is available under Analysis Systems in ANSYS Workbench. Select Geometry under Project Schematic, right click and select Properties. Select 2D Analysis Type under Advanced Geometry Options. Right click on Geometry in Project Schematic and select New DesignModeler Geometry to start DesignModeler.

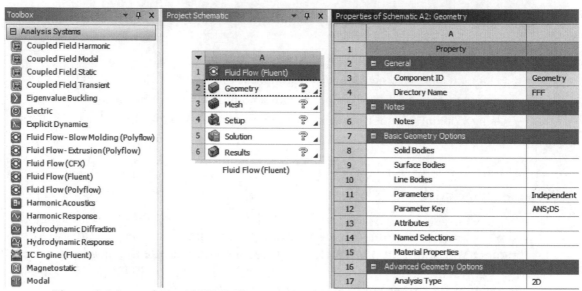

Figure 8.1 Launching ANSYS Fluent and selecting 2D analysis type

D. Launching ANSYS DesignModeler

2. Select **Units>>Millimeter** from the menu in DesignModeler. Select the XYPlane in the

    Tree Outline. Select Look At Face/Plane/Sketch [icon]. Select the Sketching tab and Rectangle. Draw a rectangle from the origin in the first quadrant of the graphics window. Make sure that the letter P is showing when you move the cursor over the origin of the coordinate system and before you start drawing the rectangle in the first quadrant.

    Select Dimensions and click on the left vertical edge of the rectangle and the lower horizontal edge. Enter **304.8 mm** as the horizontal dimension and **152.4 mm** as the vertical dimension. Right click in the graphics window and select Zoom to Fit.

    Select Concept>>Surfaces from Sketches from the menu. Select Sketch 1 under XY Plane in the Tree Outline. Apply the sketch as a Base Object in Details View. Click on Generate and close DesignModeler.

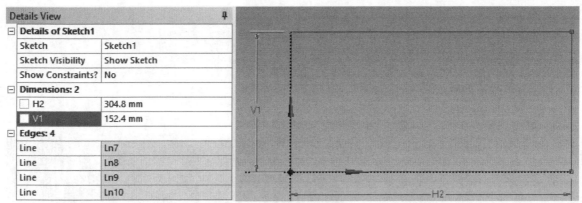

Figure 8.2a) Rectangle with dimensions

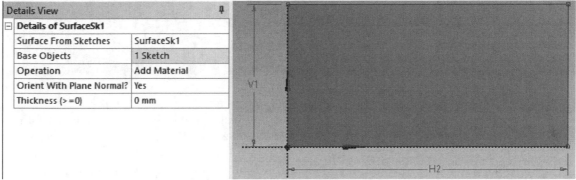

Figure 8.2b) Surface sketch for the rectangle

E. Launching ANSYS Meshing

3. Double click on Mesh under Project Schematic in ANSYS Workbench. Right-click on Mesh under Project and Model (A3) in the Meshing window and select Update. Select Mesh>>Controls>Face Meshing from the menu. Click on the rectangle in the graphics window and Apply it as Geometry in Details of Face Meshing.

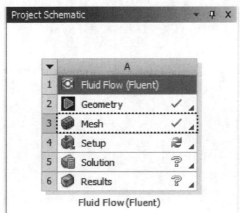

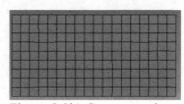

Figure 8.3a) Launching meshing window   Figure 8.3b) Coarse mesh

173

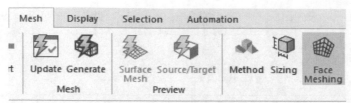

Figure 8.3c) Selecting face meshing

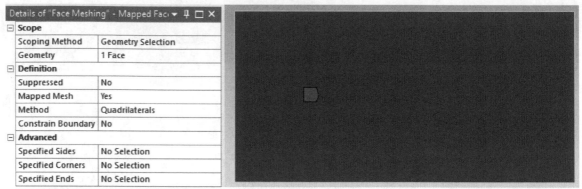

Figure 8.3d) Details of face meshing

Select Mesh>>Controls>>Sizing from the menu. Select the Edge tool 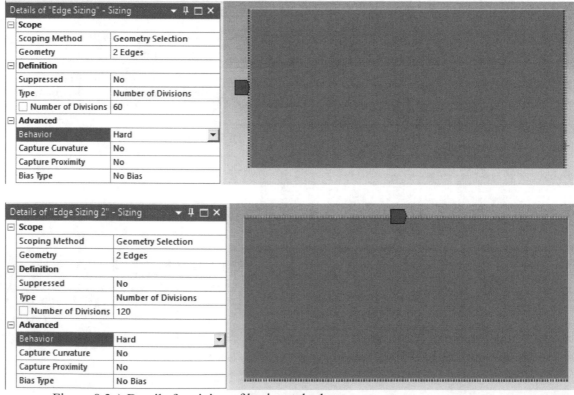, control click on the two vertical edges and Apply them as Geometry in Details of Sizing. Select Number of Divisions as Type and enter 60. Select Hard as Behavior. Repeat this step but select the two horizontal edges and 120 as the Number of Divisions and Hard as Behavior. Right click on Mesh and select Update. Open Statistics under Details of "Mesh". The number of Nodes is 7381 and the number of Elements is 7200.

Figure 8.3e) Details for sizing of horizontal edges

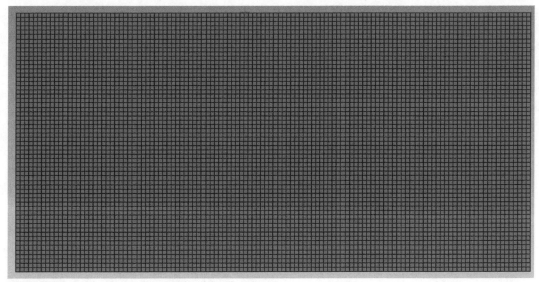

Figure 8.3g) Final mesh used in this chapter

Select the lower horizontal edge, right click and select Create Named Selection. Enter the name *axis-2* and click OK to close the window. Control-select the right vertical edge and the upper horizontal edge, right click and select Create Named Selection. Enter the name *wall-1*. Name the left vertical edge as *pressure-inlet-4*. Finally, select the face tool 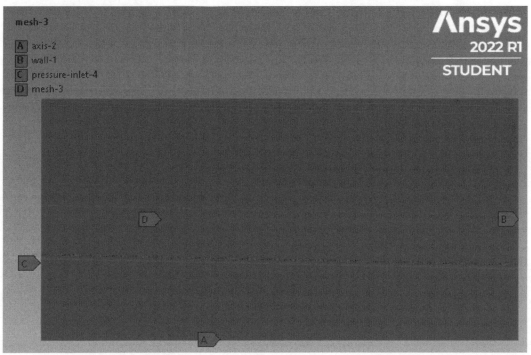, select the interior of the mesh, right click and name it *mesh-3*.

Figure 8.3h) Named selections

Select File>>Export...>>Mesh>>FLUENT Input File>>Export from the menu. Save the mesh with the name *spinning-cylinder.msh*. Select File>>Save Project from the menu and save the project with the name *Spinning Cylinder.wbpj*. Close the meshing window. Right click on Mesh in the Project Schematic and select Update.

F. Launching ANSYS Fluent

4. Double click on Setup under Project Schematic in ANSYS Workbench. Check the box for Double Precision and uncheck the box for Display Mesh After Reading. Click on Show More Options and write down the location of your *Working Directory*. You will need this information later. Select Parallel Processing Options and set the number of processes equal to the number of processor cores for your computer. To check the number of physical cores, press the Ctrl + Shift + Esc keys simultaneously to open the Task Manager. Go to the Performance tab and select CPU from the left column. You'll see the number of physical cores on the bottom-right side. Close the Task Manager window. Click on the Start button in the Fluent Launcher window. Click OK when you get the Key Behavioral Changes window.

    Click on Display under General in Mesh on the Task Page in ANSYS Fluent. Check the Edges box under Options in the Mesh Display window. Select all Surfaces and click on Display in the Mesh Display window and close the same window.

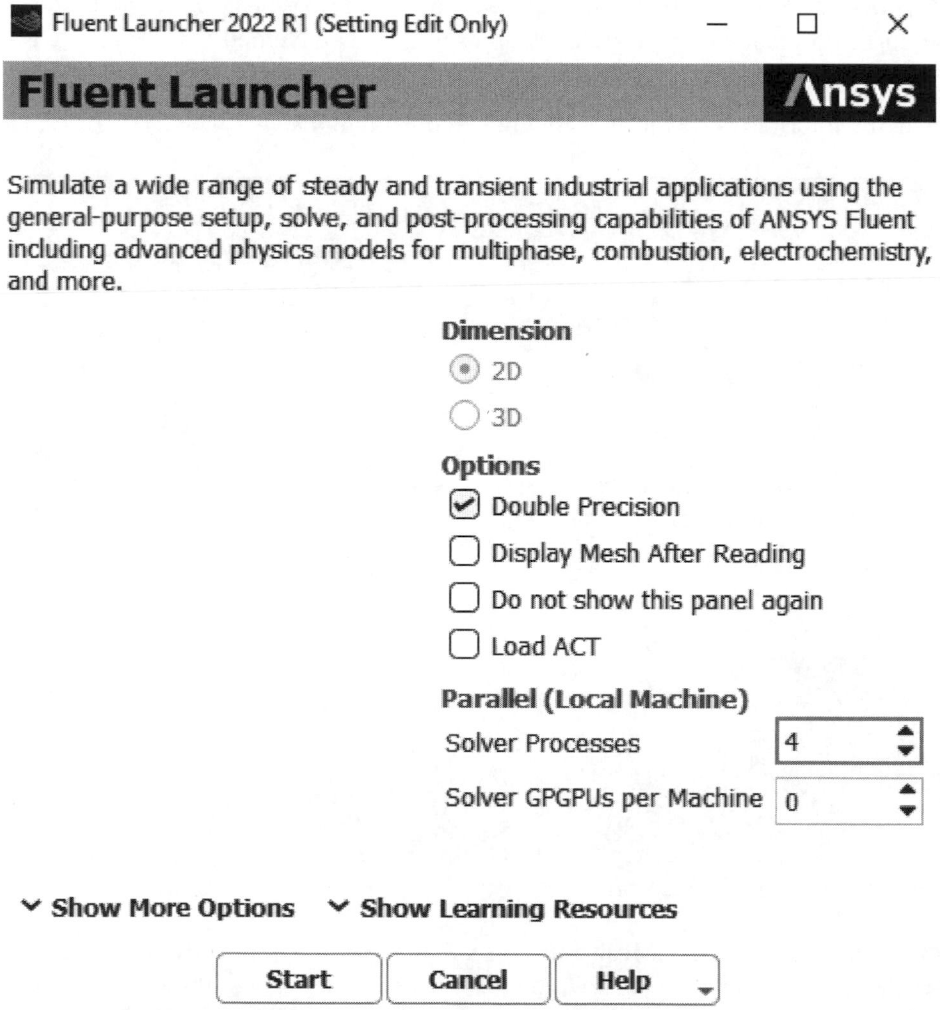

Figure 8.4a) Fluent launcher

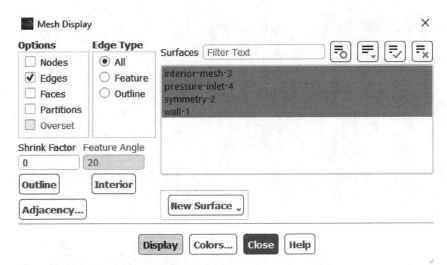

Figure 8.4b) Mesh display window

Figure 8.4c) Cylinder with mesh in ANSYS Fluent

5. Select Axisymmetric Swirl for 2D Space in the General Solver settings on the Task Page in ANSYS Fluent. Select Transient for Time in the General Solver settings.

Open Models under Setup in the Outline View and double click on Viscous (SST k-omega). Select Laminar Model and click on OK to close the Viscous Model window. Double click on Multiphase. Select the Volume of Fluid Model and check the box for Implicit Body Force under Body Force Formulation. Click on Apply and Close the Multiphase Model window.

> **Why did we select Axisymmetric Swirl and Transient for Time?** This problem is axisymmetric and has rotation. We choose Transient since we are interested in studying the shape of the free surface and how it deforms over time.

> **Why did we select Volume of Fluid as Multiphase model and Implicit Body Force?** The Volume of Fluid model is used when we have two or more immiscible fluids where the position of the interface between the fluids is of interest. The applications of the VOF Model includes free-surface flows, stratified flows and the tracking of liquid-gas interfaces in general. The VOF model is limited to the pressure based solver and doesn't allow for void regions without a fluid. Implicit Body Force is used when we have large body forces such as is the case when we have gravity, surface tension and rotation in multiphase flows. Checking the box for Implicit Body Force will make the solution more robust with better convergence.

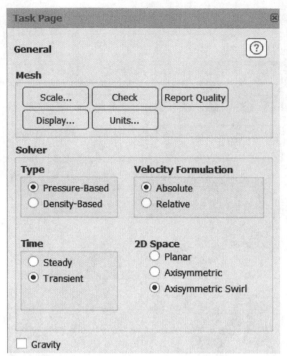

Figure 8.5a) General settings

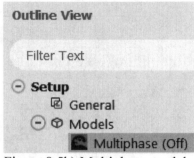

Figure 8.5b) Multiphase model

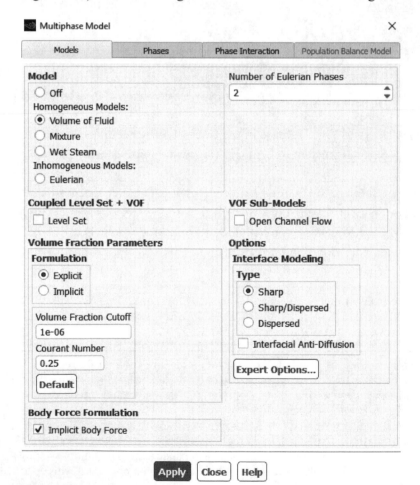

Figure 8.5c) Selecting volume of fluid model

6. Next, we double click on Materials under Setup in the Outline View. Select Fluid under Materials on the Task Page and click on Create/Edit…. Click on Fluent Database… in the Create/Edit Materials window. Scroll down in the Fluent Fluid Material section and select *water-liquid (h2o<l>)*. Click Copy at the bottom of the Fluent Database Materials window. Close the two windows.

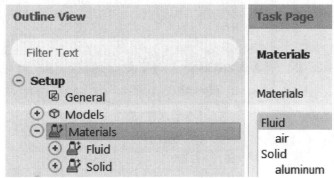

Figure 8.6a) Selecting fluid materials

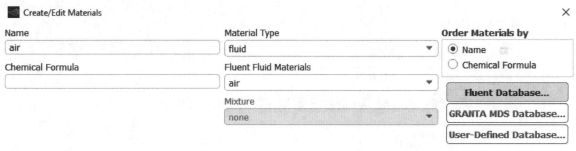

Figure 8.6b) Selecting Fluent database

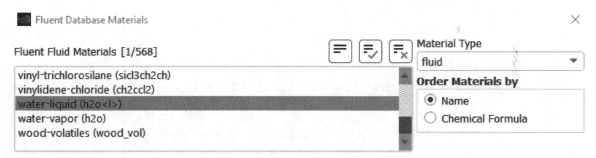

Figure 8.6c) Selection of water-liquid as fluid material

179

7. Select the Physics tab in the menu and select Models>>Multiphase Model. Select the Phases tab at the top of the Multiphase Model window. Select phase-1-Primary Phase and enter *air* as name for the Primary Phase. Select phase-2-Secondary Phase, select *water-liquid* as the Phase Material and enter *water* as the Name for the Secondary Phase. Click on Apply in the Multiphase Model window. Click on the Phase Interaction tab at the top of the Multiphase Model window. Select constant for Surface Tension Coefficients [N/m] and enter 0.07286 as the value. Click on Apply and Close the Multiphase Model window.

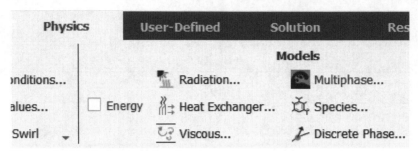

Figure 8.7a) Setting up the physics

Figure 8.7b) Editing phase-1-Primary Phase

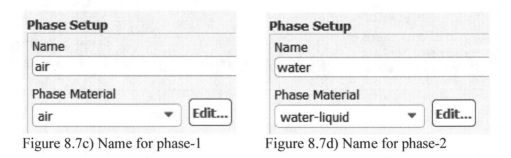

Figure 8.7c) Name for phase-1   Figure 8.7d) Name for phase-2

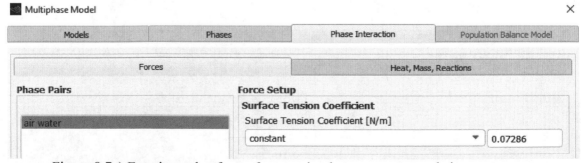

Figure 8.7e) Entering value for surface tension between water and air

8. Select the Physics tab in the menu and select Operating Conditions… under Solver. Check the box for Gravity and enter 9.81 [m/s^2] for Gravitational Acceleration in the positive X direction. Select user-input as Operating Density Method. Enter 1.225 as Operating Density [kg/m^3]. Click on the OK button to close the window.

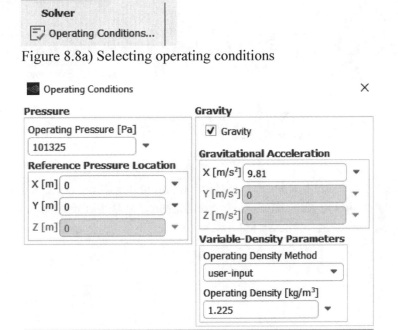

Figure 8.8a) Selecting operating conditions

Figure 8.8b) Including gravity and specified operating density

9. Double click on Boundary Conditions under Setup in the Outline View. Select *pressure-inlet-4* under Zone in Boundary Conditions on the Task Page. Select *mixture* from the Phase drop-down menu under Boundary Conditions on the Task Page. Click on the Edit… button. Leave the Gauge Total Pressure at 0 and click Apply and Close the window.

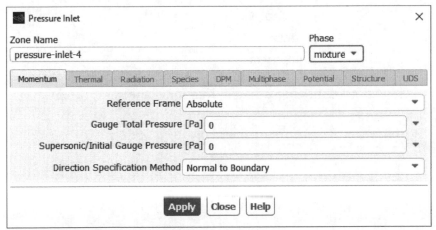

Figure 8.9a) Pressure-inlet boundary

10. For *pressure-inlet-4*, select *water* from the Phase drop-down menu under Boundary Conditions on the Task Page. Click on the Edit… button, make sure that the value for the Volume Fraction is 0, select Apply and Close the window.

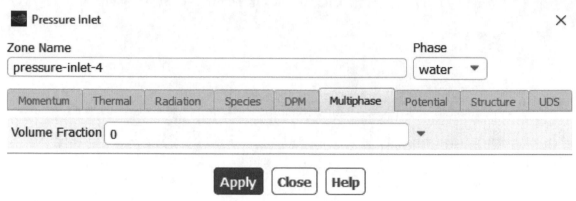

Figure 8.10 Details for pressure-inlet boundary condition

11. For *wall-1*, select *mixture* from the Phase drop-down menu under Boundary Conditions on the Task Page. Click on the Edit… button and select Moving Wall for Wall Motion. Choose Rotational Motion and set the Speed [rad/s] to 12. Click Apply and Close the Wall window.

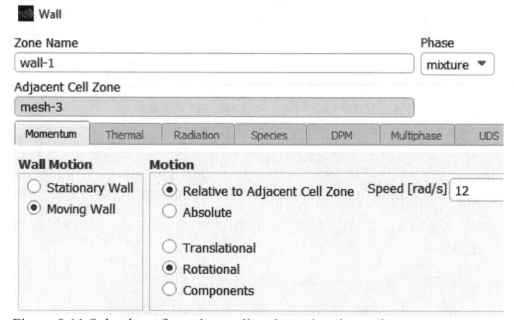

Figure 8.11 Selection of moving wall and rotational speed

12. Double click on Methods under Solution in the Outline View and choose Body Force Weighted for Pressure and First Order Upwind for Momentum under Spatial Discretization on the Task Page. Select SIMPLE as the Pressure-Velocity Coupling Scheme.

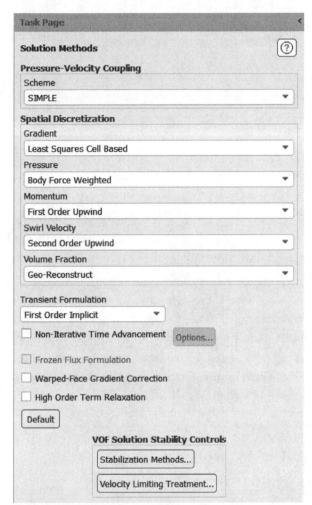

Figure 8.12 Solution methods

13. We are going to define a point close to the outer edge of the cylinder to enable the plotting of the swirl velocity over time. Select the Results tab from the menu and Create Point. Set the x [m] and y [m] coordinates to 0.2339 and 0.1372, respectively. Click Create and Close the window.

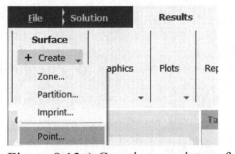

Figure 8.13a) Creating a point surface

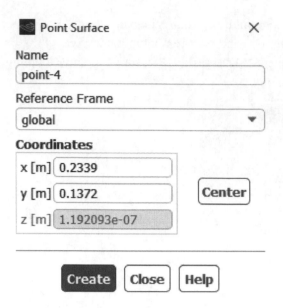

Figure 8.13b) Point surface window

14. Open Monitors and double click on Report Plots under Solution in the Outline View. Click on New… in the Report Plot Definitions window. Select New>>Surface Report>>Vertex Average… in the New Report Plot window. Check the Report File, Report Plot and Print to Console boxes under Create in the Surface Report Definition window. Select Vertex Average as Report Type. Select Velocity and Swirl Velocity as Field Variable. Select *point-4* in the Surfaces list. Enter *swirl-velocity.out* as Name. Select OK to exit the Surface Report Definition window. Click OK to close the New Report Plot window. Close the Report Plot Definitions window.

Double click on Initialization under Solution in the Tree and select Compute from *pressure-inlet-4* on the Task Page. Click on Initialize.

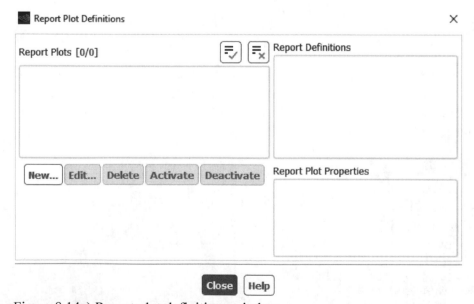

Figure 8.14a) Report plot definitions window

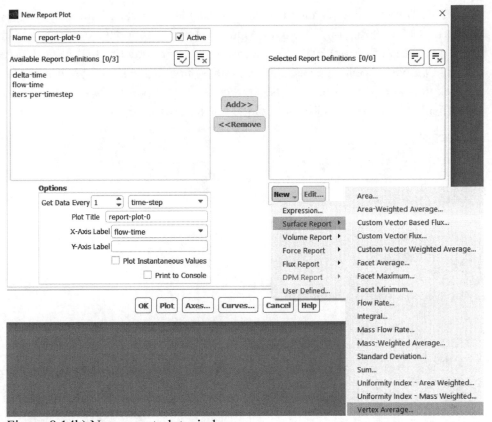

Figure 8.14b) New report plot window

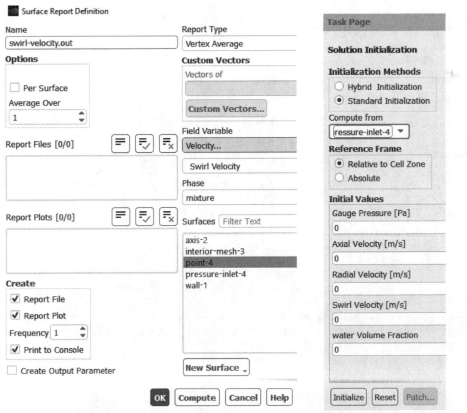

Figure 8.14c) Surface report definition   Figure 8.14d) Initialization

15. Select the *Domain* tab in the menu and select Adapt>>Manual.... Select Cell Registers>>New>>Region... from the Manual Mesh Adaption window. Set X Min [m] to 0.2058, X Max [m] to 0.3048, Y Min [m] to 0, and Y Max (m) to 0.1524. Click on Display Options.... Select red as Color and no Symbol. Check the box for Filled and uncheck the boxes for Wireframe and Marker under Options. Click on OK. Click on Save/Display in the Region Register window. Close the windows.

Figure 8.15a) Domain adaption

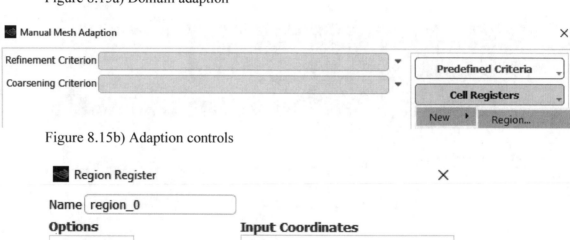

Figure 8.15b) Adaption controls

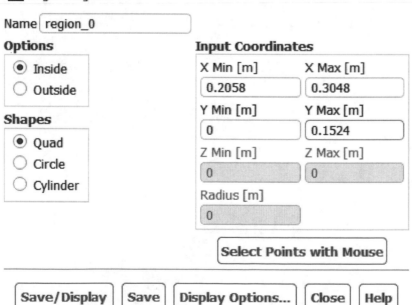

Figure 8.15c) Settings for region register

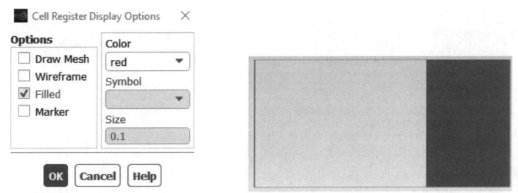

Figure 8.15d) Cell register display   Figure 8.15e) Filled bottom 33% of cylinder

Click on Copy Screenshot of Active Window to Clipboard, see Figure 8.15f). The contour plot can be pasted into a Word document.

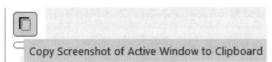

Figure 8.15f) Copying screenshot

16. Click on the Patch… button on the Solution Initialization Task Page. Select *water* as Phase in the Patch window and select Volume Fraction as Variable. Select region_0 as Registers to Patch and set the Value to 1. Click on the Patch button. You have now defined the lower third of the cylinder as filled with water. Select Mixture as Phase and choose Swirl Velocity as Variable. Set the Value [m/s] to 0 and click on the Patch button. Close the Patch window.

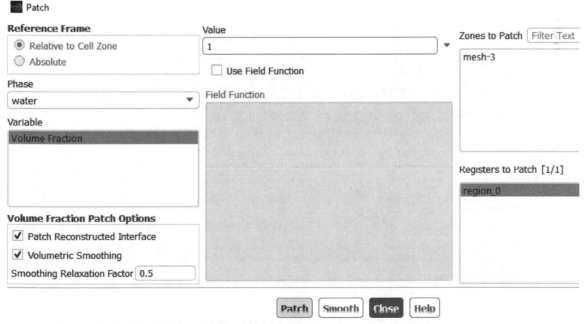

Figure 8.16a) Patching settings for water

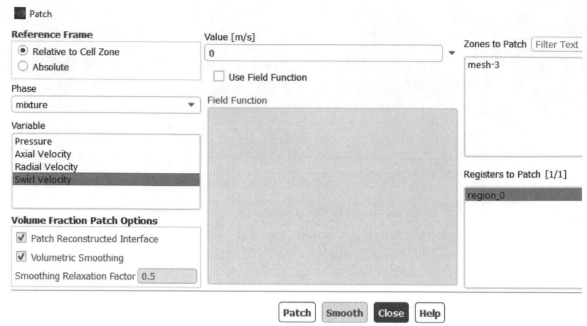

Figure 8.16b) Patching settings for mixture

17. Double click on Graphics and Contours under Results in the Outline View. Select Contours of Velocity and Swirl Velocity. Select Colormap Options... and set the Colormap Size to 30. Click on Apply and Close the Colormap window. Deselect all Surfaces and click on Save/Display. Close the Contours window.

Figure 8.17a) Contours window

Select the View tab in the menu and click on Views…. Select *axis-2* as Mirror Plane and click on Apply. Zoom out and translate the view if needed so that the entire cylinder is visible in the graphics window. Click on the Camera… button in the Views window. Use your left mouse button to rotate the dial clockwise until the cylinder rotates 90 degrees clockwise and appears upright. Close the Camera Parameters window. Click on the Save button under Actions in the Views window and Close the Views window. Click on Copy Screenshot of Active Window to Clipboard, see Figure 8.15f). The swirl velocity contour plot can be pasted into a Word document.

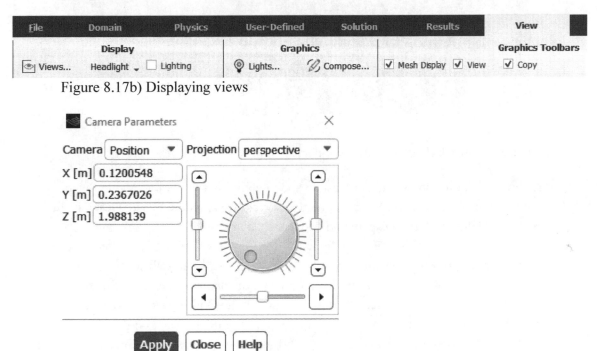

Figure 8.17b) Displaying views

Figure 8.17c) Camera parameters window

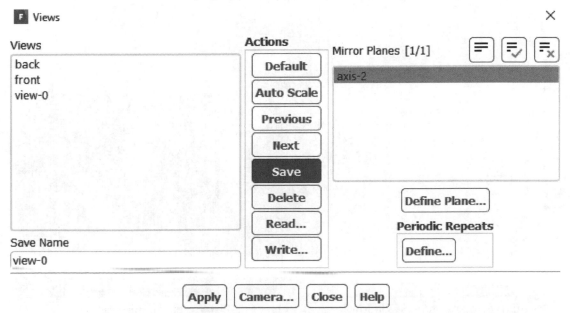

Figure 8.17d) Views window

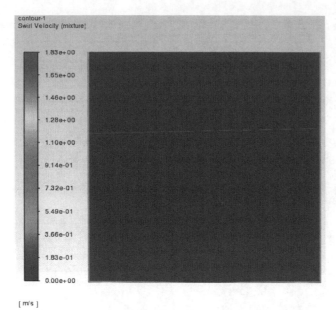

Figure 8.17e) Swirl velocity for cylinder

18. Double click on Contours under Results and Graphics in the Outline View. Select Contours of Phases… and Volume Fraction. Select *water* as the Phase. Deselect all Surfaces. Click on Save/Display and Close the Contours window. Click on Copy Screenshot of Active Window to Clipboard, see Figure 8.15f). The water volume fraction contour plot can be pasted into a Word document. Select File>>Write>>Case and Data… from the menu. Save the Case and Data File with the name Spinning Cylinder.cas.h5. Double click on Run Calculation under Solution in the Outline View and set the Time Step Size to 0.01 s. Set the Number of Time Steps to 10000. Click on the Calculate button on the Task Page. Click OK in the Information window when the calculation is complete. Click on Copy Screenshot of Active Window to Clipboard, see Figure 8.15f). The Scaled Residuals plot can be pasted into a Word document.

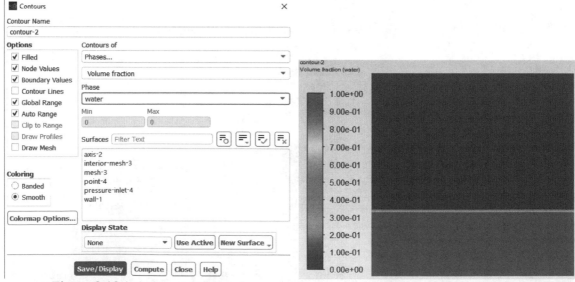

Figure 8.18a) Contours window   Figure 8.18b) Water volume fraction

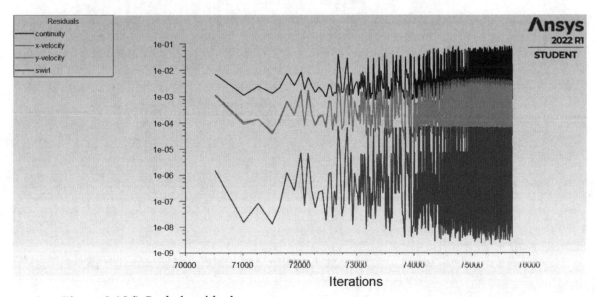

Figure 8.18c) Calculation settings

Figure 8.18d) Scaled residuals

G. Post-Processing

19. Select File>>Write>>Data… from the menu. Save the data file in the *Working Directory* folder with the name *t=100s.dat.* Select Files of type: Legacy Compressed Data Files (*.dat.gz) before you select OK in the Select File window.

    Select the Results tab from the menu and select Surface>>Create>>Iso-Surface…. Select Surface of Constant Phases… and Volume fraction. Select *water* as the Phase and set Iso-Values to 0.5 by moving the slider with the left mouse button. Select the left and right arrows on each side of the slider as needed to get exactly the value 0.5. Select mesh-3 under From Zones and mesh-3 under From Surface. Enter *free-surface-t-100s* as New Surface Name and click on Create. Close the window.

Figure 8.19a) Iso-surface settings for water phase

Select File>>Export>>Solution Data… from the menu. Select ASCII as File Type, Node under Location and Space as Delimiter. Select mesh-3 under Cell Zones and *free-surface-t-100s* under Surfaces. Select Volume fraction (water) under Quantities. Click on Write and save the ASCII File in the Working Directory with the name *free-surface-coordinates-t-100s.* Close the Export window.

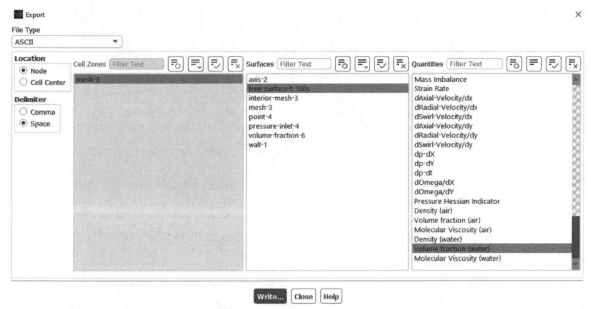

Figure 8.19b) Export settings

Open the saved file in Excel and answer Yes if you get a question. You will need to select All Files as file format in Excel to find and open the saved file. Click on Next in the Text Import Wizard – Step 1 of 3. Click on Next in Text Import Wizard – Step 2 of 3. Click on Finish in Text Import Wizard – Step 3 of 3. The opened file will contain 83 rows. If you get rows with other water-vof values than 0.5, then you will need to sort the rows by increasing or decreasing water-vof values and delete all rows except the rows with water-vof equal to 0.5. Plot the free surface in comparison with the initial undisturbed flat shape of the free surface and the final parabolic shape for the free surface, see Figures 8.19c), 8.19d). The process for this is described below and from the theory section we find the definition of the two coordinate systems (x, y) and (r, z):

$$z = Z(r,t) = H_w + H_a - x = 0.3048 - x, \; r = y \tag{8.1}$$

$$H(r,t) = Z(r,t) - H_w = z - H_w \tag{8.2}$$

Copy the y-coordinate column C in the Excel file and paste it to column E. Label this column $r\ (m)$. Create column F with the value 0.099 for all node numbers and label the column $t = 0$. This is the initial height of the water. Enter the equation =0.3048-B2 in the G column and label this column $t = 100\ s$. Enter the equation =0.099+12^2*(E2^2-0.1524^2/2)/(2*9.81) in the H column and label this column $parabola$. Select columns E – H in Excel and select the Insert tab from the menu.

Select Charts>>Scatter>>Scatter with Smooth lines in Excel, see Figure 8.19c). Click on the plus sign on the right hand side of the Excel graph and add Axis Titles, $r\ (m)$ for the horizontal axis and $Z\ (r,t)$ for the vertical axis. Double click on the horizontal axis scale on the Excel graph and set the Minimum and Maximum Bounds under Axis Options to 0 and 0.15, respectively. Set the Major Units under Axis Option to 0.01 and select the Category under Number to Number with 2 Decimal places. Double click on the vertical axis scale on the Excel graph and set the Minimum and Maximum Bounds under Axis Options to 0 and 0.2, respectively. Set the Major Units under Axis Option to 0.02 and select the Category under Number to Number with 2 Decimal places.

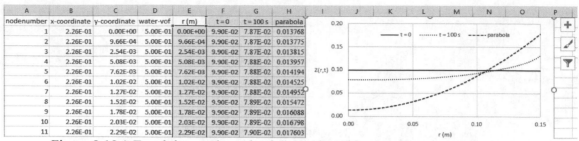

nodenumber	x-coordinate	y-coordinate	water-vof	r (m)	t = 0	t = 100 s	parabola
1	2.26E-01	0.00E+00	5.00E-01	0.00E+00	9.90E-02	7.87E-02	0.013768
2	2.26E-01	9.66E-04	5.00E-01	9.66E-04	9.90E-02	7.87E-02	0.013775
3	2.26E-01	2.54E-03	5.00E-01	2.54E-03	9.90E-02	7.87E-02	0.013815
4	2.26E-01	5.08E-03	5.00E-01	5.08E-03	9.90E-02	7.88E-02	0.013957
5	2.26E-01	7.62E-03	5.00E-01	7.62E-03	9.90E-02	7.88E-02	0.014194
6	2.26E-01	1.02E-02	5.00E-01	1.02E-02	9.90E-02	7.88E-02	0.014525
7	2.26E-01	1.27E-02	5.00E-01	1.27E-02	9.90E-02	7.88E-02	0.014952
8	2.26E-01	1.52E-02	5.00E-01	1.52E-02	9.90E-02	7.89E-02	0.015472
9	2.26E-01	1.78E-02	5.00E-01	1.78E-02	9.90E-02	7.89E-02	0.016088
10	2.26E-01	2.03E-02	5.00E-01	2.03E-02	9.90E-02	7.89E-02	0.016798
11	2.26E-01	2.29E-02	5.00E-01	2.29E-02	9.90E-02	7.90E-02	0.017603

Figure 8.19c) Excel data and graph of dimensional free surface elevation

Enter the equation =E2/0.1524 in the I column and label this column *r/R*. Create column J with the value 0 for all node numbers and label the column *t = 0*. Enter the equation =9.81*(G2-0.099)/(12*0.1524)^2 in the K column and label this column *t = 100 s*. Enter the equation =9.81*(H2-0.099)/(12*0.1524)^2 in the L column and label this column *parabola*. Select the columns I – L in Excel and select the Insert tab from the menu.

Select Charts>>Scatter>>Scatter with Smooth lines in Excel, see Figure 8.19d). Click on the plus sign on the right hand side of the Excel graph and add Axis Titles, *r/R* for the horizontal axis and *gH(r,t)/(ΩR)^2* for the vertical axis. Double click on the horizontal axis scale on the Excel graph and set the Minimum and Maximum Bounds under Axis Options to 0 and 1.0, respectively. Set the Major Units under Axis Option to 0.1 and select the Category under Number to Number with 1 Decimal place. Double click on the vertical axis scale on the Excel graph and set the Minimum and Maximum Bounds under Axis Options to -0.3 and 0.3, respectively. Set the Major Units under Axis Option to 0.1 and select the Category under Number to General.

Select Excel Workbook as Save as type:. Save the Excel file with the name *Free Surface Data and Plots for Spinning Cylinder.xlsx.*

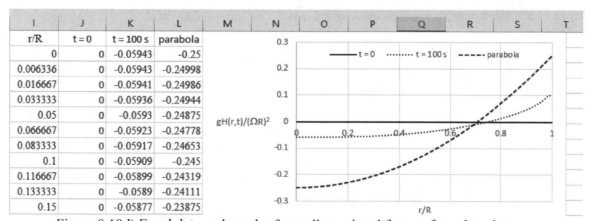

r/R	t = 0	t = 100 s	parabola
0	0	-0.05943	-0.25
0.006336	0	-0.05943	-0.24998
0.016667	0	-0.05941	-0.24986
0.033333	0	-0.05936	-0.24944
0.05	0	-0.0593	-0.24875
0.066667	0	-0.05923	-0.24778
0.083333	0	-0.05917	-0.24653
0.1	0	-0.05909	-0.245
0.116667	0	-0.05899	-0.24319
0.133333	0	-0.0589	-0.24111
0.15	0	-0.05877	-0.23875

Figure 8.19d) Excel data and graph of non-dimensional free surface elevation

Double click on contour-2 under Contours that is located under Results and Graphics in the Outline View. Select Colormap Options… and set Font Behavior to Fixed and Font Size to 12 under Font. Select Apply and Close the Colormap window. Click on Save/Display in the Contours window to display the volume fraction (water) at *t* = 100 s, see Figure 8.19e). Close the Contours window. Select the View tab in the menu and click on Views. Select *view-0*, click on Apply and Close the window.

Repeat this step but instead double click on contour-1 to display swirl velocity, see Figure 8.19e). Click on Copy Screenshot of Active Window to Clipboard, see Figure 8.15f). The water volume fraction contour plot and swirl velocity contour plot can be pasted into a Word document.

Continue to run the calculations to $t = 200$ s and $t = 300$ s, repeat step **19** and add to the free surface graphs, volume fraction of water and swirl velocity, see Figures 8.19f) to 8.19j).

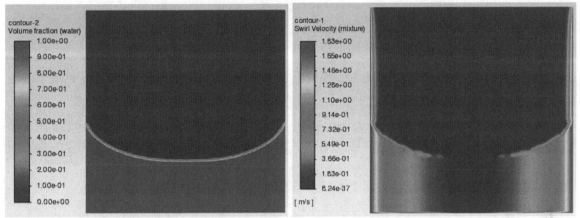

Figure 8.19e) Contours of water volume fraction and swirl velocity at $t = 100$ s

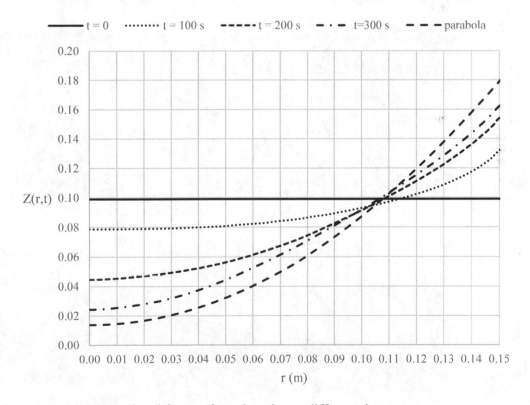

Figure 8.19f) Dimensional free surface elevation at different times

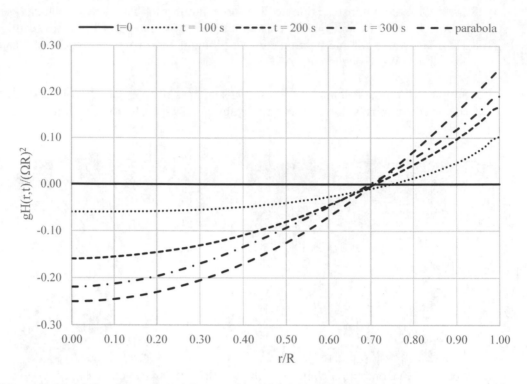

Figure 8.19g) Non-dimensional free surface elevation at different times.

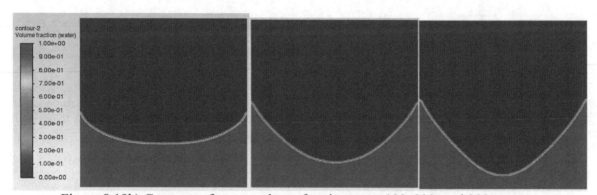

Figure 8.19h) Contours of water volume fraction at t = 100, 200, and 300 s

Figure 8.19i) Contours of swirl velocity at t = 100, 200, and 300 s

196

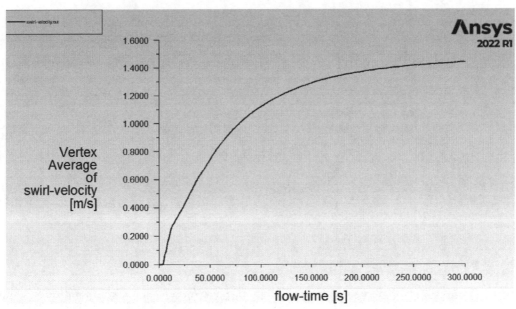

Figure 8.19j) Swirl velocity versus flow time at $(x, y) = (0.2339, 0.1372)$

H. Theory

20. We start by defining the Reynolds number for this flow case as

$$Re = \frac{\Omega R^2}{\nu} = \frac{12*0.1524^2}{1.0038*10^{-6}} = 277,650 \tag{8.3}$$

where Ω (rad/s) is the angular velocity of the cylinder, R (m) is the inside radius of the cylindrical container and ν (m^2/s) is kinematic viscosity. Another non-dimensional number that is commonly used in the Ekman number that is the inverse of the Reynolds number. The initial height of water is $H_w = 0.099$ m and the initial height of air in the cylinder is $H_a = 0.2058$ m. The initial condition for the location of the free surface without rotation can be described as $Z(r,0) = H_w$. Next, we assume that the cylinder has been rotating long enough with constant angular velocity so that the fluid has attained solid-body rotation. We can then describe the shape of the surface as a parabola with the following function for the elevation of the free surface

$$Z(r, \infty) = H_w + \frac{\Omega^2}{2g}(r^2 - \frac{R^2}{2}) \tag{8.4}$$

We can now define the deviation for the height of the free surface from the initial height H_w as

$$H(r, \infty) = Z(r, \infty) - H_w = \frac{\Omega^2}{2g}(r^2 - \frac{R^2}{2}) \tag{8.5}$$

Then, we can modify equation (8.3) and define a non-dimensional variable related to the deviation of the free surface from the initial height, see Figure 8.19f).

$$\frac{gH(r,\infty)}{\Omega^2 R^2} = \frac{1}{2}\left|\left(\frac{r}{R}\right)^2 - \frac{1}{2}\right| \tag{8.6}$$

197

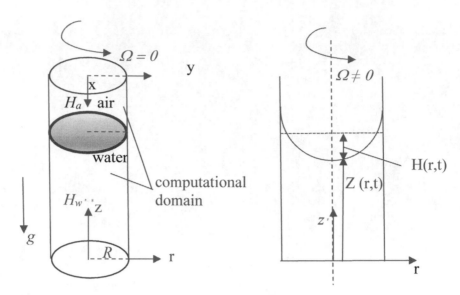

Figure 8.20 Geometry and shape of free surface for cylinder without and with solid body rotation

Generally, the incompressible Navier-Stokes equations in cylindrical coordinates (r, θ, z) with velocity components $\left(u_r(r, \theta, z, t), u_\theta(r, \theta, z, t), u_z(r, \theta, z, t)\right)$ are given by the following equations:

Continuity equation: $\quad \dfrac{1}{r}\dfrac{\partial(ru_r)}{\partial r} + \dfrac{1}{r}\dfrac{\partial u_\theta}{\partial \theta} + \dfrac{\partial u_z}{\partial z} = 0$ (8.7)

r-component: $\quad \rho\left(\dfrac{\partial u_r}{\partial t} + u_r\dfrac{\partial u_r}{\partial r} + \dfrac{u_\theta}{r}\dfrac{\partial u_r}{\partial \theta} - \dfrac{u_\theta^2}{r} + u_z\dfrac{\partial u_r}{\partial z}\right) =$
$$-\dfrac{\partial p}{\partial r} + \mu\left[\dfrac{1}{r}\dfrac{\partial}{\partial r}\left(r\dfrac{\partial u_r}{\partial r}\right) - \dfrac{u_r}{r^2} + \dfrac{1}{r^2}\dfrac{\partial^2 u_r}{\partial \theta^2} - \dfrac{2}{r^2}\dfrac{\partial u_\theta}{\partial \theta} + \dfrac{\partial^2 u_r}{\partial z^2}\right]$$ (8.8)

θ-component: $\quad \rho\left(\dfrac{\partial u_\theta}{\partial t} + u_r\dfrac{\partial u_\theta}{\partial r} + \dfrac{u_\theta}{r}\dfrac{\partial u_\theta}{\partial \theta} + \dfrac{u_r u_\theta}{r} + u_z\dfrac{\partial u_\theta}{\partial z}\right) =$
$$-\dfrac{1}{r}\dfrac{\partial p}{\partial \theta} + \mu\left[\dfrac{1}{r}\dfrac{\partial}{\partial r}\left(r\dfrac{\partial u_\theta}{\partial r}\right) - \dfrac{u_\theta}{r^2} + \dfrac{1}{r^2}\dfrac{\partial^2 u_\theta}{\partial \theta^2} + \dfrac{2}{r^2}\dfrac{\partial u_r}{\partial \theta} + \dfrac{\partial^2 u_\theta}{\partial z^2}\right]$$ (8.9)

z-component: $\quad \rho\left(\dfrac{\partial u_z}{\partial t} + u_r\dfrac{\partial u_z}{\partial r} + \dfrac{u_\theta}{r}\dfrac{\partial u_z}{\partial \theta} + u_z\dfrac{\partial u_z}{\partial z}\right) =$
$$-\dfrac{\partial p}{\partial z} + \mu\left[\dfrac{1}{r}\dfrac{\partial}{\partial r}\left(r\dfrac{\partial u_z}{\partial r}\right) + \dfrac{1}{r^2}\dfrac{\partial^2 u_z}{\partial \theta^2} + \dfrac{\partial^2 u_z}{\partial z^2}\right]$$ (8.10)

where ρ (kg/m³) is density, g (m/s²) is acceleration due to gravity and μ (kg/m-s) is dynamic viscosity. For axisymmetric flow there is no θ dependence for the velocity components so they can be expressed as $\left(u_r(r, z, t), u_\theta(r, z, t), u_z(r, z, t)\right)$. This will reduce the equations above to the following equations:

Continuity equation: $\quad \dfrac{1}{r}\dfrac{\partial(ru_r)}{\partial r} + \dfrac{\partial u_z}{\partial z} = 0$ (8.11)

r-component: $\quad \rho\left(\dfrac{\partial u_r}{\partial t} + u_r\dfrac{\partial u_r}{\partial r} - \dfrac{u_\theta^2}{r} + u_z\dfrac{\partial u_r}{\partial z}\right) =$
$$-\dfrac{\partial p}{\partial r} + \mu\left[\dfrac{1}{r}\dfrac{\partial}{\partial r}\left(r\dfrac{\partial u_r}{\partial r}\right) - \dfrac{u_r}{r^2} + \dfrac{\partial^2 u_r}{\partial z^2}\right]$$ (8.12)

θ-component:
$$\rho\left(\frac{\partial u_\theta}{\partial t} + u_r \frac{\partial u_\theta}{\partial r} + \frac{u_r u_\theta}{r} + u_z \frac{\partial u_\theta}{\partial z}\right) =$$
$$\mu\left[\frac{1}{r}\frac{\partial}{\partial r}\left(r\frac{\partial u_\theta}{\partial r}\right) - \frac{u_\theta}{r^2} + \frac{\partial^2 u_\theta}{\partial z^2}\right] \tag{8.13}$$

z-component:
$$\rho\left(\frac{\partial u_z}{\partial t} + u_r \frac{\partial u_z}{\partial r} + u_z \frac{\partial u_z}{\partial z}\right) = -\frac{\partial p}{\partial z} + \mu\left[\frac{1}{r}\frac{\partial}{\partial r}\left(r\frac{\partial u_z}{\partial r}\right) + \frac{\partial^2 u_z}{\partial z^2}\right] \tag{8.14}$$

We make the velocity components and pressure non-dimensional using the following relations:

$$u_r^* = u_r/\Omega R, u_\theta^* = u_\theta/\Omega R, u_z^* = u_z/\Omega R, p^* = p/\rho\,\Omega^2 R^2 \tag{8.15}$$

and we make the coordinates non-dimensional using the following relations:

$$r^* = \frac{r}{R},\, \theta^* = \theta,\, z^* = \frac{z}{R},\, t^* = t\,\Omega \tag{8.16}$$

The continuity equation and the different component equations in non-dimensional form after skipping the $*$ superscript symbol:

$$\frac{\partial u_r}{\partial r} + \frac{u_r}{r} + \frac{\partial u_z}{\partial z} = 0 \tag{8.17}$$

$$\frac{\partial u_r}{\partial t} + u_r \frac{\partial u_r}{\partial r} - \frac{u_\theta^2}{r} + u_z \frac{\partial u_r}{\partial z} = -\frac{\partial p}{\partial r} + \frac{1}{Re}\left[\frac{\partial^2 u_r}{\partial r^2} + \frac{1}{r}\frac{\partial u_r}{\partial r} - \frac{u_r}{r^2} + \frac{\partial^2 u_r}{\partial z^2}\right] \tag{8.18}$$

$$\frac{\partial u_\theta}{\partial t} + u_r \frac{\partial u_\theta}{\partial r} + \frac{u_r u_\theta}{r} + u_z \frac{\partial u_\theta}{\partial z} = \frac{1}{Re}\left[\frac{\partial^2 u_\theta}{\partial r^2} + \frac{1}{r}\frac{\partial u_\theta}{\partial r} - \frac{u_\theta}{r^2} + \frac{\partial^2 u_\theta}{\partial z^2}\right] \tag{8.19}$$

$$\frac{\partial u_z}{\partial t} + u_r \frac{\partial u_z}{\partial r} + u_z \frac{\partial u_z}{\partial z} = -\frac{\partial p}{\partial z} + \frac{1}{Re}\left[\frac{\partial^2 u_z}{\partial r^2} + \frac{1}{r}\frac{\partial u_z}{\partial r} + \frac{\partial^2 u_z}{\partial z^2}\right] \tag{8.20}$$

The boundary conditions excluding surface tension are the following:

$$u_r(0,z,t) = u_\theta(0,z,t) = \frac{\partial u_z(0,z,t)}{\partial r} = 0, \, , \; 0 \le z \le h(0,t) \tag{8.21}$$

$$u_r(1,z,t) = u_z(1,z,t) = 0,\, u_\theta(1,z,t) = 1, \; 0 \le z \le h(1,t) \tag{8.22}$$

$$p = \frac{\partial h}{\partial t} + u_r \frac{\partial h}{\partial r} - u_z = 0, \; z = h(r,t) \tag{8.23}$$

and the initial conditions are
$$u_r(r,z,0) = u_\theta(r,z,0) = u_z(r,z,0) = 0, \; 0 \le r \le 1, 0 \le z \le H_w/R \tag{8.24}$$

I. References

1. Greenspan, H.P. and Howard, L.N., "On a Time-Dependent Motion of a Rotating Fluid", *Journal of Fluid Mechanics,* **17**, 3, 385-404, (1963).
2. Kim, K.Y. and Hyun, J.M., "Solution for Spin-Up from Rest of Liquid with a Free Surface", *AIAA Journal,* **34**, 7, 1441-1446, (1996).
3. Van de Konijnenberg, J.A, van Heijst, G.J.F., "Nonlinear spin-up in a circular cylinder", *Phys. Fluids,* **7** (12), (1995).

J. Exercises

8.1 Use ANSYS Fluent to continue running the simulations in this chapter and include free surface elevation plots at $t = 150$ s and $t = 200$ s as shown in Figures 8.19f) and 8.19g). Include contours of water volume fraction and swirl velocity at $t = 150$ and 200 s.

8.2 Use ANSYS Fluent to rerun the calculations shown in this chapter but change the rotational speed to 8 rad/s. Include free surface elevation plots at $t = 100, 150$ and 200 s as shown in Figures 8.19f) and 8.19g). Include contours of water volume fraction and swirl velocity at $t = 100, 150$ and 200 s. Determine the Reynolds number.

8.3 Use ANSYS Fluent to rerun the calculations shown in this chapter but change the rotational speed to 16 rad/s. Include free surface elevation plots at $t = 100, 150$ and 200 s as shown in Figures 8.19f) and 8.19g). Include contours of water volume fraction and swirl velocity at $t = 100, 150$ and 200 s. Determine the Reynolds number.

8.4 Use ANSYS Fluent to rerun the calculations shown in this chapter but change the rotational speed to 10 rad/s. Also, change the radius of the cylinder to R = 0.2 m with everything else being the same. Include free surface elevation plots at $t = 100, 150$ and 200 s as shown in Figures 8.19f) and 8.19g). Include contours of water volume fraction and swirl velocity at $t = 100, 150$ and 200 s. Determine the Reynolds number.

8.5 Use ANSYS Fluent to rerun the calculations shown in this chapter but change the rotational speed to 10 rad/s. Also, change the geometry to a frustum shape with bottom radius of the cylinder $R_{bottom} = 0.1$ m and top radius $R_{top} = 0.2$ m with everything else being the same. Include free surface elevation plots at $t = 100, 150$ and 200 s as shown in Figures 8.19f) and 8.19g). Include contours of water volume fraction and swirl velocity at $t = 100, 150$ and 200 s. Determine the Reynolds number based on the average radius.

8.6 Use ANSYS Fluent to run the simulations in this chapter and determine the contour plots and graph at your own Flow Time as listed in the table below. Include contours of water volume fraction and swirl velocity at your Flow Time. You will need to change the number of time steps in Figure 8.18c) to your number of time steps as listed in the table below. Use the same time step 0.01 s as was used in this chapter. In this way the simulations will stop at $t =$ your flow time as listed in the table below.

Student Name	Number of Time Steps	Time step (s)	Flow Time (s)
Student A	5000	0.01	50
Student B	5500	0.01	55
Student C	6000	0.01	60
Student D	6500	0.01	65
Student E	7000	0.01	70
Student F	7500	0.01	75
Student G	8000	0.01	80
Student H	8500	0.01	85
Student I	9000	0.01	90
Student J	9500	0.01	95

Table 8.1 Data related to Exercise 8.6

CHAPTER 9. KELVIN-HELMHOLTZ INSTABILITY

A. Objectives

- Using ANSYS Workbench to Create Geometry and Mesh for Kelvin-Helmholtz Instability
- Inserting Wall Boundary Conditions and an Acceleration Due to Gravity Vector Corresponding to a Tilt Angle of 4 degrees for the Tube.
- Using Volume of Fluid Model for Multiphase Flow with Surface Tension
- Running Laminar Transient 2D Planar ANSYS Fluent Simulations
- Using Contour Plots of Volume Fraction and a Movie for Visualizations

B. Problem Description

The Kelvin-Helmholtz instability is a classical problem originally studied by Helmholtz[1] and Kelvin[2]. The mechanism causing the instability has been studied in detail by Lamb[3], Bachelor[4], Drazin and Reid[5], Chandrasekahr[6], Craik[7], and many others. The Kelvin-Helmholtz instability can appear at the interface of two fluid layers flowing with different velocities and with different densities. For a complete review of this instability, please see Thorpe[8]. Thorpe[9-12] found that the Kelvin-Helmholtz instability can be generated and visualized by tilting a tube that contains two fluids at different densities.

We will study the Kelvin-Helmholtz instability and analyze the problem using ANSYS Fluent. The length of the tube is 2,400 mm and the diameter 200 mm.

(fresh water) $\rho_1 = 998.2$ kg/m^3, $\mu_1 = 0.001002$ kg/ms, $\gamma_1 = 0.07274$ N/m
(salt water) $\rho_2 = 1074.9$ kg/m^3, $\mu_2 = 0.001259$ kg/ms, $\gamma_2 = 0.07422$ N/m

C. Launching ANSYS Workbench and Selecting Fluent

1. Start by launching the ANSYS Workbench. Double click on Fluid Flow (Fluent) under Analysis Systems in the Toolbox.

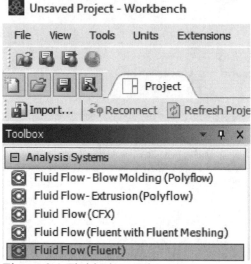

Figure 9.1 Fluid Flow (Fluent)

D. Launching ANSYS DesignModeler

2. Right click on Geometry in the Project Schematic and select Properties. Select 2D Analysis Type under Advanced Geometry Options in Properties of Schematic A2: Geometry. Right click on Geometry and select New DesignModeler Geometry. Select **Units>>Millimeter** from the menu as the length unit in DesignModeler.

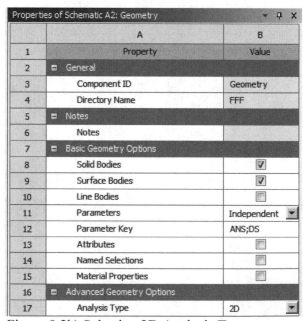

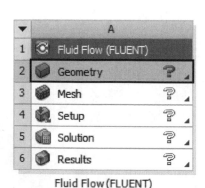

Figure 9.2a) Selecting Geometry   Figure 9.2b) Selecting 2D Analysis Type

3. Next, we will be creating the mesh for the simulation. Select XYPlane from the Tree Outline. Select Look at Sketch 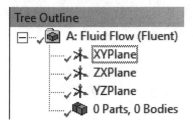. Click on the Sketching tab on the left-hand side. Select Draw and sketch a Rectangle from the origin. Make sure that the letter *P* is visible as you move the cursor over the origin and before you start drawing the rectangle. Select the Dimensions tab in the Sketching Toolboxes. Select one of the vertical edges of the rectangle and enter a length of 200 mm. Select one of the horizontal edges of the rectangle and enter a length of 2400 mm.

Figure 9.3a) Selection of XYPlane

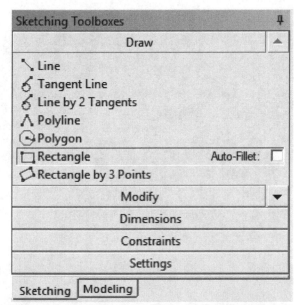

Figure 9.3b) Selection of Sketching Toolboxes and Rectangle

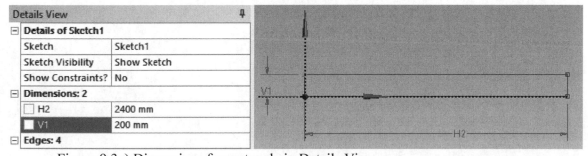

Figure 9.3c) Dimensions for rectangle in Details View

4. Click on the Modeling tab. Select Concept>>Surfaces from Sketches from the menu. Select Sketch1 under XYPlane in the Tree Outline on the left and select Apply as Base Objects in Details View. Click on Generate in the toolbar. The rectangle turns gray. Right click in the graphics window and select Zoom to Fit. Close the DesignModeler window.

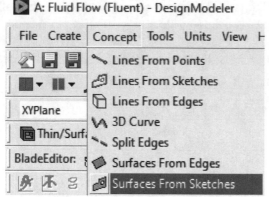

Figure 9.4a) Selecting Surfaces from Sketches

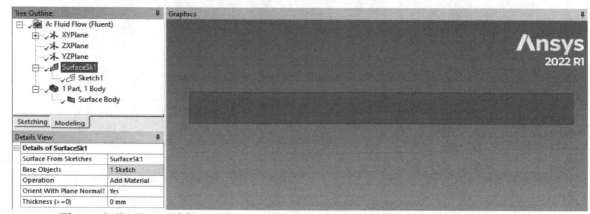

Figure 9.4b) Base Object and completed rectangle in DesignModeler

E. Launching ANSYS Meshing

5. We are now going to double click on Mesh in ANSYS Workbench to open the Meshing window. Select Mesh in the Outline. Click on Update under Mesh in the menu. A coarse mesh is created.

    Select Mesh>>Controls>>Face Meshing from the menu. Click on the rectangle in the graphics window. Select the yellow region next to Geometry that is labeled No Selection and click on the Apply button for Geometry in Details of "Face Meshing".

    Select Mesh>>Controls>>Sizing from the menu and select the Edge ⬚ selection filter. Click on the upper horizontal edge of the rectangle and control select the lower horizontal edge. Click on Apply for the Geometry in "Details of Edge Sizing". Under Definition in "Details of Edge Sizing", select Number of Divisions as Type, 480 as Number of Divisions, and Hard as Behavior.

    Repeat the selection of Mesh>>Controls>>Sizing from the menu and control-select the two vertical edges of the rectangle. Enter 40 for Number of Divisions and Hard Behavior. Choose the Bias Type —— — - — —— and enter 3.0 as Bias Factor. Select Mesh>>Update in the menu and select Mesh in the Outline. The finished mesh is shown in the graphics window.

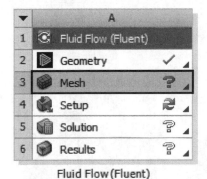

Figure 9.5a) Starting Mesh

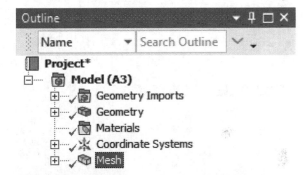

Figure 9.5b) Selection of Mesh in Outline

Figure 9.5c) Coarse mesh

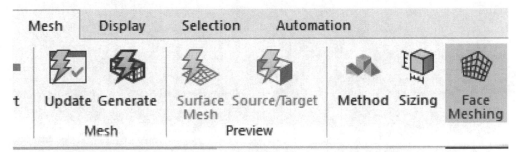

Figure 9.5d) Face Meshing

Figure 9.5e) Applied Geometry in Details

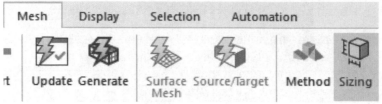

Figure 9.5f) Mesh Control Sizing

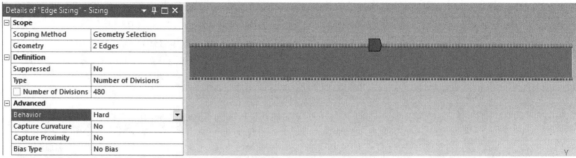

Figure 9.5g) Details of edge sizing for horizontal edges

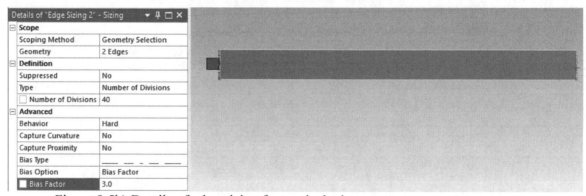

Figure 9.5h) Details of edge sizing for vertical edges

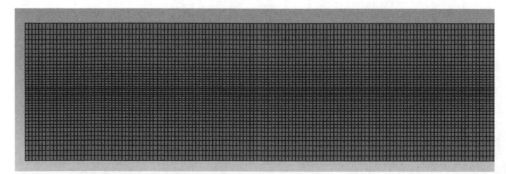

Figure 9.5i) Details of finished mesh

6. We are now going to rename the edges for the rectangle. Select the Edge filter 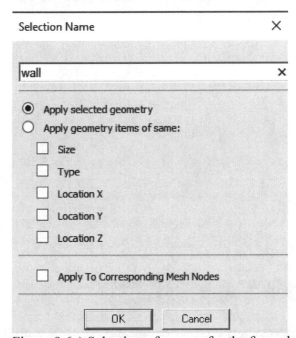 and control-select the four edges of the rectangle, right click and select Create Named Selection. Enter *wall* as the name for the selection group and click on the OK button.

    Select File>>Export…>>Mesh>>FLUENT Input File>>Export from the menu and save the mesh with the name *khi-mesh.msh* in the working directory. Select File>>Save Project from the menu and save the project with the name *Kelvin-Helmholtz Instability*. Close the Meshing window. Right click on Mesh in ANSYS Workbench in the Project Schematic and select Update.

Figure 9.6a) Selection of a name for the four edges of the rectangle

Figure 9.6b) Named selections

F. Launching ANSYS Fluent

7. Double-click on Setup under Project Schematic in ANSYS Workbench. Check the box for Double Precision. Click on Show More Options and write down the location of your *Working Directory*. You will need this information later. Set the number of processes equal to the number of processor cores for your computer. To check the number of physical cores, press the Ctrl + Shift + Esc keys simultaneously to open the Task Manager. Go to the Performance tab and select CPU from the left column. You will see the number of physical cores on the bottom-right side. Close the Task Manager window. Click on the Start button in the Fluent Launcher window. Click OK when you get the Key Behavioral Changes window.

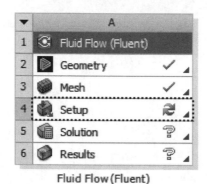

A
1 Fluid Flow (Fluent)
2 Geometry
3 Mesh
4 Setup
5 Solution
6 Results
Fluid Flow (Fluent)

Figure 9.7a) Starting Setup

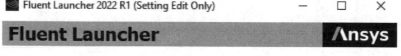

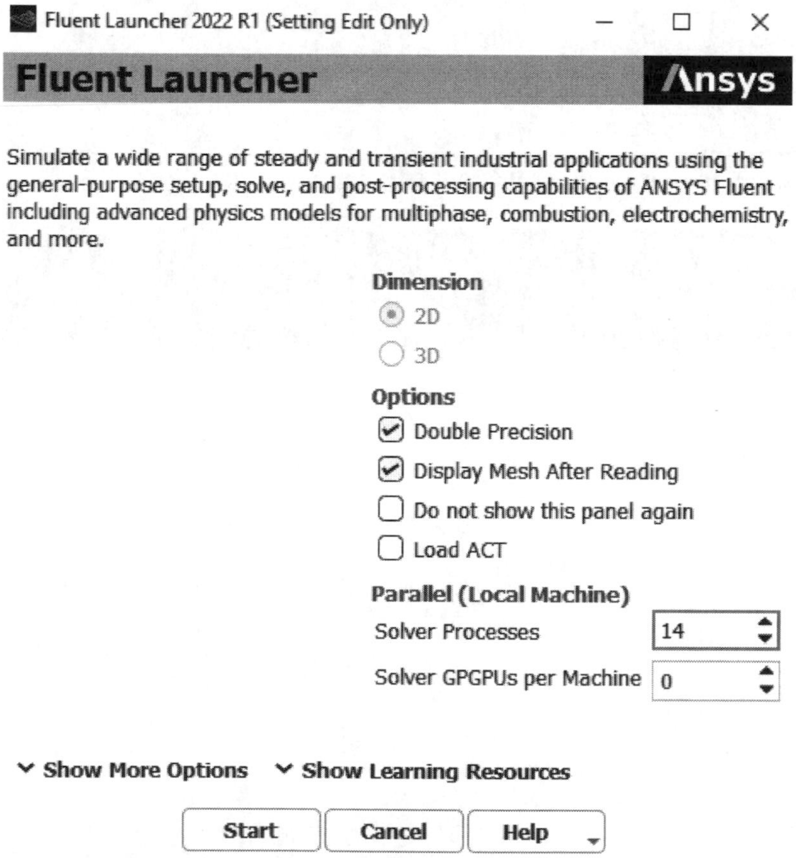

Fluent Launcher 2022 R1 (Setting Edit Only)
Fluent Launcher
Ansys
Simulate a wide range of steady and transient industrial applications using the general-purpose setup, solve, and post-processing capabilities of ANSYS Fluent including advanced physics models for multiphase, combustion, electrochemistry, and more.
Dimension
2D
3D
Options
Double Precision
Display Mesh After Reading
Do not show this panel again
Load ACT
Parallel (Local Machine)
Solver Processes
14
Solver GPGPUs per Machine
0
Show More Options
Show Learning Resources
Start
Cancel
Help

Figure 9.7b) Fluent launcher window

8. Check the mesh by selecting the Check button under Mesh in General on Task Page. Check the scale of the mesh by selecting the Scale button. Make sure that the Domain Extent is correct and close the Scale Mesh window.

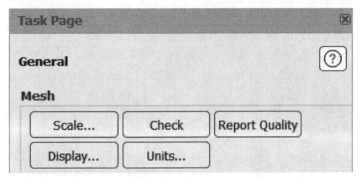

Figure 9.8a) Mesh Check

Scale Mesh

Domain Extents

Xmin [m]	0	Xmax [m]	2.4
Ymin [m]	0	Ymax [m]	0.2

Scaling

- ● Convert Units
- ○ Specify Scaling Factors

Mesh Was Created In
<Select> ▼

View Length Unit In
m ▼

Scaling Factors
X 1
Y 1

Scale   Unscale

Close   Help

Figure 9.8b) Scale Mesh window

9. Select Transient Time for the Solver in General on the Task Page. Check the Gravity box and enter -9.71453 m/s^2 as the Gravitational Acceleration in the Y direction. Enter -1.365288 m/s^2 as the Gravitational Acceleration in the X direction. These components correspond to a tilt angle of 8 degrees for the tube.

Open Models under Setup in the Outline View and double click on Viscous (SST k-omega). Select the Laminar Model and select OK to close the Viscous Model window. Double click on Multiphase (Off). Select the Volume of Fluid Model. Click on Apply and Close the Multiphase Model window.

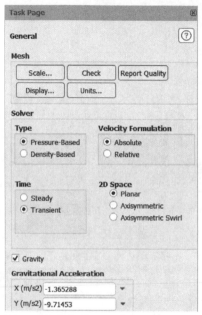

Figure 9.9a) General Setup

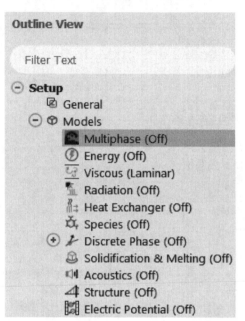

Figure 9.9b) Models Problem Setup

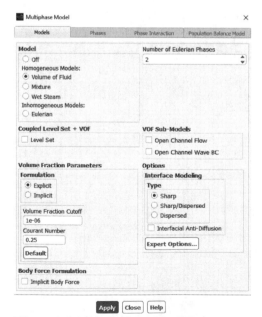

Figure 9.9c) Multiphase model

10. Double click on Materials under Setup in the Outline View. Click on the Create/Edit button for Fluid on the Task Page to open the Create/Edit Materials window. Click on Fluent Database. Scroll down in the Fluent Fluid Materials window and select *water-liquid (h2o<l>)*. Click on the Copy button and Close the Fluent Database Materials window. Click on the Change/Create button and the Close button in the Create/Edit Materials window.

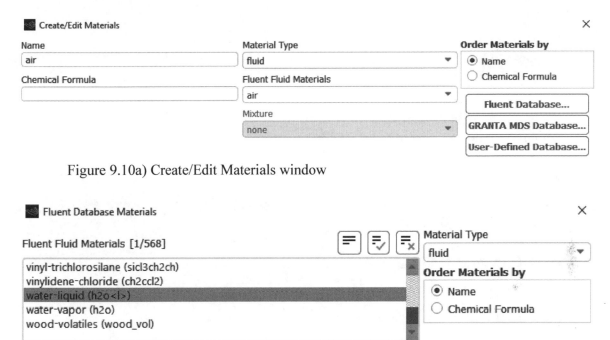

Figure 9.10a) Create/Edit Materials window

Figure 9.10b) Selecting water-liquid as FLUENT Fluid Material

Repeat the steps above and select *water-liquid*. You will get a New Material Name window when you click on the Copy button in the Fluent Database Materials window. Enter *salt-water-liquid* as the New Name and click OK. Close the Fluent Database Materials window. Click on the Change/Create button and the Close button in the Create/Edit Materials window.

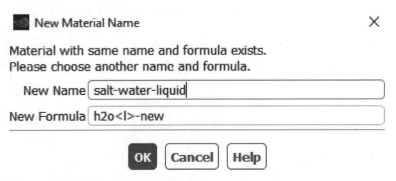

Figure 9.10c) New Material Name

11. Open Models under Setup in the Outline View, double click on Multiphase (Volume of Fluid) and select the Phases tab in the Multiphase Model window. Select phase-1-Primary Phase under Phases. Select *water-liquid* from Phase Materials and enter *fresh-water* as the name of the Primary Phase. Click on the Apply button.

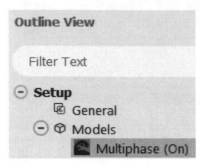

Figure 9.11a) Multiphase

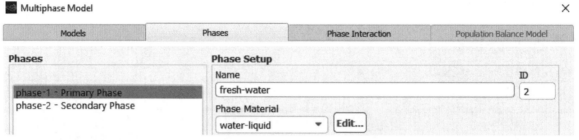

Figure 9.11b) Primary Phase Setup

Repeat these steps for the Secondary Phase to select *salt-water-liquid* as Phase Material and enter the name *salt-water* under Phase Setup. Click on the Apply button.

Select the Phase Interactions tab at the top of the Multiphase Model window. Select the Surface Tension tab, check the Surface Tension Force Modeling box, select constant from the Surface Tension Coefficients [N/m] drop down menu, enter the value 0.00148 and click on the Apply button. Close the Multiphase Model window.

Select *salt-water-liquid* under Materials and Fluid on the Task Page and click on Create/Edit. Enter 1074.9 as Density [kg/m³] and 0.001259 as Viscosity [kg/(m s)]. Select Change/Create and Close the window. For a reference of the dynamic viscosity value used for salt-water, see Sharqawy, Lienhard, and Zubair[13].

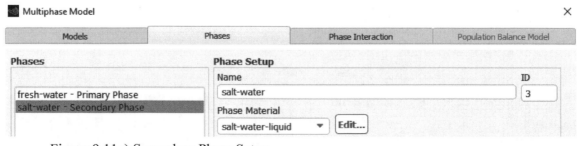

Figure 9.11c) Secondary Phase Setup

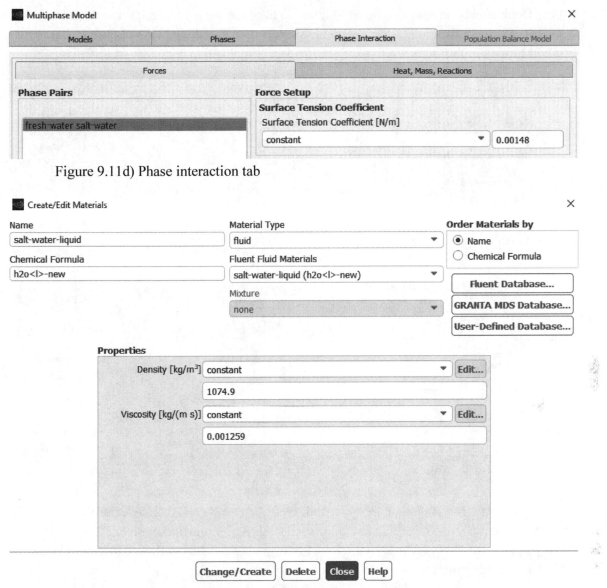

Figure 9.11d) Phase interaction tab

Figure 9.11f) Values of density and viscosity for salt-water

12. Double click on Methods under Solution in the Tree. Select Default settings on the Task Page except for First Order Upwind for Momentum. Double click on Initialization under Solution in the Outline View, select Standard Initialization and click on Initialize.

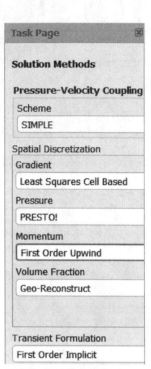

Figure 9.12a) Solution Methods

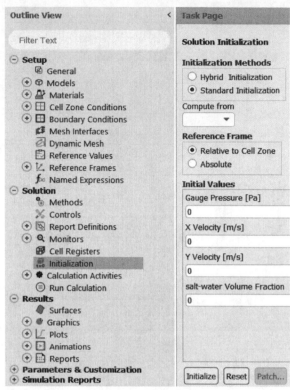

Figure 9.12b) Solution Initialization

13. Select the Domain tab in the menu and select Adapt>>Manual…. Select Cell Registers>>New>>Region in the Manual Mesh Adaption window. Enter 2.4 for X Max, and 0.1 for Y Max. Click on the Save button and Close the window. Close the Manual Mesh Adaption window.

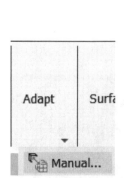

Figure 9.13a) Manual Adaption

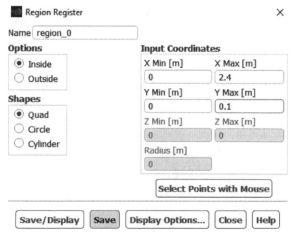

Figure 9.13b) Region Adaption Settings

Click on the Patch… button on the Task Page for Solution Initialization and select *salt-water* as the phase. Select Volume Fraction as the Variable and region_0 as Registers to Patch. Set the value to 1 and click on the Patch button.

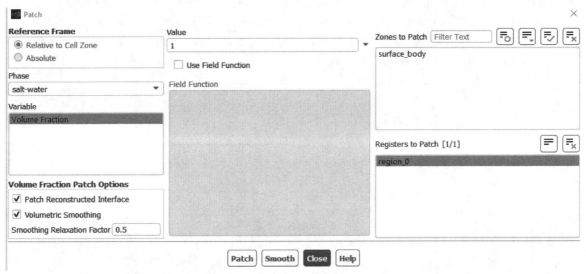

Figure 9.13c) Patch Settings for Volume Fraction

Select mixture for the phase. Select X Velocity as the Variable and region_0 as Registers to Patch. Set the value to 0 and click on the Patch button. Close the Patch window.

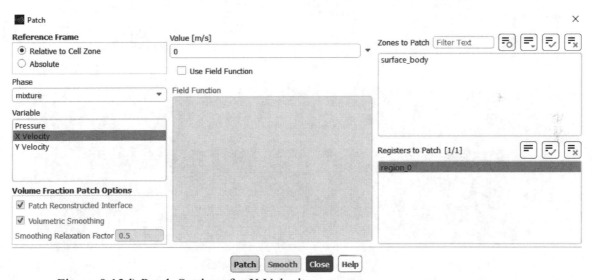

Figure 9.13d) Patch Settings for X Velocity

14. Open Graphics and double click on Contours under Results in the Outline View. Select Phases and Volume fraction under Contours of and select *fresh-water* as the Phase. Click on Colormap Options and select Top as Colormap Alignment. Set Type to float and Precision to 1 under Number Format. Set Font Behavior to Fixed and Font Size to 12 under Font. Click on Apply and Close the Colormap window. Click on the Save/Display button and Close the window. Click on Copy Screenshot of Active Window to Clipboard in the graphics window, see Figure 9.14c). The contour plot can be pasted into a Word document.

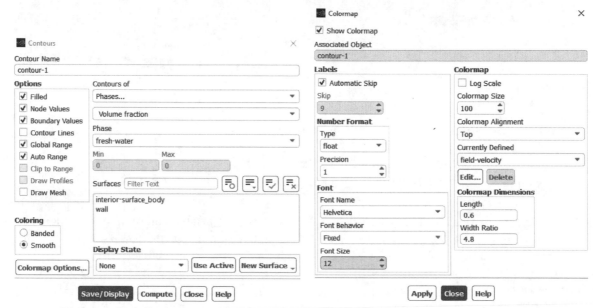

Figure 9.14a) Contours and colormap settings

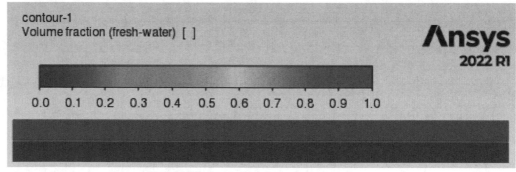

Figure 9.14b) Contours of Volume Fraction (fresh-water)

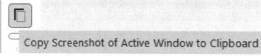

Figure 9.14c) Copying screenshot

15. Open Calculation Activities and double click on Solution Animations under Solution in the Outline View. Enter *khi-animation* as the name and select *contour-1* as Animation Object. Set Record after every to 10 time-step. Select the OK button to close the Animation Definition window.

Double click on Run Calculation under Solution, enter 0.01 as Time Step Size, and 200 for Number of Time Steps. Click on the Calculate button. Click OK when the calculation is complete. Open Graphics and double click on *contour-1* under Contours that is under Results in the Outline View. Click on the Save/Display button and Close the window.

Figure 9.15a) Animation definition window   Figure 9.15b) Running calculations

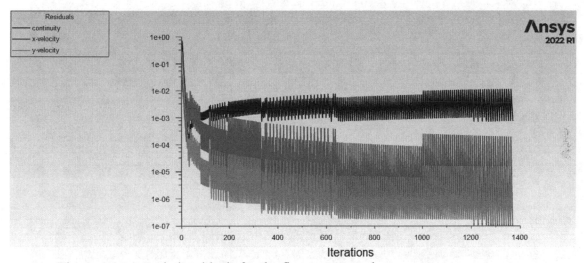

Figure 9.15c) Scaled residuals for the first two seconds.

Click on Copy Screenshot of Active Window to Clipboard in the graphics window, see Figure 9.14c). The contour plot can be pasted into a Word document.

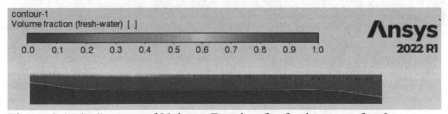

Figure 9.15d) Contour of Volume Fraction for fresh-water after 2 s

G. Post-Processing

16. Enter 100 for Number of Time Steps and continue running the calculations twice followed by entering 50 for the Number of Time Steps. Continue running the calculations six more times for a total of 7 seconds. ANSYS Fluent simulations can be compared with experiments of the instability, see Matsson and Boiselle[14].

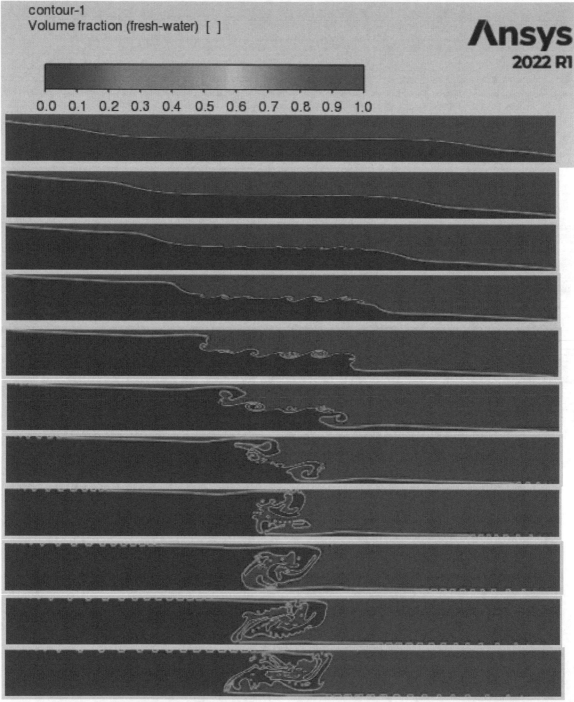

Figure 9.16a) Volume Fraction for fresh-water after 3, 4, 4.5, 5, 5.5, 6, 6.5, 7, 7.5, 8, 8.5 s

CHAPTER 9. KELVIN-HELMHOLTZ INSTABILITY

Click on Copy Screenshot of Active Window to Clipboard in the graphics window, see Figure 9.14c). The contour plots can be pasted into a Word document.

Figure 9.16b) Experiments for Kelvin-Helmholtz instability at $t = 5.0$ s (8° tilt angle), from Matsson and Boisselle[1]

Select File>>Export>Case & Data… from the menu. Save the file with the name *khi-computation.cas.h5*.

Double click on Animations and Playback under Results in the Outline View. Set the Replay Speed to low and select Play Once as Playback Mode. Select Video File as Write/Record Format. Click on the Write button. This will create a movie with the name *khi-animation.mpeg*. Close the Playback window. The movie can be viewed using for example Windows Media Player.

H. Theory

17. We will look at the inviscid theory for the Kelvin-Helmholtz instability in terms of a sinusoidal disturbance between two fluids with velocities U_1, U_2 and densities ρ_1, ρ_2.

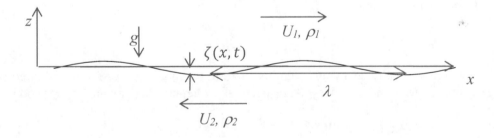

Figure 9.17 Development of Kelvin Helmholtz instability from a perturbation

The velocity vector in a Cartesian coordinate system can be described as

$$\bar{U} = u_x \bar{e}_x + u_y \bar{e}_y + u_z \bar{e}_z \tag{9.1}$$

where u_x, $u_y = 0$ and u_z are the velocity components in the streamwise, spanwise and normal directions, respectively. Euler equations of motion can be written as

$$\frac{\partial u_x}{\partial t} + u_x \frac{\partial u_x}{\partial x} + u_z \frac{\partial u_x}{\partial z} = -\frac{1}{\rho} \frac{\partial P}{\partial x} \tag{9.2}$$

$$\frac{\partial u_z}{\partial t} + u_x \frac{\partial u_z}{\partial x} + u_z \frac{\partial u_z}{\partial z} = -\frac{1}{\rho}\frac{\partial P}{\partial z} - g \qquad (9.3)$$

$$\frac{\partial u_x}{\partial x} + \frac{\partial u_z}{\partial z} = 0 \qquad (9.4)$$

The velocity components can be divided into mean flow components in the main streamwise direction plus a perturbation

$$\bar{U} = (U(z) + u)\bar{e}_x + w\bar{e}_z \qquad (9.5)$$

and the pressure is $\qquad P = P_0 - \rho g z + p \qquad (9.6)$

A perturbation is introduced in the form

$$u, w, p = [u_n(z), w_n(z), p_n(z)]e^{i(kx-\omega t)} \qquad (9.7)$$

where k is the streamwise wave number, ω is the frequency and $u_n(z), w_n(z), p_n(z)$ are eigenfunctions. The linearized Euler equations can be written as

$$i(kU - \omega)u_n + \frac{ik}{\rho}p_n = 0 \qquad (9.8)$$

$$i(kU - \omega)w_n + \frac{1}{\rho}p_n' = 0 \qquad (9.9)$$

$$iku_n + w_n' = 0 \qquad (9.10)$$

From equations (9.8) – (9.10) we can find the equation for w_n

$$w_n'' - k^2 w_n = 0 \qquad (9.11)$$

with the general solution

$$w_n = c_1 e^{-kz} + c_2 e^{kz} \qquad (9.12)$$

To get exponentially decaying solutions far from the interface, we set c_1 and c_2 equal to zero in the lower and upper layers, respectively. The interface location is defined by

$$z = \zeta(x,t) = ae^{i(kx-\omega t)} \qquad (9.13)$$

The normal velocity w at the interface is given by the material derivative of $\zeta(x,t)$

$$w = w_n(z)e^{i(kx-\omega t)} = \frac{D\zeta}{Dt} = \frac{\partial \zeta}{\partial t} + U\frac{\partial \zeta}{\partial x} =$$
$$i(kU - \omega)\,\zeta = i(kU - \omega)ae^{i(kx-\omega t)} \qquad (9.14)$$

, we find that $c_1 = i(kU_1 - \omega)a$, $c_2 = -i(kU_2 + \omega)a$

$$w_{n,1} = i(kU_1 - \omega)ae^{-kz}$$
$$w_{n,2} = -i(kU_2 + \omega)ae^{kz} \qquad (9.15)$$

We can now find the pressure eigenfunctions in the upper and lower layers from equation (9.9).

$$p_{n,1} = -\rho_1(kU_1 - \omega)^2 ae^{-kz}/k$$
$$p_{n,2} = \rho_2(kU_2 + \omega)^2 ae^{kz}/k \tag{9.16}$$

Next, we use equation (9.6) and set the pressures identical at the interface to get the dispersion relation

$$P_0 - \rho_1 g\zeta + p_{n,1}e^{i(kx-\omega t)} = P_0 - \rho_2 g\zeta + p_{n,2}e^{i(kx-\omega t)} \tag{9.17}$$

$$(\rho_2 - \rho_1)gk = \rho_2(kU_2 + \omega)^2 + \rho_1(kU_1 - \omega)^2 \tag{9.18}$$

and we find the following equation for the frequency

$$\omega = \frac{k}{\rho_1 + \rho_2}\left[(\rho_1 U_1 - \rho_2 U_2) \pm \sqrt{\frac{2(\rho_2^2 - \rho_1^2)g}{k} - \rho_1\rho_2(U_1 + U_2)^2}\right] \tag{9.19}$$

The frequency is complex with the associated growth of Kelvin-Helmholtz instability waves when the expression under the square root is negative

$$\frac{2(\rho_2^2 - \rho_1^2)g}{k} < \rho_1\rho_2(U_1 + U_2)^2 \tag{9.20}$$

Using the definition of wave number as inversely proportional to the wave-length

$$k = \frac{2\pi}{\lambda} \tag{9.21}$$

We find that waves are amplified and develop when their wave-lengths fulfill the following condition as shown in equation (9.22). A boundary below or above the interface can destabilize the flow as shown by Hazel[15].

$$\lambda < \frac{\pi\rho_1\rho_2(U_1 + U_2)^2}{g(\rho_2^2 - \rho_1^2)} \tag{9.22}$$

I. References

1. Helmholtz H. "On discontinuous movement of fluids", *Phil. Mag.* (4) **36**, 337, (1868).
2. Kelvin L. "Hydrokinetic solutions and observations", *Phil. Mag.* (4) **42**, 362, (1871).
3. Lamb H. "Hydrodynamics", *Cambridge University Press*, (1932).
4. Batchelor G.K. "An introduction to fluid dynamics", *Cambridge University Press*, (1967).
5. Drazin P.G. and Reid W.H. "Introduction to hydrodynamic instability", *Cambridge University Press*, (1981).
6. Chandrasekhar S. "Hydrodynamic and Hydromagnetic Stability", *Dover Publications, Inc.*, New York (1961).
7. Craik A.D.D. "Wave interactions and fluid flows", *Cambridge University Press*, (1985).
8. Thorpe S.A. "Transitional phenomena and the development of turbulence in stratified fluids: a review", *J. Geophys. Res.*, **92**, 5231 (1987).
9. Thorpe S.A. "A method of producing a shear flow in a stratified fluid", *J. Fluid Mech.*, **32**, 693 (1968).
10. Thorpe S.A. "Experiments on the instability of stratified shear flows: miscible

fluids", *J. Fluid Mech.*, **46**, 299 (1971).

11. Thorpe S.A. "Turbulence in stably stratified fluids: a review of laboratory experiments", *Boundary-Layer Meteorol.*, **5**, 95 (1973).

12. Thorpe S.A. "Experiments on the instability and turbulence in a stratified shear flow", *J. Fluid Mech.*, **61**, 731 (1973).

13. Sharqawy M.H, Lienhard J.H., and Zubair S.M. "Thermophysical Properties of Sea Water: A Review of existing Correlations and Data, Desalination, and Water Treatment", *Desalination and Water Treatment*, **16**, 354 (2010).

14. Matsson J. and Boisselle J. "A Senior Design Project on the Kelvin-Helmholtz Instability", 122nd ASEE Annual Conference and Exposition, June 14-17, 2015, Seattle, WA.

15. Hazel P. "Numerical Studies of the stability of inviscid stratified shear flows", *J. Fluid Mech.*, **51**, 39 (1972).

J. Exercises

9.1 Run the ANSYS Fluent simulations for a tube tilt angle of 4 degrees and show the development of the instability using contours of volume fraction for fresh water for the same fluids as used in this chapter.

9.2. Run the ANSYS Fluent simulations for a tube tilt angle of 12 degrees and show the development of the instability using contours of volume fraction for fresh water for the same fluids as used in this chapter.

9.3 Run the ANSYS Fluent simulations shown in this chapter but instead use a transient 3D geometry and mesh for the model. Use a tube tilt angle of 4 degrees and use contours of volume fraction for fresh water to visualize the flow for the same fluids as used in this chapter.

9.4 Run the ANSYS Fluent simulations shown in this chapter but instead use a transient 3D geometry and mesh for the model. Use a tube tilt angle of 8 degrees and use contours of volume fraction for fresh water to visualize the flow for the same fluids as used in this chapter.

9.5 Run the ANSYS Fluent simulations shown in this chapter but instead use a transient 3D geometry and mesh for the model. Use a tube tilt angle of 12 degrees and use contours of volume fraction for fresh water to visualize the flow for the same fluids as used in this chapter.

CHAPTER 10. RAYLEIGH-TAYLOR INSTABILITY

A. Objectives

- Using ANSYS Workbench to Create Geometry and Mesh for Rayleigh-Taylor Instability
- Inserting Symmetry and Wall Boundary Conditions
- Using Volume of Fluid Model for Multiphase Flow
- Using a User Defined Function to Trigger the Rayleigh-Taylor Instability
- Running Laminar Transient 2D Planar ANSYS Fluent Simulations
- Using Contour Plots of Volume Fractions and a Movie for Visualizations

B. Problem Description

We will study the Rayleigh-Taylor instability and will analyze the problem using ANSYS. The Geometry for Rayleigh-Taylor instability (total height 1 m, width 0.25 m) is shown below. The heavier fluid is initially on top of the lighter fluid.

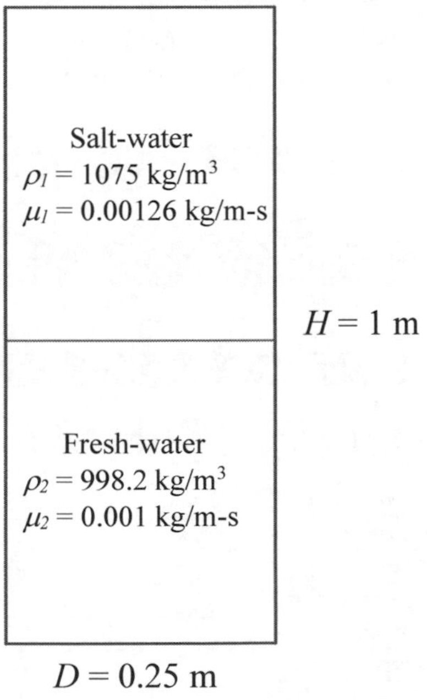

Salt-water
$\rho_l = 1075$ kg/m^3
$\mu_l = 0.00126$ kg/m-s

$H = 1$ m

Fresh-water
$\rho_2 = 998.2$ kg/m^3
$\mu_2 = 0.001$ kg/m-s

$D = 0.25$ m

C. Launching ANSYS Workbench and Selecting Fluent

1. Start by launching ANSYS Workbench. Double click on Fluid Flow (Fluent).

Figure 10.1 Selecting Fluent

D. Launching ANSYS DesignModeler

2. Right click on Geometry under Project Schematic in ANSYS Workbench and select Properties. Select 2D Analysis Type under Advanced Geometry Options in Properties of Schematic A2: Geometry. Right click on Geometry and select New DesignModeler Geometry to open ANSYS DesignModeler. **Select Units>>Millimeter from the menu as the length unit in DesignModeler.**

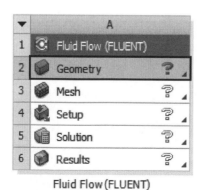

Fluid Flow (FLUENT)

Figure 10.2a) Selecting Geometry

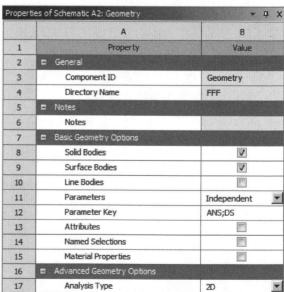

Figure 10.2b) Selecting 2D Analysis Type

3. Next, we will be creating the mesh for the simulation. Select XYPlane in the Tree Outline. Select Look at Sketch 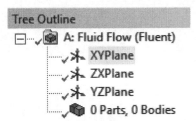 . Click on the Sketching tab and Draw a Rectangle from the origin. Make sure that you see the letter *P* as you move the cursor above the origin of the coordinate system and before you start drawing the rectangle. Select the Dimensions tab in the Sketching Toolboxes after completion of the rectangle. Click on one of the vertical edges of the rectangle and enter a vertical length of 1000 mm. Click on one of the horizontal edges and set the horizontal length of 250 mm.

Figure 10.3a) Selection of XYPlane in the Tree Outline

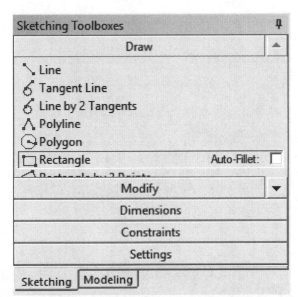

Figure 10.3b) Selection of Sketching Toolboxes and Rectangle

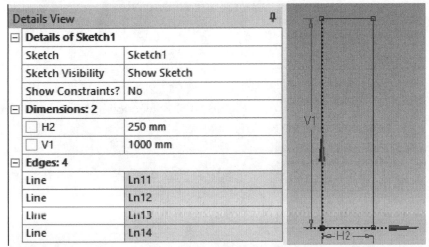

Figure 10.3c) Dimensions in Details View

225

4. Click on the Modeling tab and select Concept>>Surfaces from Sketches in the menu. Open XYPlane in the Tree Outline, select Sketch1 as Base Objects and select Apply in Details View. Click on $\boxed{\text{Generate}}$ in the toolbar. The rectangle turns gray. Right click in the graphics window, select Zoom to Fit and close DesignModeler.

Figure 10.4 Completed rectangle in DesignModeler

E. Launching ANSYS Meshing

5. We are now going to double click on Mesh in ANSYS Workbench to open the Meshing window. Select Mesh in the Outline. Click on Update. A coarse mesh is created. Select Mesh>>Controls>>Face Meshing from the menu. Click on the rectangle in the graphics window. Apply the Geometry in Details of "Face Meshing".

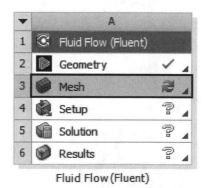

Figure 10.5a) Starting Mesh

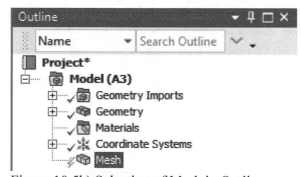

Figure 10.5b) Selection of Mesh in Outline

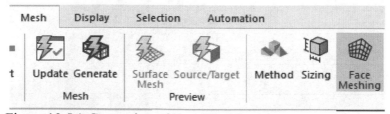

Figure 10.5c) Generation of Face Meshing

Details of "Face Meshing" - Mapped Fi ▼ ⊓ □ ×	
⊟ **Scope**	
Scoping Method	Geometry Selection
Geometry	1 Face
⊟ **Definition**	
Suppressed	No
Mapped Mesh	Yes
Method	Quadrilaterals
Constrain Boundary	No
⊟ **Advanced**	
Specified Sides	No Selection
Specified Corners	No Selection
Specified Ends	No Selection

Figure 10.5d) Applying Geometry in Details of "Face Meshing"

6. Select Mesh>>Controls>>Sizing from the menu and select Edge ▣. Control-select the four edges of the rectangle. Click on Apply for the Geometry in "Details of Edge Sizing". Under Definition in "Details of Edge Sizing", select Element Size as Type, 2.0 mm as Element Size and Hard as Behavior. Click on Update and select Mesh in the Outline. The finished mesh is shown in Figure 10.6b).

Figure 10.6a) Mesh Control Sizing

Details of "Edge Sizing" - Sizing ▼ ⊓ □ ×	
⊟ **Scope**	
Scoping Method	Geometry Selection
Geometry	4 Edges
⊟ **Definition**	
Suppressed	No
Type	Element Size
☐ Element Size	2.0 mm
⊟ **Advanced**	
Behavior	Hard
Capture Curvature	No
Capture Proximity	No
Bias Type	No Bias

Figure 10.6b) Details of "Edge Sizing"

Figure 10.6c) Details of finished mesh

7. We are now going to rename the edges for the rectangle. Use the Edge tool ⬚ , control-select the left and right edges of the rectangle, right click and select Create Named Selection. Enter *symmetry* as the name and click on the OK button.

Repeat this step for the upper and lower horizontal edges (control-select both edges) of the rectangle and name them *wall*. Select File>>Export...>>Mesh>>FLUENT Input File>>Export from the menu and save the mesh with the name *rt-mesh.msh*. Select File>>Save Project from the menu. Save the project with the name *Rayleigh-Taylor-Instability*. Select File>>Close Meshing. Right click on Mesh under Project Schematic in ANSYS Workbench and select Update.

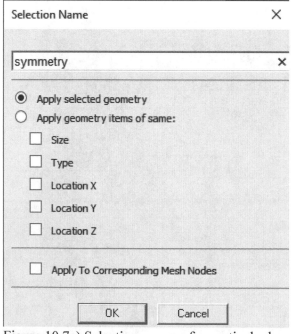

Figure 10.7a) Selecting a name for vertical edges

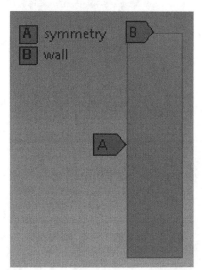

Figure 10.7b) Named selections

F. Launching ANSYS Fluent

8. Double-click on Setup under Project Schematic in ANSYS Workbench. Check the box for Double Precision. Click on Show More Options and write down the location of your *Working Directory*. You will need this information later. Set the number of processes equal to the number of processor cores for your computer. To check the number of physical cores, press the Ctrl + Shift + Esc keys simultaneously to open the Task Manager. Go to the Performance tab and select CPU from the left column. You will see the number of physical cores on the bottom-right side. Close the Task Manager window. Click on the Start button in the Fluent Launcher window.

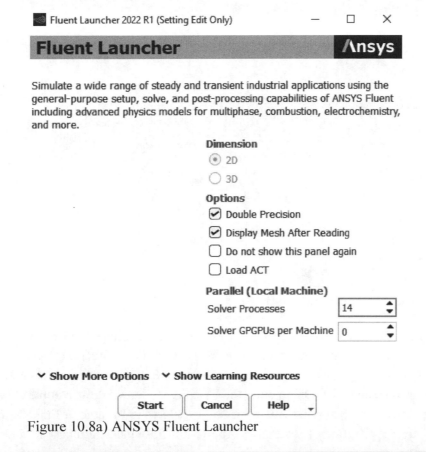

Figure 10.8a) ANSYS Fluent Launcher

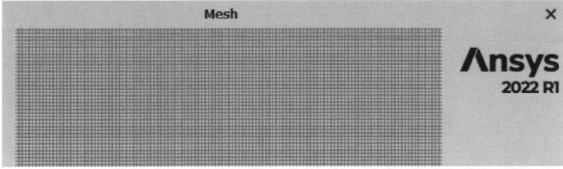

Figure 10.8b) Part of mesh in ANSYS Fluent

9. Check the mesh in ANSYS Fluent by selecting the Check button under Mesh in General Problem Setup. Check the scale of the mesh by selecting the Scale button. Make sure that the Domain Extent is correct and close the Scale Mesh window.

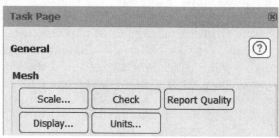

Figure 10.9a) Mesh check

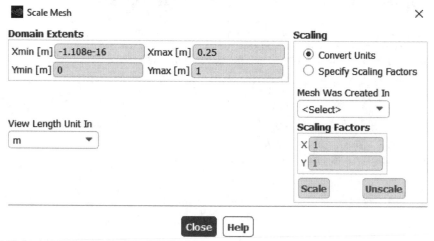

Figure 10.9b) Scale Mesh window

10. Select Transient Time for the Solver in General Problem Setup. Check the Gravity box and enter -9.81 as the Gravitational Acceleration Y [m/s^2] in the Y direction. Open Models and double click on Viscous (SST k-omega) under Setup in the Outline View. Select Laminar Model and click OK to close the Viscous Model window. Double click on Multiphase (Off) under Setup in the Outline View. Select the Volume of Fluid Model. Check the box for Implicit Body Force. Click Apply and Close the Multiphase Model window.

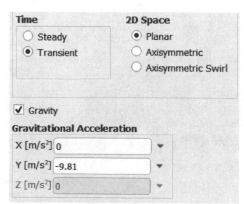

Figure 10.10a) General Setup

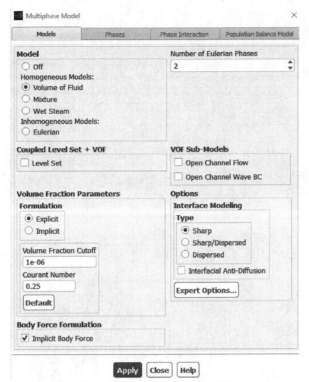

Figure 10.10b) Volume of Fluid Model

11. Double click on Materials, Fluid and air under Setup in the Outline View. Change the Name from *air* to *light-fluid*. Change the Density [kg/m³] to 998.2 and the Viscosity [kg/m s)] to 0.001002. Click on the Change/Create button and answer *Yes* to the question that appears and Close the window. Click on Create/Edit for Fluid under Materials on the Task Page. Click on Fluent Database. Scroll down in the Fluent Fluid Materials window and select *water-liquid (h2o<l>)*. Click on the Copy button and Close the Fluent Database Materials window. Change the Name from *water-liquid* to *heavy-fluid*. Change the Density [kg/m³] to 1074.9 and the Viscosity [kg/m s] to 0.001259. Erase the Chemical Formula. Click on the Change/Create button, answer *Yes* to the question that appears and Close the window.

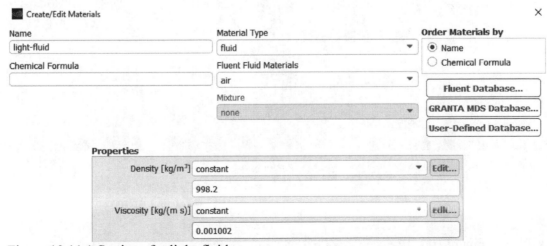

Figure 10.11a) Settings for light-fluid

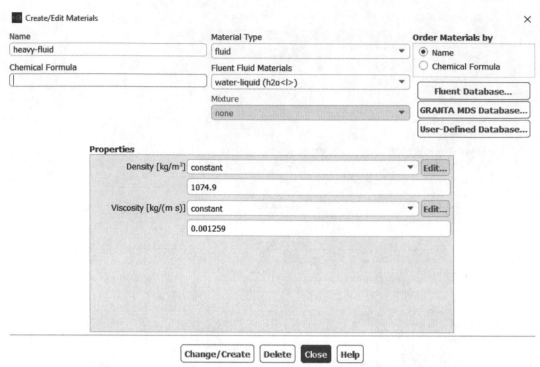

Figure 10.11b) Settings for heavy-fluid

12. Double click on Models and Multiphase (Volume of Fluid) under Setup in the Outline View. Select the Phases tab at the top of the Multiphase Model window. Select *light-fluid* as Phase Materials and enter *light-fluid* as the name of the Primary Phase. Click on Apply. Repeat this step to select and name *heavy-fluid* as the Secondary Phase. Click on Apply. Select the Phase Interaction tab at the top of the Multiphase Model window, check the Surface Tension Force Modeling box, select constant from the Surface Tension Coefficient [N/m] drop down menu and enter the value *0.00148*. Click on the Apply button and Close the Multiphase Model window.

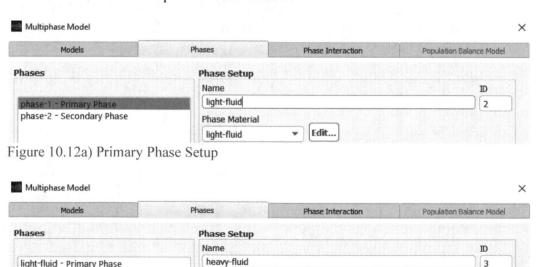

Figure 10.12a) Primary Phase Setup

Figure 10.12b) Secondary Phase Setup

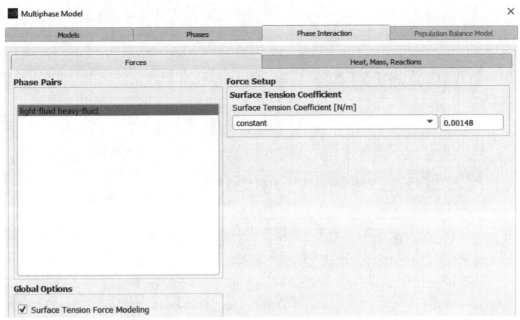

Figure 10.12c) Phase interaction toolbox

13. Copy the files *rayleigh-taylor.c* and *udfconfig.h* to your *Working Directory*. These files are available for download at *sdcpublications.com*. Select User-Defined>>Functions>>Interpreted… from the menu. Browse for the file *rayleigh-taylor.c,* select the file and click OK. Click on the Interpret button. Close the window.

```
#include "udf.h"
DEFINE_INIT(init_cos, domain)
{
cell_t c;
Thread*thread;
real xc[ND_ND];
thread_loop_c(thread,domain)
{
begin_c_loop_all(c,thread)
{
C_CENTROID(xc,c,thread);
if(xc[1] <= 0.5+0.05*cos(2*3.141592654/0.25*xc[0]))
{
C_V(c,thread)=0;
}
else
{
C_V(c,thread)=-0.1;
}
}
end_c_loop_all (c,thread)
}
}
```

Figure 10.13a) Code for the *rayleigh-taylor.c* file

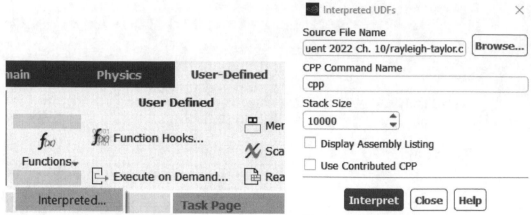

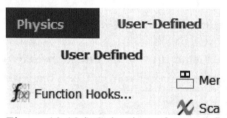

Figure 10.13b) User defined interpreted function   Figure 10.13c) Interpreted UDFs

Select User-Defined>>Functions Hook from the menu. Click on the Edit button next to Initialization. Select *init_cos* from Available Initialization Functions and click on the Add button to move it over to the Selected Initialization Functions. Click on OK buttons to close the windows.

Figure 10.13d) Selection of user defined function hooks

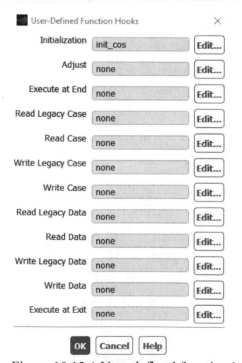

Figure 10.13e) User defined function hooks window

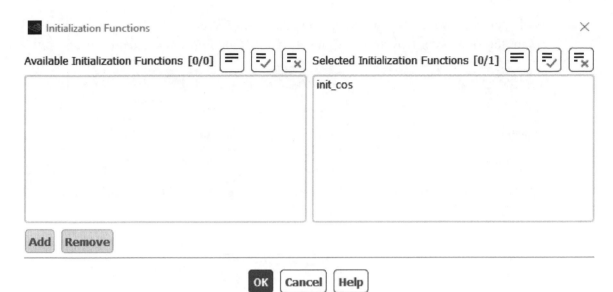

Figure 10.13f) Initializations functions window

14. Double click on Boundary Conditions under Setup in the Outline View. Click on the Operating Conditions on the Task Page in the Boundary Conditions section. Select user-input as Operating Density Method and enter the value 998.2 for Operating Density [kg/m³] (this is the value for the less dense fluid). Click OK to close the Operating Conditions window.

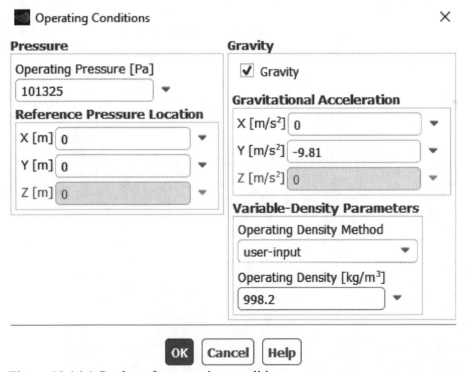

Figure 10.14a) Settings for operating conditions

Double click on Methods under Solution in the Outline View. Select Default settings except for First Order Upwind for Momentum and Body Force Weighted for Pressure.

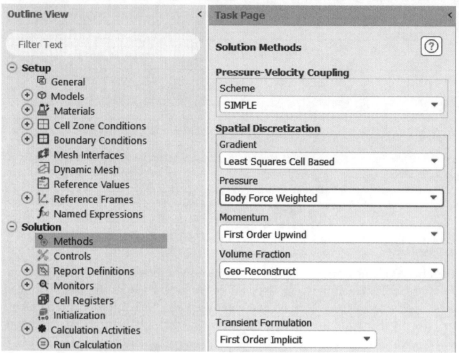

Figure 10.14b) Solution Methods Problem Setup

15. Double click on Initialization under Solution in the Outline View, select Standard Initialization under Solution Initialization on the Task Page, set the Y Velocity to -0.1 and select the Initialize button.

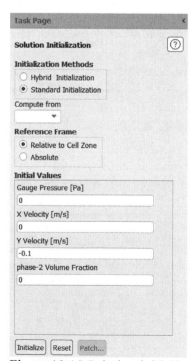

Figure 10.15 Solution initialization settings

16. Select the *Domain* tab in the menu and choose Adapt>>Manual….. Select Cell Registers>>New>>Region… in the Adaption Controls window. Enter 0.25 for X Max, 0.5 for Y Min, and 1 for Y Max. Click on the Save button and close the windows.

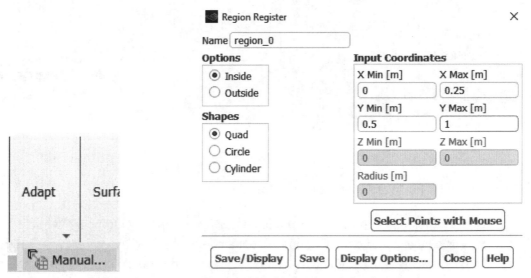

Figure 10.16a) Adaption   Figure 10.16b) Region Adaption Settings

17. Click on the Patch button on the Task Page and select *heavy-fluid* as the phase. Select Volume Fraction as the Variable and region_0 as Registers to Patch. Set the value to *1*, click on the Patch button and close the window.

Click on the Patch button on the Task Page and select *mixture* for the phase. Select Y Velocity as the Variable and region_0 as Registers to Patch. Set the value to *-0.1* and click on the Patch button and Close the window.

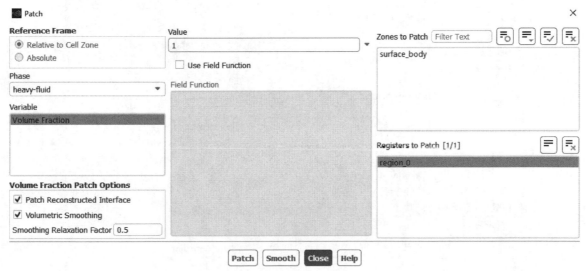

Figure 10.17a) Patch Settings for Volume Fraction

237

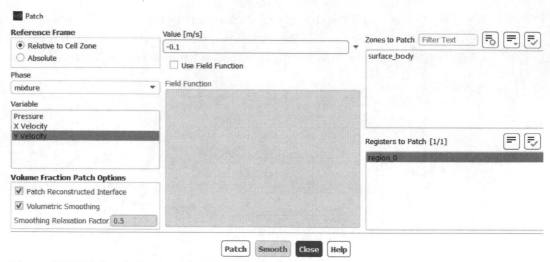

Figure 10.17b) Patch Settings for Y Velocity

18. Double click on Graphics and Contours under Results in the Outline View. Select Phases… and Volume fraction under Contours of and select *heavy-fluid* as the Phase. Uncheck the Boundary Values box under Options. Click on Colormap Options…, set Type to float and Precision to 1 under Number Format, set Font Behavior to Fixed and Font Size to 12, click on Apply and Close the Colormap window. Click on the Save/Display button and Close the Contours window. Click on Copy Screenshot of Active Window to Clipboard in the graphics window, see Figure 10.18d). The contour plot can be pasted into a Word document.

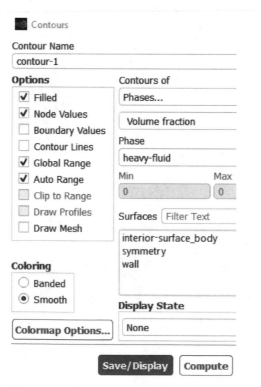

Figure 10.18a) Contours window

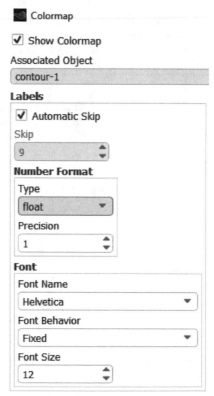

Figure 10.18b) Colormap window

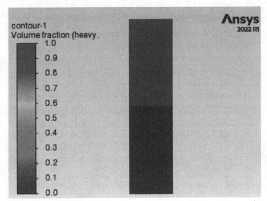

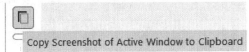

Figure 10.18c) Contours of Volume Fraction   Figure 10.18d) Copying screenshot

G. Calculations and Post-Processing

19. Double click on Calculation Activities and Solution Animations under Solution in the
Outline View. Enter *rayleigh-taylor-instability-animation* as the Name, enter Record
after every 10 and select time-step. Select contour-1 under Animation Object. Select OK
to close the Animation Definition window. Double click on Run Calculation under
Solution, enter 0.01 as Time Step Size, and 100 for Number of Time Steps. Click on the
Calculate button. Click on Copy Screenshot of Active Window to Clipboard in the
graphics window, see Figure 10.18d). The contour plot can be pasted into a Word
document. Continue running the calculations with 20 as Number of Time Steps, 14 more
times for a total of 3.8 seconds. Select File>>Export>Case & Data… from the menu.
Save the file with the name *rt-computation.cas.h5*.

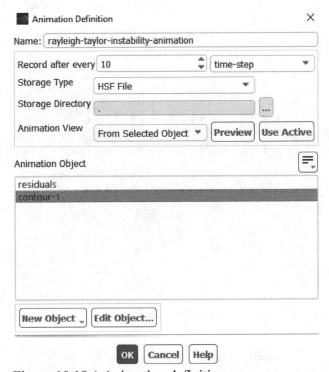

Figure 10.19a) Animation definition

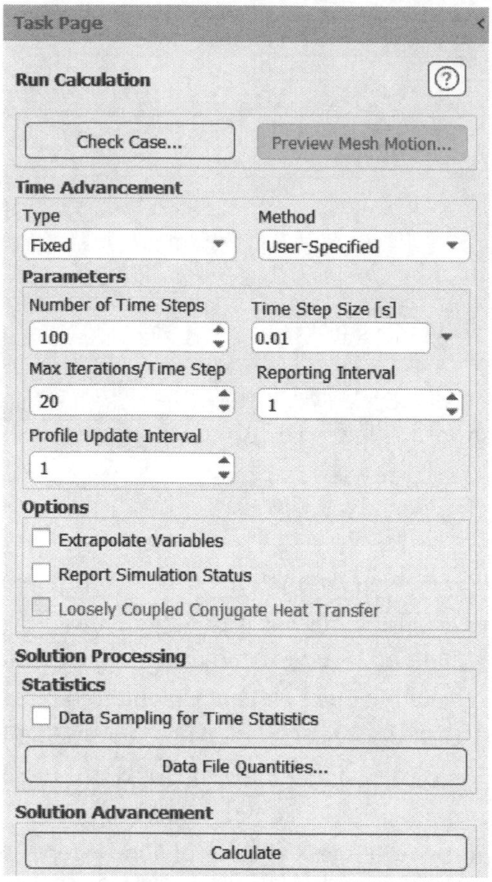

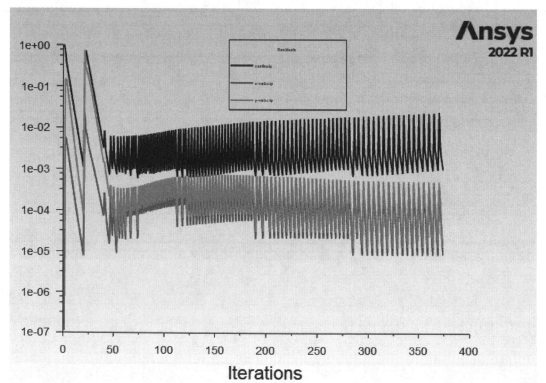

Figure 10.19b) Calculations

Figure 10.19c) Scaled residuals for the first second

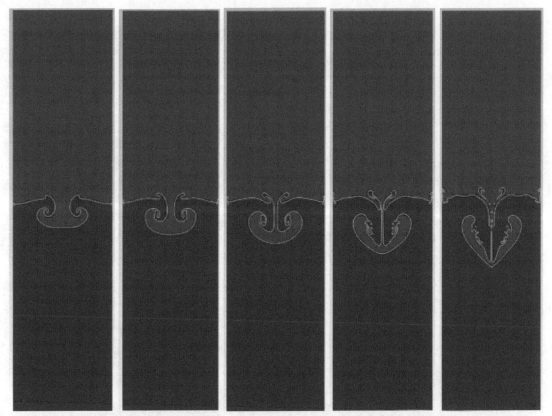

Figure 10.19d) Volume Fraction Contours after 1, 1.2, 1.4, 1.6 and 1.8 s

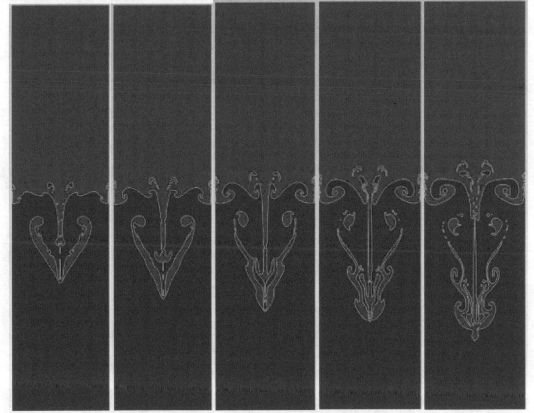

Figure 10.19e) Volume Fraction Contours at 2, 2.2, 2.4, 2.6, 2.8 s

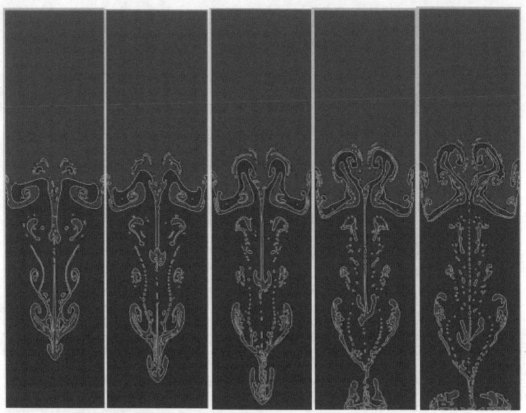

Figure 10.19f) Volume Fraction Contours at 3, 3.2, 3.4, 3.6 and 3.8 s

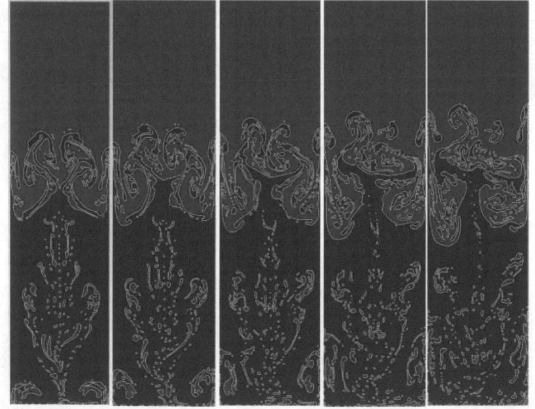

Figure 10.19g) Volume Fraction Contours at 4, 4.2, 4.4, 4.6 and 4.8 s

20. Double click on Animations and Playback under Results in the Outline View. Set the Replay Speed to low and select Play Once as Playback Mode. Select Video File as Write/Record Format. Click on the Write button. This will create the movie. Close the Playback window. The movie can be viewed using for example Windows Media Player.

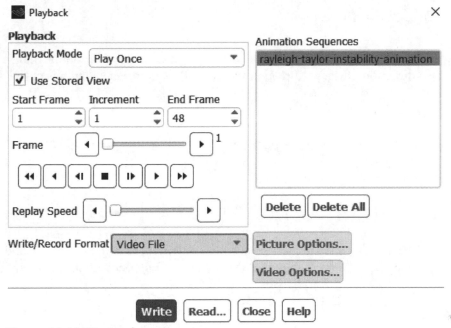

Figure 10.20 Playback settings

H. Theory

21. We will look at the inviscid theory for the Rayleigh-Taylor instability in the same way as we did for the Kelvin-Helmholz instability in terms of a sinusoidal disturbance between two fluids with densities ρ_1, ρ_2.

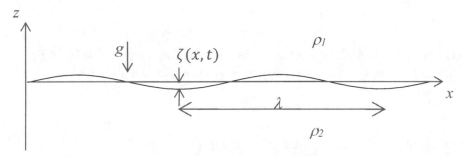

Figure 10.21 Development of Rayleigh-Taylor instability from a perturbation

The velocity vector in a Cartesian coordinate system can be described as

$$\bar{U} = u_x \bar{e}_x + u_y \bar{e}_y + u_z \bar{e}_z \qquad (10.1)$$

where $u_x, u_y = 0, u_z$ are the velocity components in the streamwise, spanwise and normal directions, respectively. Euler equations of motion can be written as

$$\frac{\partial u_x}{\partial t} + u_x \frac{\partial u_x}{\partial x} + u_z \frac{\partial u_x}{\partial z} = -\frac{1}{\rho} \frac{\partial P}{\partial x} \tag{10.2}$$

$$\frac{\partial u_z}{\partial t} + u_x \frac{\partial u_z}{\partial x} + u_z \frac{\partial u_z}{\partial z} = -\frac{1}{\rho} \frac{\partial P}{\partial z} - g \tag{10.3}$$

$$\frac{\partial u_x}{\partial x} + \frac{\partial u_z}{\partial z} = 0 \tag{10.4}$$

The velocity components can be divided into mean flow components in the main streamwise direction plus a perturbation (no mean flow components in this case)

$$\bar{U} = u\bar{e}_x + w\bar{e}_z \tag{10.5}$$

and the pressure is $\quad P = P_0 - \rho g z + p \tag{10.6}$

A perturbation is introduced in the form

$$u, w, p = [u_n(z), w_n(z), p_n(z)]e^{i(kx-\omega t)} \tag{10.7}$$

where $k = 2\pi/\lambda$ is the wave number, λ is the wave length, ω is the frequency and $u_n(z), w_n(z), p_n(z)$ are eigenfunctions. The linearized Euler equations can be written as

$$-i\omega u_n + \frac{ik}{\rho} p_n = 0 \tag{10.8}$$

$$-i\omega w_n + \frac{1}{\rho} p_n' = 0 \tag{10.9}$$

$$iku_n + w_n' = 0 \tag{10.10}$$

From equations (10.8) – (10.10) we can find the equation for w_n

$$w_n'' - k^2 w_n = 0 \tag{10.11}$$

with the general solution

$$w_n = c_1 e^{-kz} + c_2 e^{kz} \tag{10.12}$$

To get exponentially decaying solutions far from the interface, we set c_1 and c_2 equal to zero in the lower and upper layers, respectively. The interface location is defined by

$$z = \zeta(x,t) = \mathcal{R}\{ae^{i(kx-\omega t)}\} \tag{10.13}$$

where $\mathcal{R}$ stands for the real part and a is the amplitude of the initial disturbance. The normal velocity w at the interface is given by the material derivative of $\zeta(x,t)$

$$w = w_n(z)e^{i(kx-\omega t)} = \frac{D\zeta}{Dt} = \frac{\partial \zeta}{\partial t} = -i\omega\,\zeta = -i\omega a e^{i(kx-\omega t)} \tag{10.14}$$

We find the following results: $c_1 = -i\omega a$, $c_2 = -i\omega a$

$$w_{n,1} = -i\omega a e^{-kz}, \quad w_{n,2} = -i\omega a e^{kz} \tag{10.15}$$

We can now find the pressure eigenfunctions in the upper and lower layers from equation (10.9)

$$p_{n,1} = -\rho_1 \omega^2 a e^{-kz}/k \tag{10.16}$$
$$p_{n,2} = \rho_2 \omega^2 a e^{kz}/k \tag{10.17}$$

Next, we use equation (10.6) and set the pressures identical at the interface to get the dispersion relation

$$P_0 - \rho_1 g \zeta + p_{n,1} e^{i(kx-\omega t)} = P_0 - \rho_2 g \zeta + p_{n,2} e^{i(kx-\omega t)} - \sigma k^2 \zeta \tag{10.18}$$

$$(\rho_2 - \rho_1)gk = (\rho_2 + \rho_1)\omega^2 - \sigma k^3 \tag{10.19}$$

where σ is surface tension. We find the following equation for the frequency without surface tension

$$\omega = \sqrt{\frac{(\rho_2-\rho_1)gk}{\rho_1+\rho_2}} = i\sqrt{\frac{(\rho_1-\rho_2)gk}{\rho_1+\rho_2}} = i\sqrt{gkAt} \tag{10.20}$$

and including surface tension

$$\omega = \sqrt{\frac{(\rho_2-\rho_1)gk+\sigma k^3}{\rho_1+\rho_2}} = i\sqrt{\frac{(\rho_1-\rho_2)gk-\sigma k^3}{\rho_1+\rho_2}} = i\sqrt{gkAt(1-\frac{4\pi^2}{Bo})} \tag{10.21}$$

where $At = (\rho_1 - \rho_2)/\rho_1 + \rho_2$ is non-dimensional Atwood number and $Bo = (\rho_1 - \rho_2)gD^2/\sigma$ is non-dimensional Bond number. The frequency is complex with the associated growth of waves when the expression under the square root without surface tension is negative for

$$\rho_2 - \rho_1 < 0 \qquad \text{or} \qquad \rho_2 < \rho_1 \tag{10.22}$$

and including surface tension we get

$$(\rho_2 - \rho_1)gk + \sigma k^3 < 0 \qquad \text{or} \qquad \lambda > 2\pi \sqrt{\frac{\sigma}{(\rho_1-\rho_2)g}} \tag{10.23}$$

The Bond number $Bo > 4\pi^2$ will give growth of waves when $\lambda = D$. The interface location can now be written as

$$z = \zeta(x,t) = \mathcal{R}\left\{ a e^{\sqrt{gkAt(1-\frac{4\pi^2}{Bo})}\,t} e^{ikx} \right\} = a e^{\sqrt{gkAt(1-\frac{4\pi^2}{Bo})}\,t} \cos(kx) \tag{10.24}$$

and we see that surface tension has a very small but stabilizing effect. In ANSYS Fluent simulations we used $a = 0.05$ m, $k = 8\pi$ m^{-1}, $g = 9.81$ m/s^2, $\lambda = D = 0.25$ m, $\sigma = 0.00148$ N/m, $\rho_1 = 1074.9 \frac{kg}{m^3}$, $\rho_2 = 998.2 \frac{kg}{m^3}$. The Atwood and Bond numbers used in ANSYS Fluent are

$$At = \frac{1074.9-998.2}{1074.9+998.2} = 0.037 \tag{10.25}$$

$$Bo = \frac{(1074.9 - 998.2) * 9.81 * 0.25^2}{0.00148} = 31,775 \tag{10.26}$$

There are a few more non-dimensional parameters in this problem that we can define such as the aspect ratio for the rectangular computational domain $/D = 4$, the density ratio $\rho_1/\rho_2 = 1.077$, the dynamic viscosity ratio $\mu_1/\mu_2 = 1.256$ and non-dimensional time $\tau = t\sqrt{gkAt(1 - \frac{4\pi^2}{Bo})}$. The interface location is

$$z = \zeta(x,t) \approx \frac{e^{3t}}{20} \cos(8\pi x) \tag{10.27}$$

Using the inviscid theory, we find that the wave-length with surface tension included according to equation (10.23) must be $\lambda > 0.0088\, m$ to get growth of waves.

I. References

1. Andrews, M.J. and Dalziel, S.B. "Small Atwood number Rayleigh-Taylor Experiments", *Phil. Trans. R. Soc. A*, **368**, 1663-1679, (2010).
2. Chandrasekhar S. "Hydrodynamic and Hydromagnetic Stability", *Dover Publications, Inc.*, New York (1961).
3. Drazin P.G. and Reid W.H. "Introduction to hydrodynamic instability", *Cambridge University Press*, (1981).
4. Sharqawy M.H, Lienhard J.H., and Zubair S.M. "Thermophysical Properties of Sea Water: A Review of existing Correlations and Data, Desalination, and Water Treatment", *Desalination and Water Treatment*, **16**, 354 (2010).
5. Strubelj, L., and Tiselj, I., "Numerical Simulations of Basic Interfacial Instabilities with Improved Two-Fluid Model", *International Conference Nuclear Energy for New Europe*, September 14–17, Bled, Slovenia, (2009).
6. Strubelj, L., and Tiselj, I., "Numerical Simulations of Rayleigh-Taylor Instability with Two-Fluid Model and Interface Sharpening", *Proceedings of ASME Fluids Engineering Division Summer Conference*, August 10-14, Jacksonville, FL, (2008).
7. Tryggvason, G., "Numerical Simulations of the Rayleigh-Taylor Instability", *Journal of Computational Physics*, **75**, 253-282, (1988).

J. Exercises

10.1 Use ANSYS Fluent to study the same problem as we completed in this chapter but instead complete transient 2D axisymmetric simulations. Visualize the instability using contours of volume fraction for the heavy fluid.

10.2 Use ANSYS Fluent to study the same problem as we completed in this chapter but instead use the following dimensions $H = 2$ m, $D = 0.25$ m. Visualize the instability using contours of volume fraction for the heavy fluid.

10.3 Use ANSYS Fluent to study the same problem as we completed in this chapter but instead use a frustum shaped geometry with the following dimensions $H = 1$ m, $D_{bottom} = 0.25$ m and $D_{top} = 0.5$ m. Visualize the instability using contours of volume fraction for the heavy fluid.

CHAPTER 11. FLOW UNDER A DAM

A. Objectives

- Using ANSYS Workbench to Model Seepage Flow of Water under a Concrete Dam
- Inserting Boundary Conditions and a Porous Medium Zone with Viscous and Inertial Resistance
- Running Laminar Steady 2D Planar ANSYS Fluent Simulations
- Using Contour Plots for Static Pressure and Velocity Magnitude
- Using XY Plots for Visualizations of Velocity and Pressure
- Comparing ANSYS Fluent Results with Results Using Mathematica

B. Problem Description

We will study the seepage flow of water under a concrete dam using ANSYS Fluent. The dimensions and water levels are shown below.

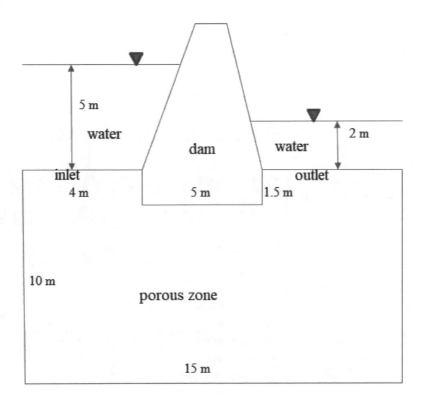

C. Launching ANSYS Workbench and Selecting Fluent

1. Start by launching ANSYS Workbench. Double click on Fluid Flow (Fluent) under Analysis Systems in the Toolbox.

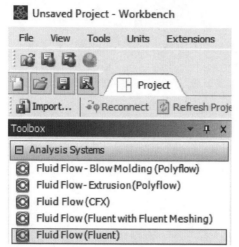

Figure 11.1 Selecting Fluid Flow (Fluent)

D. Launching ANSYS DesignModeler

2. Right click Geometry under Project Schematic in ANSYS Workbench and select Properties. In Properties of Schematic A2: Geometry, select Analysis Type 2D under Advanced Geometry Options. Right click on Geometry in the Project Schematic window and select New DesignModeler Geometry.

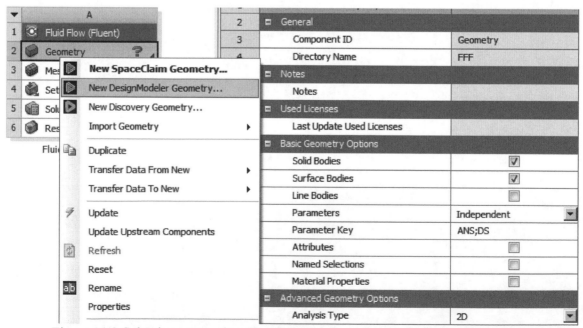

Figure 11.2 Selecting properties, 2D analysis type and new DesignModeler geometry

3. Select the Sketching tab in the Tree Outline and select Line under Draw. Click on Look at Face/Plane/Sketch . Draw a horizontal line to the right from the origin of the coordinate system in the graphics window. Make sure you have the letters *P* at the origin, *H* along the line and *C* when you end the line.

   Click on the Dimensions tab under Sketching Toolboxes and click on the new line. Enter 15 m for the length of the line. Continue adding lines with lengths as shown in Fig. 11.3a).

   Select Concepts>>Surfaces from Sketches from the menu. Click on Sketch 1 under XYPlane in the Tree Outline and select Apply for Base Objects in Details View. Click on Generate. Close DesignModeler.

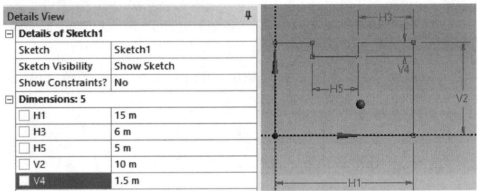

Figure 11.3a) Geometry and dimensions for flow through a porous medium

Figure 11.3b) Creating surfaces from Sketches

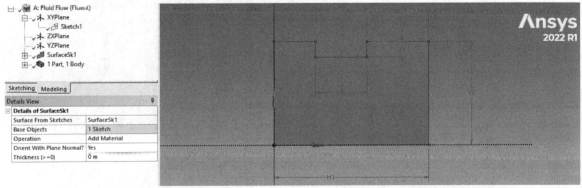

Figure 11.3c) Finished model for seepage flow under a dam

249

E. Launching ANSYS Meshing

4. Double click on Mesh under the Project Schematic in ANSYS Workbench. Select Mesh under Model (A3) and Project in the Outline. Right click on Mesh and select Update.

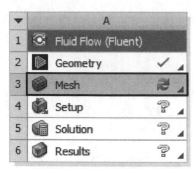

Fluid Flow (Fluent)

Figure 11.4a) Mesh in ANSYS

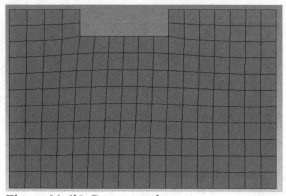

Figure 11.4b) Coarse mesh

Select Mesh>>Controls>>Face Meshing from the menu. Select the face of the mesh region. Apply the Geometry under Details of "Face Meshing".

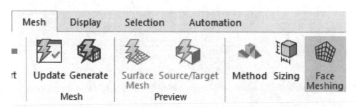

Figure 11.4c) Face meshing

Select Mesh>>Controls>>Sizing from the menu and select the face of the mesh region. Apply the Geometry under Details of "Face Sizing". Set the Element Size to 50.0 mm and set the Behavior to Hard. Right click on Mesh in the tree outline and select Generate Mesh.

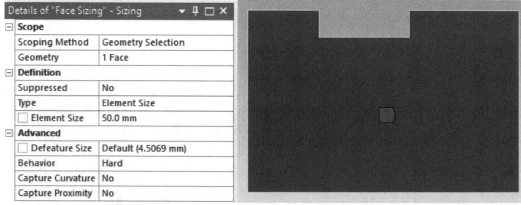

Figure 11.4d) Details of face sizing for the mesh

Figure 11.4e) A portion of the finished mesh

Select Geometry under Project and Model (A3) in the Outline. Select the Edge Selection

Filter 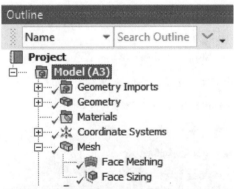. Select the upper left horizontal edge, right click and select Create Named Selection. Name the edge *pressure-inlet* and click OK. Name the upper horizontal edge to the right *pressure-outlet* and control-select all the remaining six edges and enter the name *wall*.

Select File>>Export...>>Mesh>>FLUENT Input File>>Export from the menu and save the mesh with the name *seepage-flow.msh*. Select File>>Save Project from the menu and save the project with the name *Dam-Seepage-Flow.wbpj*. Close the meshing window. Right click on Mesh and select Update under Project Schematic in ANSYS Workbench.

Figure 11.4f) Selection of Geometry

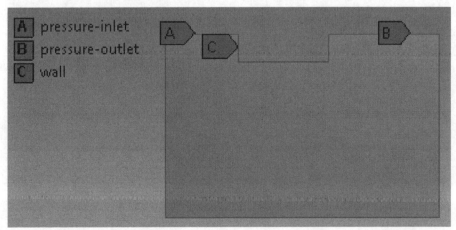

Figure 11.4g) Named selections

F. Launching ANSYS Fluent

5. Double-click on Setup under Project Schematic in ANSYS Workbench. Check the box for Double Precision. Click on Show More Options and take a note of the location of your *Working Directory*. Set the number of processes equal to the number of processor cores for your computer. To check the number of physical cores, press the Ctrl + Shift + Esc keys simultaneously to open the Task Manager. Go to the Performance tab and select CPU from the left column. You will see the number of physical cores on the bottom-right side. Close the Task Manager window. Click on the Start button in the Fluent Launcher window.

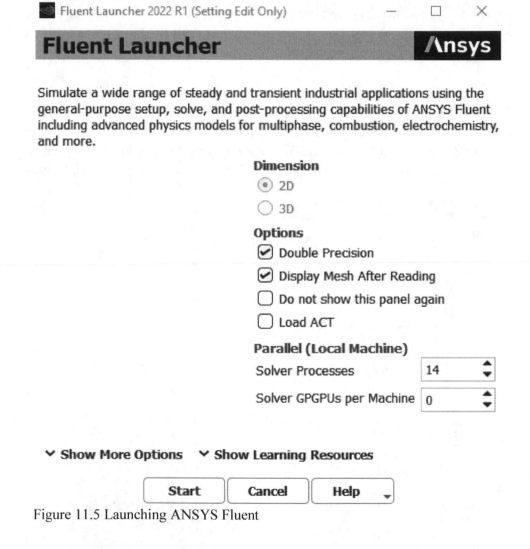

Figure 11.5 Launching ANSYS Fluent

6. Double click on Models and Viscous (SST k-omega) under Setup in the Outline View. Select Laminar Model and select OK to close the Viscous Model window. Double click on Materials under Setup in the Outline View. Click on Create/Edit for Fluid under Materials on the Task Page. Click on Fluent Database in the Create/Edit Materials window. Select *water-liquid (h2o<l>)* as the Fluent Fluid Material. Click on Copy and Close the window. Set the Density to 992.2 [kg/m^3] and the Viscosity to 0.00065 [kg/m-s]. Click Change/Create and Close the window.

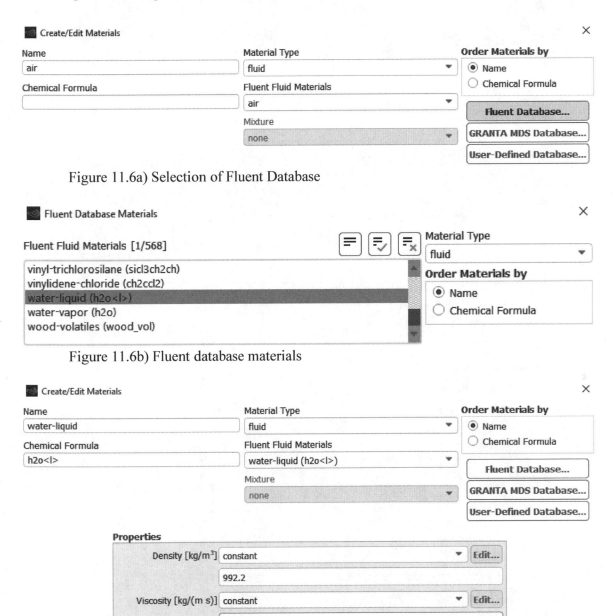

Figure 11.6a) Selection of Fluent Database

Figure 11.6b) Fluent database materials

Figure 11.6c) Properties for water-liquid

7. Double click on Cell Zone Conditions, Fluid and *surface_body* under Setup in the Outline View. Select *water-liquid* from the Material Name drop-down menu. Check the box for Porous Zone. Click on the Porous Zone tab and enter 1.98807e+06 [m^{-2}] for Viscous Resistance in Direction-1 and Direction-2. Enter 1719.4 [m^{-1}] for Inertial Resistance in Direction-1 and Direction-2. Set the Fluid Porosity to 0.5. Click on Apply button to Close the Fluid window.

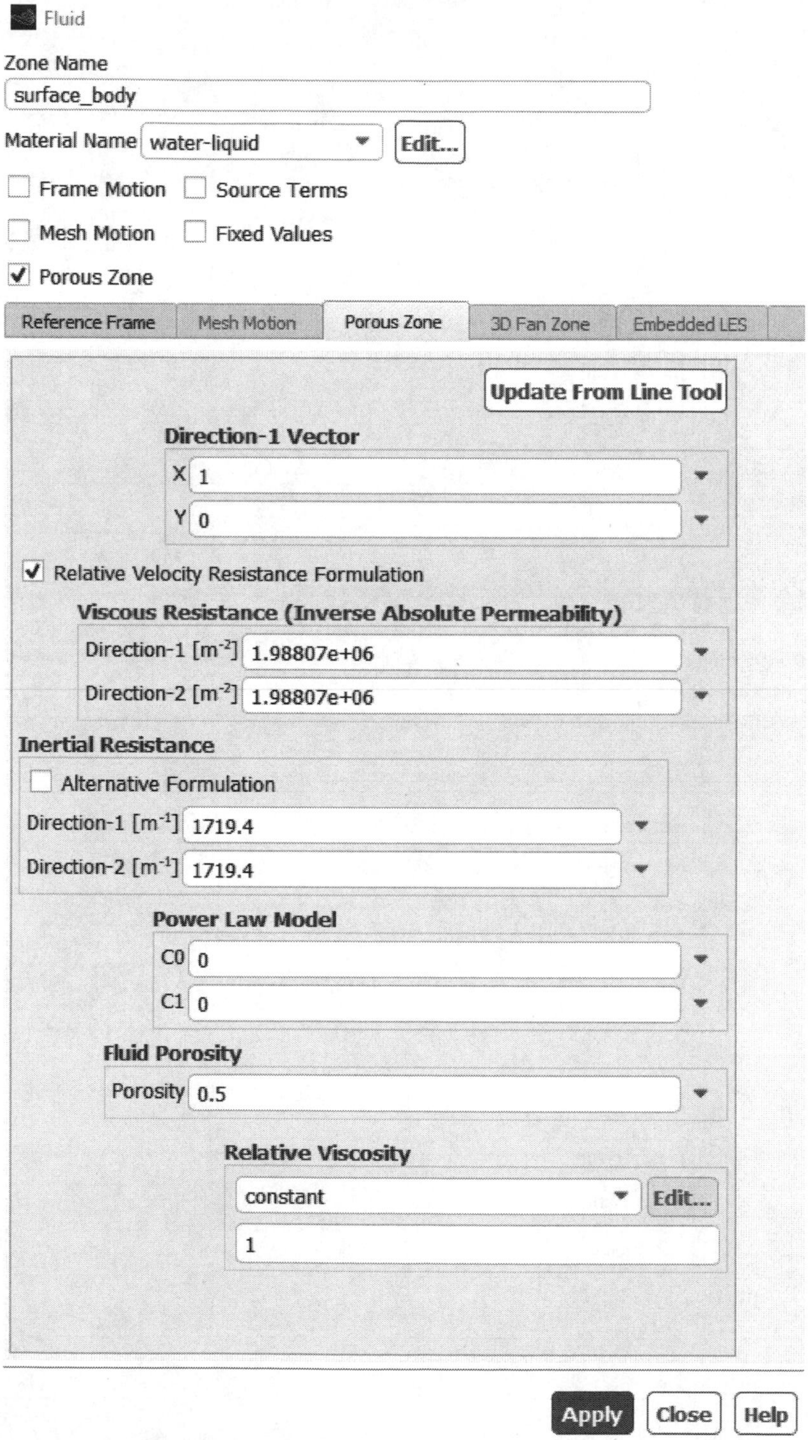

Figure 11.7 Fluid window

8. Double click on Boundary Conditions under Setup in the Outline View. Select the *pressure-inlet* Zone on the Task Page. Click on Edit.... Enter 49050 as the Gauge Total Pressure [Pa]. Click on Apply and Close the window. Select the *pressure-outlet* Zone on the Task Page. Click on Edit.... Enter 19620 as the Gauge Pressure [Pa]. Select Apply and Close the window.

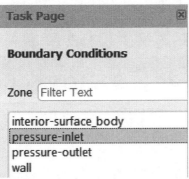

Figure 11.8a) Selection of pressure-inlet boundary condition

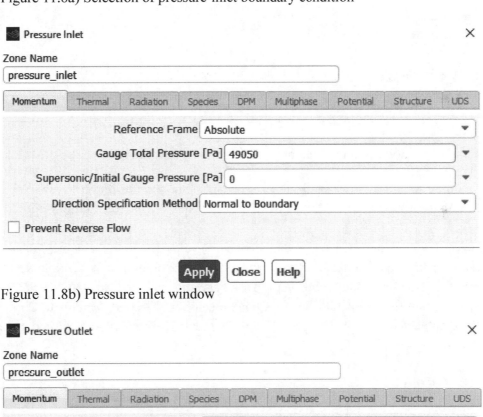

Figure 11.8b) Pressure inlet window

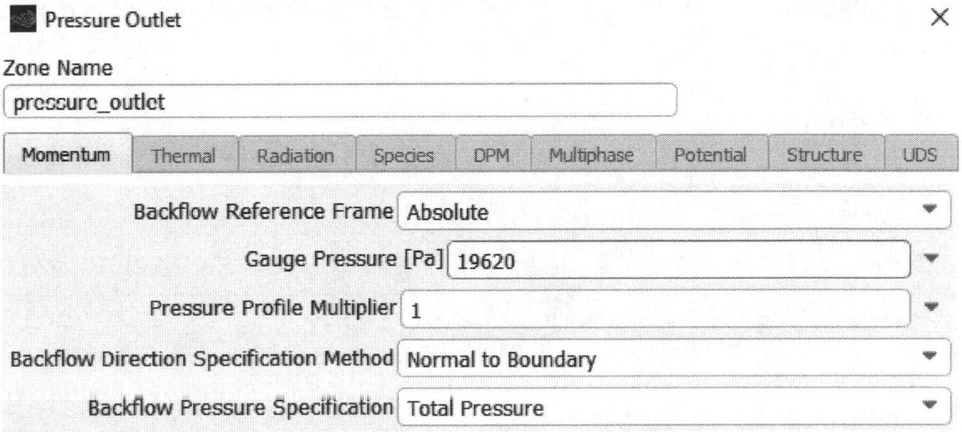

Figure 11.8c) Pressure outlet window

9. Double click on Initialization under Solution in the Outlet View. Check Standard Initialization as Initialization Method on the Task Page. Select Compute from *pressure-inlet* from the drop-down menu and click on Initialize.

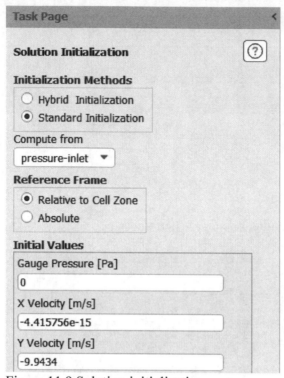

Figure 11.9 Solution initialization

10. Double click on Monitors and Residual under Solution in the Outline View. Set the Absolute Criteria to 1e-12 for all Residuals and click on OK to exit the window.

Select File>>Export>Case... from the menu. Save the file with the name *flow-under-dam.cas.h5*. Double click on Run Calculation under Solution in the Outline View, enter 1000 for Number of Iterations on the Task Page and click on Calculate. Click OK when calculations are complete. Click on Copy Screenshot of Active Window to Clipboard in the graphics window, see Figure 11.10d). The scaled residuals can be pasted into a Word document.

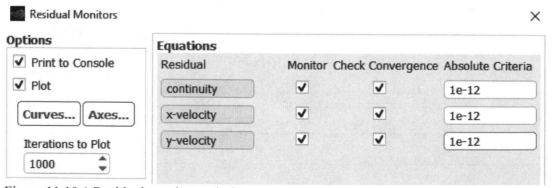

Figure 11.10a) Residual monitors window

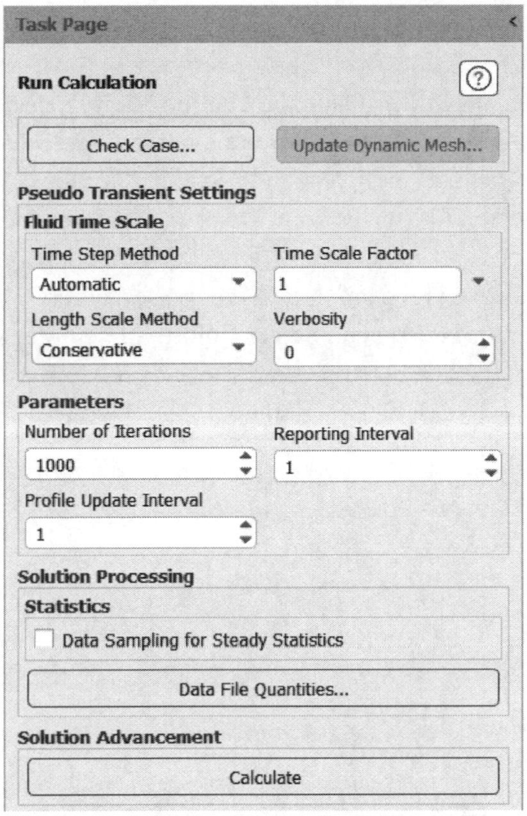

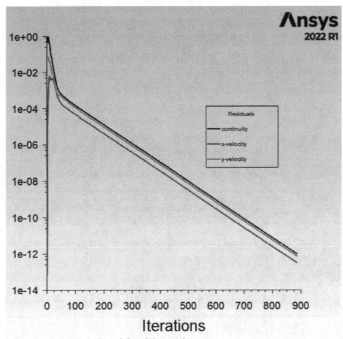

Figure 11.10b) Running calculations

Figure 11.10c) Residual iterations

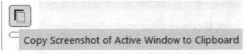

Figure 11.10d) Copying screenshot

G. Post-Processing

11. Double click on Graphics and Contours under Results in the Outline View. Select Contours of Velocity and Velocity Magnitude from the drop-down menu. Select Colormap Options…. Select Type as float and set Precision to 4 under Number Format. Set Font Behavior to Fixed and Font Size to 12 under Font. Click on Apply and Close the Colormap window. Deselect all Surfaces and click on Save/Display. Click on Copy Screenshot of Active Window to Clipboard in the graphics window, see Figure 11.10d). The contour plot can be pasted into a Word document. Select Colormap Options… and set Precision to 0 under Number Format. Click on Apply and Close the Colormap window. Select Contours of Pressure and Static Pressure and click on Save/Display. Close the window.

Figure 11.11a) Contours window

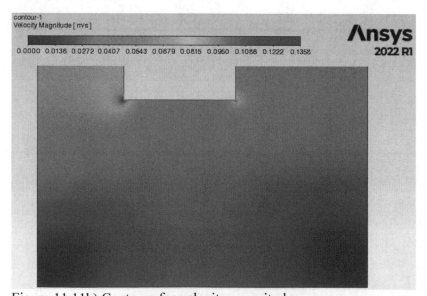

Figure 11.11b) Contours for velocity magnitude

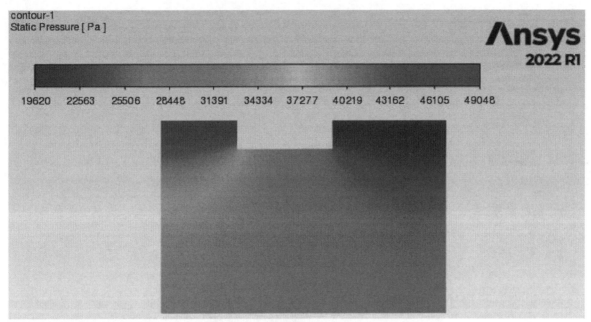

Figure 11.11c) Contours for static pressure

12. Double click on Plots and XY Plot under Results in the Outline View. Select Velocity and Velocity Magnitude for Y Axis Function and Direction Vector for X Axis Function. Select New Surface >> Line/Rake from the drop–down menu. Enter the following end points: x0 [m] = 9, y0 [m] = 10 and x1 [m] = 15, y1 [m] = 10. Click on Create and Close the window. Click on Load File… and load the file with the name *theory-flow-under-dam.dat*. This and other files are available for download at *sdcpublications.com*. Select *line-5* in the Surfaces section, select *Theory* under File Data. Click on Curves…, select Curve #0 and select the first available pattern and no symbol. Click on Apply. Select Curve # 1, select the second available pattern and no symbol as shown in Figure 11.12c). Click on Apply and Close the Curves-Solution XY Plot window. Click on Axes… in the Solution XY Plot window. Select X Axis, set Type to float, set Precision under Number Format to 0 and click on Apply. Select Y Axis, set Type to General, set Precision under Number Format to 2 and click on Apply. Close the Axes – Solution XY Plot window. Click on Save/Plot in the Solution XY Plot window. Click on Copy Screenshot of Active Window to Clipboard in the graphics window, see Figure 11.10d). XY plot can be pasted into a Word document.

Select New Surface >> Line/Rake from the drop–down menu. Enter the following end points: x0 [m] = 9, y0 [m)] = 0 and x1 [m] = 9, y1 [m] = 8.5. Click on Create and Close the window. Select *line-6* in the Surfaces section, deselect *line-5,* uncheck Position on X Axis under Options and select Mesh and Y-Coordinate as Y Axis Function. Select Pressure and Static Pressure as X Axis Function. Click on Axes… and uncheck Auto Range under Options for the X Axis. Set Minimum under Range to 25000 and Maximum to 30000. Set Precision under Number Format to 0. Select Y Axis and set the Precision to 0. Click on Apply and Close the Axes window. Click on Save/Plot. Click on Copy Screenshot of Active Window to Clipboard in the graphics window, see Figure 11.10d). XY plot can be pasted into a Word document.

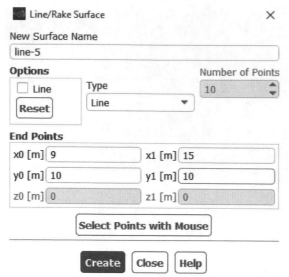

Figure 11.12a) Line/Rake Surface window

Figure 11.12b) Solution XY Plot

Figure 11.12c) Curves window   Figure 11.12d) Axes window

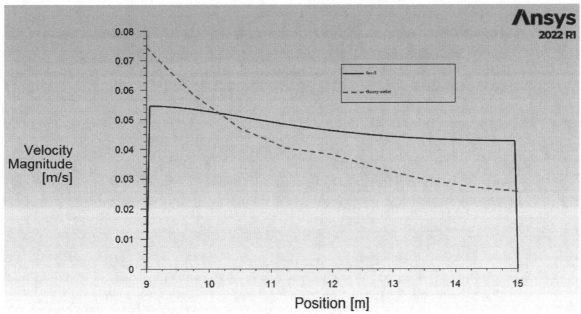

Figure 11.12e) Comparison of the variation for seepage velocity at the outlet from ANSYS Fluent simulations (full line) and finite element calculations (dashed line).

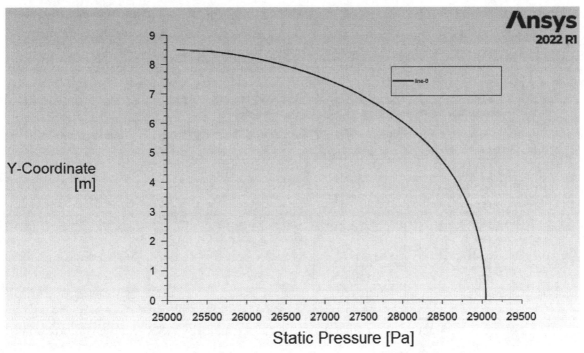

Figure 11.12f) Static pressure distribution along a vertical line under the dam

H. Theory

13. Groundwater seepage can frequently occur in soils and has many engineering applications. We can relate the mean fluid velocity v (m/s) through the porous medium to the pressure gradient dP/dx (Pa) using Darcy's law

$$v = -\frac{\kappa}{\mu}\frac{dP}{dx} \tag{11.1}$$

where κ (m^2) is permeability of the porous medium and μ (kg/m-s) is the dynamic viscosity of the fluid. Furthermore, we define the hydraulic conductivity or coefficient of permeability k (m/s) as

$$k = \kappa\frac{\rho g}{\mu} \tag{11.2}$$

Where g (m/s^2) is acceleration due to gravity and ρ (kg/m^3) is the density of the fluid. We will also use the following equation from the ANSYS Fluent user manual

$$\frac{dP}{dx} = -\left(\frac{\mu v}{\alpha} + \frac{C_2\rho v^2}{2}\right) \tag{11.3}$$

where $1/\alpha$ (1/m^2) is viscous resistance or inverse absolute permeability for the porous medium and C_2 (1/m) is inertial resistance of the porous medium. When we combine the equations we get the following expression for the permeability coefficient

$$k = \frac{g}{\frac{v C_2}{2} + \frac{\mu}{\alpha\rho}} \tag{11.4}$$

The following values were used in or determined from the ANSYS Fluent simulations.

C_2 (1/m)	1719.4
$1/\alpha$ (1/m^2)	$1.98807 \cdot 10^6$
ρ (kg/m^3)	992.2
μ (kg/m-s)	0.00065
v (m/s)	0.05
g (m/s^2)	9.81
k (m/s)	0.221508
p_1 (Pa)	29046.9
p_2 (Pa)	25124.4

Table 11.1 Parameters & coefficients used or determined by ANSYS Fluent simulations

The 2D flow through a porous medium is governed by

$$k_x\frac{\partial^2\phi}{\partial x^2} + k_y\frac{\partial^2\phi}{\partial y^2} = 0 \tag{11.5}$$

where ϕ is the hydraulic head and $k_x = k_y = k$ are coefficients of permeability in the x and y-directions. The seepage velocity components are related to the hydraulic head through

$$v_x = -k_x\frac{\partial\phi}{\partial x} \quad \text{and} \quad v_y = -k_y\frac{\partial\phi}{\partial y} \tag{11.6}$$

We follow the approach by Bhatti[3] using second order ($n = 2$) p-formulation. The equations for element **1**:

$$\begin{pmatrix}
+0.2424 & -0.2021 & -0.1212 & +0.0809 & -0.0164 & +0.1320 & +0.0164 & -0.1320 \\
-0.2021 & +0.2424 & +0.0809 & -0.1212 & -0.0164 & -0.1320 & +0.0164 & +0.1320 \\
-0.1212 & +0.0809 & +0.2424 & -0.2021 & +0.0164 & -0.1320 & -0.0164 & +0.1320 \\
+0.0809 & -0.1212 & -0.2021 & +0.2424 & +0.0164 & +0.1320 & -0.0164 & -0.1320 \\
-0.0164 & -0.0164 & +0.0164 & +0.0164 & +0.4445 & 0 & +0.2021 & 0 \\
+0.1320 & -0.1320 & -0.1320 & +0.1320 & 0 & +0.1615 & 0 & -0.0809 \\
+0.0164 & +0.0164 & -0.0164 & -0.0164 & +0.2021 & 0 & +0.4445 & 0 \\
-0.1320 & +0.1320 & +0.1320 & -0.1320 & 0 & -0.0809 & 0 & +0.1615
\end{pmatrix}
\begin{pmatrix}
\phi_1 \\ \phi_4 \\ \phi_5 \\ \phi_2 \\ \delta_1^{(1,4)} \\ \delta_1^{(4,5)} \\ \delta_1^{(2,5)} \\ \delta_1^{(1,2)}
\end{pmatrix}
=
\begin{pmatrix}
0 \\ 0 \\ 0 \\ 0 \\ 0 \\ 0 \\ 0 \\ 0
\end{pmatrix}$$

Equations for element **2**:

$$\begin{pmatrix}
+0.1902 & +0.0380 & -0.0951 & -0.1331 & -0.0932 & +0.0233 & +0.0932 & -0.0233 \\
+0.0380 & +0.1902 & -0.1331 & -0.0951 & -0.0932 & -0.0233 & +0.0932 & +0.0233 \\
-0.0951 & -0.1331 & +0.1902 & +0.0380 & +0.0932 & -0.0233 & -0.0932 & +0.0233 \\
-0.1331 & -0.0951 & +0.0380 & +0.1902 & +0.0932 & +0.0233 & -0.0932 & -0.0233 \\
-0.0932 & -0.0932 & +0.0932 & +0.0932 & +0.1521 & 0 & -0.0380 & 0 \\
+0.0233 & -0.0233 & -0.0233 & +0.0233 & 0 & +0.3233 & 0 & +0.1331 \\
+0.0932 & +0.0932 & -0.0932 & -0.0932 & -0.0380 & 0 & +0.1521 & 0 \\
-0.0233 & +0.0233 & +0.0233 & -0.0233 & 0 & +0.1331 & 0 & +0.3233
\end{pmatrix}
\begin{pmatrix}
\phi_2 \\ \phi_5 \\ \phi_6 \\ \phi_3 \\ \delta_1^{(2,5)} \\ \delta_1^{(5,6)} \\ \delta_1^{(3,6)} \\ \delta_1^{(2,3)}
\end{pmatrix}
=
\begin{pmatrix}
0 \\ 0 \\ 0 \\ 0 \\ 0 \\ 0 \\ 0 \\ 0
\end{pmatrix}$$

Equations for element **3**:

$$\begin{pmatrix}
+0.2424 & -0.2021 & -0.1212 & +0.0809 & -0.0164 & +0.1320 & +0.0164 & -0.1320 \\
-0.2021 & +0.2424 & +0.0809 & -0.1212 & -0.0164 & -0.1320 & +0.0164 & +0.1320 \\
-0.1212 & +0.0809 & +0.2424 & -0.2021 & +0.0164 & -0.1320 & -0.0164 & +0.1320 \\
+0.0809 & -0.1212 & -0.2021 & +0.2424 & +0.0164 & +0.1320 & -0.0164 & -0.1320 \\
-0.0164 & -0.0164 & +0.0164 & +0.0164 & +0.4445 & 0 & +0.2021 & 0 \\
+0.1320 & -0.1320 & -0.1320 & +0.1320 & 0 & +0.1615 & 0 & -0.0809 \\
+0.0164 & +0.0164 & -0.0164 & -0.0164 & +0.2021 & 0 & +0.4445 & 0 \\
-0.1320 & +0.1320 & +0.1320 & -0.1320 & 0 & -0.0809 & 0 & +0.1615
\end{pmatrix}
\begin{pmatrix}
\phi_4 \\ \phi_7 \\ \phi_8 \\ \phi_5 \\ \delta_1^{(4,7)} \\ \delta_1^{(7,8)} \\ \delta_1^{(5,8)} \\ \delta_1^{(4,5)}
\end{pmatrix}
=
\begin{pmatrix}
0 \\ 0 \\ 0 \\ 0 \\ 0 \\ 0 \\ 0 \\ 0
\end{pmatrix}$$

Equations for element **4**:

$$\begin{pmatrix}
+0.1902 & +0.0380 & -0.0951 & -0.1331 & -0.0932 & +0.0233 & +0.0932 & -0.0233 \\
+0.0380 & +0.1902 & -0.1331 & -0.0951 & -0.0932 & -0.0233 & +0.0932 & +0.0233 \\
-0.0951 & -0.1331 & +0.1902 & +0.0380 & +0.0932 & -0.0233 & -0.0932 & +0.0233 \\
-0.1331 & -0.0951 & +0.0380 & +0.1902 & +0.0932 & +0.0233 & -0.0932 & -0.0233 \\
-0.0932 & -0.0932 & +0.0932 & +0.0932 & +0.1521 & 0 & -0.0380 & 0 \\
+0.0233 & -0.0233 & -0.0233 & +0.0233 & 0 & +0.3233 & 0 & +0.1331 \\
+0.0932 & +0.0932 & -0.0932 & -0.0932 & -0.0380 & 0 & +0.1521 & 0 \\
-0.0233 & +0.0233 & +0.0233 & -0.0233 & 0 & +0.1331 & 0 & +0.3233
\end{pmatrix}
\begin{pmatrix}
\phi_5 \\ \phi_8 \\ \phi_9 \\ \phi_6 \\ \delta_1^{(5,8)} \\ \delta_1^{(8,9)} \\ \delta_1^{(6,9)} \\ \delta_1^{(5,6)}
\end{pmatrix}
=
\begin{pmatrix}
0 \\ 0 \\ 0 \\ 0 \\ 0 \\ 0 \\ 0 \\ 0
\end{pmatrix}$$

where $\phi_1 \ldots, \phi_9$ are hydraulic heads at the node points for the four elements and δ are the p-mode parameters, subscripts indicate mode number for the side modes and superscripts indicate node numbers on the corresponding line segments for an element. The boundary conditions are the following:

$$(\phi_1, \phi_2, \phi_3, \phi_6, \phi_9, \delta_1^{(1,2)}, \delta_1^{(3,6)}, \delta_1^{(6,9)}) - (2.98422, 2.58123, 2, 2, 2, 0, 0, 0) \tag{11.7}$$

The hydraulic head boundary conditions ϕ_1, ϕ_2 were determined from the static pressures

$$p_1, p_2 = \rho g \phi_1, \rho g \phi_2 = 29046.9, 25124.4 \, Pa \tag{11.8}$$

at nodes and 1 and 2, respectively. We can now assemble the global stiffness matrix.

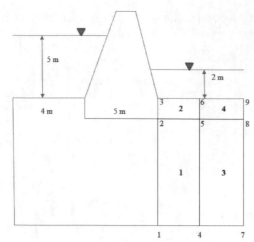

Figure 11.13a) Four element model for seepage flow

$$\begin{pmatrix} +0.485 & +0.162 & -0.202 & -0.121 & -0.016 & 0 & +0.016 & -0.264 & -0.016 & 0 & +0.016 & +0.132 & 0 \\ +0.162 & +0.865 & -0.121 & -0.164 & +0.016 & +0.023 & -0.110 & -0.264 & +0.016 & -0.047 & -0.110 & +0.132 & +0.023 \\ -0.202 & -0.121 & +0.242 & +0.081 & 0 & 0 & 0 & +0.132 & -0.016 & 0 & +0.016 & -0.132 & 0 \\ -0.121 & -0.164 & +0.081 & +0.433 & 0 & 0 & 0 & +0.132 & +0.016 & +0.023 & -0.110 & -0.132 & -0.023 \\ -0.016 & +0.016 & 0 & 0 & +0.445 & 0 & +0.202 & 0 & 0 & 0 & 0 & 0 & 0 \\ 0 & +0.023 & 0 & 0 & 0 & +0.323 & 0 & 0 & 0 & +0.133 & 0 & 0 & 0 \\ +0.016 & -0.110 & 0 & 0 & +0.202 & 0 & +0.597 & 0 & 0 & 0 & 0 & 0 & 0 \\ -0.264 & -0.264 & +0.132 & +0.132 & 0 & 0 & 0 & +0.323 & 0 & 0 & 0 & -0.081 & 0 \\ -0.016 & +0.016 & -0.016 & +0.016 & 0 & 0 & 0 & 0 & +0.445 & 0 & +0.202 & 0 & 0 \\ 0 & -0.047 & 0 & +0.023 & 0 & +0.133 & 0 & 0 & 0 & +0.647 & 0 & 0 & +0.133 \\ +0.016 & -0.110 & +0.016 & -0.110 & 0 & 0 & 0 & 0 & +0.202 & 0 & +0.597 & 0 & 0 \\ +0.132 & +0.132 & -0.132 & -0.132 & 0 & 0 & 0 & -0.081 & 0 & 0 & 0 & +0.161 & 0 \\ 0 & +0.023 & 0 & -0.023 & 0 & 0 & 0 & 0 & 0 & +0.133 & 0 & 0 & +0.323 \end{pmatrix} \begin{pmatrix} \phi_4 \\ \phi_5 \\ \phi_7 \\ \phi_8 \\ \delta_1^{(1,4)} \\ \delta_1^{(2,3)} \\ \delta_1^{(2,5)} \\ \delta_1^{(4,5)} \\ \delta_1^{(4,7)} \\ \delta_1^{(5,6)} \\ \delta_1^{(5,8)} \\ \delta_1^{(7,8)} \\ \delta_1^{(8,9)} \end{pmatrix} =$$

$$\begin{pmatrix} +0.916 \\ +1.698 \\ 0 \\ +0.456 \\ +0.007 \\ +0.060 \\ -0.139 \\ -0.735 \\ 0 \\ -0.060 \\ -0.373 \\ 0 \\ 0 \end{pmatrix}, \text{ with the solution } \begin{pmatrix} \phi_4 \\ \phi_5 \\ \phi_7 \\ \phi_8 \\ \delta_1^{(1,4)} \\ \delta_1^{(2,3)} \\ \delta_1^{(2,5)} \\ \delta_1^{(4,5)} \\ \delta_1^{(4,7)} \\ \delta_1^{(5,6)} \\ \delta_1^{(5,8)} \\ \delta_1^{(7,8)} \\ \delta_1^{(8,9)} \end{pmatrix} = \begin{pmatrix} +2.699 \\ +2.192 \\ +2.703 \\ +2.193 \\ -0.012 \\ +0.038 \\ +0.100 \\ -0.316 \\ +0.027 \\ -0.024 \\ +0.023 \\ -0.155 \\ +0.010 \end{pmatrix} \quad (11.9)$$

We can summarize the solution at the outflow for elements 2 and 4 with the following table where v_y velocity component is compared with ANSYS Fluent in Figure 11.12e).

	x	y	φ	∂φ/∂x	∂φ/∂y	v_y
2	0	10	2.	0.	−0.32565825428980166	0.07432144875149277
2	0.75	10	2.	0.	−0.2555161785079309	0.05831368410289106
2	1.5	10	2.	0.	−0.2057152107246451	0.04694814975476898
2	2.25	10	2.	0.	−0.1762553509399443	0.04022484570712651
4	3	10	2.	0.	−0.16713659915382845	0.03814377195996365
4	3.75	10	2.	0.	−0.14647549864168652	0.03342851324124527
4	4.5	10	2.	0.	−0.13050473334316898	0.029783678820436723
4	5.25	10	2.	0.	−0.1192243032582763	0.027209268697538133
4	6	10	2.	0.	−0.11263420838700836	0.025705282872549466

Table 11.2 Solution summary at the outflow region

```
(*Seepage through soil, p=q=0,
2nd order p-formulation (n=2), wi = wj = 1, kx = ky = 0.2282191462137928` *)
Clear["Global`*"]; wi = 1; wj = 1; kx = 0.2282191462137928`; ky = 0.2282191462137928`;
NT = {(1 - s) * (1 - t) / 4, (1 + s) * (1 - t) / 4, (1 + s) * (1 + t) / 4, (1 - s) * (1 + t) / 4,
     (3 * s^2 / 2 - 3 / 2) * (1 - t) / (2 * Sqrt[6]), (3 * t^2 / 2 - 3 / 2) * (1 + s) / (2 * Sqrt[6]),
     (3 * s^2 / 2 - 3 / 2) * (1 + t) / (2 * Sqrt[6]), (3 * t^2 / 2 - 3 / 2) * (1 - s) / (2 * Sqrt[6])};
gcd[x0_, y0_] := Module[{x = x0, y = y0}, {J = {{D[x, s], D[x, t]}, {D[y, s], D[y, t]}};
     BT = Inverse[J] . {D[NT, s], D[NT, t]}; k = wi * wj * Transpose[BT] . {{kx, 0}, {0, ky}} . BT * Det[J];
     k1 = k //. {s -> -0.57735, t -> -0.57735}; k2 = k //. {s -> -0.57735, t -> 0.57735};
     k3 = k //. {s -> 0.57735, t -> -0.57735}; k4 = k //. {s -> 0.57735, t -> 0.57735}; kk = k1 + k2 + k3 + k4;}]

(*Element 1: 0<x<3, 0<y<8.5, -1<s<1, -1<t<1, Element 2: 0<x<3, 8.5<y<10, -1<s<1, -1<t<1,
Element 3: 3<x<6, 0<y<8.5, -1<s<1, -1<t<1, Element 4: 3<x<6, 8.5<y<10, -1<s<1, -1<t<1 *)
gcd[3 * s / 2 + 3 / 2, 8.5 * t / 2 + 8.5 / 2]; kk1 = kk; kk1 // MatrixForm;
gcd[3 * s / 2 + 3 / 2, 3 * t / 4 + 37 / 4]; kk2 = kk; kk2 // MatrixForm;
gcd[3 * s / 2 + 9 / 2, 8.5 * t / 2 + 8.5 / 2]; kk3 = kk; kk3 // MatrixForm;
gcd[3 * s / 2 + 9 / 2, y = 3 * t / 4 + 37 / 4]; kk4 = kk; kk4 // MatrixForm;

(*Global stiffness matrix and Boundary conditions *)
kG = SparseArray[{}, {21, 21}]; r1 = {1, 4, 5, 2, 11, 15, 13, 10}; kG[[r1, r1]] = kG[[r1, r1]] + kk1;
r2 = {2, 5, 6, 3, 13, 17, 14, 12};
kG[[r2, r2]] = kG[[r2, r2]] + kk2; r3 = {4, 7, 8, 5, 16, 20, 18, 15};
kG[[r3, r3]] = kG[[r3, r3]] + kk3; r4 = {5, 8, 9, 6, 18, 21, 19, 17}; kG[[r4, r4]] = kG[[r4, r4]] + kk4;
kG[[1]] = UnitVector[21, 1]; kG[[2]] = UnitVector[21, 2]; kG[[3]] = UnitVector[21, 3];
kG[[6]] = UnitVector[21, 6]; kG[[9]] = UnitVector[21, 9]; kG[[10]] = UnitVector[21, 10];
kG[[14]] = UnitVector[21, 14]; kG[[19]] = UnitVector[21, 19];

(*Global solution and element solutions) *)
RHS = {2.984224966974819`, 2.581234546896989`, 2, 0, 0, 2, 0, 0, 2, 0, 0, 0, 0, 0, 0, 0, 0, 0, 0, 0, 0};
φ = LinearSolve[kG, RHS]
gce[x0_, y0_, s0_, t0_] :=
  Module[{x = x0, y = y0}, {J = {{D[x, s], D[x, t]}, {D[y, s], D[y, t]}}; NT1 = NT //. {s -> s0, t -> t0};
     dNTds = D[NT, s] //. {s -> s0, t -> t0}; dNTdt = D[NT, t] //. {s -> s0, t -> t0};
     dNTdsodNTdt = {dNTds, dNTdt} //. {s -> s0, t -> t0}; BT = Inverse[J] . dNTdsodNTdt;
     B = Transpose[BT]; BTx = BT[[1]]; BTy = BT[[2]]; φxy = NT1.dT; dφdx = BTx.dT; dφdy = BTy.dT;}]

(*Solution for element 2 at location (x,y) = (0,10), (0.75,10), (1.5,10), (2.25,10) *)
dT = {φ[[2]], φ[[5]], φ[[6]], φ[[3]], φ[[13]], φ[[17]], φ[[14]], φ[[12]]};
gce[3 * s / 2 + 3 / 2, 3 * t / 4 + 37 / 4, -1, 1]; mxy1 = {0, 10, φxy, dφdx, dφdy, -ky * dφdy};
gce[3 * s / 2 + 3 / 2, 3 * t / 4 + 37 / 4, -0.5, 1]; mxy2 = {0.75, 10, φxy, dφdx, dφdy, -ky * dφdy};
gce[3 * s / 2 + 3 / 2, 3 * t / 4 + 37 / 4, 0, 1]; mxy3 = {1.5, 10, φxy, dφdx, dφdy, -ky * dφdy};
gce[3 * s / 2 + 3 / 2, 3 * t / 4 + 37 / 4, 0.5, 1]; mxy4 = {2.25, 10, φxy, dφdx, dφdy, -ky * dφdy};

(*Solution for element 4 at location (x,y) = (3,10), (3.75,10), (4.5,10), (5.25,10), (6,10) *)
dT = {φ[[5]], φ[[8]], φ[[9]], φ[[6]], φ[[18]], φ[[21]], φ[[19]], φ[[17]]};
gce[3 * s / 2 + 9 / 2, 3 * t / 4 + 37 / 4, -1, 1]; mxy5 = {3, 10, φxy, dφdx, dφdy, -ky * dφdy};
gce[3 * s / 2 + 9 / 2, 3 * t / 4 + 37 / 4, -0.5, 1]; mxy6 = {3.75, 10, φxy, dφdx, dφdy, -ky * dφdy};
gce[3 * s / 2 + 9 / 2, 3 * t / 4 + 37 / 4, 0, 1]; mxy7 = {4.5, 10, φxy, dφdx, dφdy, -ky * dφdy};
gce[3 * s / 2 + 9 / 2, 3 * t / 4 + 37 / 4, 0.5, 1]; mxy8 = {5.25, 10, φxy, dφdx, dφdy, -ky * dφdy};
gce[3 * s / 2 + 9 / 2, 3 * t / 4 + 37 / 4, 1, 1]; mxy9 = {6, 10, φxy, dφdx, dφdy, -ky * dφdy};
```

265

```
(* Nodal solution summary and write to  file*)
mm = {mxy1, mxy2, mxy3, mxy4, mxy5, mxy6, mxy7, mxy8, mxy9}; MatrixForm[mm,
 TableHeadings → {{"2", "2", "2", "2", "4", "4", "4", "4", "4"}, {"x", "y", "ϕ", "∂ϕ/∂x", "∂ϕ/∂y", "vᵧ"}}]
SetDirectory["C:\\Users\\jmatsson"]; mylist = Table[
  {{9, mxy1[[6]]}, {9.75, mxy2[[6]]}, {10.5, mxy3[[6]]}, {11.25, mxy4[[6]]}, {12, mxy5[[6]]},
    {12.75, mxy6[[6]]}, {13.5, mxy7[[6]]}, {14.25, mxy8[[6]]}, {15, mxy9[[6]]}}];
TableOfValues1 = Prepend[mylist, {"((xy/key/label \"theory-outlet\")"}];
TableOfValues1 = Prepend[TableOfValues1, {""}];
TableOfValues1 = Prepend[TableOfValues1, {"(labels \"X-Coordinate\" \"Velocity\")"}];
TableOfValues1 = Prepend[TableOfValues1, {"(title \"Theory\")"}];
TableOfValues1 = Append[TableOfValues1, {")"}]; Grid[TableOfValues1]
Export["theory.dat", TableOfValues1]
```

Figure 11.13b) Mathematica code for seepage flow under a dam

I. References

1. Ochs et al., "Soil-Water Pit Heat Store with Direct Charging System", Ecostock 2005 Conference Proceedings.
2. Moaveni S., "Finite Element Analysis: Theory and Applications with ANSYS", 3rd Ed., Prentice Hall, 2007.
3. Bhatti, M.A., "Fundamental Finite Element Analysis and Applications", John Wiley & Sons, Inc., 2005.

J. Exercise

11.1 Use ANSYS Fluent to study the seepage flow of water through the porous medium under the dam as shown in Figure 11.14. Include contour plots for velocity magnitude and static pressure together with an XY Plot of the velocity magnitude at the outlet. Use the same values for viscous and inertial resistance and mesh size as was used in this chapter.

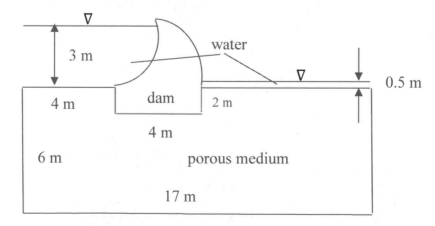

Figure 11.14 Geometry for Exercise 11.1

CHAPTER 12. WATER FILTER FLOW

A. Objectives

- Using ANSYS Workbench to Model Flow Through Water Filter
- Inserting Boundary Conditions and Porous Medium Zone for Filter with Viscous and Inertial Resistance
- Using Volume of Fluid Model for Multiphase Flow with Surface Tension
- Running Laminar Transient 2D Axisymmetric ANSYS Fluent Simulations
- Using Volume Fraction Contour Plots for Visualizations and Movie
- Comparing ANSYS Fluent Results with Theory

B. Problem Description

We will study the flow through a ceramic gravity driven clay pot water filter using ANSYS Fluent. The dimensions and initial water levels are shown below. The thickness of the filter is $d = 20$ mm and the radius $R = 100$ mm while the initial height of water $h_0 = 200$ mm. The height $h(t)$ of the water in the filter will vary over time as the fluid seeps through the filter.

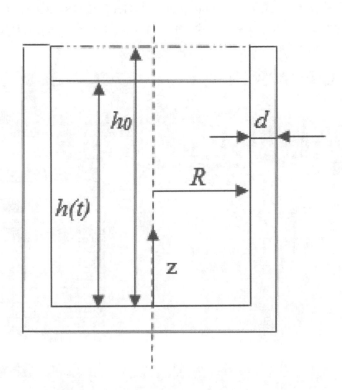

C. Launching ANSYS Workbench and Selecting Fluent

1. Start by launching ANSYS Workbench. Double click on Fluid Flow (Fluent) under Analysis Systems in the Toolbox.

Figure 12.1 Selecting Fluent

D. Launching ANSYS DesignModeler

2. Right click Geometry under Project Schematic in ANSYS Workbench and select Properties. In Properties of Schematic A2: Geometry, select Analysis Type 2D under Advanced Geometry Options. Right click on Geometry in the Project Schematic window and select New DesignModeler Geometry.

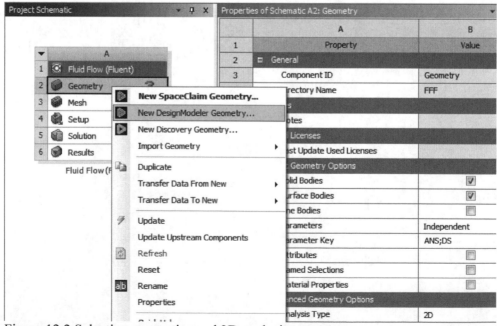

Figure 12.2 Selecting properties and 2D analysis type

3. Select Units>>Millimeter from the menu in DesignModeler. Select XY Plane in the Tree Outline and click on Look at Face/Plane/Sketch 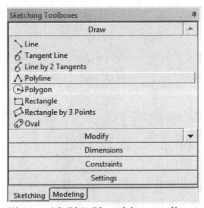. Select Sketching tab and Draw>>Polyline from Sketching Toolboxes. Start at origin with a *P* and draw the Sketch with Dimensions as shown in Figure 12.3c). Make sure that you have a *V* when drawing vertical lines and an *H* when drawing horizontal lines. Right click and select Closed End when you have completed the sketch. Enter the dimensions as shown in Figure 12.3c).

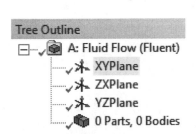

Figure 12.3a) Tree outline   Figure 12.3b) Sketching toolboxes

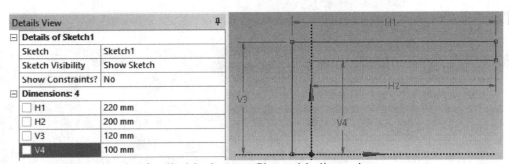

Figure 12.3c) Sketch of cylindrical water filter with dimensions

4. Select the Modeling tab and select Concepts>>Surfaces from Sketches from the menu. Select Sketch1 under XYPlane in the Tree Outline and Apply the sketch in Details View. Click on Generate.

   Select the XYPlane once again in the Tree Outline and click on New Sketch . Select the new sketch (Sketch2) under XYPlane and select the Sketching tab. Select Rectangle under Draw in Sketching Toolboxes and draw a rectangle inside the filter from the origin, see Figure 12.4. Select Concepts>>Surfaces from Sketches from the menu, select Sketch 2 and Apply the new sketch in Details View. Select Add Frozen as Operation in Details View. Click on Generate and Close DesignModeler.

Figure 12.4 Finished model for cylindrical water filter

269

E. Launching ANSYS Meshing

5. Double click on Mesh under Project Schematic in ANSYS Workbench. Select Mesh in the Outline. Set the Element Size to 2.0 mm under Defaults in Details of "Mesh". Select Yes for Capture Curvature under Sizing and set Smoothing to High under Quality. Right click on Mesh in the Outline and select Generate Mesh.

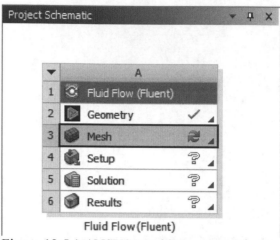

Figure 12.5a) ANSYS Meshing

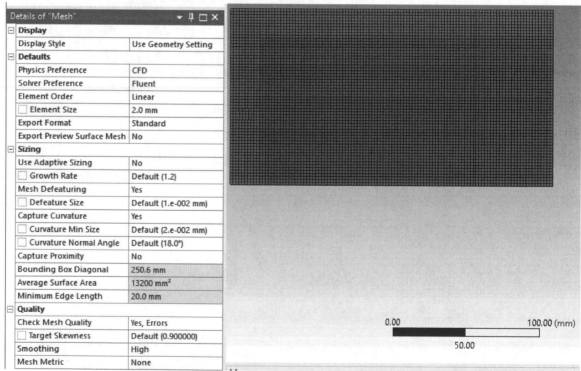

Figure 12.5b) Finished mesh for cylindrical water filter

6. Select Edge 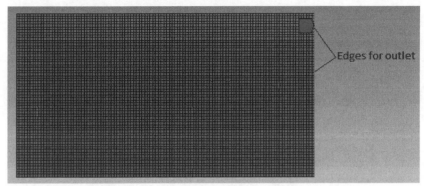 selection filter and control select the two edges on the right side of the computational domain, see Figure 12.6a). Right click on the edges and select Create Named Selection. Name the edges *outlet*.

Figure 12.6a) Named selections for outlet

Control-select the upper horizontal edge and the left outer vertical edge, see Figure 12.6b). Right click and select Create Named Selection. Name the edges *outlet-filter*.

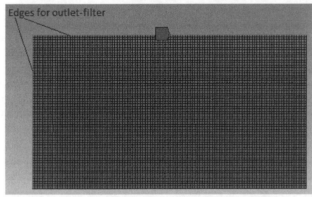

Figure 12.6b) Named selections for outlet-filter

Select Body 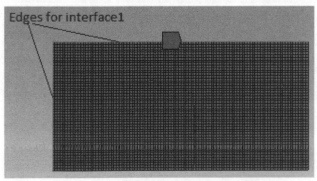 and select the outer L-shaped filter region, right click and select Hide Body. Select Edge and control-select the upper horizontal edge and the left vertical edge, see Figure 12.6c). Right click and select Create Named Selection. Name the edges *interface1*. Right click in the graphics window and select Show All Bodies.

Figure 12.6c) Named selections for interface1

Select Body , left click on the inner fluid region, right click and select Hide Body.

Select Edge 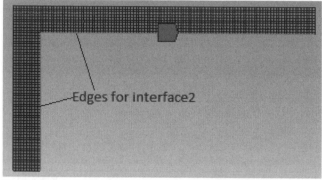 and control-select the inner horizontal edge and the inner vertical edge, see Figure 12.6d). Right click and select Create Named Selection. Name the edges *interface2*. Right click in the graphics window and select Show All Bodies.

Figure 12.6d) Named selections for interface2

Control select the two lower horizontal edges and name them *axis*, see Figure 12.6e).

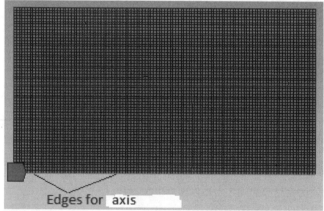

Figure 12.6e) Named selections for axis

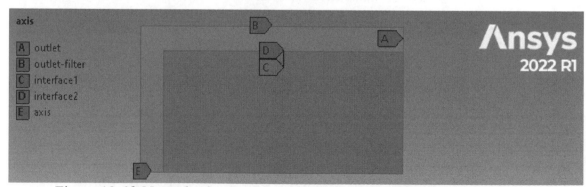

Figure 12.6f) Named selections for cylindrical water filter

Open Geometry in the Outline, select the first Surface Body, right click and select Rename. Name this body *Porous zone*. Rename the next Surface Body and call it *Inlet zone*.

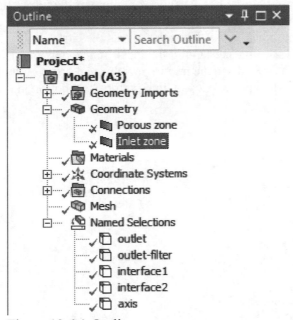

Figure 12.6g) Outline

Select File>>Export…>>Mesh>>FLUENT Input File>>Export from the menu and save the mesh with the name *cylindrical-water-filter.msh*. Select File>>Save Project from the menu and save the project with the name *Cylindrical Water Filter*. Close the ANSYS Meshing window. Right click on Mesh under Project Schematic in ANSYS Workbench and select Update.

F. Launching ANSYS Fluent

7. Double-click on Setup under Project Schematic in ANSYS Workbench. Check the box for Double Precision. Click on Show More Options and take a note of the location of the *Working Directory*. Set the number of processes equal to the number of processor cores for your computer. To check the number of physical cores, press the Ctrl + Shift + Esc keys simultaneously to open the Task Manager. Go to the Performance tab and select CPU from the left column. You will see the number of physical cores on the bottom-right side. Close the Task Manager window. Click on the Start button in the Fluent Launcher window.

Figure 12.7a) Fluent launcher settings

Select Transient Time and Axisymmetric 2D Space for the Solver under General on the Task Page. Check the box for Gravity and enter -9.81 as Gravitational Acceleration in the X direction. Double click on Models and Viscous (SST k-omega) and select Laminar Model. Click on OK to close the Viscous Model window. Double click on Models and Multiphase (Off) under Setup in the Outline View. Select Volume of Fluid model and check the box for Implicit Body Force. Click on Apply and Close the window.

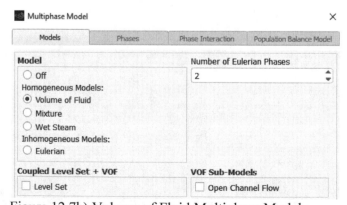

Figure 12.7b) Volume of Fluid Multiphase Model

274

8. Double click on Materials under Setup in the Outline View. Select Create/Edit for Fluid under Materials on the Task Page. Select Fluent Database and scroll down to *water-liquid (h2o<l>)* and select this as Fluent Fluid Materials. Click on Copy and Close the window. Close the Create/Edit Materials window.

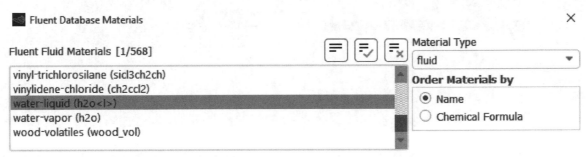

Figure 12.8 Fluent Fluid Materials

9. Double click on Models and Multiphase (Volume of Fluid) under Setup in the Outline View. Select the Phases tab at the top of the Multiphase Model window. Select phase-1 – Primary Phase under Phases. Select *water-liquid* as the Primary Phase Material and enter *water* as the name under Phase Setup. Click on Apply.

    Select phase-2 – Secondary Phase under Phases and select *air* as the Secondary Phase Material. Enter *air* as the name under Phase Setup. Click on Apply. Select the Phase Interaction tab at the top of the Multiphase Model window. Check the box for Surface Tension Force Modeling. Select constant for the Surface Tension Coefficients [N/m] and enter 0.07286 as the value. Click on Apply and Close the Multiphase Model window.

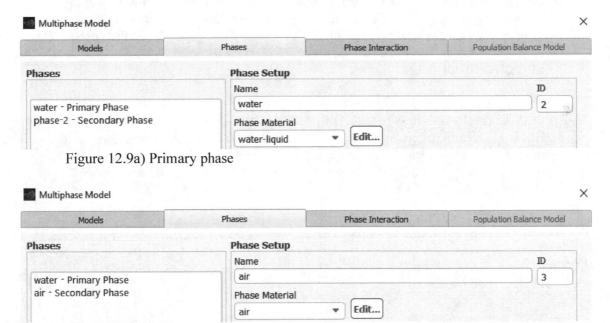

Figure 12.9a) Primary phase

Figure 12.9b) Secondary phase

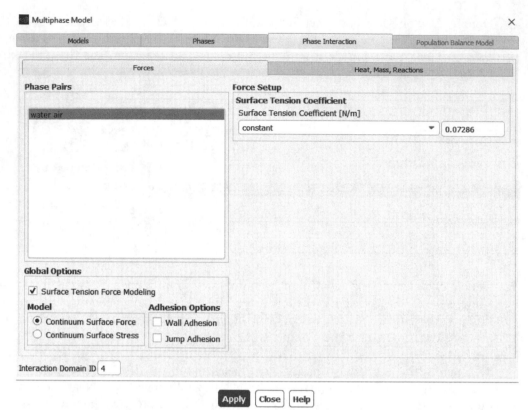

Figure 12.9c) Phase interaction

10. Double click on Cell Zone Conditions under Setup in the Outline View. Select the *porous_zone* under Cell Zone Conditions on the Task Page. Select *fluid* as Type and *mixture* as Phase and click on the Edit button on the Task Page. Check the Porous Zone box. Select the Porous Zone tab and make sure that Porosity is 0.5. Click Apply and Close the Fluid window.

Select *water* as the Phase for the *porous_zone* and click on Edit. Enter 9.76305e+08 as the Viscous Resistance in both directions. Click Apply and Close the Fluid window. Click on Operating Conditions on the Task Page under Cell Zone Conditions. Select user-input as Operating Density Method and enter the value 1.225 for Operating Density [kg/m³]. Enter X [m] 0.2 as Reference Pressure Location. Click OK to close the Operating Conditions window.

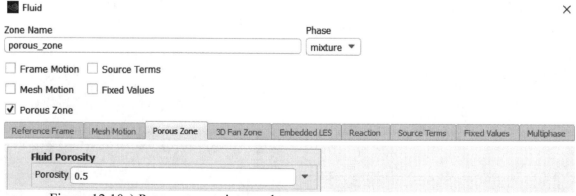

Figure 12.10a) Porous zone mixture phase

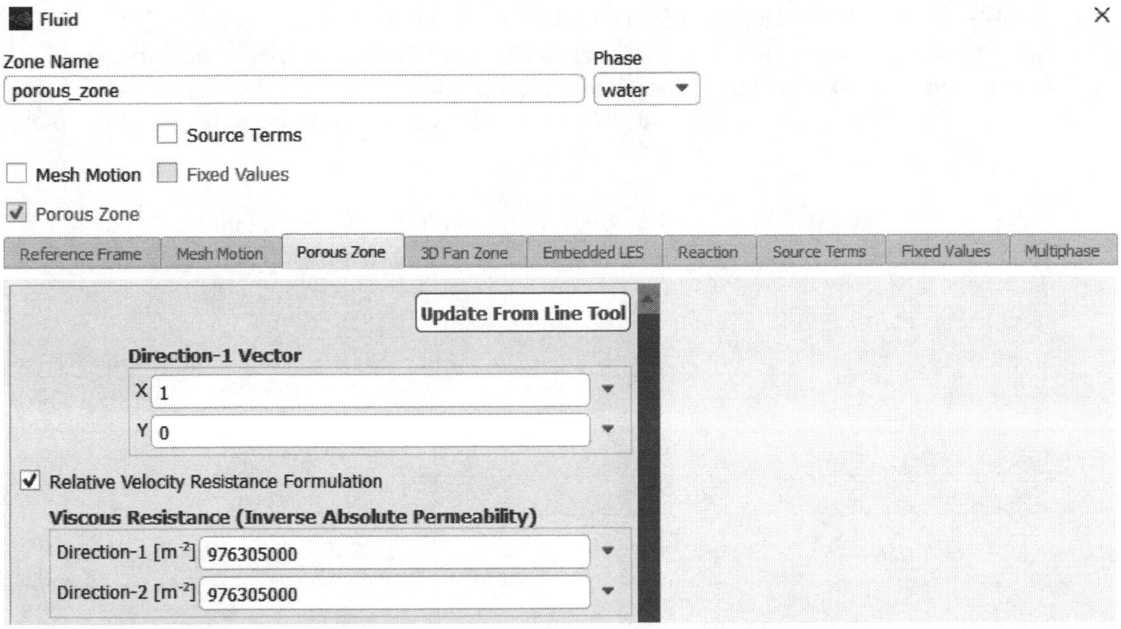

Figure 12.10b) Porous zone water phase

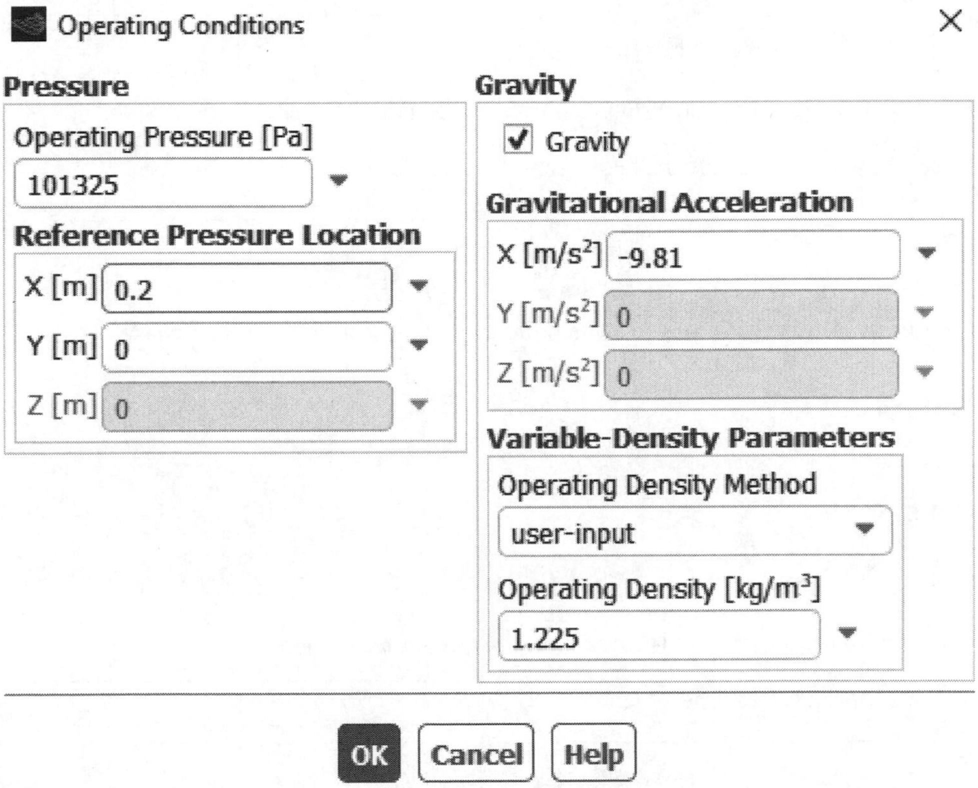

Figure 12.10c) Operating conditions

11. Double click on Boundary Conditions under Setup in the Outline View. Select the *outlet-inlet_zone* Boundary Condition on the Task Page and make sure that Type is *pressure-outlet*. Select *mixture* as Phase and click on Edit…. Make sure that Gauge Pressure [Pa] is 0 and Backflow Pressure Specification as Total Pressure. Select Apply and Close the window.

Select *air* as Phase and click on the Edit… button on the Task Page under Boundary Conditions. Set the Backflow Volume Fraction to 1, click Apply and Close the window.

Repeat this step 11 for the *outlet-porous_zone* Boundary Condition.

Figure 12.11a) Outlet-inlet_zone boundary condition for mixture phase

Figure 12.11b) Outlet-inlet_zone boundary condition for air phase

Figure 12.11c) Outlet-porous_zone boundary condition for mixture phase

Figure 12.11d) Outlet-porous_zone boundary condition for air phase

Continue with the *outlet-filter* Boundary Condition. Make sure that Type is *pressure-outlet* on the Task Page and select *mixture* as Phase and click on Edit.... Gauge Pressure [Pa] is 0 and Backflow Pressure Specification as Total Pressure. Click Apply and Close the window. Select *air* as Phase and click on the Edit... button. Select *From Neighboring Cell* as Volume Fraction Specification Method, click Apply and Close the window.

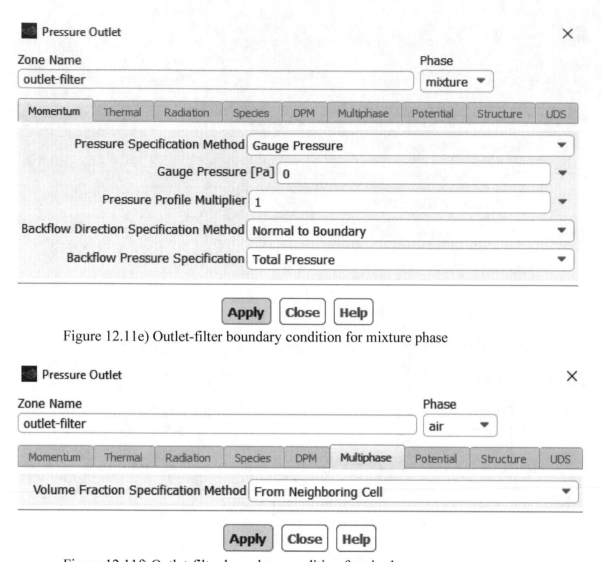

Figure 12.11e) Outlet-filter boundary condition for mixture phase

Figure 12.11f) Outlet-filter boundary condition for air phase

12. Double click on Initialization under Solution in the Outline View. Select Hybrid Initialization Method on the Task Page and click on Initialize. Click on Patch... on the Task Page under Solution Initialization. Select *air* as the Phase, select Volume Fraction as Variable, select *inlet_zone* and *porous_zone* as Zones to Patch, set the Value to 0 and click on Patch. Close the window.

Figure 12.12a) Initialization of solution

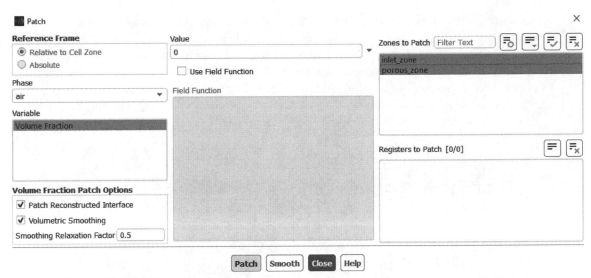

Figure 12.12b) Patch settings

13. Double click on Graphics and Contours under Results in the Outline View. Select Contours of Phases… and Volume fraction for *water* as the Phase. Scroll down and deselect all Surfaces and uncheck Auto Range Options. Set Min to 0 and Max to 1. Select Colormap Options…, set Font Behavior to Fixed and Font Size to 12 under Font. Click on Apply and Close the Colormap window. Click on Save/Display and Close the Contours window.

Figure 12.13a) Contours settings

Select the View tab in the menu and click on Views…. Select *axis-porous_zone* and *axis-inlet_zone* as Mirror Planes and click on Apply. Select the Camera… button in the Views window. Use your left mouse button to rotate the dial counter-clockwise 90 degrees until the bowl is upright. Close the Camera Parameters window. Click on the Save button under Actions in the Views window and Close the same window. Click on Copy Screenshot of Active Window to Clipboard in the graphics window, see Figure 12.13c). The contour plot can be pasted into a Word document.

Figure 12.13b) Water volume fraction after rotation and mirroring

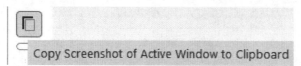

Figure 12.13c) Copying screenshot

Double click on Calculation Activities and Solution Animations under Solution in the Outline View. Enter *water-filter* as the Name, enter Record after every 10 and select time-step. Select *contour-1* as Animation Object. Select OK to close the Animation Definition window.

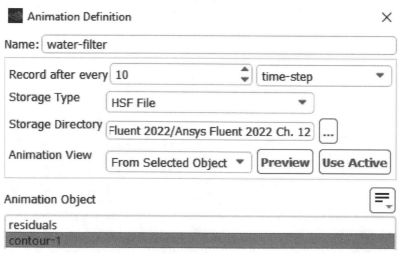

Figure 12.13d) Animation definition

14. Double click on Report Definitions under Solution in the Outline View. Select New>>Surface Report>>Volume Flow Rate…. Enter *volume-flow-rate* as the name. Select *water* as the Phase and select *outlet-filter* under Surfaces. Check the boxes for Report File, Report Plot and Print to Console under Create. Click OK to close the window.

Figure 12.14a) Surface report definition for volume flow rate

Select New>>Expression in the Report Definitions window. Select Functions>>Mathematical>>abs and select *volume-flow-rate* as Report Definition, see Figure 12.14b). Enter *absolute-volume-flow-rate* as Name and check the boxes for Report File, Report Plot and Print to Console under Create. Click on OK to close the window. Close the Report Definitions window. Select File>>Write>Case & Data... from the menu. Save the file with the name *water-filter-flow.cas.h5*.

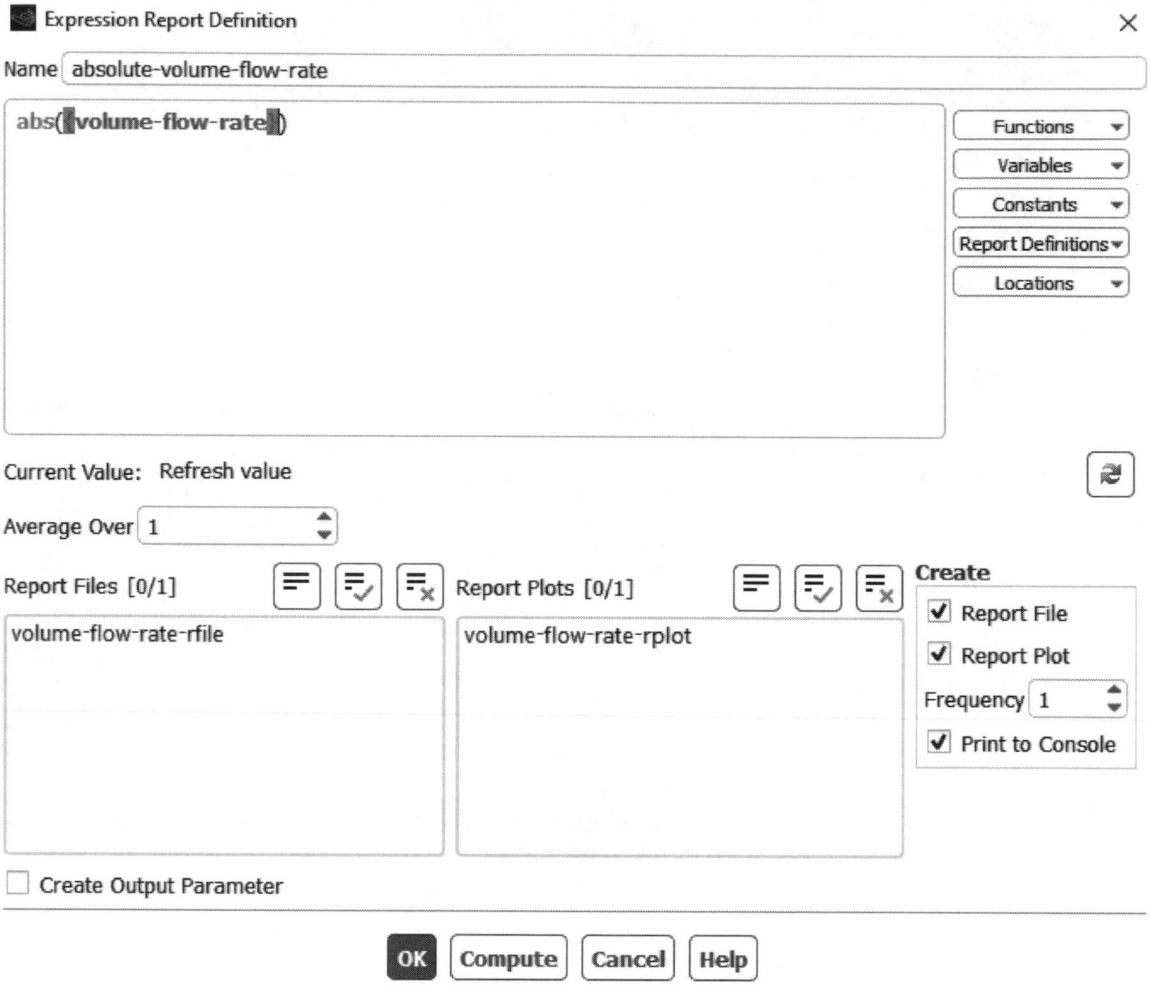

Figure 12.14b) Expression report definition for absolute flow rate

15. Double click on Run Calculation under Solution in the Outline View and set the Time Step Size [s] to 0.005. Set the Number of Time Steps to 50 and Max Iterations/Time Step to 30. Click on the Calculate button on the Task Page. Click OK in the Information window when the calculations have finished. Click on Copy Screenshot of Active Window to Clipboard in the graphics window, see Figure 12.13c). The scaled residuals can be pasted into a Word document.

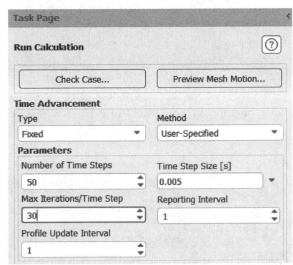

Figure 12.15a) Calculation settings

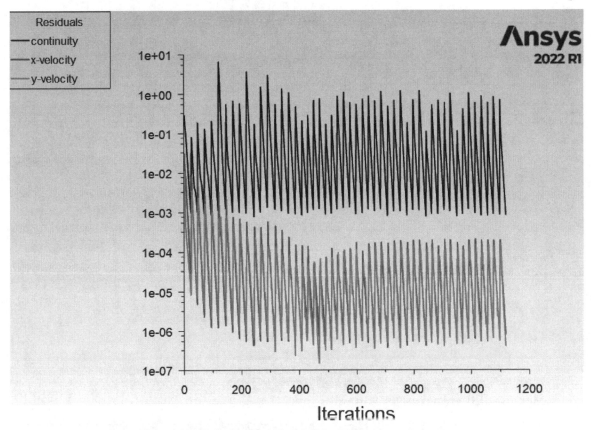

Figure 12.15b) Residuals during the first 0.25 s

G. Post-Processing and Continued Simulations

16. Select the tab Contours of Volume fraction (water) in the graphics window. Select the View tab in the menu and click on Views…. Select *view-0* in Views, click on Apply in the Views window and Close the window. Click on Copy Screenshot of Active Window to Clipboard in the graphics window, see Figure 12.13c). The scaled residuals can be pasted into a Word document.

Select the Results tab from the menu and select Surface>>Create>>Iso-Surface…. Select Surface of Constant Phases… and Volume Fraction. Select *water* as the Phase and set Iso-Values to 0.5. Use the horizontal slider to set the Iso-Value. Select *inlet_zone* under From Zones. Enter *free-surface-t=0.25s* as New Surface Name and click on Create. The new surface will now be available under From Surface, see Figure 12.16a). Close the Iso-Surface window.

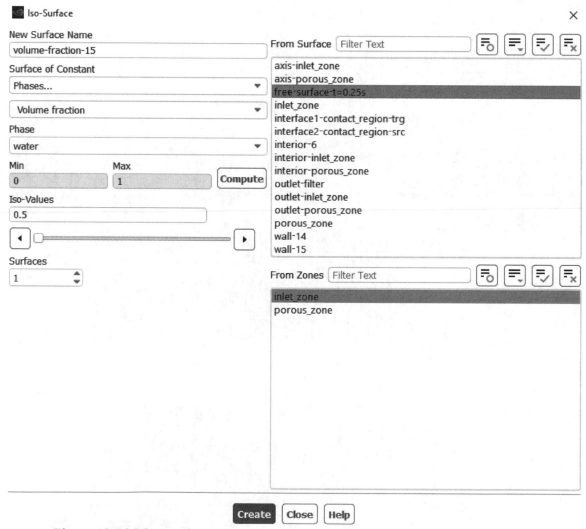

Figure 12.16a) Iso-surface settings

Select File>>Export>>Solution Data… from the menu. Select ASCII as File Type, Node under Location and Space as Delimiter. Select *inlet_zone* and *porous_zone* under Cell Zones and *Iso-surface>> free-surface-time-t=0.25s* under Surfaces. Select *Volume fraction (water)* under Quantities. Click on Write and save the ASCII File in the *Working Directory* with the name *free-surface-coordinates-t=0.25s*. Close the Export window.

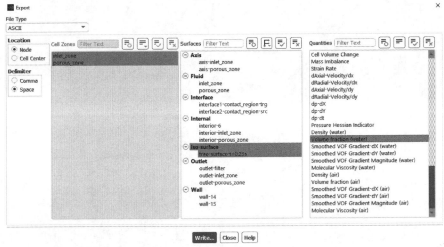

Figure 12.16b) Export settings

Open the saved file *free-surface-coordinates-t=0.25s* in Excel. Click on Next twice in the Text Import Wizard and click on Finish. Copy the 50 first values from the *C* column and paste the data in the *F* column. Label the new column *r (m)*. Copy the 50 first values from the *B* column and paste the data in the *G* column. Label the new column *h (m)*. Plot the free surface and take the average free surface height for the first 50 points. Repeat steps **15 – 16** seven more times for a total of 2 seconds.

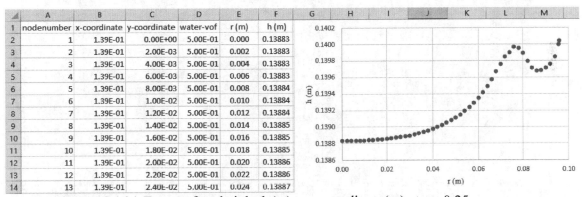

Figure 12.16c) Free surface height *h* (m) versus radius *r* (m) at *t* = 0.25 s

t (s)	Average *h* (m), ANSYS Fluent	*h* (m), Theory
0	0.2	0.2
0.25	0.13927	0.1429
0.5	0.09868	0.1080
0.75	0.07290	0.08457
1	0.05487	0.06788
1.25	0.04156	0.05548
1.5	0.03185	0.04597
1.75	0.02417	0.03849
2	0.01769	0.03249

Table 12.1 Data for average free surface height over time

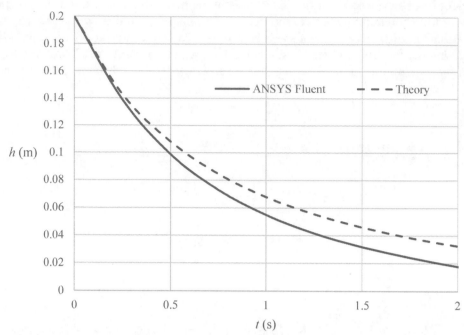

Figure 12.16d) Average free surface height h (m) versus time

Figure 12.16e) Water volume fraction at t = 0 – 2 s with an increment of 0.25 s

Double click on Animations and Playback under Results in the Outline View. Uncheck the box for Use Stored View, set the Replay Speed to low and select Play Once as Playback Mode. Select Video File as Write/Record Format. Click on the Write button for your chosen Animation Sequence. This will create the movie. Close the Playback window. The movie can be viewed using for example Windows Media Player.

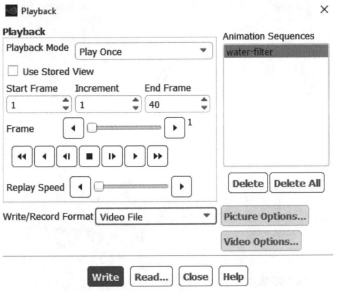

Figure 12.16f) Playback settings

Double click on Data Sources under Results and Plots in the Outline View. Select Load File... under Data Sources and load the file *absolute-volume-flow-rate-rfile.out*. Select *flow-time* as X Axis Variable and *absolute-volume-flow-rate* as Y Axis Variable.

Next, load the file *volume-flow-rate.dat*. This and other files are available for download from *sdcpublications.com*. Select *Time(s)* as X Axis Variable and *Volume Flow Rate (m3/s)* as Y Axis Variable. Click on Axes... at the bottom of the Plot Data Sources window. Uncheck Auto Range under Options for X Axis. Set Minimum to 0 and Maximum to 2 under Range. Click on Apply and select Y Axis. Select float as Type under Number Format and set Precision to 3. Click on Apply and Close the Axes – Plot Data Sources window. Change the X Axis Label under Plot to *t* and the Y Axis Label to *dV/dt* (m3/s). Delete the Title and Legend Label under Plot.

Click on Curves... and select the first available Pattern under Line Style for Curve # 0. Select no (blank) Symbol under Marker Style and click on Apply. Select Curve # 1, select the second available Pattern under Line Style. Select no (blank) Symbol under Marker Style and click on Apply. Close the Curves – Plot Data Sources window.

Click on Plot at the bottom of the Plot Data Sources window. Click on Copy Screenshot of Active Window to Clipboard in the graphics window, see Figure 12.13c). The curves can be pasted into a Word document.

Open the saved file *absolute-volume-flow-rate-rfile.out* in Excel. Select Delimited instead of Fixed width in Text Import Wizard – Step 1 of 3. Click on Next. Uncheck Tab as Delimiters and check Space as Delimiters in Text Import Wizard – Step 2 of 3. Click on Next. Click on Finish in Text Import Wizard – Step 3 of 3. Use the trapezoidal rule to find the area under the volume flow rate curve (2nd column *B* in Excel file) for each time interval (3rd column *C*) and add these areas in order to show the volume produced by the filter.

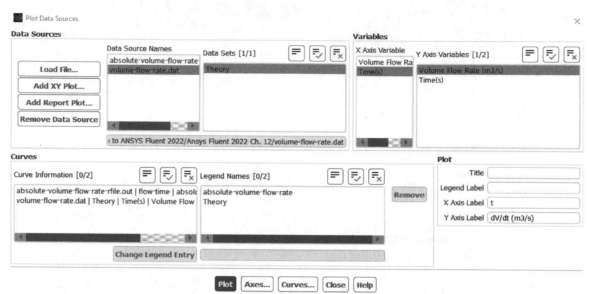

Figure 12.16g) Plot data sources settings

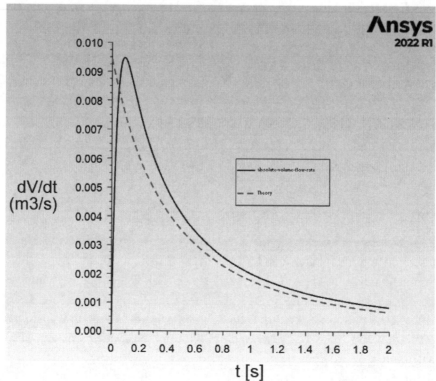

Figure 12.16h) Comparing ANSYS Fluent and theory-dashed curve for volume flow rate

This is accomplished by first deleting the title rows in the file and relabeling the three first columns $A - C$ as shown in Figure 12.16i).

Next, add the D column by inserting the equation =(B2+B3)/2 in $D3$ and drag this equation down the whole column D. Add the title *(f(a)+f(b))/2* as the title to column D.

Add the E column by inserting the equation =C3-C2 in $E3$ and drag this equation down the whole column E. Add the title *b-a* as the title to column E.

Add the F column by inserting the equation =D3*E3 in $F3$ and drag this equation down the whole column. Add the title *(b-a)*(f(a)+f(b))/2* as the title to column F.

Copy column C to column G. Add the title *t* as the title to column G.

Enter 0 in $H2$. Enter the equation =H2+F3 in $H3$ and drag this equation down the whole column. Add the title *V(t)* as the title to column H.

Enter equation (12.8) in column I and add equation (12.10) in column J as shown in the theory section. Equation (12.8) is =0.2*0.1/((0.1+0.2)*Exp(0.01*G2/0.02)-0.2) that is entered in cell $I2$. Drag this equation down the whole column I. Add the title *h(t)* as the title to column I.

Equation (12.10) is =PI()*0.1^2*(0.2-I2) that is entered in cell $J2$. Drag this equation down the whole column J. Add the title *V(t)* as the title to column J.

A	B	C	D	E	F	G	H	I	J
	dV/dt	t	(f(a)+f(b))/2	b-a	(b-a)*(f(a)+f(b))/2	t	V(t)	h(t)	V(t)
0	0	0				0	0	0.2	8.71967E-19
1	0.001527555	0.005	0.000763778	0.005	3.81889E-06	0.005	3.81889E-06	0.198509	4.68312E-05
2	0.002794584	0.01	0.00216107	0.005	1.08053E-05	0.01	1.46242E-05	0.197037	9.3084E-05
3	0.003907453	0.015	0.003351019	0.005	1.67551E-05	0.015	3.13793E-05	0.195583	0.000138769
4	0.004881165	0.02	0.004394309	0.005	2.19715E-05	0.02	5.33509E-05	0.194146	0.000183897
5	0.005712971	0.025	0.005297068	0.005	2.64853E-05	0.025	7.98362E-05	0.192727	0.000228477
6	0.006435774	0.03	0.006074373	0.005	3.03719E-05	0.03	0.000110208	0.191325	0.000272519
7	0.007042481	0.035	0.006739128	0.005	3.36956E-05	0.035	0.000143904	0.18994	0.000316033
8	0.007563954	0.04	0.007303217	0.005	3.65161E-05	0.04	0.00018042	0.188572	0.000359028
9	0.007994146	0.045	0.00777905	0.005	3.88952E-05	0.045	0.000219315	0.187219	0.000401513
10	0.00835847	0.05	0.008176308	0.005	4.08815E-05	0.05	0.000260197	0.185883	0.000443497

Figure 12.16i) Data for absolute volume flow rate

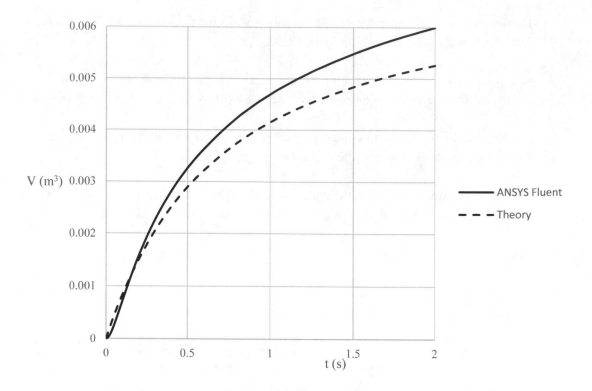

Figure 12.16j) Comparison ANSYS Fluent and theory for volume produced

H. Theory

17. We can relate the mean fluid velocity v (m/s) through the porous medium to the pressure gradient dP/dx (Pa) using Darcy's law in the same way as we did in the former chapter

$$v = -\frac{\kappa}{\mu}\frac{dP}{dx} \qquad (12.1)$$

where κ (m^2) is permeability of the porous medium and μ (kg/m-s) is the dynamic viscosity of the fluid. Furthermore, we define the hydraulic conductivity or coefficient of permeability k (m/s) as

$$k = \kappa\frac{\rho g}{\mu} \qquad (12.2)$$

Where g (m/s^2) is acceleration due to gravity and ρ (kg/m^3) is the density of the fluid. We will also use the following equation from the ANSYS Fluent user manual

$$\frac{dP}{dx} = -(\frac{\mu v}{\alpha} + \frac{C_2\rho v^2}{2}) \qquad (12.3)$$

where $1/\alpha$ (1/m^2) is viscous resistance or inverse absolute permeability for the porous medium and C_2(1/m) is inertial resistance of the porous medium. When we combine these equations we get this expression for the permeability coefficient

$$k = \frac{g}{\frac{vC_2}{2} + \frac{\mu}{\alpha\rho}} \qquad (12.4)$$

CHAPTER 12. WATER FILTER FLOW

The following values were used in the ANSYS Fluent simulations in this chapter.

Parameter	Value
C_2 (1/m)	0
$1/\alpha$ (1/m^2)	$9.76305 \cdot 10^8$
ρ (kg/m^3)	998.2
μ (kg/m-s)	0.001003
g (m/s^2)	9.81
k (m/s)	0.01
R (m)	0.1
h_0 (m)	0.2
d (m)	0.02
γ	2

Table 12.2 Parameters used in ANSYS Fluent simulations

For a cylindrical filter with a constant thickness d and constant radius R we have the schematic diagram as shown in Figure 12.17a).

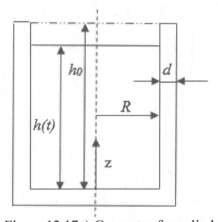

Figure 12.17a) Geometry for cylindrical ceramic water filter

The height of the free surface $h(t)$ will decrease over time as water is filtered through the filter. The volumetric flow rate through the filter can be described by

$$\dot{V}(t) = \frac{k}{d}\left[\pi R^2 h + R \int_0^{2\pi}\int_0^h (h-z)dzd\theta\right] = \frac{k\pi Rh}{d}(R+h) \tag{12.5}$$

We can alternatively describe the volumetric flow rate as

$$\dot{V}(t) = -\pi R^2 \frac{dh}{dt} \tag{12.6}$$

Combining these two equations we get the following first order ordinary differential equation

$$\frac{dh}{dt} + \frac{kh}{d}\left(1 + \frac{h}{R}\right) = 0 \tag{12.7}$$

with the following solution where we have used the initial condition $h(0) = h_0$

293

$$h(t) = \frac{h_0 R}{e^{kt/d}(R+h_0)-h_0}$$ (12.8)

The volume flow rate using equation (12.6) will be

$$\dot{V}(t) = \frac{(R+h_0)\pi R^3 k h_0 e^{kt/d}}{d(e^{kt/d}(R+h_0)-h_0)^2}$$ (12.9)

The total water volume produced can be expressed as

$$V(t) = \pi R^2(h_0 - h(t))$$ (12.10)

and with the use of equation (12.8) we get

$$V(t) = \pi R^2 h_0 \left(1 - \frac{R}{e^{kt/d}(R+h_0)-h_0}\right)$$ (12.11)

We can alternatively make equation (12.7) non-dimensional using non-dimensional variables $h^* = \frac{h}{h_0}$ and $t^* = \frac{kt}{d}$ where h_0 is the initial height of fluid in the cylinder. This will transform equation (12.7) to the following equation after skipping[*]

$$\frac{dh}{dt} + h(1 + \gamma h) = 0$$ (12.12)

where $\gamma = \frac{h_0}{R} = 2$ is the aspect-ratio.

Using initial condition $h(0) = 1$ we get the following solution

$$h(t) = \frac{1}{e^t(1+\gamma)-\gamma} = \frac{1}{3e^t-2}$$ (12.13)

Now we can express the volumetric flow rate in non-dimensional form as (after introducing $V^* = \frac{V}{\pi R^2 h_0}$ and skipping[*])

$$\dot{V}(t) = -\frac{dh}{dt} = \frac{e^t(1+\gamma)}{[\gamma - e^t(1+\gamma)]^2} = \frac{3e^t}{(2-3e^t)^2}$$ (12.14)

The total water volume produced can be expressed as

$$V(t) = \pi R^2(h_0 - h(t))$$ (12.15)

and in non-dimensional form this expression will be the following after skipping[*]

$$V(t) = 1 - h(t) = 1 - \frac{1}{e^t(1+\gamma)-\gamma} = \frac{3e^t-3}{3e^t-2}$$ (12.16)

For a frustum filter with a constant thickness d and varying radius r we have the schematic diagram as shown in Figure 12.17b).

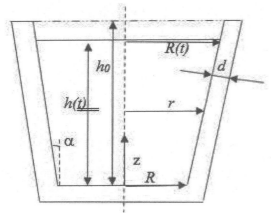

Figure 12.17b) Geometry for frustum water filter

The volumetric flow rate in this case can be described as

$$\dot{V}(t) = \frac{k}{d}\left[\pi R^2 h + \int_0^{2\pi}\int_0^h (h-z)rdzd\theta\right] \tag{12.17}$$

Using $r = R + z\,tan\alpha$, we can express equation (12.13) as

$$\dot{V}(t) = \frac{k\pi h}{d}\left(R^2 + hR + \frac{h^2 tan\alpha}{3}\right) \tag{12.18}$$

and equation (12.6) for a frustum shaped filter will be modified to

$$\dot{V}(t) = -\pi R(t)^2 \frac{dh}{dt} \tag{12.19}$$

where $R(t) = R + h(t)\,tan\alpha$. The differential equation for $h(t)$ will then be

$$\frac{dh}{dt} + \frac{kh}{d}\frac{\left(R^2 + hR + \frac{h^2 tan\alpha}{3}\right)}{(R+htan\alpha)^2} = 0 \tag{12.20}$$

and in non-dimensional form after skipping* we have

$$\frac{dh}{dt} + \frac{h}{(1+\delta h)^2}\left(1 + \gamma h + \frac{1}{3}\delta\gamma h^2\right) = 0 \tag{12.21}$$

where $\delta = \gamma\,tan\alpha$. There is no analytical solution available except in the case $\delta = 0$ corresponding to the cylindrical geometry. Equation (11.17) can be solved numerically in the general case $\delta \neq 0$ using the initial condition $h(0) = 1$.

We express the volumetric flow rate in non-dimensional form after skipping[*]

$$\dot{V}(t) = -(1 + \delta h)^2 \frac{dh}{dt} \tag{12.22}$$

The total water volume produced can be expressed as

$$V(t) = \pi\left[R^2 + R(h_0 + h)tan\alpha + \left(h_0^2 + h_0 h + h^2\right)\frac{tan^2\alpha}{3}\right](h_0 - h) \tag{12.23}$$

and in non-dimensional form this expression will be the following after skipping[*]

$$V(t) = (1 - h) + \delta(1 - h^2) + \frac{\delta^2}{3}(1 - h^3) \qquad (12.24)$$

I. References

1. Kelly, A.C., "Finite Element Modeling of Flow Through Ceramic Pot Filters.", *Master's Thesis in Civil and Environmental Engineering*, MIT, (2013).
2. Schweitzer, R.W., Cunningham, J.A., Mihelcic, J.R., "Hydraulic Modeling of Clay Ceramic Water Filters for Point-of-Use Water Treatment.", *Environ. Sci. Technol.*, **47**, 429-435, (2013).
3. Wald, I., "Modeling Flow Rate to Estimate Hydraulic Conductivity in a Parabolic Ceramic Water Filter.", *Undergraduate Journal of Mathematical Modeling*, **4**, 2, (2012).

J. Exercise

12.1 Complete ANSYS Fluent simulation of the flow through a frustum-shaped filter using the values listed in Table 12.3. Include graphs of your results showing a comparison between ANSYS Fluent and Theory for free-surface height, volume flow rate and volume produced by the filter. Use a time step size of 0.005 s and 200 time-steps with 50 time steps for each run. Include contours of volume fraction for water at t = 0, 0.25, 0.5, 0.75 and 1 second.

Parameter	Value
C_2 (1/m)	0
$1/\alpha$ (1/m^2)	$9.76305 \cdot 10^8$
ρ (kg/m^3)	998.2
μ (kg/m-s)	0.001003
g (m/s^2)	9.81
k (m/s)	0.01
R (m)	0.1
h_0 (m)	0.3
d (m)	0.01
γ	3
α (degrees)	10

Table 12.3 Parameter values for frustum shaped filter

CHAPTER 13. MODEL ROCKET FLOW

A. Objectives

- Model Geometry and Mesh
- Run Simulations for Turbulent Steady 3D Flow Past a Model Rocket in Ansys Fluent
- Determine Drag Coefficient for Model Rocket
- Visualize Flow Around Model Rocket

B. Problem Description

We will study the flow past the model rocket Estes Firestreak SST with a free stream velocity of 30 m/s.

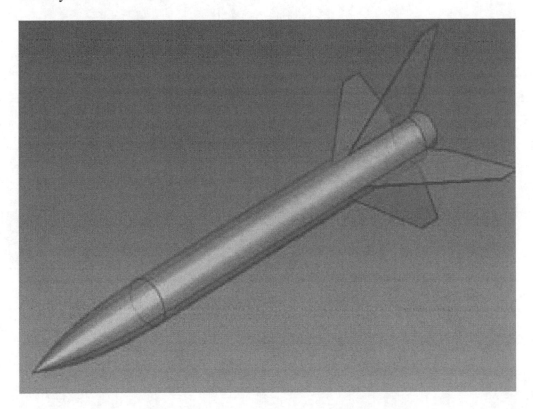

C. Launching Ansys Workbench and Importing the Geometry

1. Start by launching Ansys Workbench. Select File>>Import from the menu. Select Geometry File as format and open the file *Estes Model Rocket.IGS*. You will need to select Geometry File as file format in order to open the file. Download this and other files from *sdcpublications.com*.

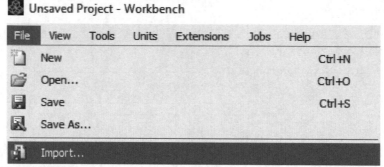

Figure 13.1 Import of geometry file

D. Launching Ansys DesignModeler

2. Right click on the imported Geometry in Ansys Workbench and select Edit Geometry in DesignModeler. Click on Generate in DesignModeler.

   Select Tools>>Enclosure from the menu. Set the Number of Planes to 2 in Details View. Select ZXPlane from the Tree Outline as Symmetry Plane 1 and Apply the plane in Details View. Select YZPlane from the Tree Outline as Symmetry Plane 2 and Apply it in Details View. Click on Generate. Right click in graphics window and select Zoom to Fit.

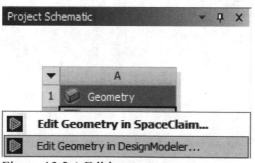

Figure 13.2a) Editing geometry

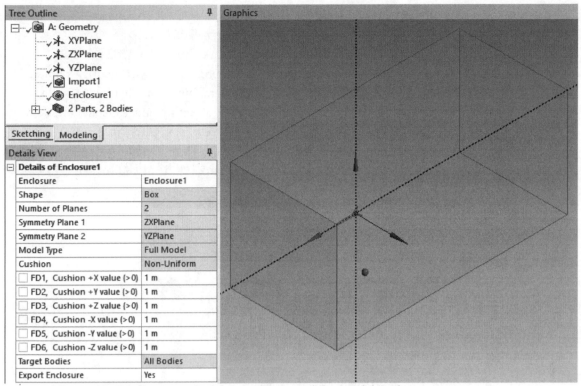

Figure 13.2b) Computational domain around the model rocket

3. Select Create>>Boolean from the menu. Select Subtract as Operation. Click on the plus sign next to 2 Parts, 2 Bodies in the Tree Outline. Select the second Solid and Apply it as Target Body in Details of Boolean1. Select the first Solid and Apply it as Tool Body in Details of Boolean1. Click on Generate. Right click on the Solid in the Tree Outline and Rename it to *Air*. Select File>>Save Project with the name "Model Rocket Flow Study" and close DesignModeler.

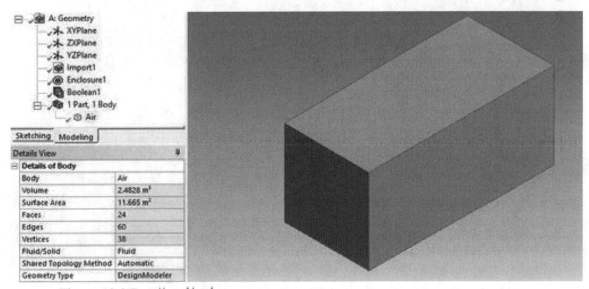

Figure 13.3 Details of body

E. Launching Ansys Meshing

4. Double click on Fluid Flow (Fluent) under Analysis Systems in Toolbox in Ansys Workbench. Drag the Geometry to the Geometry under Fluid Flow (Fluent). Double click on Mesh under Fluid Flow (Fluent). In the Meshing window, right click on Mesh in the Outline and select Update. Click on the plus sign next to Sizing under Details of Mesh. Change Capture Proximity and Capture Curvature both to Yes. Click on the plus sign next to Quality under Details of Mesh. Set Smoothing to High. Update the mesh once again. Rotate the Mesh to see the details of the mesh around the model rocket.

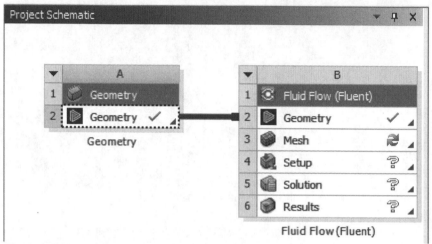

Figure 13.4a) Shared geometry with Fluent

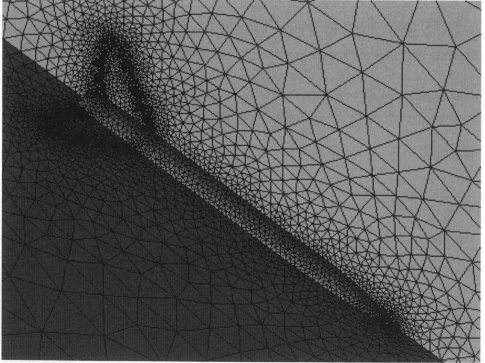

Figure 13.4b) Refined mesh around model rocket

5. Right click on Mesh in the Outline, select Insert>> Sizing. Select Face 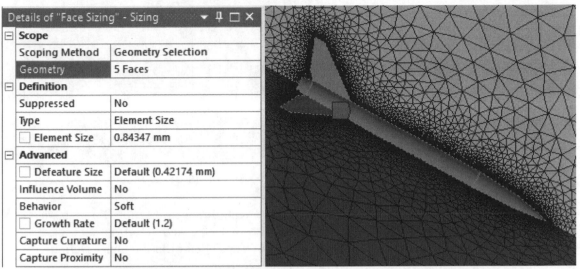 and control select the 5 faces of the model rocket as shown in figure 13.5a). Apply the faces as the Geometry under Scope in Details of "Face Sizing". Select the Home tab in the menu and Tools>>Units>>Metric (mm, kg, …). Set the Element Size to 0.84347 mm. Update the mesh once again.

Details of "Face Sizing" - Sizing	
Scope	
Scoping Method	Geometry Selection
Geometry	5 Faces
Definition	
Suppressed	No
Type	Element Size
☐ Element Size	0.84347 mm
Advanced	
☐ Defeature Size	Default (0.42174 mm)
Influence Volume	No
Behavior	Soft
☐ Growth Rate	Default (1.2)
Capture Curvature	No
Capture Proximity	No

Figure 13.5a) Face sizing for the five faces of the model rocket

Figure 13.5b) Further refinement of mesh around model rocket

6. Click on the plus sign next to Inflation in Details of "Mesh". Set Use Automatic Inflation to Program Controlled. Update the mesh once again. Select Box Select from the menu of the graphics window, see Figure 13.6a). Select Face and make a Box Select around the model rocket, see Figure 13.6a). This will select all surfaces, see Figure 13.6b).

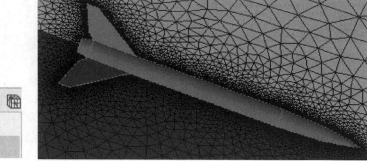

Figure 13.6a) Box Select   Figure 13.6b) Selection of surfaces

Right click in the graphics window and select Create Named Selection. Enter the name *model-rocket* and click OK. Select *model-rocket* under Named Selections and select *Include* from the drop-down menu for Program Controlled Inflation under Definition in Details of "model rocket". Update the mesh once again.

Details of "model rocket"	
Scope	
Scoping Method	Geometry Selection
Geometry	18 Faces
Definition	
Send to Solver	Yes
Protected	Program Controlled
Visible	Yes
Program Controlled Inflation	Include
Statistics	
Type	Manual
Total Selection	18 Faces
Surface Area	5864.7 mm²
Suppressed	0
Used by Mesh Worksheet	No

Figure 13.6c) Details for model rocket named selection

Right click in the graphics window and select View>>Front. Select Single Select, see Figure 13.6a), and select the square velocity-inlet face for the computational domain, see Figure 13.6d). Right click in the graphics window and select Create Named Selection. Enter the name *velocity-inlet* and click OK.

Right click in the graphics window and select View>>Back. Select Face and select the square pressure-outlet face. Right click in the graphics window and select Create Named Selection. Enter the name *pressure-outlet* and click OK.

Right click in the graphics window and select View>>Left. Select Face ▣ and select the rectangular symmetry-left face. Right click in the graphics window and select Create Named Selection. Enter the name *symmetry-left* and click OK.

Right click in the graphics window and select View>>Bottom. Select Face ▣ and select the rectangular symmetry-bottom face. Right click in the graphics window and select Create Named Selection. Enter the name *symmetry-bottom* and click OK.

Right click in the graphics window and select View>>Top. Select Face ▣ and select the rectangular symmetry-top face. Right click in the graphics window and select Create Named Selection. Enter the name *symmetry-top* and click OK.

Right click in the graphics window and select View>>Right. Select Face ▣ and select the rectangular symmetry-right face. Right click in the graphics window and select Create Named Selection. Enter the name *symmetry-right* and click OK.

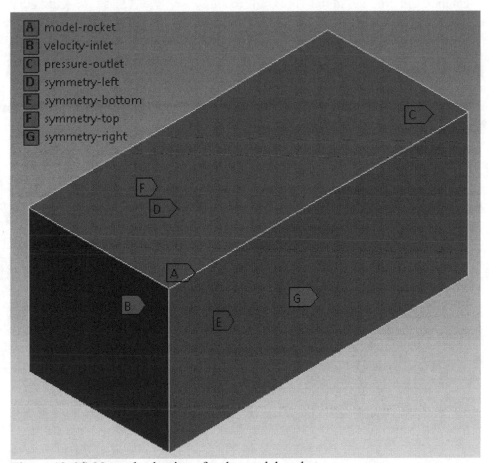

Figure 13.6d) Named selections for the model rocket

7. Go back to DesignModeler by double clicking on the Geometry under A in the Project Schematic in Ansys Workbench. Select Bottom view and select YZ Plane under Tree Outline in DesignModeler. Right click on YZ Plane and select Look At. Create a new sketch ⬚. Use your mouse wheel to zoom in on the model rocket.

Click on the Sketching tab in the Tree Outline and select Rectangle. Draw a rectangle around the model rocket. Make sure you start drawing the rectangle on the vertical axis and that you get a *C* when you move the cursor on the vertical axis. This means that one of the sides will be positioned on the same vertical axis. Select Dimensions and click on the horizontal and vertical sides of the new rectangle. Enter 0.6375 m as the vertical height of the rectangle and 0.1275 m as the horizontal width. Select the Vertical dimensioning tool, click on the horizontal coordinate axis and the lower side of the rectangle and enter 0.275 m as the vertical dimension.

Select the Modeling tab in the Tree Outline and select ⬚ Extrude. Select Sketch1 under YZ plane in the Tree Outline and Apply it as Geometry in Details of Extrude. Enter 0.1275 m as the FD1, Depth (>0) of the Extrusion. Click on Generate. Go to 2 Parts, 2 Bodies in the Tree Outline, right click on Solid and select Rename. Rename the Solid to *ModelRocketBox*.

Go back to Ansys Workbench, double click on Mesh and answer *Yes* to the question whether you would like to read upstream data.

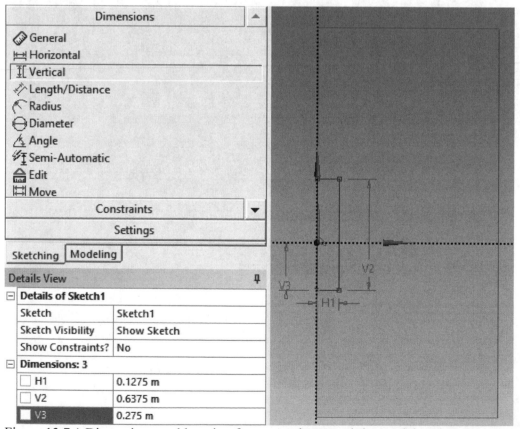

Figure 13.7a) Dimensions and location for rectangle around the model rocket

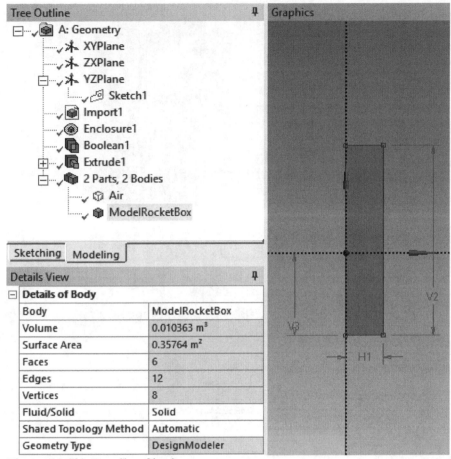

Figure 13.7b) Details of body

8. Right click Mesh under Project in Outline and select Update. Right click Mesh under Project in Outline, select Insert>>Sizing. Select Body [icon] and select the larger computational domain in the graphics window. Apply the computational domain in Geometry under Scope for Details of Body Sizing. Select *Body of Influence* as Type under Definition.

Select Bottom view in the graphics window, zoom in on the model rocket, select the *ModelRocketBox* in the graphics window and Apply it as the Body of Influence. You will need to select the second blue small rectangle in the lower left corner of the graphics window as shown in Figure 13.8a). *ModelRocketBox* turns from green to red when you apply it as the *Body of Influence*. Update the Mesh.

Right click on *ModelRocketBox* under Geometry in Project Outline and select Hide Body. Select Mesh in the Outline and click on the plus sign next to Quality in Details of "Mesh". Select Skewness as Mesh Metric. The max skewness should not be above 0.9. In this study the max value is 0.89998, see Figure 13.8b).

Select File>>Save Project. Select File>>Export…>>Mesh>>FLUENT Input File>>Export and save the mesh with the name "model-rocket-flow-mesh.msh". Close the Meshing window.

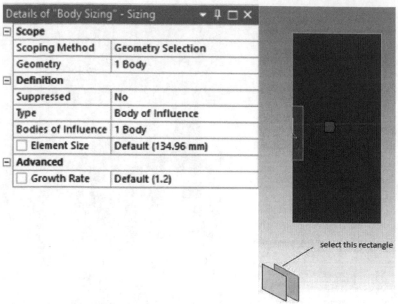

Figure 13.8a) Selection of ModelRocketBox as Body of Influence

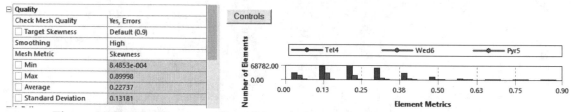

Figure 13.8b) Details on quality of the mesh

F. Launching Ansys Fluent

9. Double-click on Setup under Project Schematic in Ansys Workbench. Check the box for Double Precision. Click on Show More Options and take a note of the location of the *Working Directory*. Set the number of Solver Processes equal to the number of processor cores for your computer. Ansys Student is limited to 4 Solver Processes. To check the number of physical cores, press the Ctrl + Shift + Esc keys simultaneously to open the Task Manager. Go to the Performance tab and select CPU from the left column. You will see the number of physical cores on the bottom-right side. Close the Task Manager window. Click on the Start button in the Fluent Launcher window. Select OK when you get the Key Behavioral Changes window.

Click on Display… under Mesh and General on the Task Page. Select all Surfaces in the Mesh Display window. Check the boxes for Edges and Faces under Options. Select All as Edge Type and click on Display.

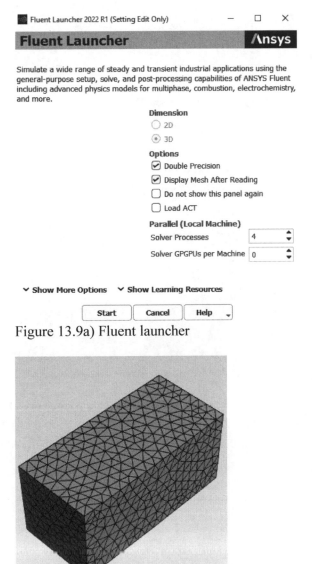

Figure 13.9a) Fluent launcher

Figure 13.9b) Displayed mesh with inflow and outflow

Double click on Models and Viscous (SST k-omega) under Setup in the Outline View. Set the k-omega Model to Standard and click OK to close the Viscous Model window.

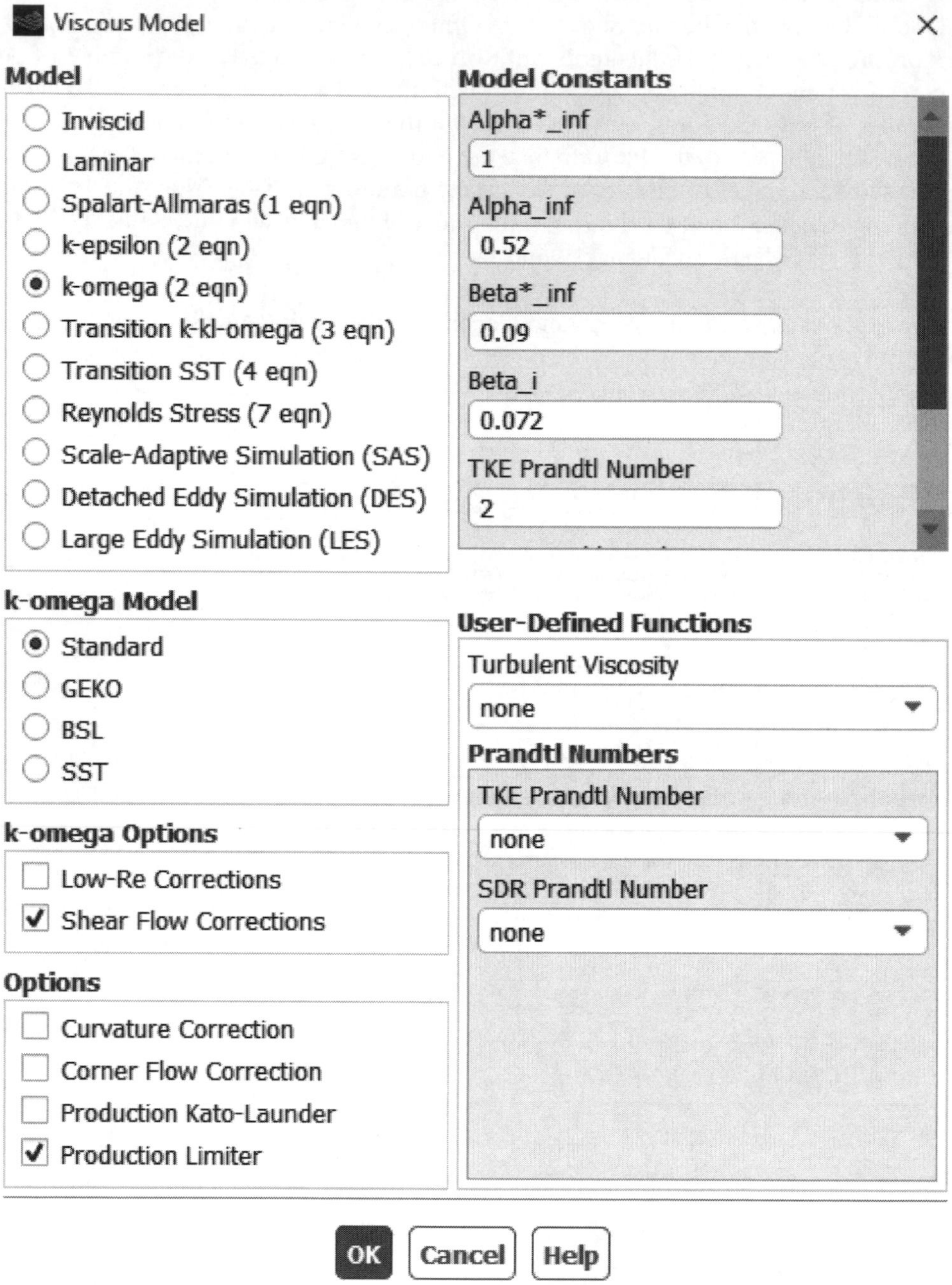

Figure 13.9c) Details for the viscous model

10. Double click on Boundary Conditions under Setup in the Tree and select velocity-inlet under Zone for Boundary Conditions on the Task Page. Click on the Edit button. Select Components as Velocity Specification Method. Enter -30 as Z-Velocity (m/s). Set the Turbulent Intensity to 1%. Select Apply and Close the Velocity Inlet window.

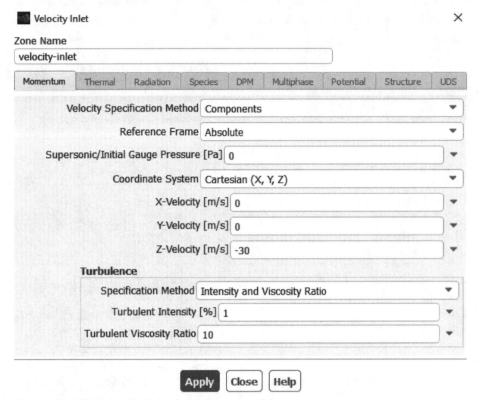

Figure 13.10 Velocity inlet boundary condition

11. Double click Projected Areas under Results and Reports in the Outline View. Select *model-rocket* as the Surface, enter 0.0001 as Min Feature Size [m], select Z as Projection Direction and click on Compute. The computed frontal Area [m²] is 0.00010233. Close the window.

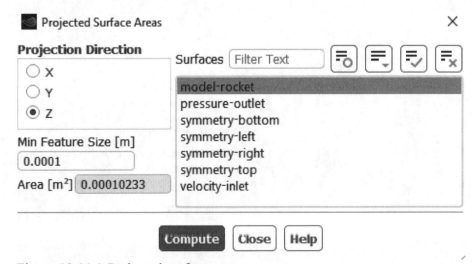

Figure 13.11a) Projected surface areas

Double click on Reference Values under Setup in the Outline View and change the Area [m²] to 0.00010233. Select Compute from *velocity-inlet* and set Reference Zone to *air*.

Double click on Methods under Solution in the Outline View and set the Pressure-Velocity Coupling scheme to Coupled. Set the Spatial Discretization Gradient to Least Squares Cell Based, and Pressure to Standard. Set the remaining three (Momentum, Turbulent Kinetic Energy and Specific Dissipation Rate) to First Order Upwind.

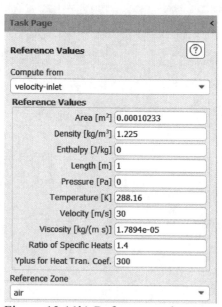

Figure 13.11b) Reference values

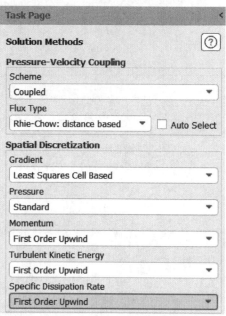

Figure 13.11c) Solution Methods

12. Double click on Controls under Solution in the Outline View and set the Pressure and Momentum Pseudo Transient Explicit Relaxation Factors to 0.25. Set the Turbulent Viscosity Explicit Relaxation Factor to 0.8. Click on Limits... and set the Maximum Turb. Viscosity Ratio to 1e+07. Click on the OK button to close the Solution Limits window.

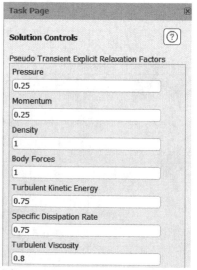

Figure 13.12a) Solution controls

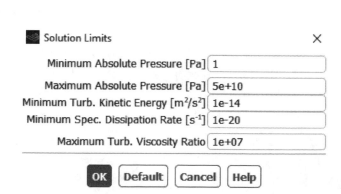

Figure 13.12b) Solution limits

13. Double click on Monitors and Residual under Solution. Check the box for Show Advanced Options. Set the Convergence Criterion to none. Click on OK to close Residual Monitors.

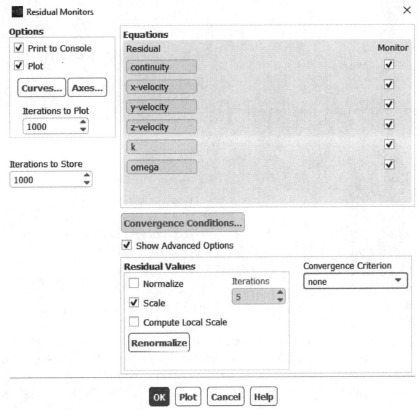

Figure 13.13 Settings for Residual Monitors

14. Double-click on Report Definitions under Solution in the Outline View. Select New>>Force Report>>Drag… from the drop-down menu. Select *model-rocket* as Zone. Check the boxes for Report File, Report Plot and Print to Console under Create. Set the Force Vector to X = 0, Y = 0, and Z = -1. Click on the OK button to close the Drag Report Definition window. Close the Report Definitions window.

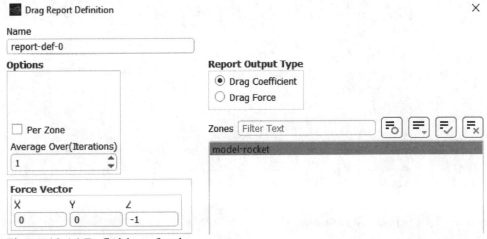

Figure 13.14 Definitions for drag report

311

15. Double click on Initialization under Solution in the Outline View and select *Hybrid Initializa*tion. Click on the Initialize button. Double click on Calculation Activities under Solution in the Outline View and click on Edit… for Autosave Every (Iterations). Set Save Data File Every (Iterations) to 100. Click OK to close the Autosave window.

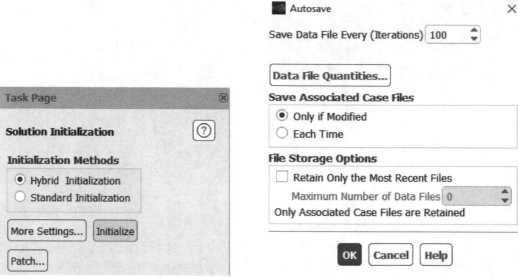

Figure 13.15a) Solution initialization   Figure 13.15b) Autosave window

Double click on Run Calculation under Solution in the Outline View. Set the Number of Iterations to 100. Click on Calculate. When calculations are complete, click on Copy Screenshot of Active Window to Clipboard in the graphics window, see Figure 13.15d). The scaled residuals and drag coefficient plots can in this way be pasted into a Word document.

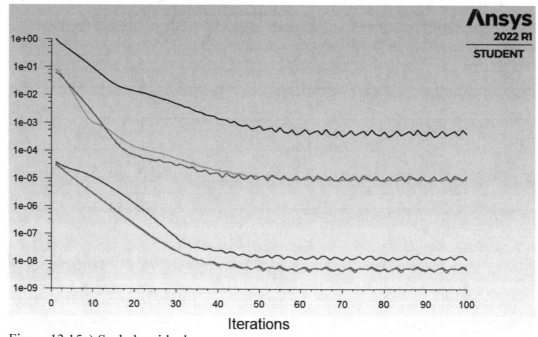

Figure 13.15c) Scaled residuals

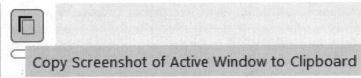

Figure 13.15d) Copying screenshot

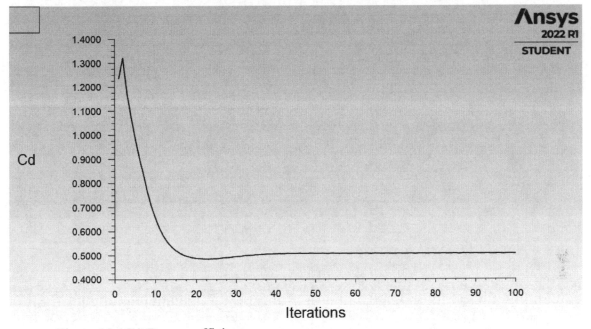

Figure 13.15e) Drag coefficient

G. Post-Processing

16. Double click on Methods under Solution in the Outline View and replace First Order Upwind with Second Order Upwind for Momentum, Turbulent Kinetic Energy and Specific Dissipation Rate.

   Double click on Controls under Solution and set the Turbulent Viscosity value to 0.95. Double click on Run Calculation under Solution and set Number of Iterations to 300. Click on Calculate and Click OK in the window that appears.

   Double-click on Reports under Results in the Outline View and double-click on Forces under Reports. Set Direction Vector for X to 0, Y to 0, and Z to -1. Check the *model-rocket* Wall Zone and click on the Print button. The total drag-coefficient is 0.43538. Close the Force Reports window.

Console						
	Forces [N]			Coefficients		
Zone	Pressure	Viscous	Total	Pressure	Viscous	Total
model-rocket	0.0038786808	0.020680884	0.024559565	0.068759462	0.36662117	0.43538063
Net	0.0038786808	0.020680884	0.024559565	0.068759462	0.36662117	0.43538063

Figure 13.16a) Drag-coefficient from console

When calculations are complete, click on Copy Screenshot of Active Window to Clipboard in the graphics window, see Figure 13.15d). The scaled residuals and drag coefficient plots can be pasted into a Word document.

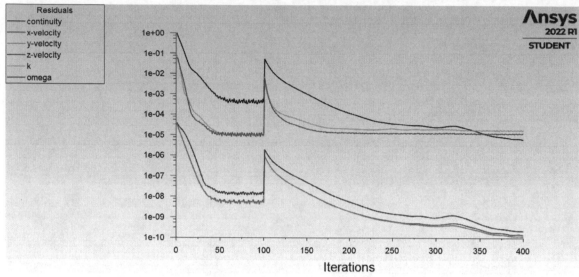

Figure 13.16b) Scaled residuals

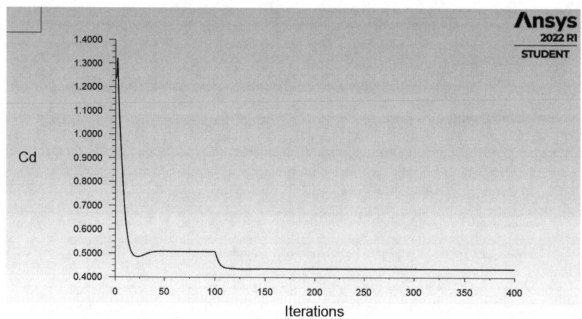

Figure 13.16c) Drag coefficient after 400 iterations

17. Double click on Graphics and Contours under Results. Select the settings as shown in
    Figure 13.17a) and Apply the Colormap settings as shown in Figure 13.17b). Click on
    Save/Display. Select the Y-Z plane view in the graphics window. Zoom in on the rocket.
    Close the Contours window.

Figure 13.17a) Contours settings

Figure 13.17b) Colormap settings

Figure 13.17c) Velocity magnitude for the flow around the model rocket at 30 m/s

Double click on Pathlines under Results and Graphics in the Outline View. Select the settings and Apply the same Colormap Options and settings as shown in Figures 13.17b) and 13.17d). Click on Save/Display. Close the Pathlines window.

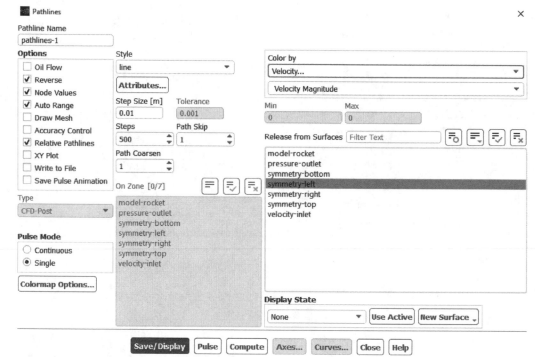

Figure 13.17d) Settings for pathlines

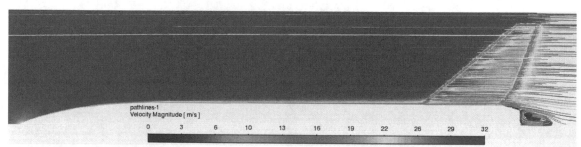

Figure 13.17e) Pathlines around model rocket fin at 30 m/s

Double click on Contours under Results and Graphics in the Outline View. Select the settings as shown in Figure 13.17f) for static pressure. Apply the same Colormap settings as shown in Figures 13.17b). Click on Save/Display. Close the Contours window.

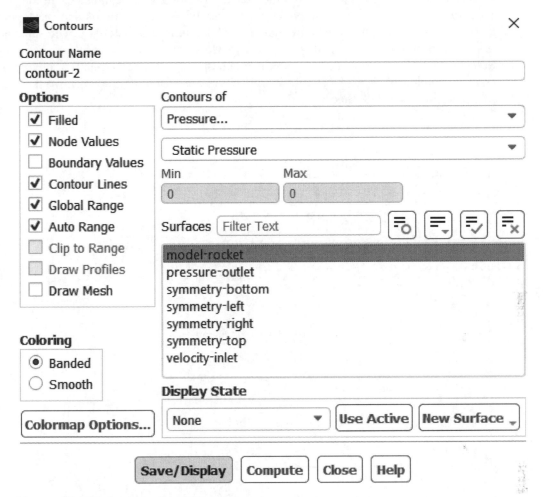

Figure 13.17f) Settings for static pressure

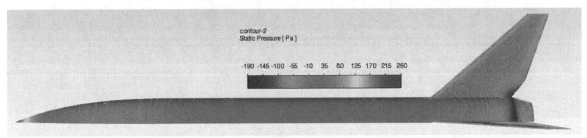

Figure 13.17g) Static pressure for the flow past the model rocket at 30 m/s

H. Batch Jobs

18. You will need to create the case and scheme files to complete a batch job. Select File>>Write>>Case & Data… from the menu. Save the case and data files in the *Working Directory* folder with the name ***model_rocket.cas.h5***. Open Notebook and create and save the scheme file with the name ***model_rocket.scm*** for the batch job. The scheme file is shown in Figure 13.18a) and is available for download from *sdcpublications.com*. Select File>>Read>>Scheme… from the menu in Ansys Fluent. Open the scheme file to start the batch job. Table 13.1 and Figure 13.18b) are showing the results from the batch job in comparison with theory.

```
(Do ((x 70 (- x 5))) ((< x 5))
(begin
  (Ti-menu-load-string (format #f "file read-case model-rocket.cas.h5"))
  (Ti-menu-load-string (format #f "define boundary-conditions velocity-inlet velocity-inlet n y y n 0 y n 0 n 0 n -~a n n y 1 10" x))
  (Ti-menu-load-string (format #f "report reference-values velocity ~a" x))
  (Ti-menu-load-string (format #f "solve set discretization-scheme mom 0"))
  (Ti-menu-load-string (format #f "solve set discretization-scheme k 0"))
  (Ti-menu-load-string (format #f "solve set discretization-scheme epsilon 0"))
  (Ti-menu-load-string (format #f "solve set under-relaxation turb-viscosity 0.8"))
  (Ti-menu-load-string (format #f "solve initialize hyb-initialization yes"))
  (Ti-menu-load-string (format #f "solve iterate 100"))
  (Ti-menu-load-string (format #f "solve set discretization-scheme mom 1"))
  (Ti-menu-load-string (format #f "solve set discretization-scheme k 1"))
  (Ti-menu-load-string (format #f "solve set discretization-scheme epsilon 1"))
  (Ti-menu-load-string (format #f "solve set under-relaxation turb-viscosity 0.95"))
  (Ti-menu-load-string (format #f "solve iterate 300"))
  (Ti-menu-load-string (format #f "report forces wall-forces y 0 0 -1 y model_rocket_drag_coefficient_data_U=~amps no ok" x))
  (Ti-menu-load-string (format #f "display objects display contour-1"))
  (Ti-menu-load-string (format #f "display views restore-view right"))
  (Ti-menu-load-string (format #f "display save-picture model_rocket_static_pressure_contour_plot_U=~amps.jpg yes" x))
  (Ti-menu-load-string (format #f "display objects display contour-2"))
  (Ti-menu-load-string (format #f "display views restore-view right"))
  (Ti-menu-load-string (format #f "display save-picture model_rocket_velocity_magnitude_contour_plot_U=~amps.jpg yes" x))
  (Ti-menu-load-string (format #f "display objects display pathlines-1"))
  (Ti-menu-load-string (format #f "display views restore-view right"))
  (Ti-menu-load-string (format #f "display save-picture model_rocket_path_lines_U=~amps.jpg yes" x))
  (Ti-menu-load-string (format #f "plot residuals y y y y y y"))
  (Ti-menu-load-string (format #f "display save-picture model_rocket_residuals_plot_U=~amps.jpg yes" x))
  (Ti-menu-load-string (format #f "plot file report-def-0-rfile.out"))
  (Ti-menu-load-string (format #f "display save-picture model_rocket_drag_coefficient_plot_U=~amps.jpg yes" x))
))
```

Figure 13.18a) Scheme file for model rocket batch job

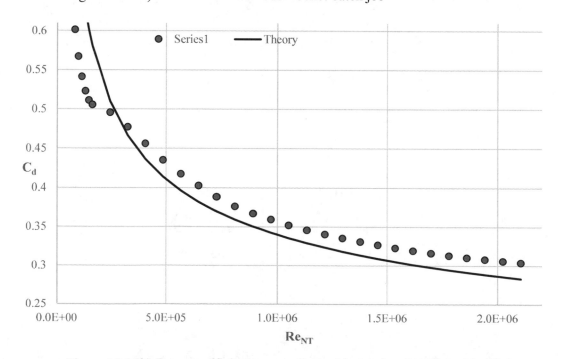

Figure 13.18b) Drag coefficient versus Reynolds number for Ahmed body

I. Theory

19. The total drag-coefficient for a model rocket[1,2] can be expressed using equation (13.1).

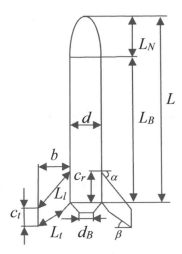

Figure 13.18 Geometry for model rocket

$$C_{D,T} = C_{D,NT} + C_{D,B} + C_{D,F} + C_{D,I} + C_{D,L} \tag{13.1}$$

, where

$$C_{D,NT} = 1.02 \frac{C_{f,turb} A_{w,NT}}{A_c} \left(1 + \frac{3}{2}\left(\frac{L}{d}\right)^{-3/2}\right) \tag{13.2}$$

$$C_{D,B} = \frac{0.029}{\sqrt{C_{D,NT}}} \left(\frac{d_b}{d}\right)^3 \tag{13.3}$$

$$C_{D,F} = \frac{2C_{f,lam} A_{w,F}}{A_c} \left(1 + 2\frac{t}{c_r}\right) \tag{13.4}$$

$$C_{D,I} = \frac{C_{f,lam} n d c_r}{A_c} \left(1 + 2\frac{t}{c_r}\right) \tag{13.5}$$

The different drag coefficients and parameters are explained in Table 13.1. We define the Reynolds number for the nose cone and body tube as

$$Re_{NT} = \frac{UL\rho}{\mu} = \frac{30*0.2365*1.225}{0.000017894} = 485{,}714 \tag{13.6}$$

, and the Reynolds number for the fin as

$$Re_F = \frac{U c_{ave} \rho}{\mu} = \frac{30*0.0271525*1.225}{0.000017894} = 55{,}765 \tag{13.7}$$

, where $c_{ave} = S/b$. The surface area for the fin can be determined from the geometry in Figure 13.19.

$$S = (c_t + L_l \sin\alpha)L_l \cos\alpha - \frac{1}{2}\{L_l^2 \sin\alpha \cos\alpha + L_t^2 \sin\beta \cos\beta + (L_l \cos\alpha - L_t \cos\beta)[(c_t - c_r + L_l \sin\alpha - L_t \sin\beta) + 2L_t \sin\beta]\} \tag{13.8}$$

319

The skin friction coefficient can be approximated assuming fully turbulent flow on nose cone and body tube

$$C_{f,turb} = \frac{0.0315}{Re_{NT}^{1/7}} = \frac{0.0315}{485,714^{1/7}} = 0.00485255 \tag{13.9}$$

, and for a laminar boundary layer on the fin as

$$C_{f,lam} = \frac{1.328}{Re_F^{1/2}} = \frac{1.328}{55,765^{1/2}} = 0.00562365 \tag{13.10}$$

For an ogive nose cone we have the following relation

$$\frac{A_{w,NT}}{A_c} = \frac{8L_N + 12L_B}{3d} = 39.2012 \tag{13.11}$$

The different drag coefficient and the total value becomes

$$C_{D,NT} = 1.02 * 0.00485255 * 39.2012 \left(1 + \frac{3}{2}(10.8986)^{-\frac{3}{2}}\right) = 0.2021 \tag{13.12}$$

$$C_{D,B} = \frac{0.029}{\sqrt{0.202}} \left(\frac{0.0164}{0.0217}\right)^3 = 0.0278 \tag{13.13}$$

$$C_{D,F} = \frac{2*0.00562*0.00419}{0.00037} \left(1 + 2\frac{0.0007}{0.0038}\right) = 0.1321 \tag{13.14}$$

$$C_{D,I} = \frac{0.00562*0.038*4*0.0217}{0.00037} \left(1 + 2\frac{0.0007}{0.0038}\right) = 0.0520 \tag{13.15}$$

$$C_{D,T} = 0.2021 + 0.0278 + 0.1321 + 0.0520 = 0.414 \tag{13.16}$$

The value for the total drag coefficient 0.414 according to the theory can be compared with the value 0.4354 from Ansys Fluent simulations, a difference of 5.2 %.

$A_c\ (m^2)$: cross sectional area for body tube 0.000369836 m^2
$A_{w,F}\ (m^2)$: total wetted surface area for all fins 0.00418879 m^2
$A_{w,NT}(m^2)$: wetted surface area for nose and body tube 0.014498 m^2
α : fin leading edge sweep angle 50°
β : fin trailing edge angle 15°
$b\ (m)$: span for the fin 0.0385673 m
$c_{ave}\ (m)$: average fin chord length 0.0271525 m
$c_r\ (m)$: root fin chord length 0.038 m
$c_t\ (m)$: tip fin chord length 0.010 m
$C_{D,NT}$: nose and body tube drag coefficient 0.2021
$C_{D,B}$: base drag coefficient 0.0278
$C_{D,F}$: fin drag coefficient 0.1321
$C_{D,I}$: interference drag coefficient 0.0520
$C_{D,T}$: total drag coefficient 0.414
$C_{f,lam}$: skin friction coefficient for laminar boundary layer 0.00562365
$C_{f,turb}$: skin friction coefficient for turbulent boundary layer 0.00485255
$d\ (m)$: body tube diameter 0.0217 m

d_b (m): base diameter 0.0164 m
L (m): length of nose and body 0.2365 m
L_B (m): length of body tube 0.165 m
L_l (m): length of leading edge of fin 0.060 m
L_N (m): length of nose 0.0715 m
L_t (m): length of trailing edge of fin 0.033 m
n: number of fins 4
Re_F: Reynolds number for the fin 55,765
Re_{NT}: Reynolds number for nose cone and body tube 485,714
S (m^2): area for fin 0.0010472 m^2
t (m): thickness of fin 0.0007 m
$U(m/s)$: free stream velocity 30 m/s
$\rho(kg/m^3)$: density of air 1.225 kg/m^3
$\mu(kg/ms)$: dynamic viscosity of air 0.000017894kg/m^3

Table 13.1 Drag coefficients and parameters

J. References

1. DeMar, J.S.,"Model Rocket Drag Analysis Using a Computerized Wind Tunnel", NARAM-37, 1995.
2. Gregorek, G.M., "Aerodynamic Drag of Model Rockets.", Estes Industries, Penrose, CO, 1970.
3. Cannon, B.,"Model Rocket Simulation with Drag Analysis", BYU, 2004.
4. Milligan, T.V, "Determining the Drag Coefficient of a Model Rocket Using Accelerometer Based Payloads", NARAM-54, 2012.

K. Exercise

13.1 Rerun ANSYS Fluent simulations in this chapter at 10, 20, 40 and 50 m/s and compare the drag coefficient with values using the theory section. Include a plot of total drag coefficient versus Reynolds number Re$_{NT}$ corresponding to free stream velocities 10, 20, 30, 40 and 50 m/s and fill out Table 12.2 below.

U (m/s)	Re_{NT}	Re_F	$C_{D,NT}$	$C_{D,B}$	$C_{D,F}$	$C_{D,I}$	$C_{D,T}$	$C_{D,Fluent}$	% Diff.
10									
20									
30	485714	55765	0.2021	0.0278	0.1321	0.0520	0.414	0.437	5.6
40									
50									

Table 13.2 Drag coefficients at different Reynolds numbers

L. Notes

CHAPTER 14. AHMED BODY

A. Objectives

- Model the Flow Past an Ahmed Body
- Run Simulations for Turbulent Steady 3D Flow Past the Ahmed Body
- Determine the Drag Coefficient for the Ahmed Body
- Visualize the Flow Around Ahmed Body using Pathlines, Pressure and Velocity

B. Problem Description

Studying the flow around the Ahmed body is a standard flow test case for wind tunnel experiments and CFD codes[1-6]. The geometry in *mm* that is being used for the Ahmed body in this chapter is shown below.

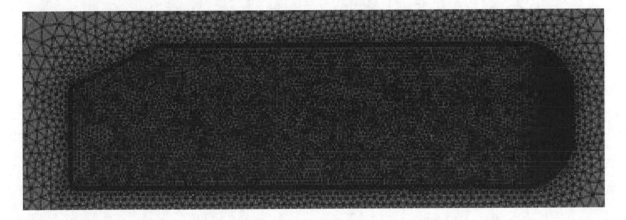

C. 3D Ahmed Body Design

1. Start by launching Ansys Workbench. Double click on Fluid Flow (Fluent) under Analysis Systems in the Toolbox. Right-click on Geometry in the Project Schematic window and select New DesignModeler Geometry. Select Units>>Millimeter from the menu as the desired length unit.

   Select Look At Face ![icon]. Select XY Plane in the Tree Outline and New Sketch ![icon]. Select the Sketching tab, click on Draw and select Rectangle. Draw a rectangle from the origin in the first quadrant. Select Dimensions and set the height of the rectangle to 288 mm and the length to 1044 mm. Right click in the graphics window and select Zoom to Fit.

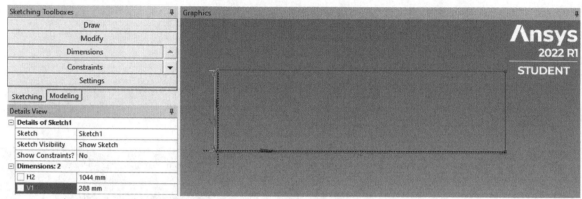

Figure 14.1 First step in the creation of the 3D Ahmed body

2. Select Modify and Fillet under Sketching Toolboxes. Set the Radius of the Fillet to 100 mm. Click on the two left hand side corners of the rectangle.

   Click on Chamfer under Modify. Set the Length of the Chamfer to 201.2 mm. Click on the upper right corner of the rectangle. The rear slant angle is 25 degrees. Select Dimensions and set the Angle to 25 degrees. You may need to right-click and select alternate angle when you define the angle. Select General under Dimensions and select the chamfer to set the length to 201.2 mm, see Figure 14.2a).

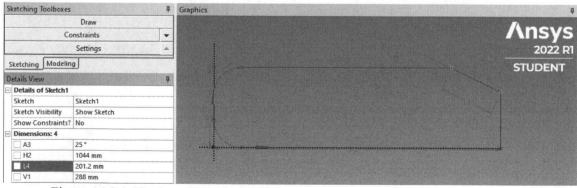

Figure 14.2a) 2D shape for Ahmed body

Click on ![Extrude icon] Extrude and select Sketch 1 in the Tree Outline under XYPlane. Apply the sketch in Details View next to Geometry. Set the Direction for the Extrude to Both -

Symmetric in Details View and set the depth of the extrusion to 194.5 mm. Click on Generate. Select Blend>>Fixed Radius from under the menu. Set the Radius to 100 mm. Control select the two vertical edges at the front of the Ahmed body and Apply the edges next to Geometry in Details View. Click on Generate.

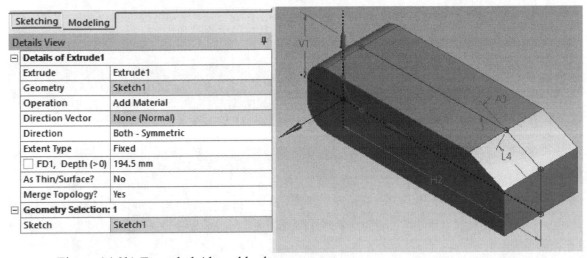

Figure 14.2b) Extruded Ahmed body

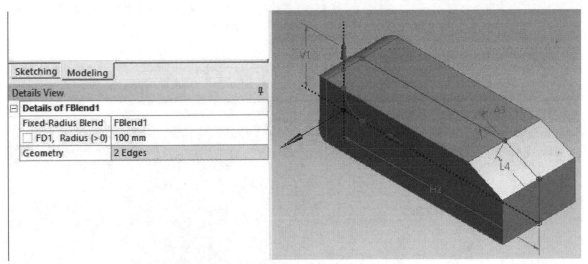

Figure 14.2c) Extruded and blended Ahmed body

3. Select ZXPlane in the Tree Outline. Right click in the graphics window and select View>>Bottom View. Select the Sketching tab in Tree Outline and select Circle. Draw a circle and use General under Dimensions to define a 30 mm diameter circle that you position 202 mm from the front of the Ahmed body and 163.5 mm from the center plane.

Create three more circles with the same diameter and position them according to Figure 14.3a).

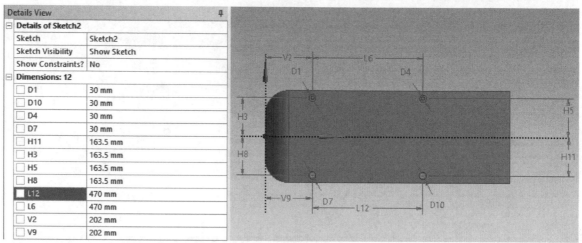

Figure 14.3a) Locations for the four cylindrical supports

Click on ⌐ Extrude and select Sketch 2 in the Tree Outline under ZXPlane. Apply the sketch in Details View next to Geometry. Set the Direction under Details View to Reversed and set the depth of the extrusion to 50 mm. Click on Generate.

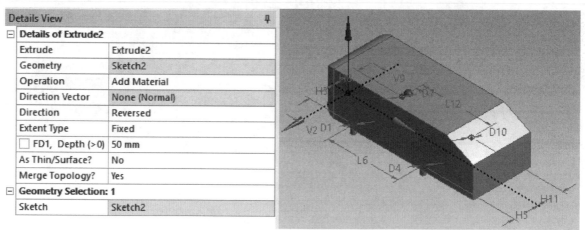

Figure 14.3b) Details view for extrusion and finished Ahmed model

Select Create>> New Plane from the menu. Select in Details View From Plane as Type, ZXPlane as the Base Plane and Offset Z as Transform 1 (RMB).
Set the Offset Z value to -50 mm. Click on Generate.

Select Tools>>Enclosure from the menu. Select Box as the Shape in Details View and set the Number of Planes to 2. Select XYPlane as Symmetry Plane 1, Plane4 as Symmetry Plane 2 and Apply them in Details View. Set the Model Type to Full Model and the Cushion to Non-Uniform. Set the different Cushion values FD1 – FD4 as listed in Figure 14.3c). Click on Generate.

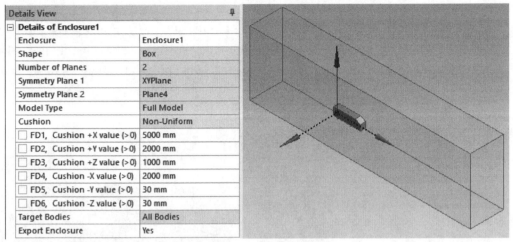

Figure 14.3c) Details for enclosure

4. Select Create>>Boolean from the menu. Select Subtract as Operation. Click on the plus sign next to 2 Parts, 2 Bodies in the Tree Outline. Select the second Solid and Apply it as Target Body in Details of Boolean1. Select the first Solid and Apply it as Tool Body in Details of Boolean1. Click on Generate. Right click on the Solid in the Tree Outline and Rename it to Air. Select File>>Save Project and name it "3D Ahmed Body Study" and close DesignModeler.

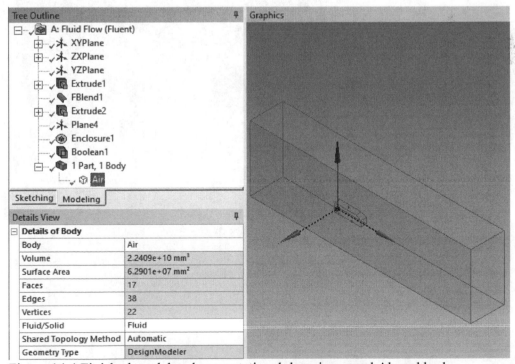

Figure 14.4 Finished model and computational domain around Ahmed body

327

D. 3D Ahmed Body Mesh

5. Double click on Mesh under Project Schematic in Ansys Workbench. In the Meshing window, right click on Mesh in the Outline and select Update. Select the Home tab in the menu. Select Layout>>Reset Layout under the Home tab.

    Click on the plus sign next to Sizing under Details of "Mesh". Change Capture Proximity and Capture Curvature both to Yes. Click on the plus sign next to Quality under Details of "Mesh". Set Smoothing under Quality in Details of Mesh to High. Update the mesh once again. Right click in the graphics window and select View>>Back.

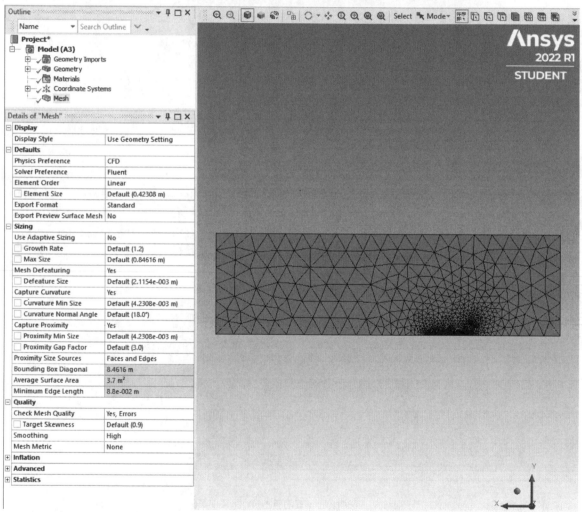

Figure 14.5 Refined mesh around Ahmed body

6. Right click on Mesh in the Outline, select Insert>> Sizing. Right click in the graphics window and select Cursor Mode>>Face. Control select all 11 faces of the Ahmed body as shown in figure 14.6a). You will need to rotate and zoom in on the model in order to select all faces. Apply the faces as the Geometry under Scope in Details of "Face Sizing". Set the Element Size to 0.0075 m. Update the mesh once again.

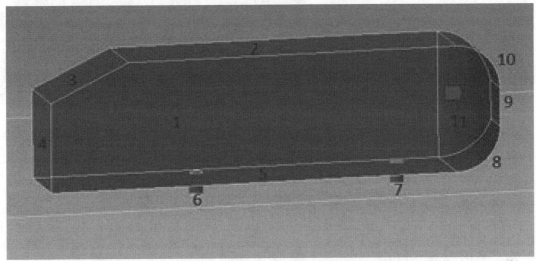

Figure 14.6a) Face sizing for the eleven faces of the Ahmed body

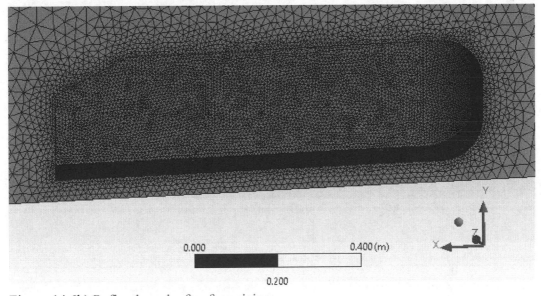

Figure 14.6b) Refined mesh after face sizing

7. Click on the plus sign next to Inflation in Details of Mesh. Set Use Automatic Inflation to Program Controlled. Right click in the graphics window and select Cursor Mode>>Vertex. Control select the 14 vertices as shown in Figure 14.7a).

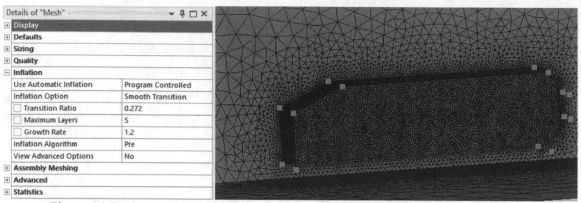

Figure 14.7a) Program controlled inflation and selected vertices for Ahmed body

Right click in the graphics window and select Create Named Selection. Enter the name "ahmed body" and click OK. Select *ahmed body* under Named Selections and select Include from the drop-down menu for Program Controlled Inflation under Definition in Details of "ahmed body", see Figure 14.7b).

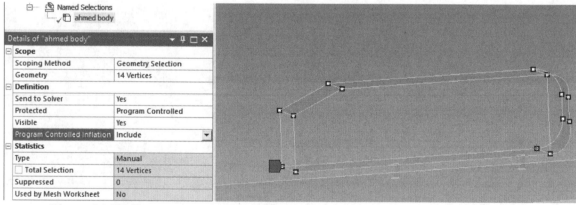

Figure 14.7b) Program controlled inflation set to include for Ahmed body

8. Right click in the graphics window and select View>>Left. Right click in the graphics window and select Cursor Mode>>Face. Select the face of the computational domain that is available from the Left View. Right click in the graphics window and select Create Named Selection. Enter the name "velocity-inlet" and select OK.

    Right click in the graphics window and select View>>Right. Select the face of the computational domain that is available from the Right View. Right click in the graphics window and select Create Named Selection. Enter the name "pressure-outlet" and select OK.

    Right click in the graphics window and select View>>Back. Select the face of the computational domain that is available from the Back View. Right click in the graphics window and select Create Named Selection. Enter the name "symmetry1" and select OK.

    Right click in the graphics window and select View>>Bottom. Select the face of the computational domain that is available from the Bottom View. Right click in the graphics window and select Create Named Selection. Enter the name "street" and click OK.

    Right click in the graphics window and select View>>Top. Select the face of the computational domain that is available from the Top View. Right click in the graphics window and select Create Named Selection. Enter the name "symmetry-top" and click OK.

    Right click in the graphics window and select View>>Front. Select the face of the computational domain that is available from the Front View. Right click in the graphics window and select Create Named Selection. Enter the name "symmetry-side" and click OK. Update the mesh.

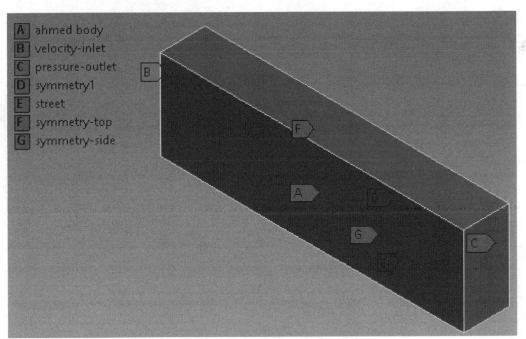

Figure 14.8 Named selections for the Ahmed body

9. Go back to DesignModeler by double clicking on Geometry under Project Schematic in Ansys Workbench. Right-click in the graphics window and select Back View. Select

XYPlane in the Tree Outline and create a new Sketch 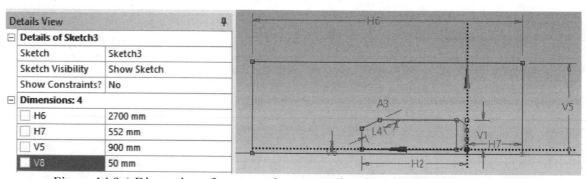.

Click on the Sketching tab in the Tree Outline and select Rectangle. Draw a rectangle around the Ahmed body from the bottom of the computational domain. Select Dimensions and click on the vertical and horizontal sides of the new rectangle. Enter 900 mm as the vertical height of the rectangle and 2700 mm as the width. Set the horizontal distance from the front of the Ahmed body to the right-hand side vertical edge of the rectangle to 552 mm. Set the vertical distance between the lower horizontal edge of the rectangle and the bottom of the Ahmed body to 50 mm.

Select the Modeling tab in the Tree Outline and select Extrude. Select the new Sketch3 under XYPlane in the Tree Outline and Apply it as Geometry in Details of Extrude3. Enter 450 mm as the FD1, Depth (>0) of the Extrusion. Click on Generate. Click on plus sign next to 2 Parts, 2 Bodies in the Tree Outline, right click on Solid and select Rename. Rename Solid to "AhmedBodyBox". Close the DesignModeler window.

Go back to Ansys Workbench, double click on Mesh and answer *Yes* to the question whether you would like to read upstream data.

Figure 14.9a) Dimensions for rectangle surrounding the Ahmed body

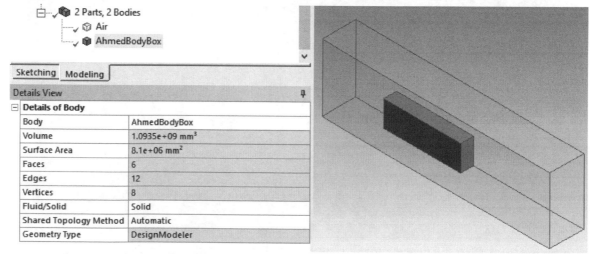

Figure 14.9b) Details of body

10. Right click Mesh under Project in Outline, select Insert>>Sizing. Right click in the graphics window and select Cursor Mode>>Body. Select the computational domain in the graphics window. Apply the computational domain in Geometry under Scope in Details of "Sizing". Select Body of Influence as Type under Definition.

Select Back view in the graphics window, zoom in on the Ahmed body, select the AhmedBodyBox and Apply it as the Body of Influence. You will need to select the second small rectangle in the lower left corner in Figure 14.10a) before you can select the AhmedRocketBox that turns from green to red when you apply it as the body of influence. Update the Mesh. If you click on the plus sign next to Statistics in Details of Mesh you will find that the number of Elements for the mesh is 482440 and the number of Nodes is 139169.

Right click on AhmedBodyBox under Geometry in Outline and select Hide Body. Select Mesh in the Outline and click on the plus sign next to Quality in Details of Mesh. Select Skewness as Mesh Metric. The max skewness should not be above 0.9 (max is 0.77402). Select File>>Save Project. Select File>>Export>>Mesh>>FLUENT Input File>>Export and save the mesh with the name "ahmed-body-flow-mesh.msh". Close the Meshing window.

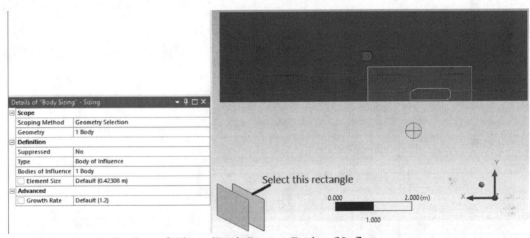

Figure 14.10a) Selection of AhmedBodyBox as Body of Influence

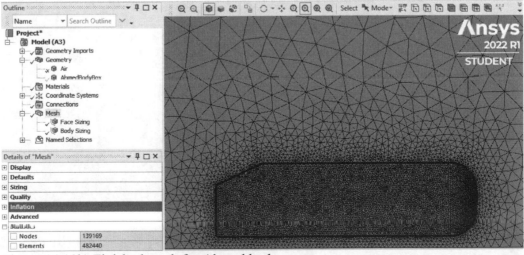

Figure 14.10b) Finished mesh for Ahmed body

E. Launching Ansys Fluent for 3D Ahmed Body

11. Double click on Setup in Ansys Workbench. Check Double Precision under Options and set the number of Processes equal to the number of cores for the computer. Select Show More Options. Select the Working Directory. Click Start to launch Fluent.

    Double click on Models under Setup in the Outline View and double click on Viscous – SST k-omega under Models on the Task Page. Select the k-epsilon (2 eqn) turbulence model. Select Realizable as the k-epsilon model and check Non-Equilibrium Wall Functions as Near-Wall Treatment. Click OK to exit the Viscous Model window.

    Double click on Boundary Conditions under Setup in the Outline View and select *velocity-inlet* under Zone for Boundary Conditions on the Task Page. Click on the Edit… button. Select Components as Velocity Specification Method. Enter 30 as X-Velocity (m/s). Click Apply and close the Velocity Inlet window.

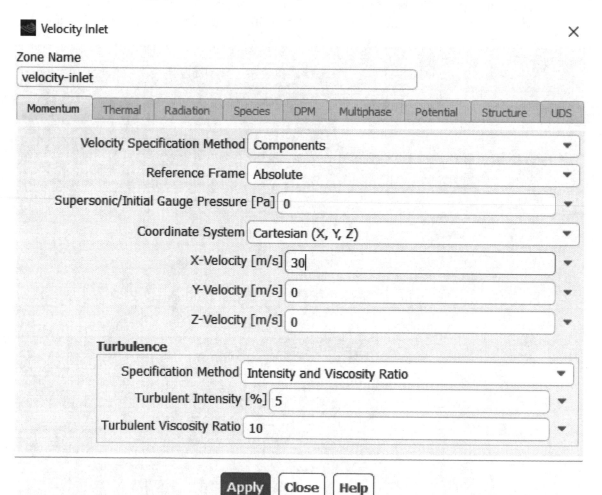

Figure 14.11 Velocity inlet specifications for X-Velocity

12. Double click Projected Areas under Results and Reports in the Outline View. Select *wall-air* as the Surface, enter 0.030 as Min Feature Size (m), select X as Projection Direction and click on Compute. The computed frontal area is 0.06175669 m2. Close the window.

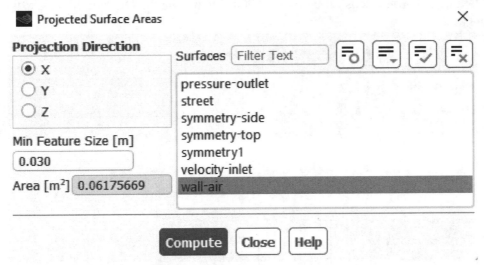

Figure 14.12a) Projected surface area

Double click on Reference Values under Setup in the Outline View and change the Area (m2) to 0.06175669. Select Compute from velocity-inlet and set Reference Zone to air.

Double click on Methods under Solution in the Outline View and set the Pressure-Velocity Coupling scheme to Coupled. Set the Spatial Discretization Gradient to Least Squares Cell Based, and Pressure to Standard. Set the remaining three for Momentum, Turbulent Kinetic Energy and Turbulent Dissipation Rate to First Order Upwind. Uncheck the box for Pseudo Transient.

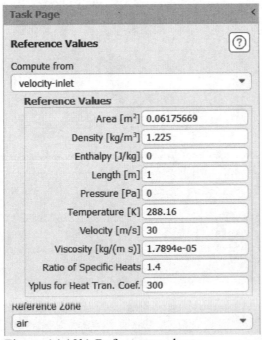

Figure 14.12b) Reference values   Figure 14.12c) Solution methods

13. Double click on Controls under Solution in the Outline View, set the Flow Courant Number to 50, set the Momentum and Pressure Explicit Relaxation Factors to 0.25 and set the Turbulent Viscosity Under-Relaxation Factor to 0.8. Click on Limits… and set the Maximum Turb. Viscosity Ratio to 10000000. Click on the OK button to close the window.

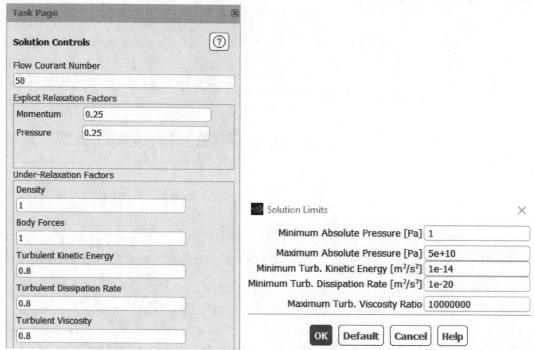

Figure 14.13a) Solution controls   Figure 14.13b) Solution limits

Double click on Monitors and Residual under Solution in the Outline View. Check the box for Show Advanced Options and set the Convergence Criterion to none. Click on OK to close Residual Monitors.

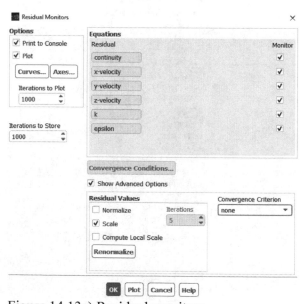

Figure 14.13c) Residual monitors

Double-click on Report Definitions on the left-hand side under Solution in the Tree. Select New>>Force Report>>Drag... from the drop-down menu. Select *wall-air* as Wall Zone. Check the boxes for Report File, Report Plot and Print to Console under Create. Set the Force Vector to X = 1, Y = 0, and Z = 0. Click on the OK button to close the window. Close the Report Definitions window.

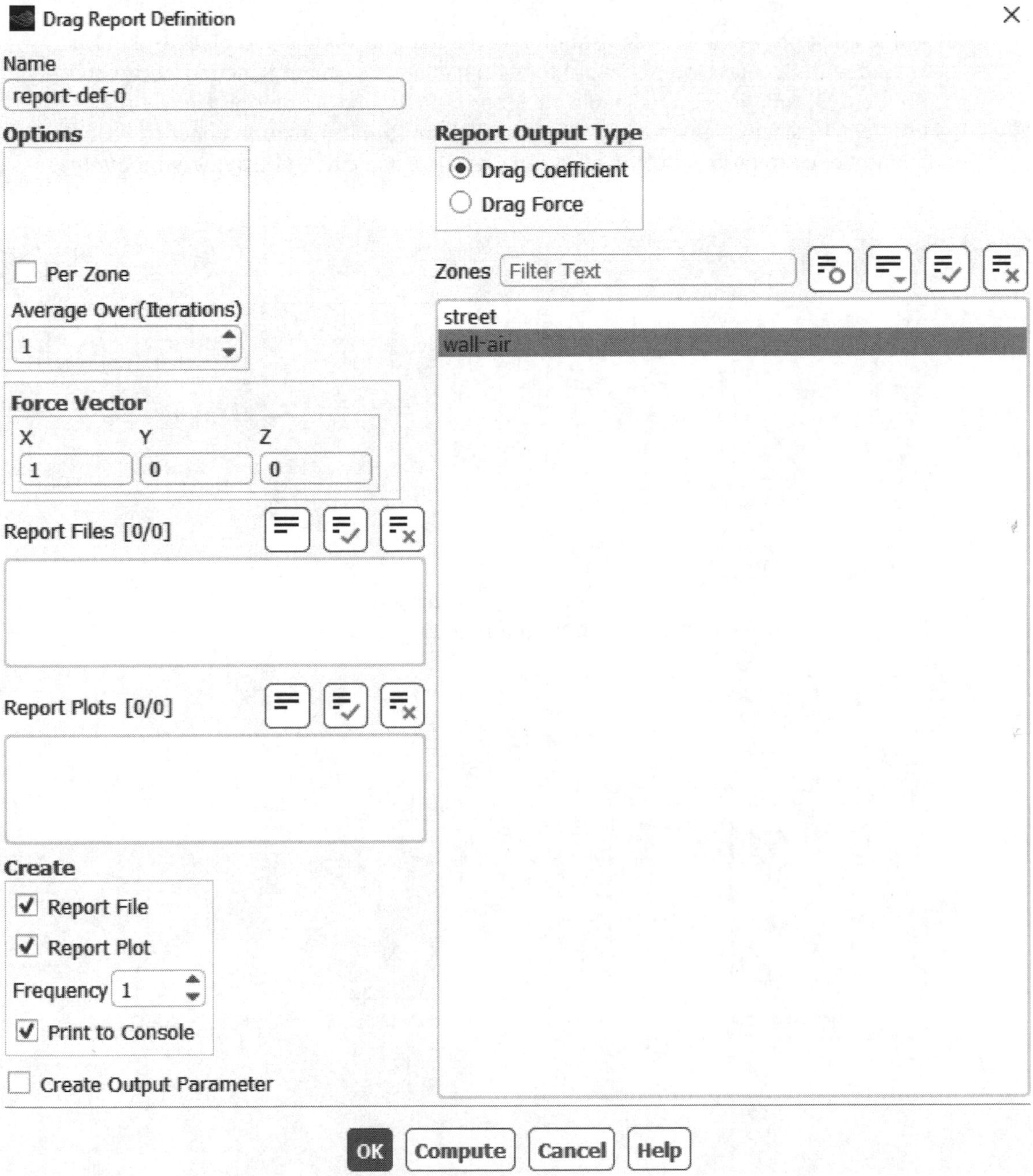

Figure 14.13d) Drag report definition

14. Double click on Initialization under Solution in the Outline View and select *Hybrid Initialization*. Click on the Initialize button. Double click on Calculation Activities under Solution in the Outline View and click on Edit… for Autosave Every (Iterations). Set Save Data File Every (Iterations) to 100. Click OK to close the Autosave window. Double click on Run Calculation under Solution in the Outline View. Set the Number of Iterations to 200. Click on Calculate.

Double click on Methods under Solution in the Outline View and replace First Order Upwind with Second Order Upwind for Momentum, Turbulent Kinetic Energy and Turbulent Dissipation Rate. Double click on Controls under Solution and set the Turbulent Viscosity value to 0.95. Double click on Run Calculation under Solution and set Number of Iterations to 500. Click on Calculate and Click OK in the window that appears.

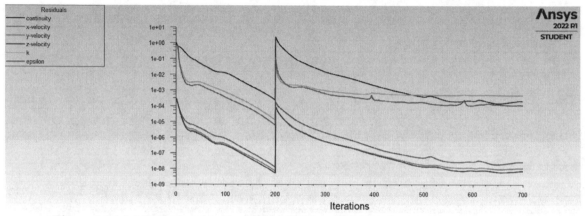

Figure 14.14a) Scaled residuals during the first 700 iterations

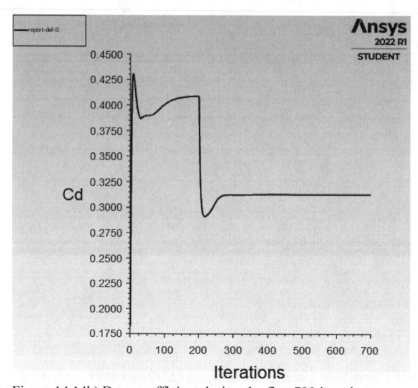

Figure 14.14b) Drag coefficient during the first 700 iterations

F. Post-Processing for 3D Ahmed Body

15. Double-click on Reports under Results in the Outline View and double-click on Forces under Reports. Set Direction Vector for X to 1, Y to 0, and Z to 0. Select the *wall-air* Wall Zone, deselect *street* Wall Zone and click on the Print button. The total drag coefficient is 0.3127421. Close the Force Reports window.

Double-click on Graphics and Contours under Results in the Outline View. Select Contours of Pressure... and Static Pressure. Select *symmetry1* as Surfaces and click on Save/Display.

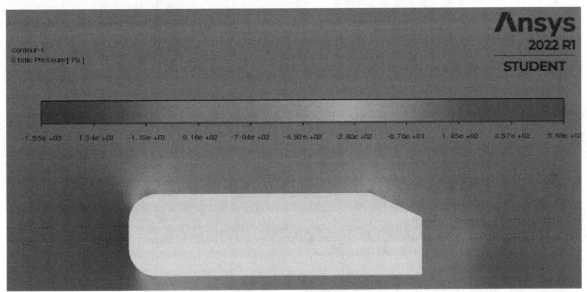

Figure 14.15a) Static pressure for Ahmed body at 30 m/s.

Double-click on Graphics and Contours under Results in the Outline View. Select Contours of Velocity... and Velocity Magnitude. Select *symmetry1* as the Surface and click on Save/Display.

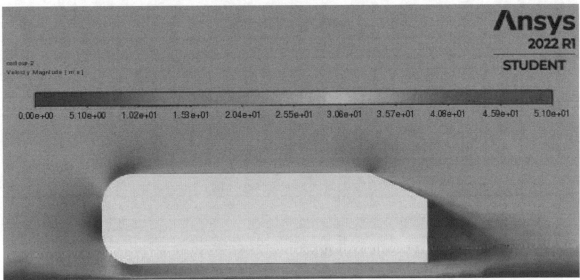

Figure 14.15b) Velocity magnitude around Ahmed body at 30 m/s.

Double-click on Graphics and Pathlines under Results in the Outline View. Select the settings as shown in Figure 14.15c). Select *symmetry1* as the Surface and click on Save/Display.

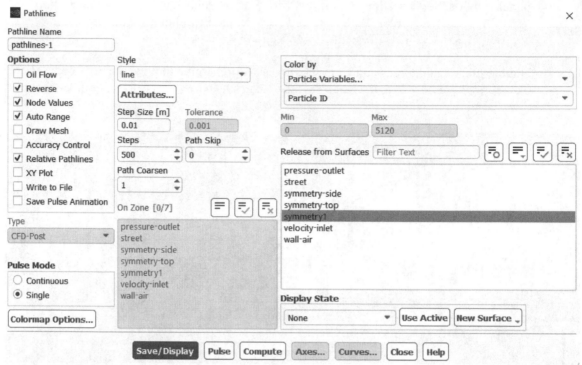

Figure 14.15c) Settings for pathlines

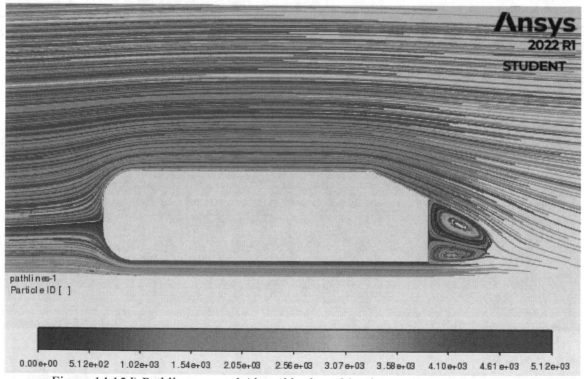

Figure 14.15d) Pathlines around Ahmed body at 30 m/s.

G. Batch Jobs

16. You will need to create the case and scheme files to complete a batch job. Select File>>Write>>Case & Data… from the menu. Save the case and data files in the *Working Directory* folder with the name ***ahmed_body.cas.h5***. Open Notebook and create and save the scheme file with the name ***ahmed_body.scm*** for the batch job. The scheme file is shown in Figure 14.16a) and is available for download from *sdcpublications.com*. Select File>>Read>>Scheme… from the menu in Ansys Fluent. Open the scheme file to start the batch job. Table 14.1 and Figure 14.16b) are showing the results from the batch job in comparison with Meile et al. (2011).

```
(Do ((x 70 (- x 5))) ((< x 5))
(begin
   (Ti-menu-load-string (format #f "file read-case ahmed_body.cas.h5"))
   (Ti-menu-load-string (format #f "define boundary-conditions velocity-inlet velocity-inlet n n y y n ~a n 0 n n y 5 10" x))
   (Ti-menu-load-string (format #f "report reference-values velocity ~a" x))
   (Ti-menu-load-string (format #f "solve set discretization-scheme mom 0"))
   (Ti-menu-load-string (format #f "solve set discretization-scheme k 0"))
   (Ti-menu-load-string (format #f "solve set discretization-scheme epsilon 0"))
   (Ti-menu-load-string (format #f "solve set under-relaxation turb-viscosity 0.8"))
   (Ti-menu-load-string (format #f "solve initialize hyb-initialization yes"))
   (Ti-menu-load-string (format #f "solve iterate 200"))
   (Ti-menu-load-string (format #f "solve set discretization-scheme mom 1"))
   (Ti-menu-load-string (format #f "solve set discretization-scheme k 1"))
   (Ti-menu-load-string (format #f "solve set discretization-scheme epsilon 1"))
   (Ti-menu-load-string (format #f "solve set under-relaxation turb-viscosity 0.95"))
   (Ti-menu-load-string (format #f "solve iterate 500"))
   (Ti-menu-load-string (format #f "report forces wall-forces n 1 () 1 0 0 y ahmed_body_drag_coefficient_data_U=~amps no ok" x))
   (Ti-menu-load-string (format #f "display objects display contour-1"))
   (Ti-menu-load-string (format #f "display views restore-view front"))
   (Ti-menu-load-string (format #f "display save-picture ahmed_body_static_pressure_contour_plot_U=~amps.jpg yes" x))
   (Ti-menu-load-string (format #f "display objects display contour-2"))
   (Ti-menu-load-string (format #f "display views restore-view front"))
   (Ti-menu-load-string (format #f "display save-picture ahmed_body_velocity_magnitude_contour_plot_U=~amps.jpg yes" x))
   (Ti-menu-load-string (format #f "display objects display pathlines-1"))
   (Ti-menu-load-string (format #f "display views restore-view front"))
   (Ti-menu-load-string (format #f "display save-picture ahmed_body_path_lines_U=~amps.jpg yes" x))
   (Ti-menu-load-string (format #f "plot residuals y y y y y"))
   (Ti-menu-load-string (format #f "display save-picture ahmed_body_residuals_plot_U=~amps.jpg yes" x))
   (Ti-menu-load-string (format #f "plot file report-def-0-rfile.out"))
   (Ti-menu-load-string (format #f "display save-picture ahmed_body_drag_coefficient_plot_U=~amps.jpg yes" x))
))
```

Figure 14.16a) Scheme file for batch job

U (m/s)	Re	Cd, Fluent	Cd, Meile *et al.*	Percent difference
2	142,942	0.44510	0.36330	22.5
3	214,413	0.39679	0.36000	10.2
4	285,884	0.370385	0.35684	3.8
5	357,354	0.354278	0.35380	1.1
6	428,825	0.341247	0.35088	2.7
8	571,767	0.33100	0.34536	4.2
10	714,709	0.326152	0.34027	4.2
15	1,072,063	0.320129	0.32919	2.8
20	1,429,418	0.316644	0.32010	1.1
25	1,786,772	0.314312	0.31265	0.5
30	2,144,127	0.311152	0.30655	2.0
35	2,501,481	0.311152	0.30155	3.2
40	2,858,835	0.309861	0.297444	4.2
45	3,216,190	0.309340	0.294082	5.2
50	3,573,544	0.308944	0.291326	6.0
55	3,930,899	0.308347	0.289067	6.7
60	4,288,253	0.307925	0.287216	7.2
65	4,645,607	0.308202	0.285698	7.9
70	5,002,962	0.307899	0.28445	8.2

Table 14.1 Comparison between ANSYS Fluent and established data

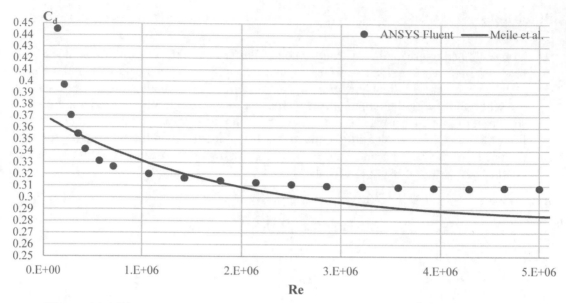

Figure 14.16b) Drag coefficient versus Reynolds number for Ahmed body

H. Theory

We start by defining the Reynolds number for the 3D case outlined in this chapter

$$Re = \frac{UL\rho}{\mu} \tag{14.1}$$

where U (m/s) is inlet velocity, L (m) is the length of the Ahmed body, ρ (kg/m³) is the density of the fluid and μ (m²/s) is dynamic viscosity of the fluid.
Next, we define the drag coefficient C_d as

$$C_d = \frac{F_d}{\frac{1}{2}\rho U^2 A} \tag{14.2}$$

where F_d(N) is the drag force and A (m²) is the frontal area of the Ahmed body. Meile *et al.*[6] determined the following relationship between the drag coefficient and the Reynolds number for the Ahmed body.

$$C_d = 0.2788 + 0.0915e^{-Re/1797100} \tag{14.3}$$

I. References

1. Ahmed, S.R., Ramm, G.,"Some Salient Features of the Time-Averaged Ground Vehicle Wake", *SAE-Paper 840300*, 1984.
2. Dogan, T., Conger, M., Kim, D-H., Mousaviraad, M., Xing, T., Stern, F.,"Simulation of Turbulent Flow over the Ahmed Body", *ME: 5160 Intermediate Mechanics of Fluids CFD Lab 4*, (2016).
3. Kalyan, D.K., Paul, A.R., "Computational Study of Flow around a Simplified 2D Ahmed Body", *International Journal of Engineering Science and Innovative Technology (IJESIT)*, **2**, 3, (2013).

4.   Khan, R.S, Umale, S.,"CFD Aerodynamic Analysis of Ahmed Body", *International Journal of Engineering Trends and Technology (IJETT)*, **18**, 7, (2014).

5.   Lienhart, H., Stoots, C. and Becker, S.,"Flow and Turbulence Structures in the Wake of a Simplified Car Model (Ahmed Model), *Notes on Numerical Fluid Mechanics*, **77**, 6, (2002).

6.   Meile, W., Brenn, G., Reppenhagen, A., Lechner, B., Fuchs, A. "Experiments and numerical simulations on the aerodynamics of the Ahmed body", *CFD Letters*, **3**, 1, (2011).

J. Exercise

14.1   Model the 3D flow past the Ahmed body with the dimensions as shown and units in mm from Figure 14.17. Follow the same steps as shown in this chapter. Find the drag curve corresponding to Figure 14.16 and compare the two rear slant angles 25 and 35 degrees in the same graph.

14.2   Model the 3D flow past the Ahmed body with the dimensions as shown and units in mm from Figure 14.18. Follow the same steps as shown in this chapter. Find the drag curve corresponding to Figure 14.16.

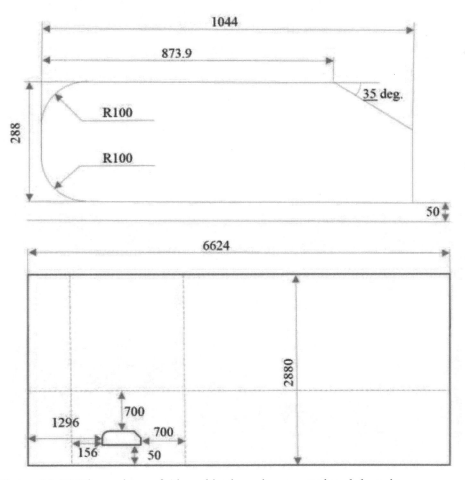

Figure 14.17 Dimensions of Ahmed body and computational domain

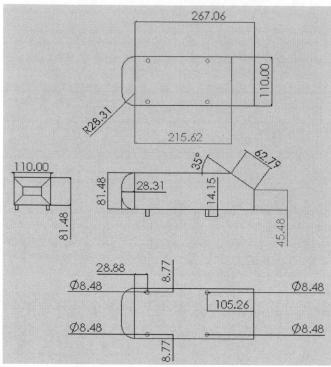

Figure 14.18 Dimensions of Ahmed body

CHAPTER 15. HOURGLASS

A. Objectives

- Using ANSYS Workbench to Create Geometry and Mesh for the Hourglass
- Inserting Boundary Conditions
- Running the Calculations
- Using Contour Plots for Volume Fractions

B. Problem Description

We will study the flow in the hourglass and will analyze the problem using ANSYS Fluent.

C. Launching ANSYS Workbench and Selecting Fluent

1. Start by launching ANSYS Workbench. Double click on Fluid Flow (Fluent) under Analysis Systems in the Toolbox.

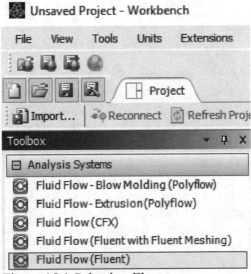

Figure 15.1 Selecting Fluent

D. Launching ANSYS DesignModeler

2. Right click Geometry under Project Schematic in ANSYS Workbench and select Properties. In Properties of Schematic A2: Geometry, select Analysis Type 2D under Advanced Geometry Options. Right click on Geometry in the Project Schematic window and select New DesignModeler Geometry.

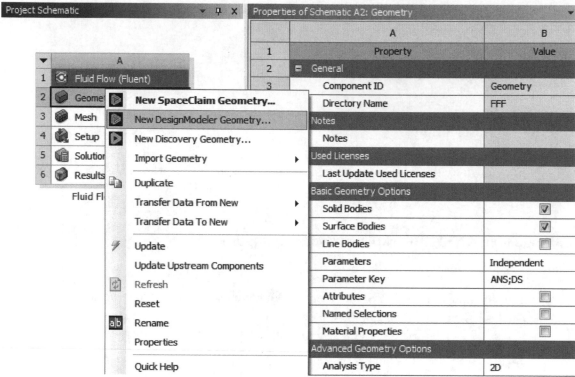

Figure 15.2 Selecting properties and 2D analysis type

3. Select Units>>Millimeter from the menu in DesignModeler. Select the XY Plane in the Tree Outline and click on Look at Face/Plane/Sketch ![icon]. Select the Sketching tab and select Polyline under Draw from Sketching Toolboxes. Start at the origin (make sure that letter *P* is shown when the cursor is close to the origin and that letters *V*, *H* and *C* are showing when you are drawing vertical, horizontal and coinciding lines) and draw the sketch with the dimensions as shown in Figure 15.3a). Right click and select *Closed End* after drawing the last line. Select General under Dimensions and select the eight lines (H1, V2-V5 and L6-L9) as shown in Figure 15.3a). Select Angle under Dimensions and create angles A10 and A11 by selecting the lines next to the angle. You may need to right click and select *Alternate Angle* when creating the angle.

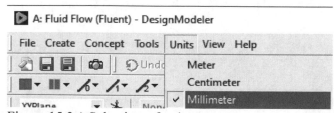

Figure 15.3a) Selection of units

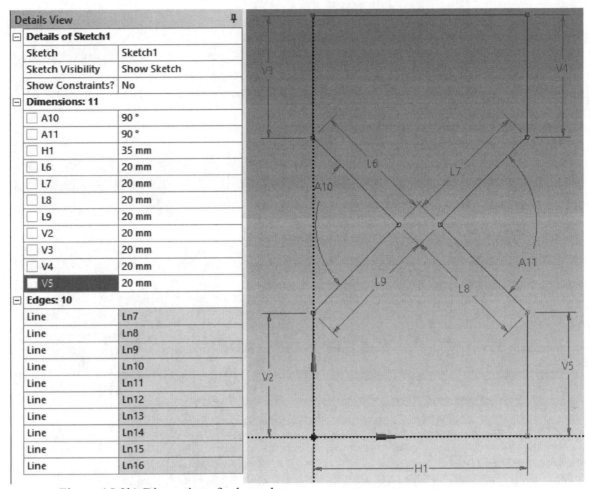

Figure 15.3b) Dimensions for hourglass

Select Concept>>Surfaces from Sketches from the menu. Select Sketch1 under XYPlane in Tree Outline and Apply the sketch as the Base Object in Details View. Click on Generate and close DesignModeler.

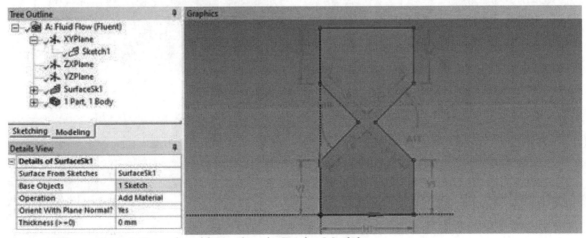

Figure 15.3c) Completed hourglass in DesignModeler

E. Launching ANSYS Meshing

4. Double click on Mesh under Project Schematic in ANSYS Workbench. Set the Element Size under Defaults in Details of Mesh to 0.5 mm. Right click on Mesh under Project in Outline and select Generate Mesh. Click on the plus sign next to Statistics in Details of "Mesh". The number of Nodes is 8317 and the number of Elements is 8098.

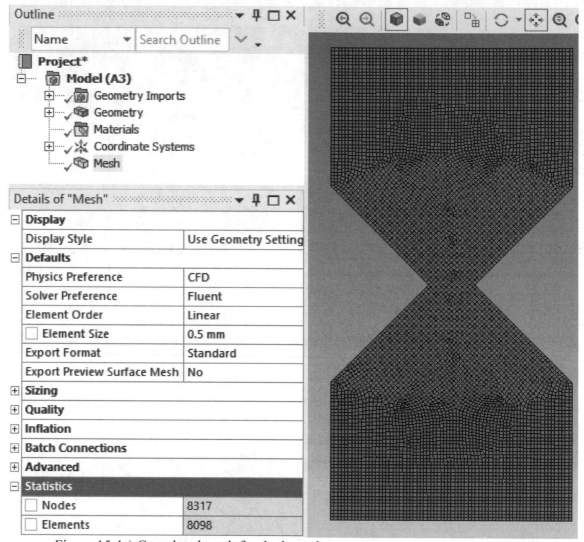

Figure 15.4a) Completed mesh for the hourglass

Right click in the graphics window and select Cursor Mode>> Edge and select the upper horizontal edge for the mesh. The edge turns green. Right click on the edge and select Create Named Selection. Enter *outlet* as the name and click OK to close the Selection Name window. Control select the remaining nine edges of the mesh, right click and select Create Named Selection. Name these edges *wall*.

Select File>>Export>>Mesh>>FLUENT Input File>>Export from the menu and save the mesh with the name *hourglass-mesh.msh*. Select File>>Save Project from the menu and save the project with the name *Hour-Glass.wbpj*. Close the Meshing window and right-click on Mesh under Project Schematic in ANSYS Workbench and select Update.

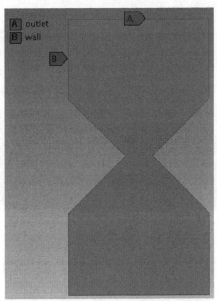

Figure 15.4b) Named selections for the hourglass

F. Launching ANSYS Fluent

5. Double-click on Setup under Project Schematic in ANSYS Workbench. Check the box for Double Precision under Options in the ANSYS Fluent Launcher. Click on Show More Options and take a note of the location of the *Working Directory*. Set the number of processes equal to the number of processor cores for your computer. To check the number of physical cores, press the Ctrl + Shift + Esc keys simultaneously to open the Task Manager. Go to the Performance tab and select CPU from the left column. You will see the number of physical cores on the bottom-right side. Close the Task Manager window. Click on the Start button in the Fluent Launcher window.

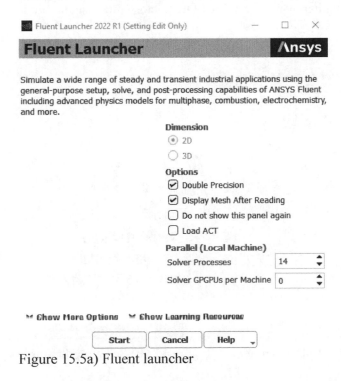

Figure 15.5a) Fluent launcher

Click on Display… under Mesh and General on the Task Page. Select all Surfaces in the Mesh Display window. Check the boxes for Edges and Faces under Options. Select All as Edge Type, click on Display and Close the Mesh Display window. Select Transient under Time on the Task Page. Check the box for Gravity and enter -9.81 as the value for Y [m/s^2].

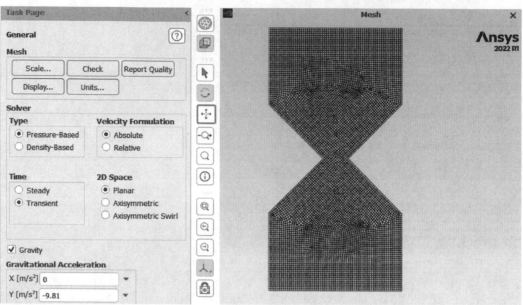

Figure 15.5b) General settings in ANSYS Fluent

6. Double click on Models and Multiphase under Setup in the Outline View. Select Eulerian Inhomogeneous Model. Set the Number of Eulerian Phases to 2 and Implicit Formulation for Volume Fraction Parameters. Click Apply and Close close the Multiphase Model window. Double click on Models and Viscous (SST, k-omega, Mixture) under Setup in the Outline View. Select Laminar Model and click OK to close the Viscous Model window.

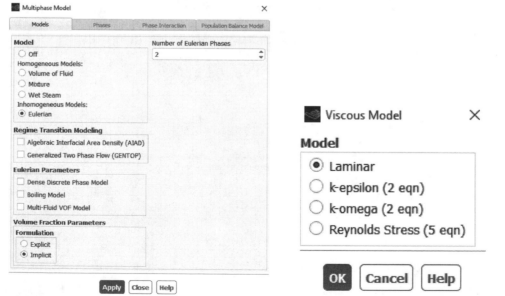

Figure 15.6a) Settings for the Eulerian model   Figure 15.6b) Selecting laminar model

7. Open Materials and Fluid under Setup in the Outline View and select *air*. Right click on *air* and select copy. Enter the name *wood* and set the Density [kg/m³] to 850. Click on Change/Create and answer Yes to the question that appears. Close the Create/Edit Materials window.

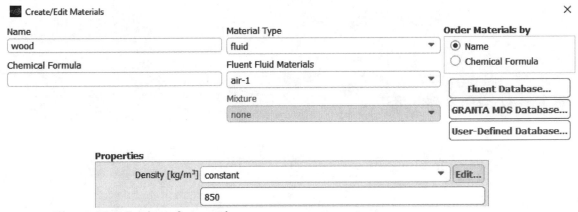

Figure 15.7 Settings for wood

8. Double click on Multiphase under Models in the Setup and select the Phases tab at the top of the Multiphase Model window. Select phase- 1 – Primary Phase under Phases. Select *air* as Phase Material and enter *air* as the Name for the Primary Phase. Click on Apply.

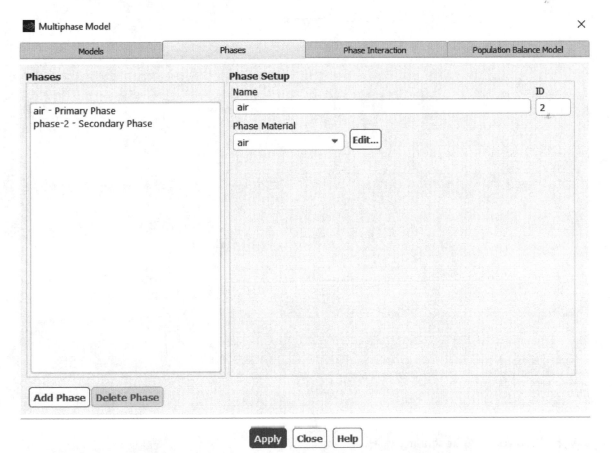

Figure 15.8a) Primary phase

Select phase- 2 – Secondary Phase under Phases. Select *wood* as Phase Material and enter *wood* as the Name for the Secondary Phase. Check the box for Granular and enter 0.0002 as Diameter [m]. Set the Granular Properties as shown in Table 15.1 and Figure 15.8b). Click on Apply and Close the window.

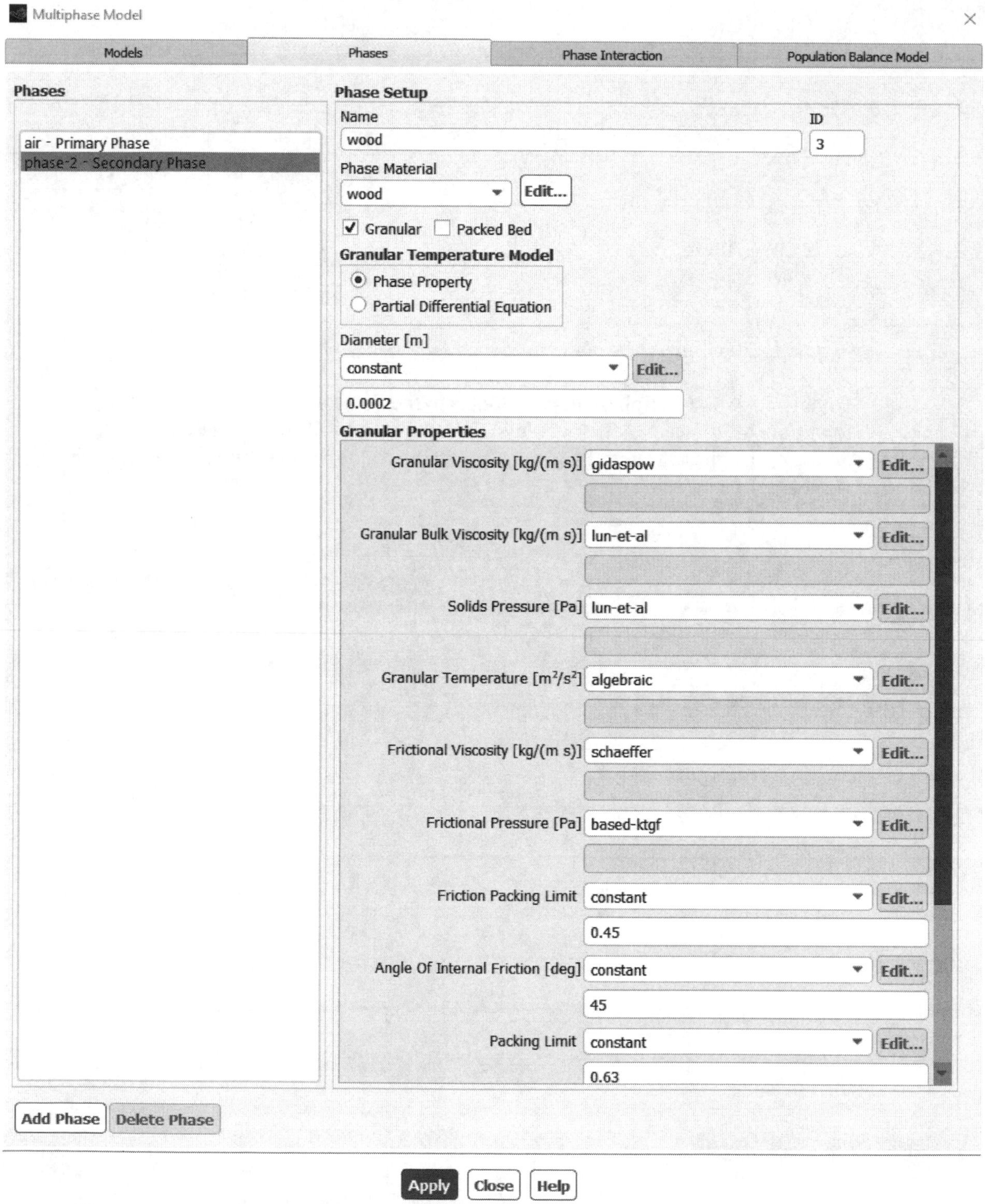

Figure 15.8b) Secondary phase settings

Diameter [m]	0.0002
Granular Viscosity [kg/m s]	gidaspow
Granular Bulk Viscosity [kg/m s]	lun-et-al
Frictional Viscosity [kg/m s]	schaeffer
Angle Of Internal Friction [deg]	45
Frictional Pressure [Pa]	based-ktgf
Frictional Modulus [Pa]	derived
Friction Packing Limit	0.45
Granular Temperature [m²/s²]	algebraic
Solids Pressure [Pa]	lun-et-al
Radial Distribution	lun-et-al
Elasticity Modulus [Pa]	derived
Packing Limit	0.63

Table 15.1 Settings for secondary phase

9. Select the Phase Interaction tab at the top of the Multiphase Model window. Select *air wood* under Phase Pairs and set Drag Coefficient, Lift Coefficient, Surface Tension Coefficient and Virtual Mass Coefficient to *none*. Select Apply and *wood wood* under Phase Pairs and set the Restitution Coefficient to 0.001. Select Apply and Close the Multiphase Model window.

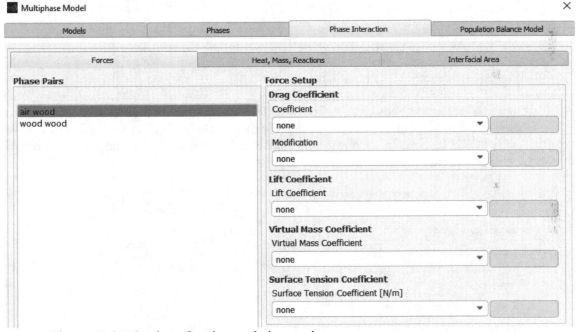

Figure 15.9a) Settings for air-wood phase pair

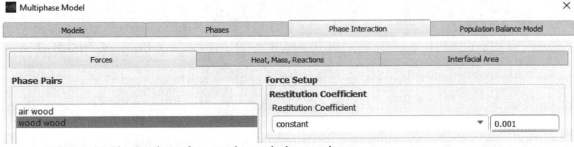

Figure 15.9b) Settings for wood-wood phase pair

353

10. Double-click on Boundary Conditions under Setup and click on Operating Conditions under Boundary Conditions on the Task Page. Select user-input as Operating Density Method. Set Reference Pressure Location X [m] to 0.0175 and Y [m] to 0.06828427. Click OK to close the Operating Conditions window.

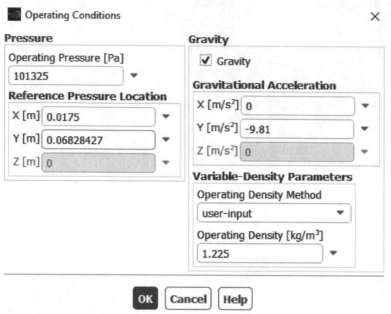

Figure 15.10 Settings for operating conditions

11. Double-click on Methods under Solution and set the Spatial Discretization for Momentum to Second Order Upwind. Set the Spatial Discretization for Pressure to Second Order. Double-click on Controls under Solution and set the Under-Relaxation Factor for Momentum to 0.3.

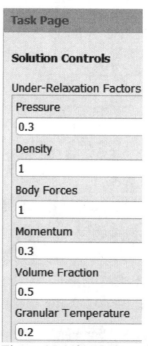

Figure 15.11a) Solution methods   Figure 15.11b) Under-relaxation factors

Double click on Residual under Monitors and Solution in the Outline View. Set Absolute Criteria for continuity, u-air and v-air to 1e-05. Set the other Absolute Criteria to 0.001 and click OK to close the Residual Monitors window.

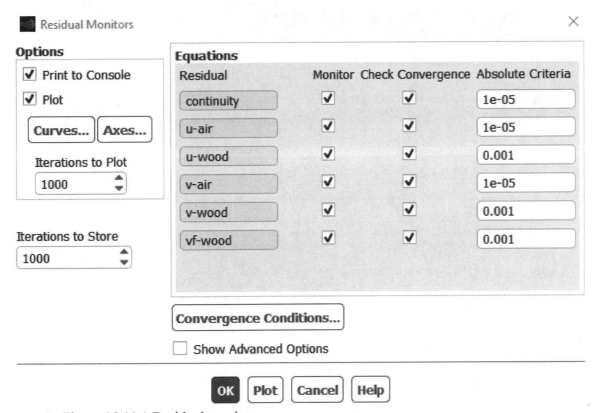

Figure 15.11c) Residual monitors

12. Double-click on Initialization under Solution in the Outline View and set Initializations Method to Hybrid Initialization on the Task Page. Click on Initialize, see Figure 15.12a).

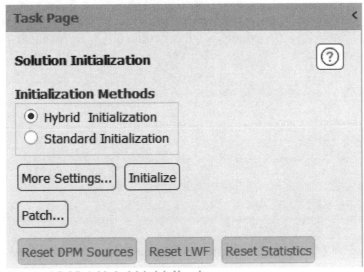

Figure 15.12a) Hybrid initialization

Select Adapt>>Automatic... from the Domain tab in the menu, see Figure 15.12b). Select Cell Registers>>New>>Field Variable... in the Automatic Mesh Adaption window, see Figure 15.12c).

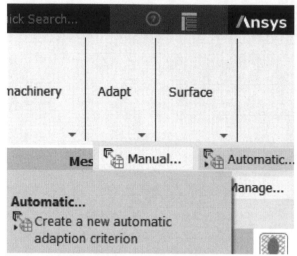

Figure 15.12b) Automatic adaption

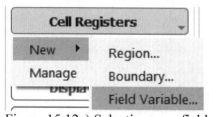

Figure 15.12c) Selecting new field variable

In the Field Variable Register, select Compute.

In the Field Variable Register, select Gradient as Derivative Option, select *wood* as Phase and select Gradient of Phases... and Volume Fraction. Select Cells in Range as Type, set the Gradient-Min to 0.001 and Gradient-Max to 0.01, see Figure 15.12d). Click on Save and Close the Field Variable Register window. Select gradient_0 as Refinement Criterion and set the Frequency (time-step) to 1 in the Automatic Mesh Adaption window, see Figure 15.12e).

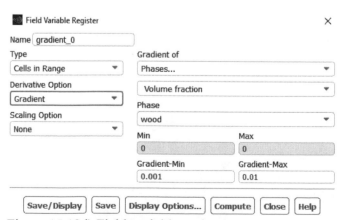

Figure 15.12d) Field variable register

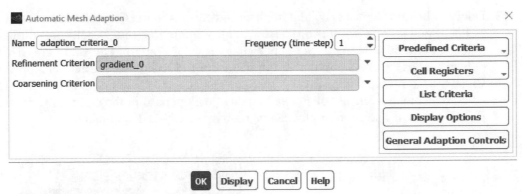

Figure 15.12e) Automatic mesh adaption

Select Cell Registers>>New>>Region… in the Adaption Controls window. Set the X Max [m] value to 0.035 and Y Min [m] value to 0.034 and Y Max [m] value to 0.04. Click on Save and Close the Region Register window. Click OK to close the Automatic Mesh Adaption window. Click OK in the Information window.

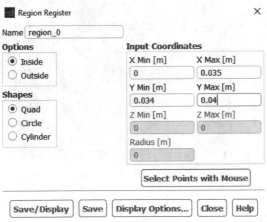

Figure 15.12f) Region register

Click on Patch… under Solution Initialization on the Task Page. Select wood as the Phase, select Volume Fraction as Variable, region_0 as Registers to Patch and Value to 0.63. Select Patch and Close the Patch window.

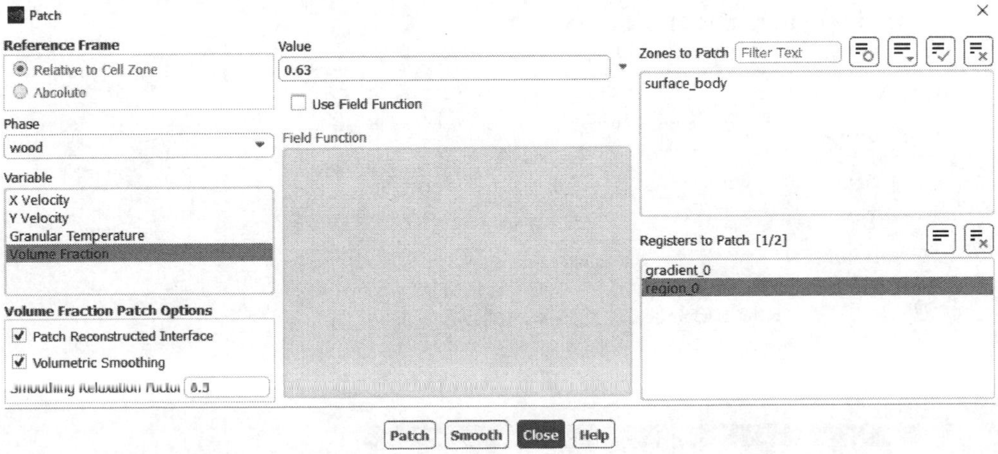

Figure 15.12g) Patch window

357

13. Double-click on Contours under Graphics and Results in the Outline View. Select wood as Phase and select Contours of Phases… and Volume fraction. Deselect all Surfaces. Select Colormap Options…. Set Font Behavior under Font to Fixed and Font Size to 12. Select float as Type under Number Format and set Precision to 2. Click on Apply and Close the Colormap window. Click on Save/Display and Close the Contours window. Click on Copy Screenshot of Active Window to Clipboard in the graphics window, see Figure 15.13c). The contour plot can be pasted into a Word document.

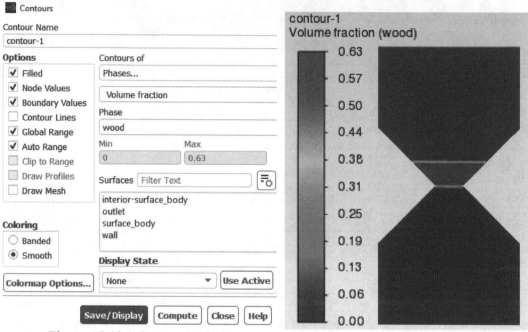

Figure 15.13a) Contours window   Figure 15.13b) Volume fraction (wood)

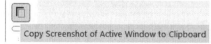

Figure 15.13c) Copying screenshot

14. Double-click on Solution Animations under Solution and Calculation Activities in the Outline View. Select Record after every 1 time-step. Select HSF File as Storage Type. Select contour-1 as Animation Object. Enter *hour-glass* as Name and click OK to close the Animation Definition window.

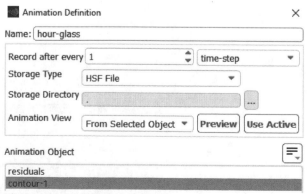

Figure 15.14 Contours window

15. Open Cell Registers under Solution in the Outline View and double click on gradient_0. Select Compute in the Field Variable Register window. Copy the computed Min and Max values and enter these as Gradient-Min and Gradient-Max, see Figure 15.15a). Select Save/Display in the Field Variable Register window to see the cells in the graphics window that have a gradient in this region. Close the Field Variable Register window.

	Field Variable Register	✕

Name: gradient_0

Type
Cells in Range ▼

Derivative Option
Gradient ▼

Scaling Option
None ▼

Gradient of
Phases... ▼

Volume fraction ▼

Phase
wood ▼

Min
4.930381e-32

Max
0.3964469

Gradient-Min
4.930381e-32

Gradient-Max
0.3964469

[Save/Display] [Save] [Display Options...] [Compute] [Close] [Help]

Figure 15.15a) Computed gradient min and max

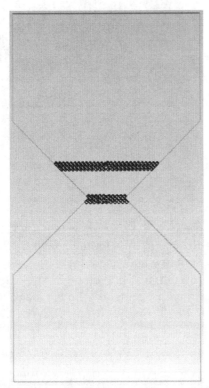

Figure 15.15b) Cells with gradients in between computed min and max values

16. Double-click on Run Calculation under Solution in the Outline View. Enter 0.001 as Time Step Size [s] under Parameters in Time Advancement on the Task Page. Enter 40 as Number of Time Steps, set Max Iterations/Time Step to 30, see Figure 15.16a). Select File>>Export>>Case & Data… from the menu and save the Case/Data File with the name *hour-glass.cas.h5*. Click on Calculate under Run Calculation on the Task Page. After the first set of calculations are complete, continue running the calculations four more times with the same settings, see Figure 15.16b). Click on Copy Screenshot of Active Window to Clipboard in the graphics window, see Figure 15.13c). The contour plot can be pasted into a Word document in between calculations.

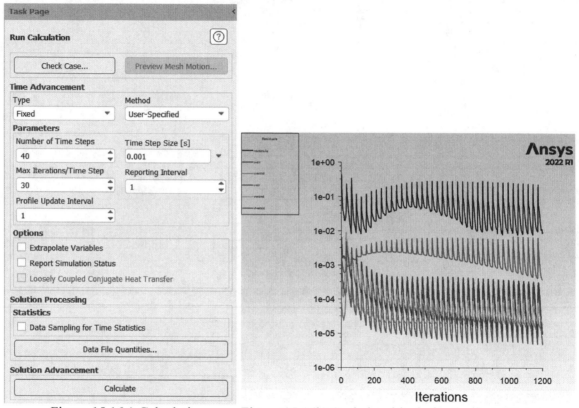

Figure 15.16a) Calculations   Figure 15.16b) Scaled residuals for the first 40 ms.

G. Post-Processing

Double click on Animations and Playback under Results in the Outline View. Uncheck the box for Use Stored View, set the Replay Speed to low and select Play Once as Playback Mode. Select Video File as Write/Record Format. Click on the Write button for your chosen Animation Sequence. This will create the movie. Close the Playback window. The movie can be viewed using, for example, Windows Media Player.

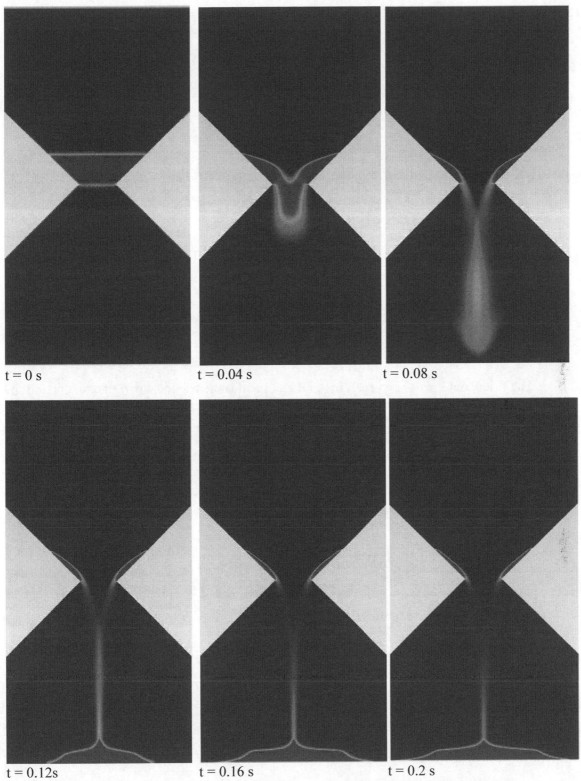

t = 0 s   t = 0.04 s   t = 0.08 s

t = 0.12s   t = 0.16 s   t = 0.2 s

Figure 15.16b) Volume fraction (wood) at t = 0, 0.04, 0.08, 0.12, 0.16 and 0.2 s.

H. Theory

17. We start with continuity, momentum and energy equations.

$$\frac{\partial}{\partial t}(\alpha_s \rho_s) + \nabla \cdot (\alpha_s \rho_s \bar{u}_s) = \dot{m}_{fs} \tag{15.1}$$

$$\frac{\partial}{\partial t}(\alpha_s \rho_s \bar{u}_s) + \nabla \cdot (\alpha_s \rho_s \bar{u}_s \bar{u}_s) = -\alpha_s \nabla p_f + \nabla \cdot \tau_s + \sum_{s=1}^{n}(\bar{R}_{fs} + \dot{m}_{fs}\bar{u}_{fs}) + \bar{F}_s \tag{15.2}$$

$$\frac{3}{2}\left[\frac{\partial}{\partial t}(\alpha_s \rho_s \theta_s) + \nabla \cdot (\alpha_s \rho_s \bar{u}_s \theta_s)\right] = \tau_s : \nabla \bar{u}_s + \nabla \cdot (\kappa_\theta \nabla \theta_s) - \gamma_s + \phi_{lm} + \phi_{fs} \tag{15.3}$$

where

$$\tau_s = -P_s \bar{I} + 2\alpha_s \mu_s \bar{S} + \alpha_s(\lambda_s - \frac{2}{3}\mu_s)\nabla \cdot \bar{u}_s \bar{I} \tag{15.4}$$

$$\bar{S} = \frac{1}{2}[\nabla \cdot \bar{u}_s + (\nabla \cdot \bar{u}_s)^T] \tag{15.5}$$

The total granular solids phase pressure is the sum of kinetic and frictional pressures

$$P_s = P_{s,friction} + P_{s,kinetic} \tag{15.6}$$

The Frictional Pressure (pascal) in ANSYS Fluent is the frictional part of equation (15.6). The following expression was given by Johnson and Jackson[4].

$$P_{s,friction} = \frac{\alpha_s}{10}\frac{(\alpha_s - \alpha_{s,min})^2}{(\alpha_{s,max} - \alpha_s)^5} \tag{15.7}$$

Alternatively, we have the expression by Syamlal et al[7].

$$P_{s,friction} = 2\alpha_s^2 \rho_s g_0 \theta_s (1 + e_s) \tag{15.8}$$

where the radial distribution function

$$g_0 = \frac{1}{1 - (\frac{\alpha_s}{\alpha_{s,max}})^{1/3}} \tag{15.9}$$

The Solids Pressure (pascal) in ANSYS Fluent is the kinetic portion of granular solids phase pressure and can be described using the following relation by Lun et al[5].

$$P_{s,kinetic} = \alpha_s \rho_s \theta_s [1 + 2\alpha_s g_0 (1 + e_s)] \tag{15.10}$$

The granular phase shear viscosity can be described as the sum of the collisional, kinetic and frictional parts.

$$\mu_s = \mu_{s,collision} + \mu_{s,kinetic} + \mu_{s,friction} \tag{15.11}$$

The collisional part of the solids phase shear viscosity was modeled by Gidaspow et al.[1] and Syamlal et al[7].

$$\mu_{s,collision} = \frac{4}{5}\alpha_s^2\rho_s d_s g_0(1+e_s)\sqrt{\frac{\theta_s}{\pi}}$$

(15.12)

The Granular Viscosity (kg/m-s) in ANSYS Fluent is the kinetic part of equation (15.11). The following expression was given by Gidaspow et al[1].

$$\mu_{s,kinetic} = \frac{10\sqrt{\pi\theta_s}\rho_s d_s}{96(1+e_s)g_0}\left[1+\frac{4}{5}(1+e_s)g_0\alpha_s\right]^2$$

(15.13)

Alternatively, we have the expression by Syamlal et al[7].

$$\mu_{s,kinetic} = \frac{\alpha_s\sqrt{\pi\theta_s}\rho_s d_s}{6(3-e_s)}\left[1+\frac{2}{5}(1+e_s)(3e_s-1)g_0\alpha_s\right]$$

(15.14)

The Frictional Viscosity (kg/m-s) in ANSYS Fluent is the frictional part of equation (15.11). The equation is given by Schaeffer[6]

$$\mu_{s,friction} = \frac{P_{s,friction}\sin\varphi}{2\sqrt{I_2}}$$

(15.15)

The Granular Bulk Viscosity (kg/m-s) in ANSYS Fluent accounts for resistance of the granular material to compression and expansion. The equation is given by Lun et al[5].

$$\lambda_s = \frac{4}{3}\alpha_s^2\rho_s d_s g_0(1+e_s)\sqrt{\frac{\theta_s}{\pi}}$$

(15.16)

α_s	volume fraction for granular material
$\alpha_{s,min}$	friction packing limit
$\alpha_{s,max}$	packing limit
φ	internal angle of friction
γ_s	energy dissipation
κ_θ	granular temperature conductivity
λ_s	bulk viscosity
μ_s	granular shear viscosity
$\mu_{s,collision}$	collisional viscosity
$\mu_{s,friction}$	frictional viscosity
$\mu_{s,kinetic}$	kinetic viscosity
ϕ_{fs}	energy exchange between fluid and granular phase
ϕ_{lm}	energy exchange between granular phases
ρ_s	density
θ_s	temperature
τ_s	stress tensor
d_s	particle diameter
e_s	coefficient of restitution
g_0	radial distribution function
I_2	second invariant of the deviatoric stress tensor
$\dot{m}_{fs}$	unidirectional mass transfer

p_f	fluid pressure
P_s	granular pressure
$P_{s,friction}$	frictional pressure
$P_{s,kinetic}$	kinetic pressure
$\bar{S}$	strain rate
t	time
$\bar{u}_s$	granular velocity vector

Table 15.2 Parameters used in theory section

I. References

1. Gidaspow, D., Bezburuah, R., Ding, J. "Hydrodynamics of Circulating Fluidized Beds: Kinetic Theory Approach", *7th Fluidization Conference*, May 3, (1992).
2. Hirt, C.W., "A Flow-3D Continuum Model for Granular Media.", Flow Science Report 02-13, (2013).
3. Jansson, M., "CFD Simulations of Silos Content, An Investigation of Flow Patterns and Segregation Mechanisms.", *Master's Thesis within the Innovative and Sustainable Chemical Engineering Programme*, Chalmers, (2014).
4. Johnson, P.C., Jackson. R., "Frictional-Collisional Constitutive Relations for Granular Materials, with Application to Plane Shearing". *J. Fluid Mech.* **176**. (1987).
5. Lun, C.K.K., Savage, S.B., Jeffrey, D.J., Chepurniy, N., "Kinetic Theories for Granular Flow: Inelastic Particles in Couette Flow and Slightly Inelastic Particles in a General Flow Field". *J. Fluid Mech.* **140**, (1984).
6. Schaeffer, D.G., "Instability in the Evolution Equations Describing Incompressible Granular Flow.", *Journal of Differential Equations*, **66**, (1987).
7. Syamlal, M., Rogers, W., O'Brien, T.J. "MFIX Documentation Theory Guide", DOE/METC-94/1004 (1993).

J. Exercises

15.1 Use ANSYS Fluent to simulate an axisymmetric cylindrical heap of sand that deforms due to gravity into a conical pile. Use an initial 100 mm in diameter and 100 mm high heap of sand resting on a horizontal surface with a volume fraction packing limit of 0.63, see Hirt[2]. Use the density 1638 kg/m3 for the sand, a grain diameter of 0.45 mm and an internal angle of friction of 34 degrees. Determine the angle of repose for the conical pile of sand and show contours for the volume fraction of sand over time as it slumps.

15.2 Use ANSYS Fluent to simulate a 2-D hourglass. The shape of the hourglass is 200 mm wide at the top and the bottom, a total height of 500 mm and sides with a slope of 40.5 degrees. The opening at the waist of the hourglass is 10 mm. Start the simulation with sand (1638 kg/m3) and a grain diameter of 0.45 mm in the upper half of the hourglass with a packing limit of 0.63 for the volume fraction of sand. Use an internal angle of friction of 34 degrees. Determine the angle of repose for the final conical pile of sand in the bottom half of the hourglass and show contours for the volume fraction of sand over time as it moves due to gravity from the upper half to the lower half of the hourglass. See Hirt[2] for a similar simulation using Flow-3D.

15.3 Use ANSYS Fluent to simulate the flow of wood when filling a 2-D axisymmetric silo, see Jansson[3]. The shape of the silo is shown in Figure 15.16 with the following dimensions: $v1 = 9$ m, $h1 = 10$ m, $v2 = 17$ m, $h2 = 0.5$ m, $v3 = 0.5$ m, $v4 = 1.5$ m, $h3 = 0.5$ m. Use an inlet volume fraction for wood of 0.4 and other values as listed in Table 15.2.

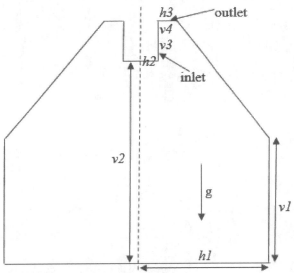

Figure 15.16 Geometry for silo

$\alpha_{s,min} = 0.45$	friction packing limit
$\alpha_{s,max} = 0.63$	packing limit
$\varphi = 45°$	internal angle of friction
$\rho_s = 850 \; kg/m3$	wood density
$d_s = 0.02 \; m$	particle diameter
$e_s = 0.001$	coefficient of restitution

Table 15.3 Parameters used in Exercise 15.3

K. Notes

CHAPTER 16. BOUNCING SPHERES

A. Objectives

- Use the Macroscopic Particle Model
- Create Pictures of the Resulting Motion of the Spheres

B. Problem Description

In this chapter we will be simulating a number of spheres falling through an hourglass shaped geometry. We will let the balls bounce with each other and they will bounce with the walls.

C. Launching ANSYS Workbench and Selecting Fluent

1. Start by launching ANSYS Workbench. Double click on Fluid Flow (Fluent) under Analysis Systems in the Toolbox.

Figure 16.1 Selecting Fluent

D. Launching ANSYS DesignModeler

2. Right click on Geometry in the Project Schematic window in ANSYS Workbench and select New DesignModeler Geometry. Select Units>>Millimeter from the menu in DesignModeler. Select the XY Plane in the Tree Outline and select Look at Face/Plane/Sketch ![icon]. Select the Sketching tab and select Polyline under Draw from Sketching Toolboxes. Start at the origin and draw the sketch, right click and select *Closed End* and enter the dimensions as shown in Figure 16.2a). When you start drawing from the origin (make sure that letter *P* is shown when the cursor is close to the origin and that letters *V*, *H* and *C* are showing when you are creating vertical, horizontal and coinciding lines) and draw the sketch with the dimensions as shown in Figure 16.2a).

Select General under Dimensions and click on the eight lines (H1-H2, V3-V4 and L5-L6) as shown in Figure 16.2a). Select Angle under Dimensions and create angle A7 by selecting the lines next to the angle. You may need to right click and select *Alternate Angle* when creating the angle. Select ![Revolve icon] Revolve and apply Sketch1 under XY Plane in the Tree Outline as Geometry in Details View. Select the left vertical side of the sketch and apply it as the Axis in Details View. Click on Generate and close DesignModeler.

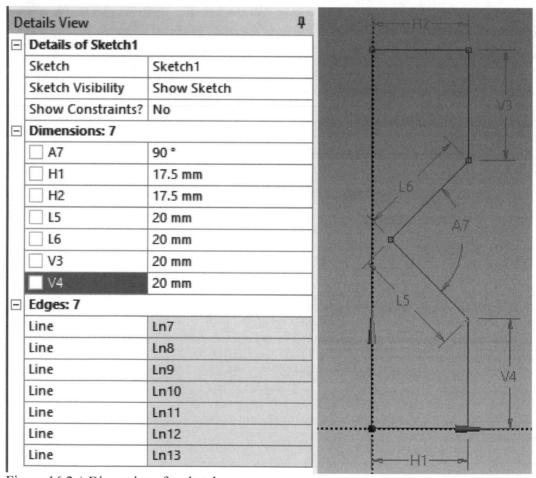

Figure 16.2a) Dimensions for sketch

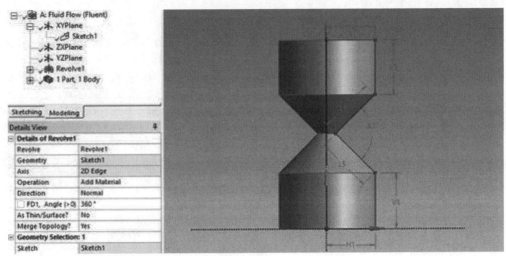

Figure 16.2b) Revolved sketch

E. Launching ANSYS Meshing

3. Double click on Mesh under Project Schematic in ANSYS Workbench. Select Mesh in the Outline and Set the Element Size under Defaults in Details of Mesh to 1.0 mm. Right click on Mesh under Project in Outline and select Generate Mesh. Under Statistics in Details of "Mesh" we can see that this mesh has 111,860 Nodes and 107,352 Elements. Right click in the graphics window and select Cursor Mode>>Face. Control select the 6 faces (push the mouse wheel and drag the mouse to turn around the model to select the bottom face) of the mesh, right click and select Create Named Selection. Enter the name *wall* and click on OK to close the Selection Name window.

   Select File>>Export…>>Mesh>>FLUENT Input File>>Export from the menu and save the mesh with the name *3d-enclosure-mesh.msh*. Select File>>Save Project from the menu and save the project with the name *3D-Enclosure.wbpj*. Close the Meshing window, right-click on Mesh under Project Schematic in ANSYS Workbench and select Update.

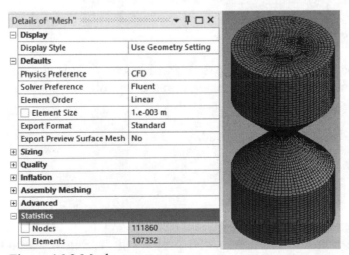

Figure 16.3 Mesh

F. Launching ANSYS Fluent

4. Double-click on Setup under Project Schematic in ANSYS Workbench. Check the box for Double Precision under Options in the ANSYS Fluent Launcher. Click on Show More Options and take a note of the location of the *Working Directory*. You will need this information later in step 9. Set the number of processes equal to the number of processor cores for your computer. To check the number of physical cores, press the Ctrl + Shift + Esc keys simultaneously to open the Task Manager. Go to the Performance tab and select CPU from the left column. You will see the number of physical cores on the bottom-right side. Close the Task Manager window. Click on the Start button in the Fluent Launcher window.

Figure 16.4a) Launching ANSYS Fluent

Click Enter on the keyboard with the cursor in the Console section that is located under the graphics window. Enter the following commands in the Console to load the Macroscopic Particle Model (MPM), see Figure 16.4b).

>**define**
/define> **models**
/define/models/> **addon**
Enter Module Number: [0] **10**

Click on Display under Mesh in General on the Task Page. Check the boxes for Edges and Faces under Options in the Mesh Display window. Select wall under Surfaces, click on Display and Close the window, see Figure 16.4c).

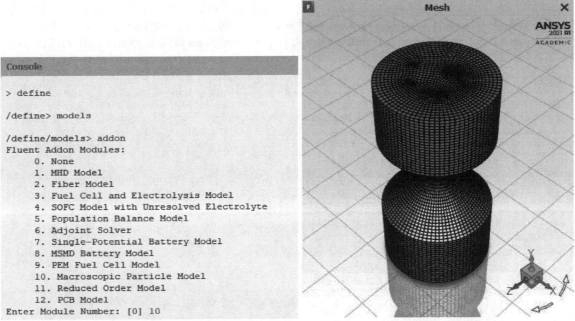

Figure 16.4b) Console commands

Figure 16.4c) Mesh in ANSYS Fluent

5. Double click on General under Setup in the Outline View and select Transient under Time on the Task Page. Check the box for Gravity and enter -9.81 as the value for Y [m/s²]. Click on Scale… under Mesh in General on the Task Page.

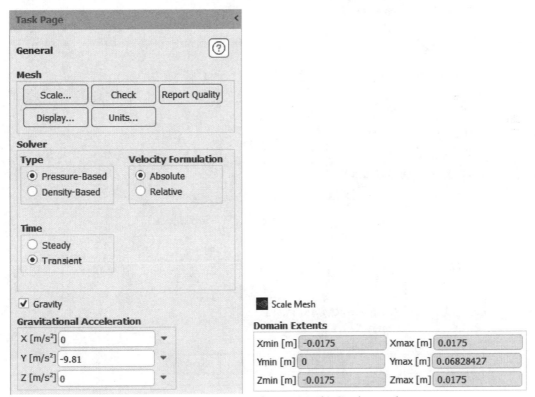

Figure 16.5a) General settings

Figure 16.5b) Scale mesh

6. Double click on Models and Macroscopic Particles under Setup in the Outline View and check the box to Enable Macroscopic Particle Model. Select the Drag tab and select Disable Drag Calculations under Drag Law Options. Select the Injections tab and choose Create…. Select plane as Injection Type. Set the Diameter [m] to 0.003. Set the Plane Shape to Circular and select Y as Axis Direction. Set Y [m] to 0.06828427 and Radius [m] to 0.0175 under Center Location/Radius of Circular Plane. Click on Create/Modify.

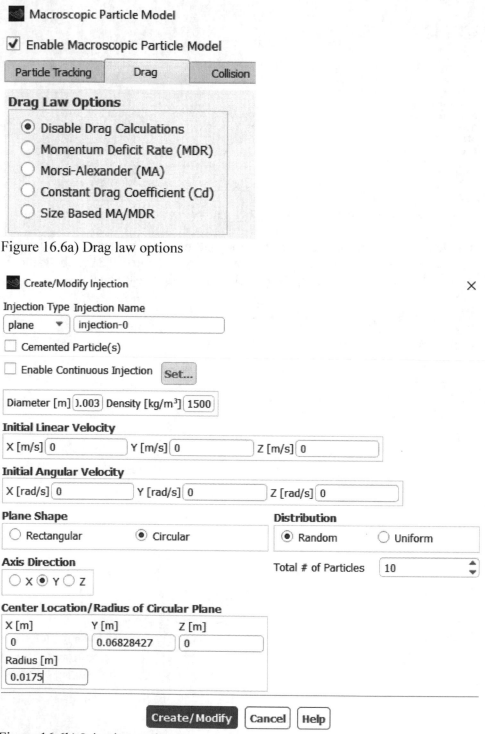

Figure 16.6a) Drag law options

Figure 16.6b) Injection options

Select the Initialize MPM tab in the MPM window. Check the box for Print Warning Messages and click on Initialize MPM Functions. Click on Initialize Particles.

Select Display Mesh 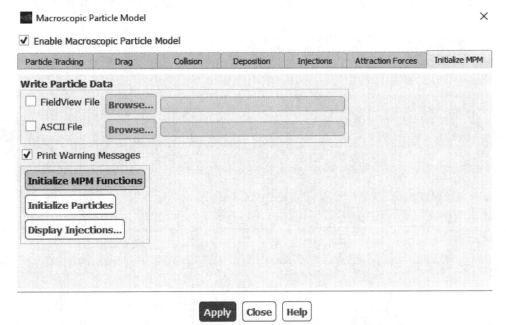 next to the graphics window to open the Mesh Display window. Select *wall* under Surfaces in the Mesh Display window. Uncheck the box for Faces under Options and select Outline as Edge Type, see Figure 16.6d). Click on Display and Close the Mesh Display window.

Click on Display Injections in the Macroscopic Particle Model MPM window. Select Particle ID under Display Particles Colored by, click on Display and Close the Display Particles window. Click on Apply in the MPM window and Close the same window.

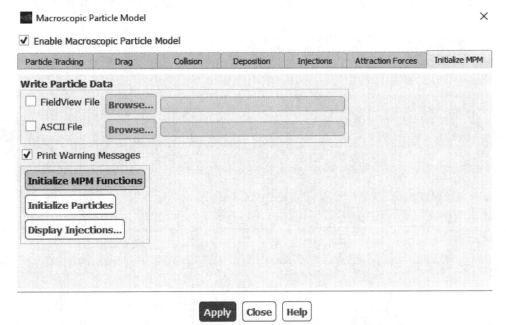

Figure 16.6c) Macroscopic Particle Model

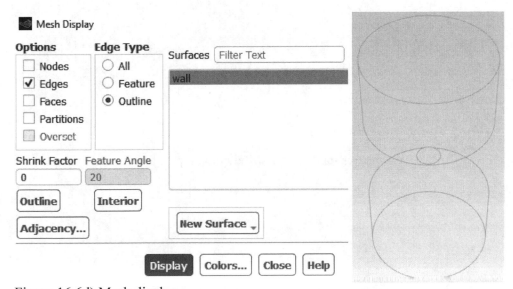

Figure 16.6d) Mesh display

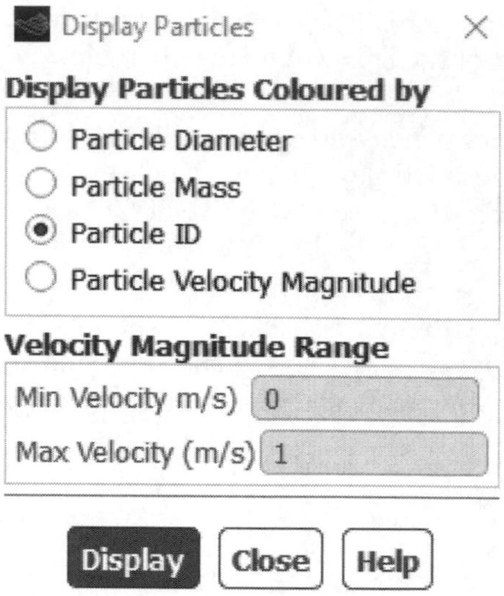

Figure 16.6e) Display particles settings

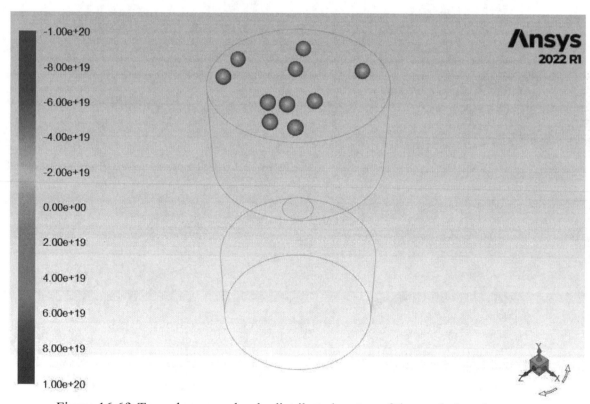

Figure 16.6f) Ten spheres randomly distributed on top of the mesh domain

7. Double click on Methods under Solution in the Outline View and select Second Order Implicit as Transient Formulation under Solution Methods on the Task Page. Select First Order Upwind for Turbulent Kinetic Energy and Specific Dissipation Rate under Spatial Discretization.

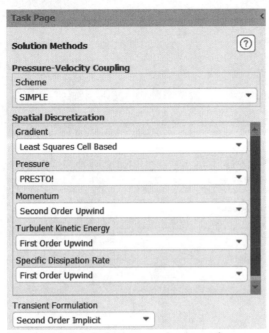

Figure 16.7a) Solution methods settings

Open Cell Zone Conditions, Fluid and solid (fluid, id=2) under Setup in the Outline View and uncheck the box for Source Terms in the Fluid window. Click on Apply and Close the Fluid window.

Double click on Monitors and Residual under Setup in the Outline View and uncheck the box for Plot under Options. Check the box for Show Advanced Options, select none as Convergence Criterion and click on OK to close the Residual Monitors window.

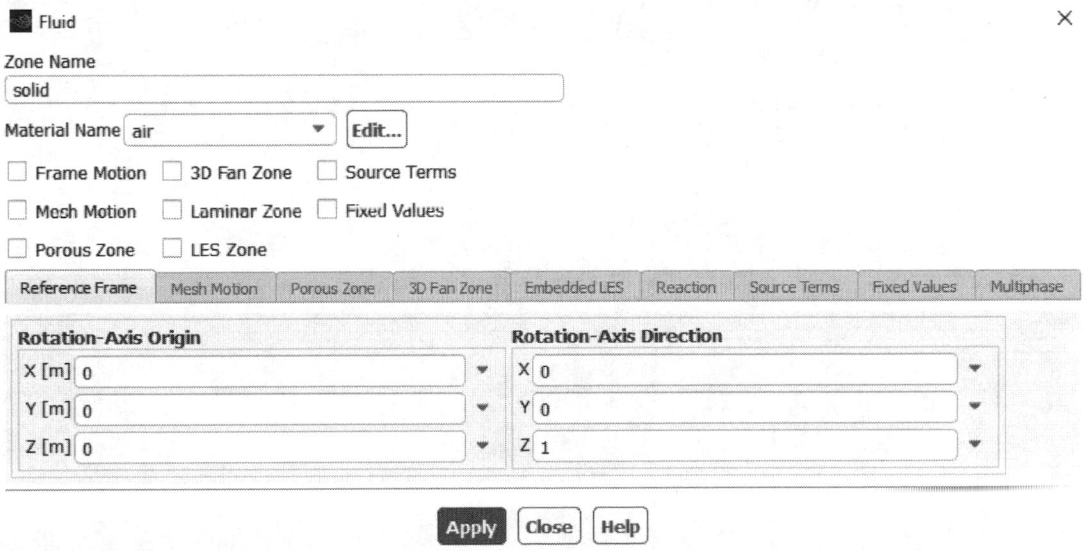

Figure 16.7b) Fluid window

Figure 16.7c) Residual monitors window

8. Double click on Initialization under Solution in the Outline View and select Hybrid Initialization. Click on Initialize. Answer OK when you get a question.

Figure 16.8 Solution initialization

9. Create a new folder in your *Working Directory* called *jpg-files*. Double click on Calculation Activities and Execute Commands under Solution in the Outline View. Enter the commands listed below and as shown with settings in Figure 16.9. You will need to change the second command so that it is in line with the name and location of your *Working Directory*. Click OK to close the window.

(display-mpm-injections 'particle-id 0 10)
dis hard C:/Users/Instructor/Desktop/An Introduction to ANSYS Fluent 2022/Ansys Fluent 2022 Ch. 16/jpg-files/mpm-%t.jpg

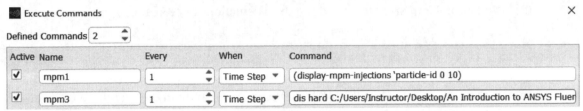

Figure 16.9 Executive commands

10. Select File>>Export>>Case & Data… from the menu and save the Case/Data File with the name *bouncing-spheres.cas.h5*. Double click on Run Calculation under Solution in the Outline View and set the Time Step Size [s] to 0.01 and Number of Time Steps to 5. Set Max Iterations/Time Step to 1 and click on Calculate.

Figure 16.10b) Running the calculations

G. Post-Processing

11. Select the *YZ* plane by clicking on the red *X* axis of the coordinate system in the lower right corner of the graphics window as shown in Figure 16.11a). Select Copy Screenshot of Active Window to Clipboard in the graphics window, see Figure 16.11b). The particle traces can be pasted into a Word document.

    Continue running the calculations two more times with the same settings. Next, change the Number of Time Steps to 10 and run the calculations once again.

    Next, change the Number of Time Steps to 25 and run the calculations once again. Change the Number of Time Steps to 50 and run the calculations once again. Answer OK to the question that you get. Next, change the Number of Time Steps to 100 and run the calculations. Change the Number of Time Steps to 300 and run the calculations.

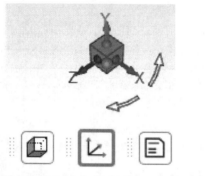

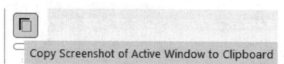

Figure 16.11a) Coordinate system   Figure 16.11b) Copying screenshot

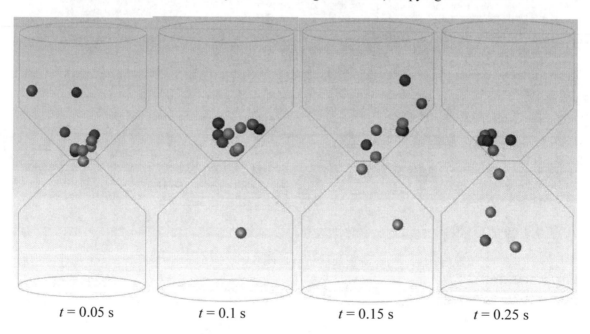

$t = 0.05$ s $t = 0.1$ s $t = 0.15$ s $t = 0.25$ s

Figure 16.11c) Distribution of spheres during the first 0.25 s.

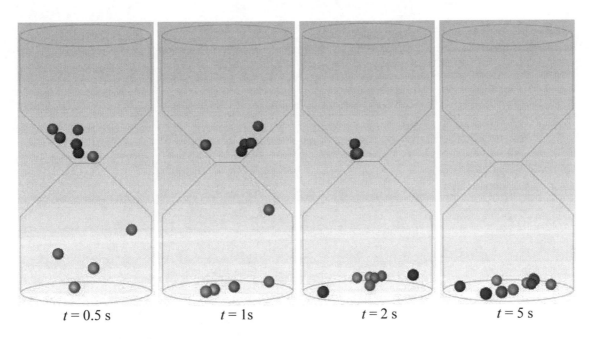

$t = 0.5$ s $t = 1$s $t = 2$ s $t = 5$ s

Figure 16.11d) Distribution of spheres over time up to 5 s.

H. Reference

1. Agrawal, M., Bakker, A., Prinkey, M.T. "Macroscopic Particle Model Tracking Big Particles in CFD", *AIChE Annual Meeting, Particle Technology Forum, Paper 268b, November 7-12, Austin, TX*, (2004).

I. Exercise

16.1 Use ANSYS Fluent and the macroscopic particle model to perform a billiard table simulation of the first shot to break the balls including the collision of the cue ball with the 15 object balls initially racked together in a triangular pattern on a 50 in. x 100 in. billiard table. The diameter of the billiard balls is 57 mm.

J. Notes

CHAPTER 17. FALLING SPHERE

A. Objectives

- Using the Macroscopic Particle Model Module
- Determine the Terminal Velocity for a Falling Sphere

B. Problem Description

A 2 mm diameter sphere with a density of 1695 kg/m^3 is falling in a tube with a diameter of 20 mm and a length of 60 mm.

C. Launching ANSYS Workbench and Selecting Fluent

1. Start by launching ANSYS Workbench. Double click on Fluid Flow (Fluent) under Analysis Systems in the Toolbox. Select **Units>>Metric (tonne, mm, s, …)** from the menu.

Figure 17.1 Selecting Fluent

D. Launching ANSYS DesignModeler

2. Right click on Geometry in the Project Schematic window in ANSYS Workbench and select New DesignModeler Geometry. Select **Units>>Millimeter** from the menu in DesignModeler. Select the XY Plane in the Tree Outline and click on Look at Face/Plane/Sketch ![icon], Select the Sketching tab and select Rectangle under Draw from Sketching Toolboxes. Start at the origin (make sure you see letter P when the cursor is at the origin and you are about to start drawing) and draw the rectangle in the first quadrant.

Click on Dimensions and click on the lower horizontal edge of the rectangle. Enter a length of 10 mm for the horizontal edge. Click on the right vertical edge of the rectangle. Enter a length of 60 mm for the vertical edge.

Select 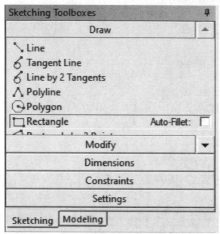 Revolve and apply Sketch1 under XYPlane in the Tree Outline as Geometry in Details View. Select the left vertical side of the rectangle sketch and apply it as the Axis in Details View. Click on Generate and close DesignModeler.

Figure 17.2a) Rectangle drawing tool

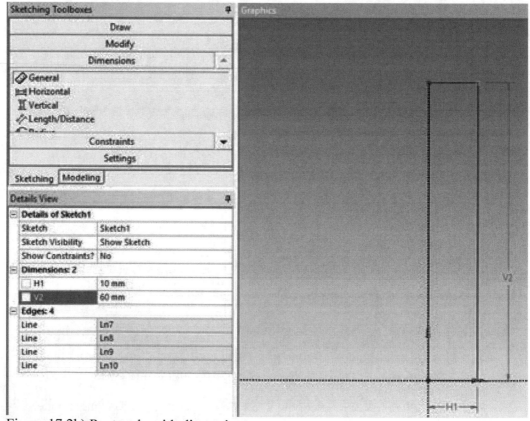

Figure 17.2b) Rectangle with dimensions

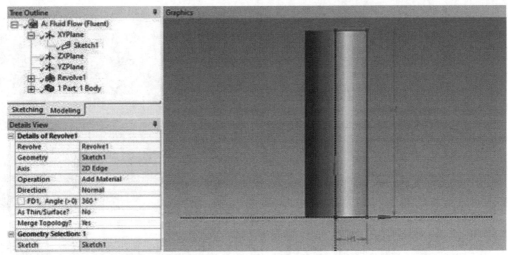

Figure 17.2c) Finished cylinder

E. Launching ANSYS Meshing

3. Double click on Mesh under Project Schematic in ANSYS Workbench. Select Mesh in the Outline under Project. Select Home>>Tools>>Units>>Metric (mm, kg, N, …) from the menu. Right click on Mesh under Project in Outline and select Generate Mesh. A coarse mesh is generated.

   Set the Element Size under Defaults in Details of Mesh to 0.4 mm. Right click on Mesh under Project in Outline and select Generate Mesh. Under Statistics in Details of Mesh we can see that this mesh has 503283 Nodes and 489450 Elements.

   Right click in the graphics window and select Cursor Mode>>Face. Control select the three faces of the mesh, right click and select Create Named Selection. Enter the name *wall* and click on OK to close the Selection Name window.

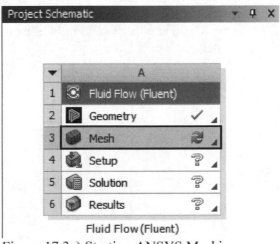

Figure 17.3a) Starting ANSYS Meshing

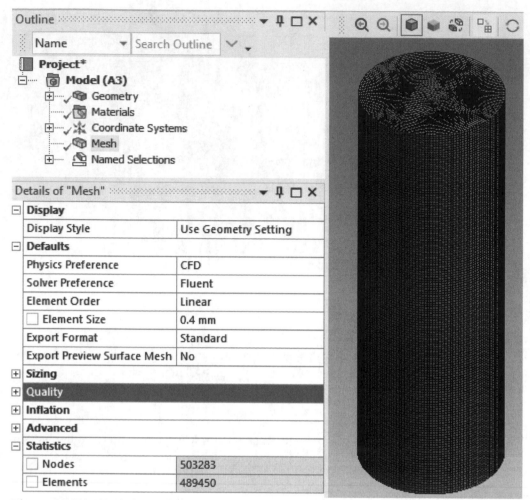

Figure 17.3b) Mesh for cylinder

Select File>>Export...>>Mesh>>FLUENT Input File>>Export from the menu and save the mesh in the *Working Directory* with the name *cylinder-mesh-lambda-0.1.msh*. Select File>>Save Project from the menu and save the project with the name *Cylinder Mesh Falling Sphere Lambda = 0.1.wbpj*. Close the Meshing window and right-click on Mesh under Project Schematic in ANSYS Workbench and select Update.

F. Launching ANSYS Fluent

4. Double-click on Setup under Project Schematic in ANSYS Workbench. Check the box for Double Precision under Options in the ANSYS Fluent Launcher. Click on Show More Options and take a note of the location of the *Working Directory*. You will need this information later. Set the number of processes equal to the number of processor cores for your computer. To check the number of physical cores, press the Ctrl + Shift + Esc keys simultaneously to open the Task Manager. Go to the Performance tab and select CPU from the left column. You will see the number of physical cores on the bottom-right side. Close the Task Manager window. Click on the Start button in the Fluent Launcher window.

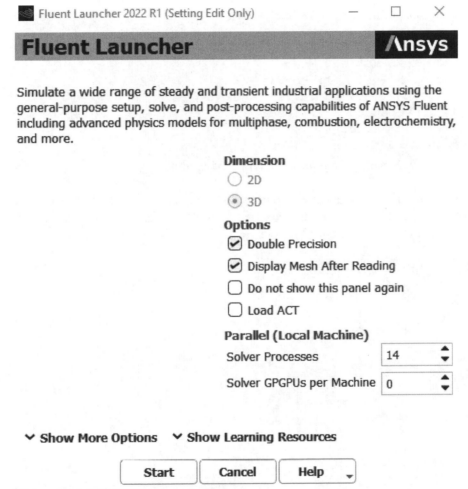

Figure 17.4a) ANSYS Fluent launcher

Enter the following commands in the Console that is located below the graphics window to load the Macroscopic Particle Model (MPM). Click in the Console window and select Enter on the keyboard to get to the prompt.

>**define**
/define> **models**
/define/models/> **addon**
Enter Module Number: [0] **10**

```
Console

> define

/define> models

/define/models> addon
Fluent Addon Modules:
     0. None
     1. MHD Model
     2. Fiber Model
     3. Fuel Cell and Electrolysis Model
     4. SOFC Model with Unresolved Electrolyte
     5. Population Balance Model
     6. Adjoint Solver
     7. Single-Potential Battery Model
     8. MSMD Battery Model
     9. PEM Fuel Cell Model
     10. Macroscopic Particle Model
     11. Reduced Order Model
     12. PCB Model
Enter Module Number: [0] 10
```

Figure 17.4b) Console commands to load MPM module

5. Double click on General under Setup in the Outline View and select Transient under Time on the Task Page. Check the box for Gravity and enter -9.8 as the value for Y [m/s²]. Click on Scale… under Mesh in General on the Task Page.

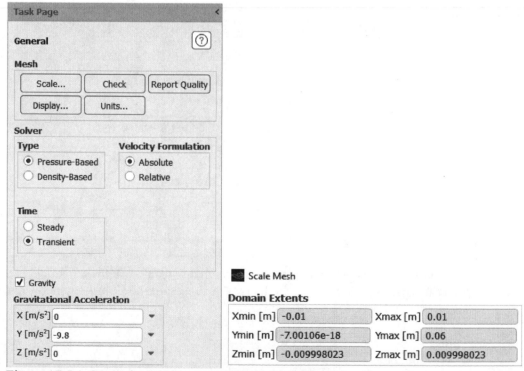

Figure 17.5a) General settings   Figure 17.5b) Scale mesh

6. Double click on Models and Viscous (SST k-omega) under Setup in the Outline View. Select Laminar Model and click OK to close the Viscous Model window. Double click on Materials under Setup in the Outline View. Click on Create/Edit for Fluid under Materials on the Task Page. Click on Fluent Database in the Create/Edit Materials window. Scroll down and select *water-liquid (h2o<l>)* as the Fluent Fluid Material. Click on Copy and Close the Fluent Database Materials window. Set the Density [kg/m³] to 1020 and Viscosity [kg/m s] to 0.49. Click on Change/Create and Close the Create/Edit Materials window.

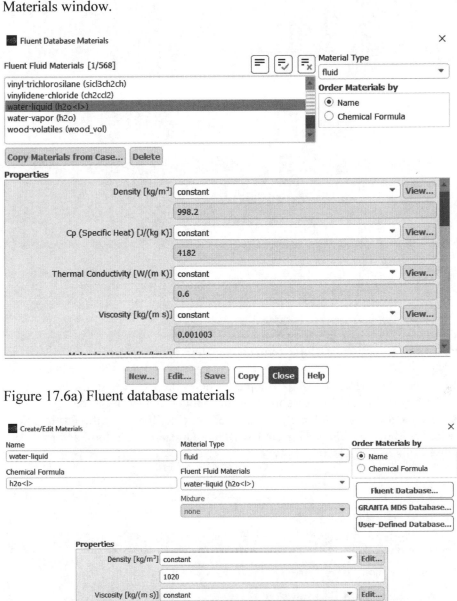

Figure 17.6a) Fluent database materials

Figure 17.6b) Setting the density and viscosity for the fluid

Open Cell Zone Conditions and Fluid under Setup in the Outline View. Double click on *solid (fluid, id=2)* under Cell Zone Conditions and Fluid in the Outline View. Select *water-liquid* as Material Name and enter *water-liquid* as Zone Name. Check the box for Source Terms and click on the Source Terms tab. Click on Edit for X Momentum. Set Number of X Momentum Sources to 1, select *udf x_mom::mpm* and click OK. Repeat this for Y Momentum, Z Momentum and Mass. Select Apply and Close the Fluid window.

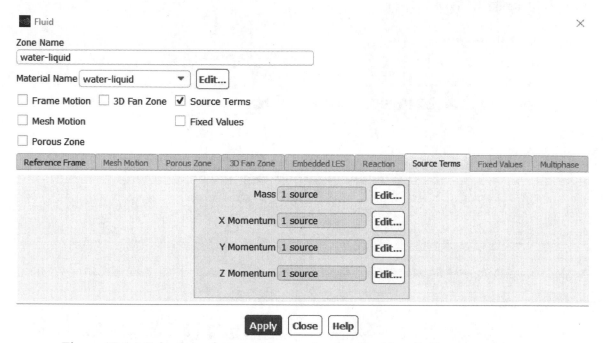

Figure 17.6c) Selection of source terms and water-liquid as fluid

Double click on Models and Macroscopic Particles under Setup in the Outline View and check the box to Enable Macroscopic Particle Model. Select the Particle Tracking tab and check the box for Enable Interaction with Continuous Phase. Set the Particle Sub-Timesteps within each Flow Timestep to 10 and Under-Relaxation Factor to Velocities in Touched Cells to 1. Click on Apply.

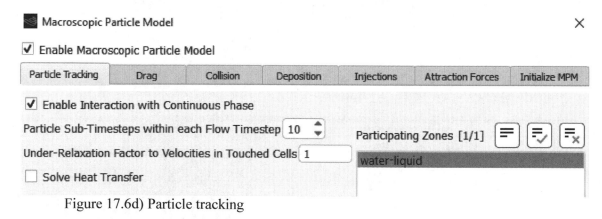

Figure 17.6d) Particle tracking

Select the Drag tab and select Momentum Deficit Rate (MDR) under Drag Law Options. Click on Apply.

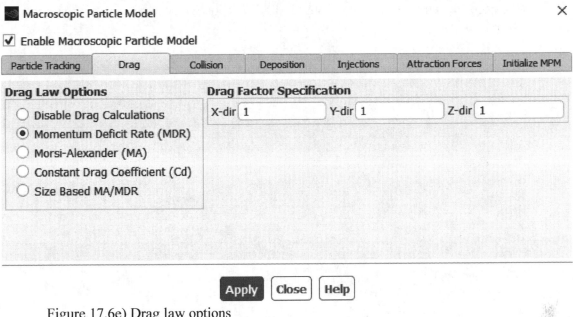

Figure 17.6e) Drag law options

Select the Collision tab and uncheck Enable Particle-Particle Collision and uncheck Enable Particle-Wall Collision. Click on Apply.

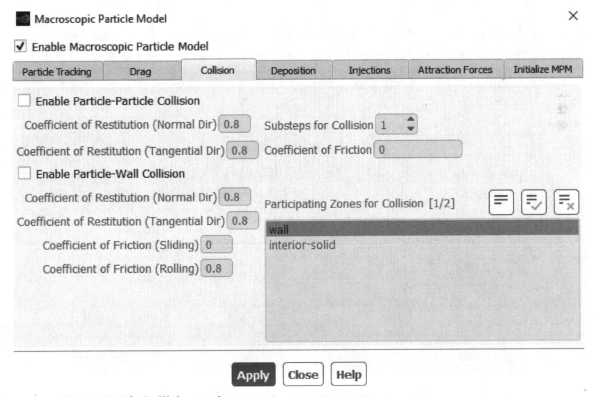

Figure 17.6f) Collision options

Select the Injections tab and click on Create. Select *point* as Injection Type. Set the Diameter [m] to 0.002 and set the Density [kg/m^3] to 1695. Check the box for Cylindrical Coordinates and select Y as Axis Direction. Set Axial [m] to 0.03 under Initial Location. Click on Create/Modify in the Create/Modify Injection window. Select Apply in the MPM window.

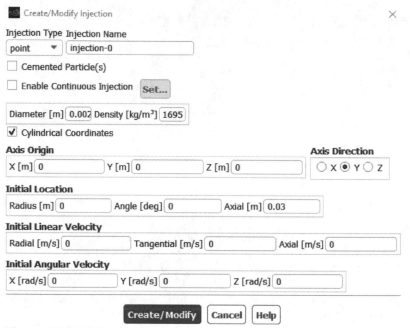

Figure 17.6g) Injection options

Select the Initialize MPM tab and check the box for ASCII File under Write Particle Data. Click on Browse, enter the file name *Ansys Fluent 2022 Ch.csv*, save the file in the *Working Directory* and click on OK. Check the box for Print Warning Messages. Select Initialize MPM Functions. Click on Initialize Particles. Select Apply in the MPM window.

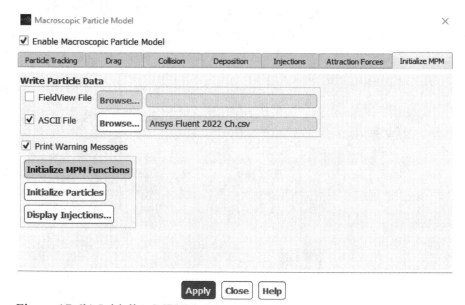

Figure 17.6h) Initialize MPM

Select Display Mesh  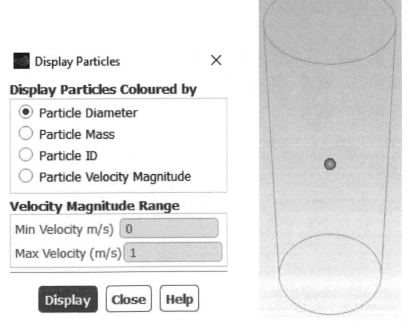 next to the graphics window. Select *wall* under Surfaces. Check the box for Edges and uncheck the box for Faces under Options and select Outline as Edge Type, see Figure 17.6i). Click on Display and Close the Mesh Display window. Click on Apply in the MPM window.

Mesh Display ✕

Options
- ☐ Nodes
- ☑ Edges
- ☐ Faces
- ☐ Partitions
- ☐ Overset

Edge Type
- ○ All
- ○ Feature
- ● Outline

Shrink Factor
`0`

Feature Angle
`20`

[Outline] [Interior]

[Adjacency...]

Surfaces | Filter Text | [🗒] [🗒] [🗒] [🗒]

wall

[New Surface ▾]

[Display] [Colors...] [Close] [Help]

Figure 17.6i) Mesh display

Click on Display Injections. Select Display Particles Colored by Particle Diameter. Click on Display and Close the Display Particles window. Click on Apply and close the Macroscopic Particle Model window.

Display Particles ✕

Display Particles Coloured by
- ● Particle Diameter
- ○ Particle Mass
- ○ Particle ID
- ○ Particle Velocity Magnitude

Velocity Magnitude Range

Min Velocity m/s) `0`

Max Velocity (m/s) `1`

[Display] [Close] [Help]

Figure 17.6j) Display particles

Figure 17.6k) Sphere in the middle of cylinder

7. Double click on Methods under Solution in the Outline View and select Second Order Implicit as Transient Formulation under Solution Methods on the Task Page. Check the box for Higher Order Term Relaxations and select Options. Click on Options…. Select Flow Variables Only and set the Relaxation Factor to 0.75. Click OK to close the window.

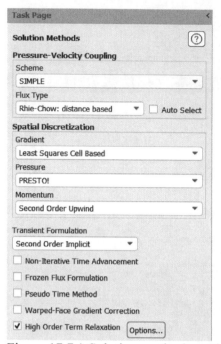

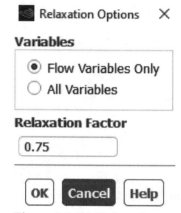

Figure 17.7a) Solution methods settings   Figure 17.7b) Relaxation options

Double click on Monitors and Residual under Setup in the Outline View. Check the boxes for Print to Console and Plot under Options. Click on OK to close the Residual Monitors window.

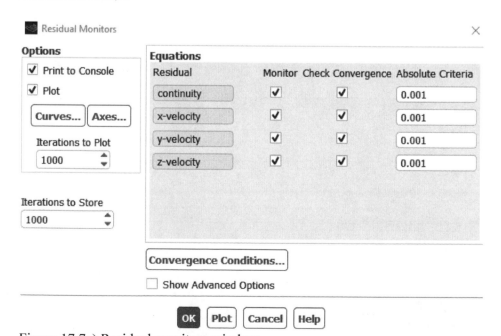

Figure 17.7c) Residual monitors window

8. Double click on Initialization under Solution in the Outline View and select Hybrid Initialization. Click on Initialize. Answer OK to the question that you get.

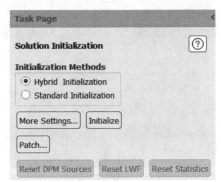

Figure 17.8 Solution initialization

9. Create a folder in your *Working Directory* named *falling-sphere-lambda-0.1-jpg-files*. Double click on Calculation Activities and Execute Commands under Solution in the Outline View. Copy and enter the commands listed below and as shown with settings in Figure 17.9a). You will need to change the command so that it is in line with the name and location of your *Working Directory*. Click OK to close the window.

(display-mpm-injections 'particle-id 0 20)
dis hard C:/Users/Instructor/Desktop/An Introduction to ANSYS Fluent 2022/Ansys Fluent 2022 Ch. 17/falling-sphere-lambda-0.1-jpg-files/mpm-%t.jpg

Figure 17.9a) Executive commands

Double click on Results>>Graphics>>Particle Tracks in the Outline View. Set the Track Style to *sphere* and Color by Particle Variables… and Particle Diameter. Select *injection-0* under Release from Injections. Click on Attributes under Track Style. Check the Constant box under Options and set the Diameter to 0.002. Click on Apply and Close the window. Check the box for Draw Mesh under Options. Select *wall* under Surfaces. Click on Display and Close the Mesh Display window. Click on Save/Display and Close the Particle Tracks window.

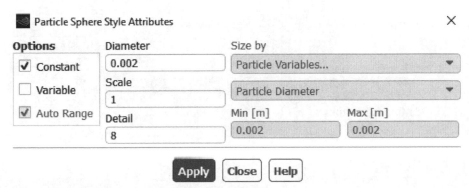

Figure 17.9b) Particle sphere style attributes

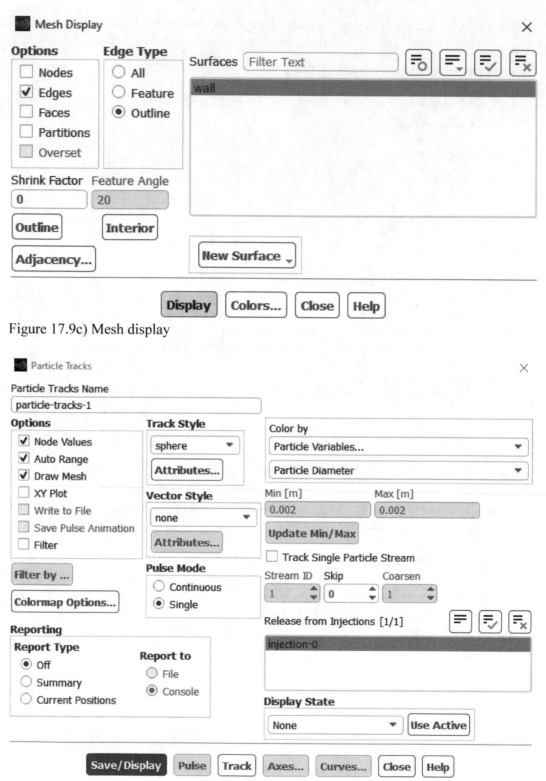

Figure 17.9c) Mesh display

Figure 17.9d) Particle tracks

10. Select File>>Save Project from the menu. Select File>>Export>>Case & Data... from the menu. Save the Case/Data File in the *Working Directory* with the name *falling-sphere-lambda-0.1.cas.h5*.

Double click on Run Calculation under Solution in the Outline View and set the Time Step Size (s) to 2e-5 and Number of Time Steps to 2000. Set Max Iterations/Time Step to 20 and click on Calculate.

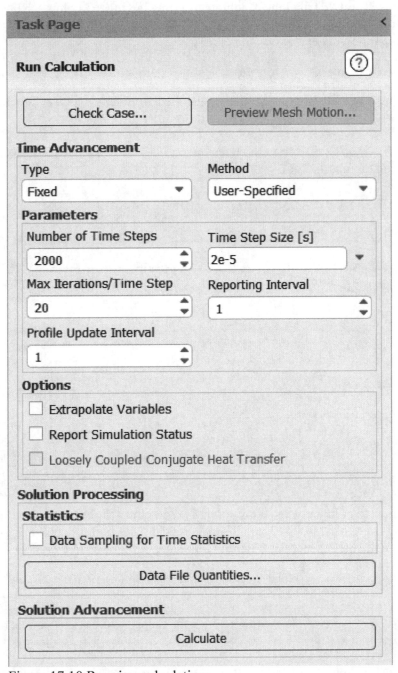

Figure 17.10 Running calculations

G. Post-Processing

11. Open Excel and open the file *Ansys Fluent 2022 Ch.csv*. You will need to select All Files format to open the file in Excel. Save the file as *Ansys Fluent 2022 Ch.xlsx*.

    Copy the first column *A* to column *N*. In position *O2*, enter =ABS(G2) and drag the *O3* value down the whole *O* column to get the absolute *y*-velocity component. Select the data values in columns *N* and *O*. Label the *N* column as *t* (m/s) and the *O* column as *Ansys Fluent*. The theoretical solution for velocity is given by equation (17.7) in the theory section. Enter the following expression in cell *P2*:

    = (0.791576*9.81*(1-(1020/1695)) / (18*0.49/ (1695*0.002^2))) *(1-Exp(-(18*0.49/ (1695*0.002^2)) *N2/0.791576))

    Drag the *P2* value down the whole *P* column to get the theoretical velocity. Label the *P* column as *Theory*.

    Select columns *N* – *P* and select from the Excel menu Insert>>Charts>>Scatter with Smooth Lines, see Figures 17.11a).

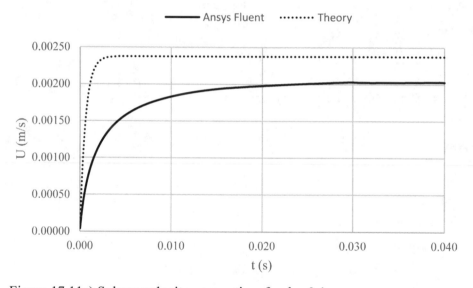

Figure 17.11a) Sphere velocity versus time for $\lambda = 0.1$

    Copy the column *N* to column *Q* and copy column *D* to column *R*. Label the *Q* column as *t* (m/s) and the *R* column as *Ansys Fluent*. The theoretical solution for position is given by equation (17.8) in the theory section. Enter the following expression in cell *S2*:

    = 0.03+(0.791576*9.81*(1-(1020/1695)) / (18*0.49/ (1695*0.002^2)) ^2) * (0.791576*(1-Exp(-(18*0.49/ (1695*0.002^2)) *N2/0.791576))- (18*0.49/ (1695*0.002^2)) *N2)

    Drag the *S2* value down the whole *S* column to get the theoretical velocity. Label the *S* column as *Theory*.

    Select columns *Q* – *S* and select from the Excel menu Insert>>Charts>>Scatter with Smooth Lines, see Figures 17.11b).

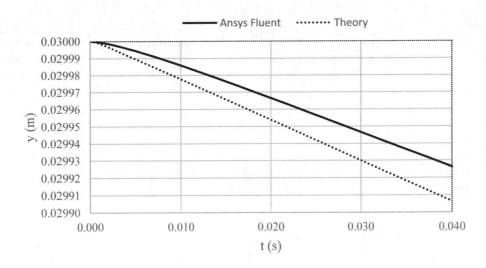

Figure 17.11b) Sphere position versus time for $\lambda = 0.1$

Copy the column Q to column T and copy column L to column U. Label the T column as t (m/s) and the U column as *Ansys Fluent*. The theoretical solution for position is given by equation (17.9) in the theory section. Enter the following expression in cell $V2$:

$$= 3*PI()*0.49*0.002*(9.81*(1-(1020/1695)) / (18*0.49/ (1695*0.002^2))) * (1-Exp(-(18*0.49/ (1695*0.002^2)) *N2/0.791576))$$

Drag the $V2$ value down the whole V column to get the theoretical velocity. Label the V column as *Theory*.

Select columns $T - V$ and select from the Excel menu Insert>>Charts>>Scatter with Smooth Lines, see Figures 17.11c).

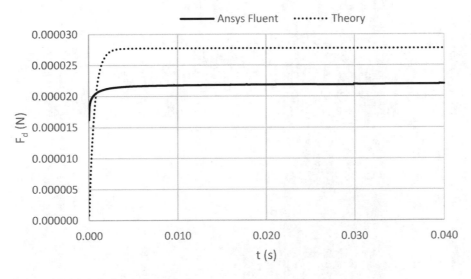

Figure 17.11c) Sphere drag force versus time for $\lambda = 0.1$

Double-click on Graphics and Contours under Results in the Outline View. Select Contours of Velocity… and Velocity Magnitude. Select New Surface>>Plane…. Select YZ Plane as Method and select Create. Close the Plane Surface window. Select *plane-1* as the Surface in the Contours window. Check and uncheck the boxes as shown in Figure 17.11d). Click on Save/Display. The velocity magnitude for the sphere at t = 0.04 s is shown in Figure 17.11e).

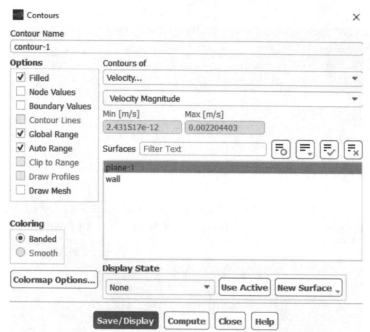

Figure 17.11d) Settings for contours

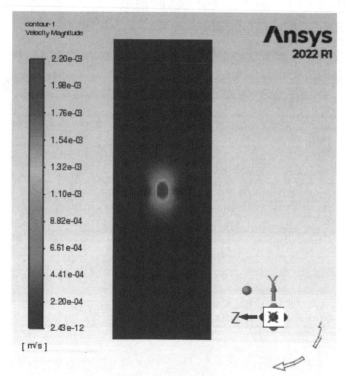

Figure 17.11e) Sphere velocity magnitude at *t* = 0.04 s.

diameter of sphere	d (m)	0.002
diameter of mesh	D (m)	0.02
diameter ratio	$\lambda = d/D$	0.1
height of mesh	H (m)	0.06
wall factor	f	0.791576
density of sphere	ρ_s (kg/m3)	1695
density of fluid	ρ_f (kg/m3)	1020
dynamic viscosity of fluid	μ (Pa-s)	0.49
terminal velocity	U_∞ (m/s)	0.003
modified terminal velocity	$U = f U_\infty$ (m/s)	0.002375
terminal drag force	F_d (N)	0.0000277
Reynolds number	Re	0.009887
upper limit for Reynolds number	Re_{max}	0.027
terminal drag coefficient	C_d	3067
relaxation time	τ_∞ (s)	0.001
upper limit for time step size	Δt (s)	0.00002
element size used to create mesh	Element Size (mm)	0.4
number of nodes for mesh	#Nodes	503,283
number of elements for mesh	#Elements	489,450
mesh count	N	489,450
volume of sphere	V_s (m^3)	4.1888E-09
total volume of mesh	V_{mesh} (m^3)	1.8850E-05
number of cells per particle volume	N_{CPV} (m^{-3})	108
number of cells per particle diameter	N_{CPD} (m^{-1})	5.9
modified terminal velocity from ANSYS Fluent	$f U_\infty$ (m/s), Fluent	0.001942
percent difference between ANSYS Fluent and Theory	% Difference	18
wall factor from ANSYS Fluent	f, Fluent	0.647239
terminal drag force on sphere ANSYS Fluent	F_d (N)	0.0000211
terminal drag coefficient for sphere ANSYS Fluent	C_d	3751

Table 17.1 Parameter values for ANSYS Fluent simulation.

H. Theory

12. For Stokes flow or creeping flow with Reynolds number $\ll 1$ we can define the terminal velocity U_∞ for a sphere falling in an infinite quiescent Newtonian fluid under the influence of gravity as

$$U_\infty = \frac{(\rho_s - \rho_f)d^2 g}{18\mu} \tag{17.1}$$

, where $d \, (m)$ is the diameter of the sphere, ρ_s (kg/m³) is the density of the sphere, ρ_f (kg/m³) is the density of the fluid, μ (kg/m-s) is the dynamic viscosity of the fluid, and g (m/s²) is acceleration due to gravity. The relaxation time for an infinite fluid is defined by

$$\tau_\infty = \frac{\left(\rho_s + \frac{\rho_f}{2}\right)d^2}{18\mu} \tag{17.2}$$

The terminal velocity will be reduced by a factor f by the presence of a surrounding cylindrical wall, see Figure 7.12a). Using Newton's second law we get

$$W - F_d - F_b = m\frac{d^2 Y}{dt^2} = m\frac{dU}{dt} \tag{17.3}$$

, where m is the mass of the sphere, W is the weight of the sphere, F_d is the drag force on the sphere and F_b is the buoyancy force. Using Stokes law in combination with the wall effect factor, we replace the drag force with $F_d = 3\pi\mu dU/f$. We also use the buoyancy force on the sphere $F_b = \rho_f g V_s$ and the weight of the sphere $W = \rho_s g V_s$ to get

$$\rho_s g V_s - 3\pi\mu dU/f - \rho_f g V_s = m\frac{dU}{dt} \tag{17.4}$$

After expressing the volume of the sphere in terms of the diameter of the sphere we get the following

$$\rho_s g \frac{\pi d^3}{6} - 3\pi\mu dU/f - \rho_f g \frac{\pi d^3}{6} = \rho_s \frac{\pi d^3}{6}\frac{dU}{dt} \tag{17.5}$$

This equation can be rewritten as an ordinary first order differential equation for the velocity of a falling sphere in a cylinder

$$\frac{dU}{dt} + bU/f - a = 0 \tag{17.6}$$

, where $a = g\left(1 - \rho_f/\rho_s\right)$ and $b = \frac{18\mu}{\rho_s d^2}$. The solution to equation 17.6 is of the form

$$U = \frac{fa}{b}\left(1 - e^{-bt/f}\right) \tag{17.7}$$

When $t \to \infty$, $U \to fU_\infty = fa/b$, equation 17.7 gives equation 17.1. The location of the sphere can be expressed as

$$y = y_0 + \frac{fa}{b^2}\left[f\left(1 - e^{-bt/f}\right) - bt\right] \tag{17.8}$$

The drag force can now be expressed as

$$F_d = \frac{3\pi\mu dU}{f} = 3\pi\mu d\frac{a}{b}\left(1 - e^{-bt/f}\right) \tag{17.9}$$

When $t \to \infty$, $F_d = \frac{3\pi\mu da}{b} = 3\pi\mu dU_\infty$. The terminal drag coefficient can be expressed as

$$C_d = \frac{F_d}{\frac{1}{2}\rho_f U^2 A} = \frac{3\pi\mu dU}{f\frac{1}{2}\rho_f U^2 \pi d^2/4} = \frac{24\mu}{f\rho_f Ud} = \frac{24}{fRe} \tag{17.10}$$

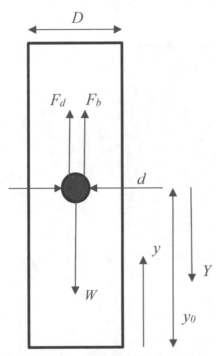

Figure 7.12a) Geometry for falling sphere

The wall factor for very low Reynolds numbers is according to Haberman and Sayre (1958)

$$f = \frac{1 - 2.105\lambda + 2.0865\lambda^3 - 1.7068\lambda^5 + 0.72603\lambda^6}{1 - 0.75857\lambda^5} \tag{17.11}$$

, where $\lambda = d/D$ and D is the mesh diameter or inner diameter of the clear glass graduated cylinder. The modified terminal velocity is defined as $U = fU_\infty$. The upper level of the Reynolds number for the Stokes regime and the wall factor is shown versus λ in Table 17.2 and Figure 17.12, see Chhabra *et al.* (2003).

λ	f	Re_{max}
0.1	0.792	0.027
0.2	0.595	0.040
0.3	0.422	0.050
0.4	0.279	0.083
0.5	0.170	0.18
0.6	0.094	0.52
0.7	0.047	2.10
0.8	0.020	8.40
0.9	0.008	25.17

Table 17.2 Values of wall factor and max Re for different λ

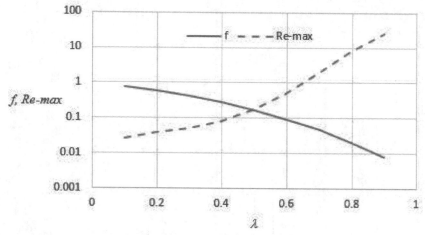

Figure 17.12b) Wall factor f and maximum Reynolds number versus λ

The Reynolds number is defined as

$$Re = \frac{Ud\rho_f}{\mu} = \frac{fU_\infty d\rho_f}{\mu} \tag{17.12}$$

The time step size used in ANSYS Fluent simulations is described by the following

$$\Delta t = \frac{2\tau_\infty}{100} \tag{17.13}$$

The volume for the spherical particle is given by

$$V_s = \frac{\pi d^3}{6} \tag{17.14}$$

The total volume of the mesh including the volume of the sphere used in ANSYS Fluent simulations

$$V_{mesh} = \frac{\pi H D^2}{4} \tag{17.15}$$

, where H is the height of the mesh. This height needs to be a minimum of $H = 3D$ in the ANSYS Fluent simulations. We can now determine the number of cells per particle volume

$$N_{CPV} = \frac{NV_s}{V_{mesh}} \tag{17.16}$$

, where N is the mesh count or number of elements for the mesh. The number of cells per particle diameter is defined as

$$N_{CPD} = \left(\frac{6N_{CPV}}{\pi}\right)^{\frac{1}{3}} = \left(\frac{6NV_s}{\pi V_{mesh}}\right)^{\frac{1}{3}} \tag{17.17}$$

, and was chosen to be at least 5 in the ANSYS Fluent simulations. The modified terminal velocity will depend on this parameter in simulations.

I. References

1. Agrawal M., Bakker A. and Prinkey M.T. "Macroscopic particle model – tracking big particles in CFD", *Proceedings of AIChE 2004 Annual Meeting, Austin, TX,* (2004).
2. Agrawal M.,Ookawara S. and Ogawa K. "Drag Force Formulation in Macroscopic Particle Model and its validation", *Proceedings of AIChE 2009 Annual Meeting,* (2009).
3. Chhabra R.P., Agarwal S. and Chaudhary K. "A note on wall effect on the terminal falling velocity of a sphere in quiescent Newtonian media in cylindrical tubes", *Powder Technology*, **129**, 53 - 58, (2003).
4. Haberman W.L and Sayre R.M. "David Taylor Model Basin Report No. 1143", *Department of Navy, Washington, DC* (1958).
5. Ookawara, S., M. Agrawal, D. Street and K. Ogawa. Modeling the motion of a sphere falling in a quiescent Newtonian liquid in a cylindrical tube by using the macroscopic particle model. In: The Seventh World Congress of Chemical Engineering. Glasgow, Scotland. C39-004, (2005).

J. Exercises

17.1 Run ANSYS Fluent Simulations for $\lambda = 0.1 - 0.9$ as shown in Table 17.3. Fill out the missing results in the table based on your simulations. Use the element size as listed in the table when you create the mesh for each λ. Change the #Nodes, #Elements and Mesh Count in the table to the values that you get when creating the mesh for each λ. The value for Mesh Count is equal to the value for #Elements. #Cells/Particle Volume and #Cells/Particle Diameter will need to be recalculated based on your actual values. Plot the modified terminal velocity versus λ from ANSYS Fluent simulations and compare with Haberman and Sayre (1958).

d (m)	0.002	0.004	0.006	0.008	0.01	0.012	0.014	0.016	0.018
D (m)	0.02	0.02	0.02	0.02	0.02	0.02	0.02	0.02	0.02
$\lambda = d / D$	0.1	0.2	0.3	0.4	0.5	0.6	0.7	0.8	0.9
H (m)	0.06	0.06	0.06	0.06	0.06	0.06	0.06	0.06	0.06
f	0.79157616	0.595336803	0.421995126	0.279200952	0.170357599	0.09440553	0.046675065	0.020398637	0.008245803
ρ_s(kg/m3)	1695	1695	1695	1695	1695	1695	1695	1695	1695
ρ_f(kg/m3)	1020	1020	1020	1020	1020	1020	1020	1020	1020
μ(Pa-s)	0.49	1.96	4.41	7.84	12.25	17.64	24.01	31.36	39.69
U_∞(m/s)	0.003	0.003	0.003	0.003	0.003	0.003	0.003	0.003	0.003
fU_∞(m/s)	0.00237473	0.00178601	0.001265985	0.000837603	0.000511073	0.000283217	0.000140025	6.11959E-05	2.47374E-05
Re	0.00988662	0.003717818	0.001756878	0.000871791	0.000425546	0.000196518	8.32803E-05	3.18469E-05	1.14432E-05
Re_{max}	0.027	0.04	0.05	0.083	0.18	0.52	2.1	8.4	25.17
τ_∞ (s)	0.001	0.001	0.001	0.001	0.001	0.001	0.001	0.001	0.001
Time Step Size (s)	0.00002	0.00002	0.00002	0.00002	0.00002	0.00002	0.00002	0.00002	0.00002
#Nodes	346,060	47,141	13,720	6,450	3,059	1,890	1,092	980	624
#Elements	335,787	44,776	12,753	5,829	2,684	1,600	840	741	465
Element Size(mm)	0.465	1.045	1.555	2.065	2.705	3.16	317	320	420
Mesh Count	335,787	44,776	12,753	5,829	2,684	1,600	840	741	465
Particle Volume (m^3)	4.1888E-09	3.3510E-08	1.1310E-07	2.6808E-07	5.2360E-07	9.0478E-07	1.4368E-06	2.1447E-06	3.0536E-06
Total Volume (m^3)	1.8850E-05	1.8850E-05	1.8850E-05	1.8850E-05	1.8850E-05	1.8850E-05	1.8850E-05	1.8850E-05	1.8850E-05
#Cells/Particle Volume	74.62	79.60	76.52	82.90	74.56	76.80	64.03	84.31	75.33
#Cells/Particle Diameter	5.2	5.3	5.3	5.4	5.2	5.3	5.0	5.4	5.2
fU_∞ (m/s), Fluent									
% Diff									
f, Fluent									

Table 17.3 Data for $\lambda = 0.1 - 0.9$

17.2 Run ANSYS Fluent Simulations for $\lambda = 0.1$ as shown in Table 17.3 for different values of #Cells/Particle Diameter from 2 – 10. Include a plot of f versus #Cells/Particle Diameter.

CHAPTER 18. FLOW PAST A SPHERE

A. Objectives

- Using ANSYS DesignModeler to Model the Sphere
- Inserting Boundary Conditions and Free Stream Velocity
- Running the Calculations
- Using Velocity and Pressure Plots for Visualizations
- Determining the Drag Coefficient

B. Problem Description

We will study the axisymmetric flow past a sphere and we will analyze the problem using ANSYS Fluent. A 50 mm diameter sphere will be used and the Reynolds number will be set to $Re = 100$.

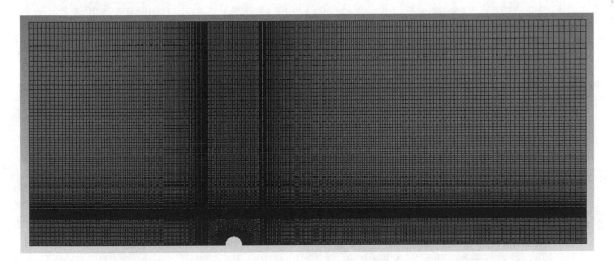

C. Launching ANSYS Workbench and Selecting Fluent

1. Start by launching ANSYS Workbench. Double click on Fluid Flow (Fluent) under Analysis Systems in the Toolbox. Select Units>>Metric (tonne, mm, s, …) from the menu.

Figure 18.1 Selecting Fluent

D. Launching ANSYS DesignModeler

2. Right click Geometry and select Properties. In Properties of Schematic A2: Geometry, select Analysis Type 2D under Advanced Geometry Options. Right click on Geometry in the Project Schematic window and select New DesignModeler Geometry. Select millimeter as the desired length unit from the menu (Units>>Millimeter).

Select the Sketching tab and click on Arc by Center. Click on Look at Face/Plane/Sketch

. Draw in the upper two quadrants a half-circle centered at the origin of the coordinate system in the graphics window. Click on the Dimensions tab and select Radius. Click on the half-circle in the graphics window and enter 25 mm for the radius. Draw a line between the two end-points of the half-circle.

Select Concepts>>Surfaces from Sketches from the menu. Control select the half-circle and the line in the graphics window, and select Apply for Base Objects in Details View. Click on Generate.

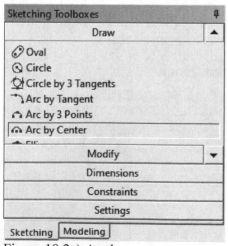

Figure 18.2a) Arc by center

Details View	⊒
⊟ **Details of Sketch1**	
Sketch	Sketch1
Sketch Visibility	Show Sketch
Show Constraints?	No
⊟ **Dimensions: 1**	
☐ R1	25 mm

Figure 18.2b) Half-circle and line

Details View	⊒
⊟ **Details of SurfaceSk1**	
Surface From Sketches	SurfaceSk1
Base Objects	1 Sketch
Operation	Add Material
Orient With Plane Normal?	Yes
Thickness (>=0)	0 mm

Figure 18.2c) Finished half-circle

Next, we will be creating the 2D axisymmetric mesh region around the half-sphere. Click on the modeling tab and select the XYPlane in the Tree Outline. Click on to generate a new sketch. Select Rectangle by 3 Points from the sketching tools. Click anywhere at the upper left corner of the graphics window, followed by clicking anywhere on the left-hand plane on the horizontal axis, and finally click on the right-hand plane on the horizontal axis. Select Dimensions, select the Horizontal dimensioning tool and click on the vertical edges of the rectangle. Enter 1575 mm as the horizontal length of the rectangle. Click on the vertical axis and the vertical edge of the rectangle on the left-hand side. Enter 575 mm as the length. Select the Vertical dimensioning tool and click on the horizontal edges of the rectangle. Enter 625 mm as the vertical height of the rectangle.

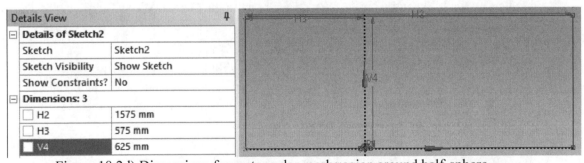

Details View	⊒
⊟ **Details of Sketch2**	
Sketch	Sketch2
Sketch Visibility	Show Sketch
Show Constraints?	No
⊟ **Dimensions: 3**	
☐ H2	1575 mm
☐ H3	575 mm
☐ V4	625 mm

Figure 18.2d) Dimensions for rectangular mesh region around half-sphere

3. Select Concept>>Surface from Sketches. Click on Sketch2 in the Tree Outline and select Apply for Base Objects in Details View. Select Operation>>Add Frozen in Details View followed by Generate. Select Create>>Boolean from the menu. Select Subtract Operation from the Details View. Select the mesh region in the graphics window and click on Apply for Target Body. Select the Sphere as the Tool Body and click on Generate.

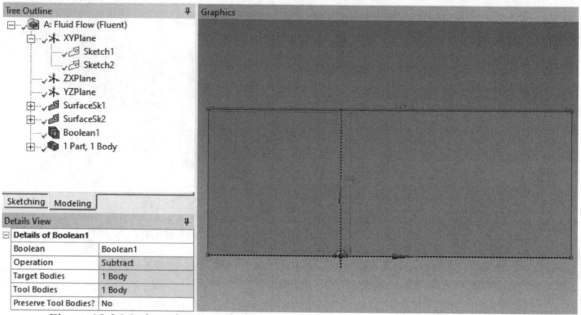

Figure 18.3 Mesh region around sphere

4. Select XYPlane in the tree outline and create a new sketch 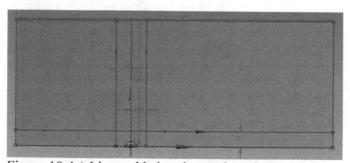. Select the Sketching tab and select Line. Draw a vertical line from top to bottom and to the left of the sphere. The vertical line will intersect the entire mesh region. Position the line 75 mm from the vertical axis. Select Concepts>>Lines from Sketches from the menu. Select the newly created vertical line and Apply it as a Base Object in the Details View. Click on Generate.

Create another sketch in the XYPlane with a vertical line 75 mm to the right of the vertical axis and repeat all the different steps as listed above. Repeat these steps one more time with a horizontal line from left to right intersecting the mesh region and positioned 75 mm above the half-sphere centerline.

Figure 18.4a) Lines added to the mesh region

Create another sketch in the XYPlane with a vertical line towards the half-sphere as shown in Figure 18.4b). Repeat the steps completed earlier. Finally, create the last sketch with two diagonal lines at 45 degrees towards the half-sphere as shown in Figure 18.4b). Use the Angle tool under Dimensions to set the angle to 45 degrees for both lines.

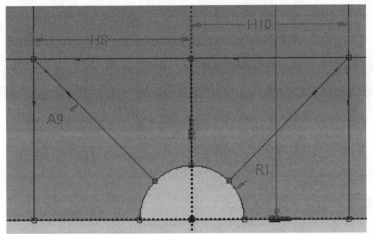

Figure 18.4b) Lines added to the mesh region close to the half-sphere

5. Select Tools>>Projection from the menu. Select Edges on Face as type in Details View.

    Use the Edge selection filter 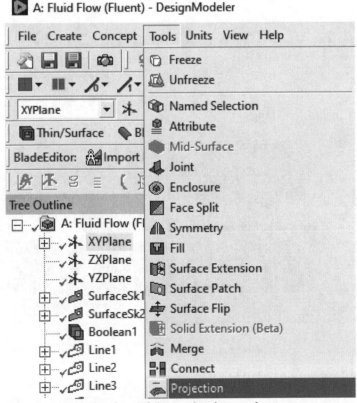 and control-select all 11 line-segments from the 6 lines that you have created and Apply as the Edges in the Details View. Select the mesh region and Apply it as the Target under Details View. Click Generate.

Figure 18.5 Selection of the projection tool

You should now have 9 different regions for meshing. Select File>>Save Project from the menu and save the project with the name *Axisymmetric Sphere Flow Study.wbpj*. Close DesignModeler.

E. Launching ANSYS Meshing

6. Double click on Mesh under Project Schematic in ANSYS Workbench. Select Mesh in the Outline under Project. Select Home>>Tools>>Units>>Metric (mm, kg, N, ...) from the menu. Right click on Mesh under Project in Outline and select Generate Mesh. A coarse mesh is generated.

    Select Mesh under Model (A3) in Project under Outline on the left-hand side. Select the Mesh tab in the menu and Controls>>Face Meshing. Control-select all 9 faces of the mesh. Apply the Geometry under Details of "Face Meshing".

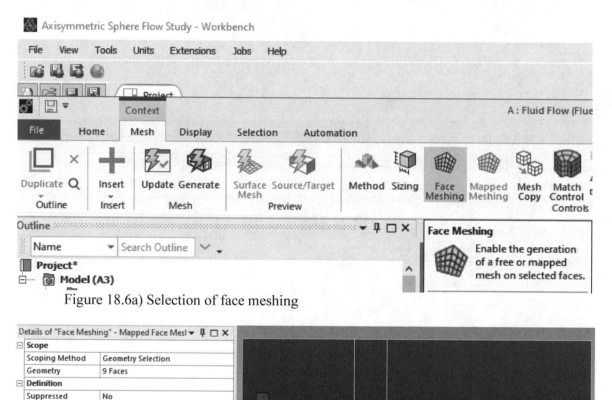

Figure 18.6a) Selection of face meshing

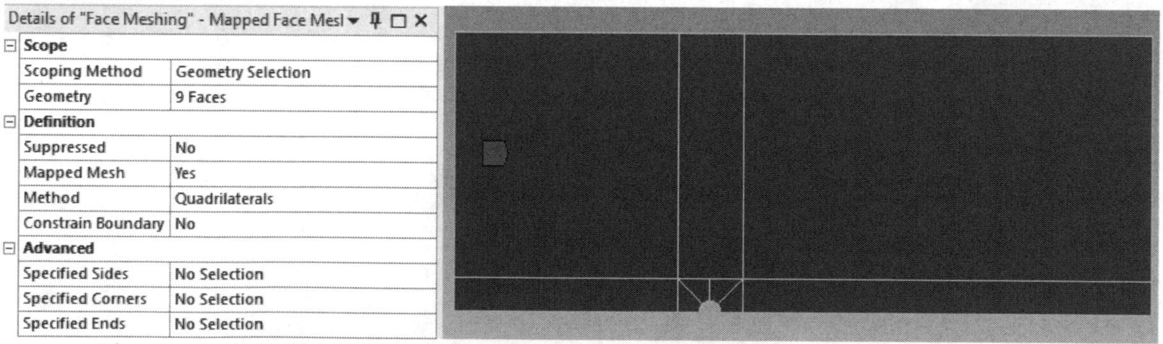

Figure 18.6b) Details of face meshing

Select Mesh>>Controls>>Sizing from the menu, right-click in the graphics window and select Cursor Mode>>Edge. Control-select the two edges 8 and 19, see Figure 18.6c).

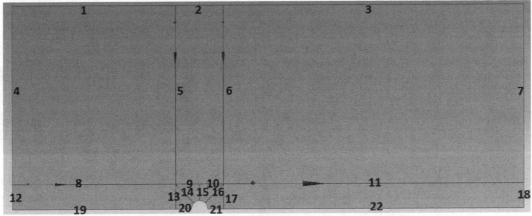

Figure 18.6c) Numbered edges for mesh sizing

Apply the Geometry under Details of "Sizing". Select Number of Divisions as Type under Details of "Sizing". Set the number of Divisions to 100. Set the Behavior to Hard. Select the first Bias Type from the drop-down menu. Enter a Bias Factor of 10.

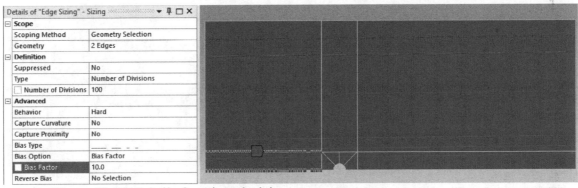

Figure 18.6d) Details for 1st mesh sizing

Select Mesh Control>>Sizing from the menu. Control-select edges 11 and 22. Apply the Geometry under Details of "Sizing". Select Number of Divisions as Type under Details of "Sizing". Set the number of Divisions to 150. Set the Behavior to Hard. Select the second Bias Type from the drop-down menu. Enter a Bias Factor of 10.

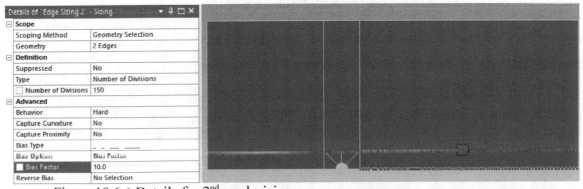

Figure 18.6e) Details for 2nd mesh sizing

Select Mesh Control>>Sizing from the menu. Select edge 1. Apply the Geometry under Details of "Sizing". Select Number of Divisions as Type under Details of "Sizing". Set the number of Divisions to 100. Set the Behavior to Hard. Select the second Bias Type from the drop-down menu. Enter a Bias Factor of 10.

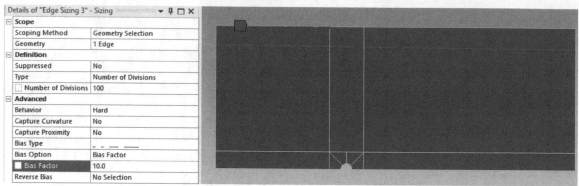

Figure 18.6f) Details for 3rd mesh sizing

Select Mesh Control>>Sizing from the menu. Select edge 3. Apply the Geometry under Details of "Sizing". Select Number of Divisions as Type under Details of "Sizing". Set the number of Divisions to 150. Set the Behavior to Hard. Select the first Bias Type from the drop-down menu. Enter a Bias Factor of 10.

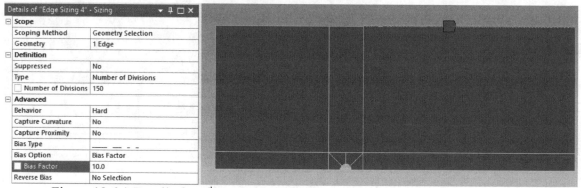

Figure 18.6g) Details for 4th mesh sizing

Select Mesh Control>>Sizing from the menu. Select edge 2. Apply the Geometry under Details of "Sizing". Select Number of Divisions as Type under Details of "Sizing". Set the number of Divisions to 54. Set the Behavior to Hard.

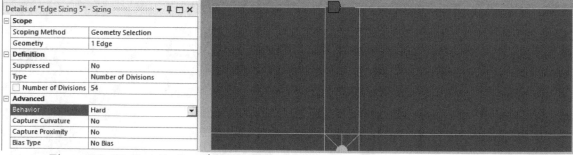

Figure 18.6h) Details for 5th mesh sizing

412

Select Mesh Control>>Sizing from the menu. Control-select edges 9, 10, 12, 13, 17, 18, and the four parts of the half-circle. Apply the Geometry under Details of "Sizing". Select Number of Divisions as Type under Details of "Sizing". Set the number of Divisions to 27. Set the Behavior to Hard.

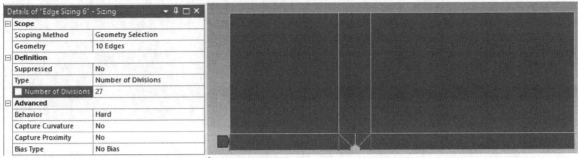

Figure 18.6i) Details for 6th mesh sizing

Select Mesh Control>>Sizing from the menu. Control-select the three edges 4, 5, and 6. Apply the Geometry under Details of "Sizing". Select Number of Divisions as Type under Details of "Sizing". Set the number of Divisions to 100. Set the Behavior to Hard. Select the first Bias Type from the drop-down menu. Enter a Bias Factor of 20.

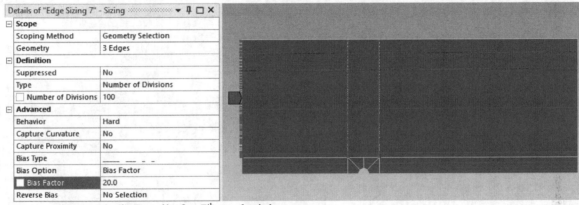

Figure 18.6j) Details for 7th mesh sizing

Select Mesh Control>>Sizing from the menu. Select edge 7. Apply the Geometry under Details of "Sizing". Select Number of Divisions as Type under Details of "Sizing". Set the number of Divisions to 100. Set the Behavior to Hard. Select the second Bias Type from the drop-down menu. Enter a Bias Factor of 20.

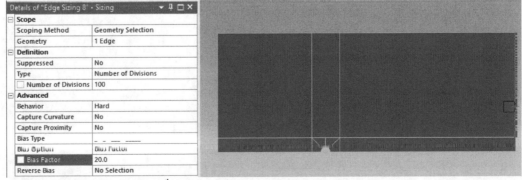

Figure 18.6k) Details for 8th mesh sizing

Select Mesh Control>>Sizing from the menu. Select edges 14, 15, 16, 20, and 21. Apply the Geometry under Details of "Sizing". Select Number of Divisions as Type under Details of "Sizing". Set the number of Divisions to 18. Set the Behavior to Hard.

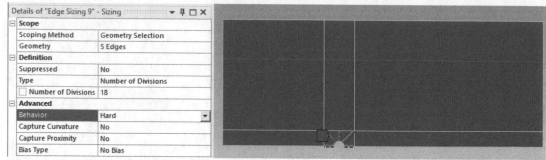

Figure 18.6l) Details for 9th mesh sizing

Right click on Mesh in the tree outline and select Generate Mesh.

Figure 18.6m) Details of mesh around half-sphere

7. Select geometry in the Outline. Right-click in the graphics window and select Cursor Mode>>Edge. Control select the two vertical edges on the right-hand side, right click and select Create Named Selection. Name the edges *outlet* and click OK.

    Name the two vertical edges to the left *inlet* and the four edges of the sphere with the name *sphere*.

    Name the three horizontal edges at the top *symmetry* and name the four horizontal edges at the bottom *axisymmetry*.

    Export the mesh and save it with the name *flow-past-sphere-mesh.msh*.

    Open Geometry in the Outline and select Line Body. Right click on Line Body and select Suppress Body.

    Close the meshing window. Right click on Mesh under Project Schematic in Ansys Workbench and select Update.

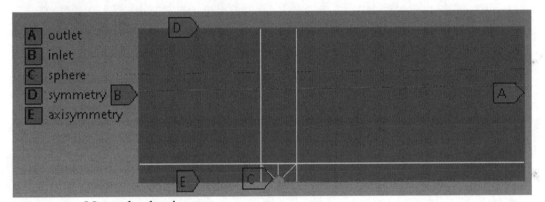

Figure 18.7 Named selections

F. Launching ANSYS Fluent

8. Double click on Setup under Project Schematic in Ansys Workbench. Check the Options box Double Precision. Set the number of processes equal to the number of processor cores for your computer. To check the number of physical cores, press the Ctrl + Shift + Esc keys simultaneously to open the Task Manager. Go to the Performance tab and select CPU from the left column. You will see the number of physical cores on the bottom-right side. Close the Task Manager window. Click on Show More Options and write down the location of the Working Directory. Click on the Start button in the Fluent Launcher window. Select 2D Space Axisymmetric under General on the Task Page.

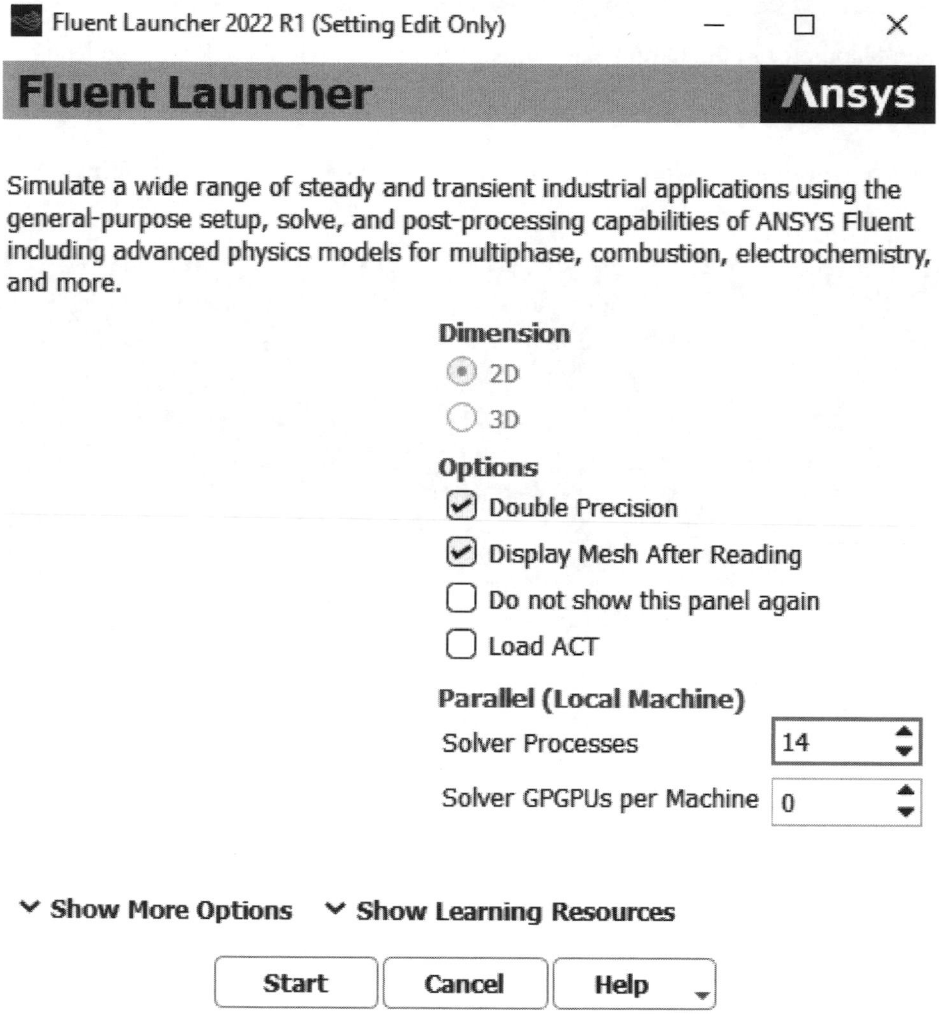

Figure 18.8a) Launching Fluent

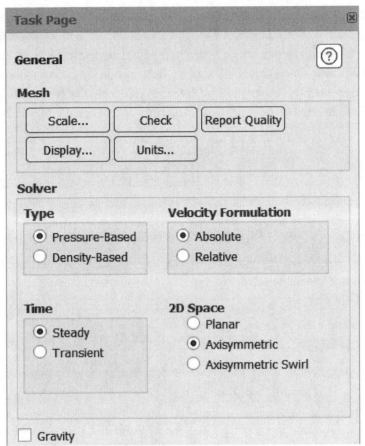

Figure 18.8b) General settings

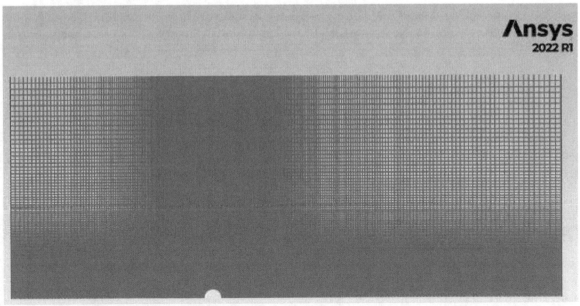

Figure 18.8c) Geometry in Fluent

9. Double click on Models under Setup in the Outline View. Double click on Viscous Model and select Laminar Model. Click on OK to close the Viscous Model window. Double click on Boundary Conditions under Setup in the Outline View. Select the *inlet* Zone under Boundary Conditions on the Task Page. Click on Edit…. Select Components as Velocity Specification Method. Enter 0.02921469 as Axial-Velocity [m/s]. This is corresponding to Reynolds number *Re* = 100. Select Apply and Close the Velocity Inlet window.

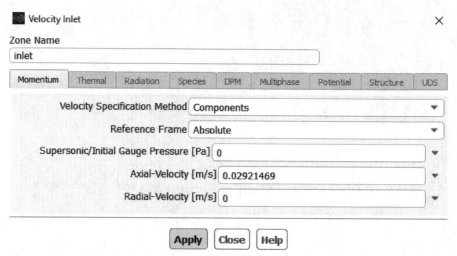

Figure 18.9 Velocity inlet settings

10. Double click on Initialization under Solution in the Outline View. Select Standard Initialization. Select Compute from inlet. Click on Initialize. Double click on Reference Values under Setup in the Outline View. Select Compute from inlet and select Initialize. Double click on Reference Values under Setup in the Outline View. Select Compute from inlet and set the reference values for Area [m^2] to 0.001963496, Length [m] to 0.05 and Velocity [m/s] to 0.02921469, see Figure 18.10b). Set the Reference Zone to *solid-surface_body*.

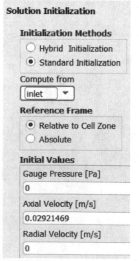
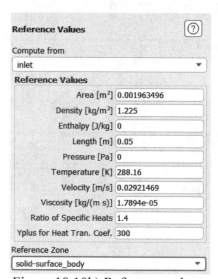

Figure 18.10a) Solution initialization   Figure 18.10b) Reference values

418

Double-click on Report Definitions under Solution in the Outline View. Select New>>Force Report>>Lift… from the drop-down menu. Select *sphere* as Wall Zones. Check the boxes for Report File, Report Plot and Print to Console under Create. Enter *lift-report-def* as Name. Click on the OK button to close the Lift Report Definition window. Repeat this step for the Drag and set the Force Vector for X to 1, Y to 0 for the Drag Report Definition. Enter *drag-report-def* as Name. Close the Report Definitions window.

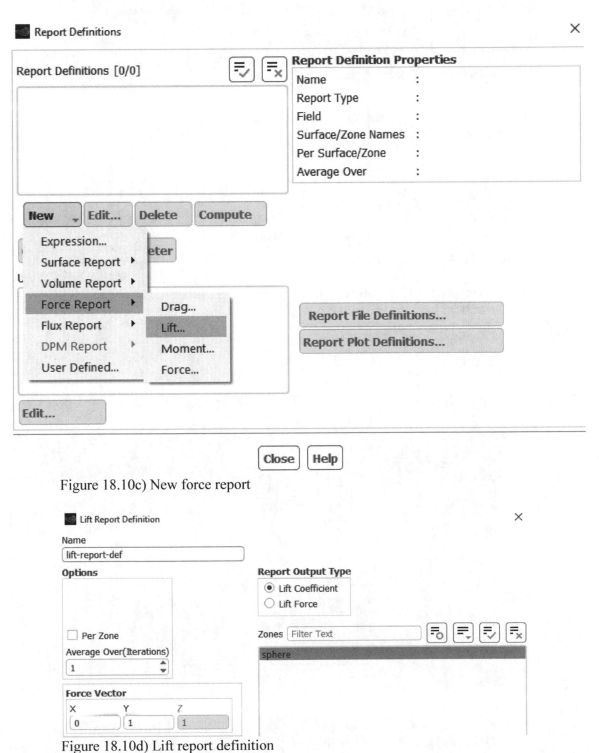

Figure 18.10c) New force report

Figure 18.10d) Lift report definition

11. Double click on Monitors and Residual under Solution in the Outline View. Check the boxes for Print to Console and Plot. Set all three Absolute Criteria to 1e-12. Click on OK to exit the Residual Monitors window.

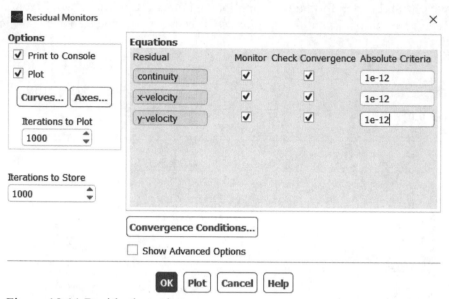

Figure 18.11 Residual monitors

Double click on Run Calculation under Solution in the Outline View. Enter 300 for the Number of Iterations and click on Calculate.

G. Post-Processing

12. The value of the drag coefficient at $Re = 100$ for the first coarse mesh is $C_d = 1.08837$.

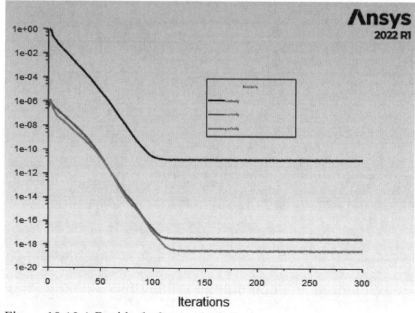

Figure 18.12a) Residuals from calculations

Double-click on Contours in Graphics under Results in the Outline View. Enter *static-pressure* as Contour Name in the Contours window. Select Contours of Pressure... and Static Pressure. Check the box Filled under Options. Deselect all Surfaces and click on Save/Display. Close the Contours window.

Figure 18.12b) Contours plot for static pressure

Double-click on Contours in Graphics under Results in the Outline View. Select Contours of Velocity and Axial Velocity. Enter *axial-velocity* as Contour Name in the Contours window. Check the box Filled under Options. Deselect all surfaces and click on Save/Display. Close the Contours window. Find the coefficient of drag by double-clicking on Reports under Results in the Outline View. Double-click on Forces under Reports. Select sphere under Wall Zones. Select Print to get the value for the total drag coefficient that you have calculated: 1.0883686.

Figure 18.12c) Contours plot for axial velocity

13. Select the Domain tab in the menu. Select Adapt>>Manual…. under the Domain tab from the menu. Select Cell Registers>>New>>Region… in the Manual Mesh Adaption window and enter -0.575 for X Min [m], 1 for X Max [m], 0 for Y Min [m], and 0.625 for Y Max [m]. Select Save and Close the Region Register window. Select General Adaption Controls and set the Maximum Refinement Level to 10. Select *region_0* as Refinement Criterion in the Manual Mesh Adaption window and select Adapt. Select OK in the Information window. This increases the number of cells by a factor of 4 from 39094 to 156376 cells. Close the Manual Mesh Adaption window. Select File>>Save Project from the menu. Select File>>Export>>Case & Data… from the menu. Save the file in the Working Directory as Files of type: CFF Case/Data Files (*.cas.h5 *.dat.h5) with the name *Axisymmetric Coarse Mesh Flow Past a Sphere at Re = 100.*

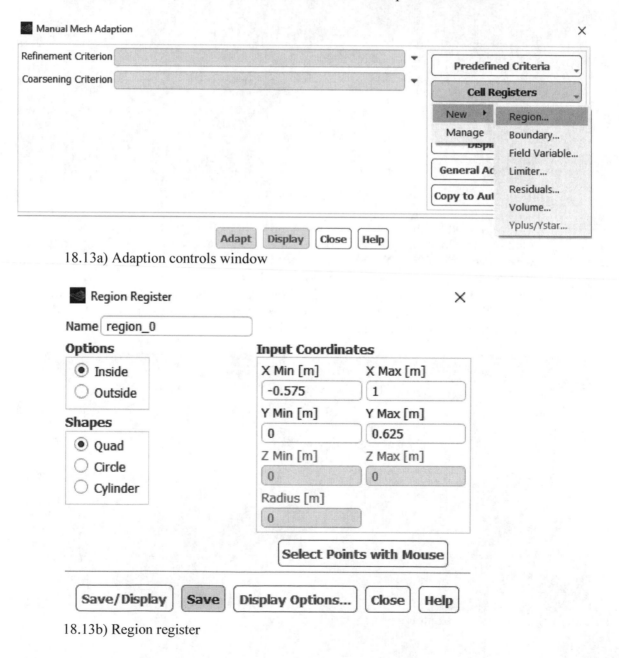

18.13a) Adaption controls window

18.13b) Region register

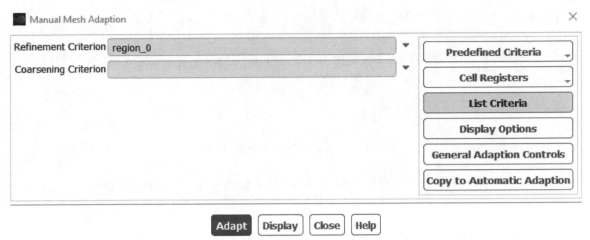

18.13c) Region_0 as refinement criterion

Initialize the flow and run the calculations once again for this new mesh. Enter 300 for the Number of Iterations and click on Calculate. The value of the drag coefficient at $Re = 100$ for a refined mesh is $C_d = 1.0955586$. Continue refining the mesh further to see how the value for the drag coefficient is affected.

Number of Cells	$C_{d,\ Fluent}$	$C_{d,\ Experiment}$	% difference
39094	1.0883686	1.1855	8.2
156376	1.0953921	1.1855	7.6
625504	1.0892230	1.1855	8.1
2502016	1.0888105	1.1855	8.2
10008064	1.0891897	1.1855	8.1

Table 18.1 Drag coefficient with a refined mesh at $Re = 100$.

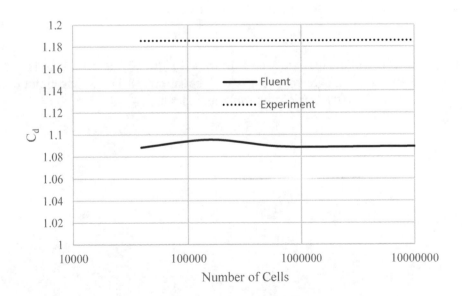

18.13d) Drag coefficient for sphere versus number of cells at $Re = 100$.

H. Theory

14. The drag coefficient from experiments for a smooth sphere is found using the following curve-fit formula:

$$C_{D,Exp} = \frac{24}{Re} + \frac{6}{1+Re^{1/2}} + 0.4 \qquad 0 \le Re \le 200,000 \tag{18.1}$$

We determine the Reynolds number for the flow around the sphere that is defined as

$$Re = Ud/v = 0.02921469 * 0.05 / 1.46 * 10^{-5} = 100 \tag{18.2}$$

, where U is the magnitude of the free stream velocity, d is the diameter of the sphere, and v is the kinematic viscosity of air at room temperature. The difference between Ansys Fluent and experiments for the drag coefficient is 8.1%.

I. References

1. Clift, R., Grace, J.R. and Weber, M.E. "Bubbles, Drops and Particles", Academic Press, New York, (1978).
2. Jones, D.A. and Clarke, D.B. "Simulation of Flow Past a Sphere using the Fluent Code", Australian Government Department of Defence, Defence Science and Technology Organisation, Maritime Platforms Division, DSTO-TR-2232, (2008).
3. Roos, F.W. and Willmarth, W.W. "Some Experimental Results on Sphere and Disk Drag", AIAA Journal, **9**(2), 285-291, (1971).
4. Tabata, M. and Itakura, K., "A Precise Computation of Drag Coefficients of a Sphere", Int. J. Comput. Fluid Dynam. **9**, 303, (1998).

J. Exercises

18.1 Run ANSYS Fluent Simulations for Re = 20, 40, 60, 80, 100, 120, 140, 160, 180 and 200 and compare your results with equation (18.1). Include a plot of drag coefficient versus Reynolds number in this range.

CHAPTER 19. TAYLOR-COUETTE FLOW

A. Objectives

- Using ANSYS Meshing to Create the Mesh for 3D Taylor-Couette Flow
- Inserting Wall Boundary Conditions for Rotating Inner Wall
- Running Laminar Steady 3D ANSYS Fluent Simulations
- Using Contour Plots for Visualizations
- Compare Results with Neutral Stability Theory

B. Problem Description

We will study the flow between two vertical cylinders where the inner wall is rotating at 1.5 rad/s. The inner and outer cylinders have radii of 30 mm and 35 mm, respectively. The height of both cylinders is 100 mm. The centrifugal instability will cause the appearance of counter-rotating vortices at low rotation speeds for the inner wall with a fixed outer wall.

C. Launching ANSYS Workbench and Selecting Fluent

1. Start by launching ANSYS Workbench. Double click on Fluid Flow (Fluent) under Analysis Systems in the Toolbox. Select Units>>Metric (tonne, mm, s, …) from the menu.

Figure 19.1 Selecting Fluent

D. Launching ANSYS DesignModeler

2. Right click on Geometry in the Project Schematic window and select New DesignModeler Geometry. Select millimeter as the desired length unit from the menu (Units>>Millimeter) in DesignModeler. Select the ZX Plane in the Tree Outline and select Look at Face/Plane/Sketch ⬚. Select the Sketching tab and click on Circle. Draw two circles centered at the origin. Make sure that you see a *P* next to the cursor before you start drawing the circles from the origin. Click on the Dimensions tab and select Radius. Click on the circles in the graphics window and enter 30 mm and 35 mm for the radii of the circles in Details View.

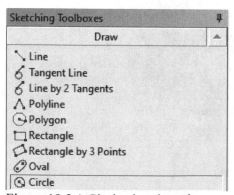

Figure 19.2a) Circle sketch tool

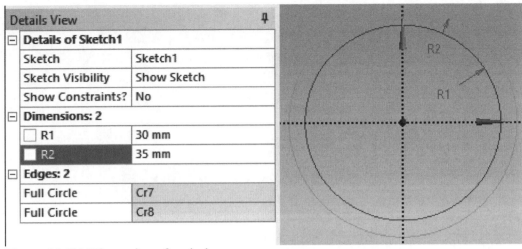

Figure 19.2b) Dimensions for circles

Select 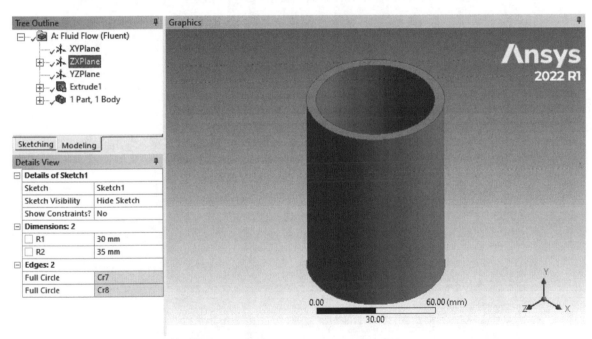 Extrude and apply Sketch1 under ZXPlane in the Tree Outline as Geometry in Details View. Enter 100 mm as the FD1, Depth (>0) of the extrusion. Click on Generate and select isometric view in the graphics window.

Figure 19.2c) Details for extrusion

Create a new sketch in the ZX plane. Select the ZXPlane in the Tree Outline, select New Sketch and Look At Face/Plane/Sketch. Select the new Sketch2 and select the Sketching tab and Line under Draw. Draw a line on the vertical axis through the extruded tube, see Figure 19.2d). Make sure you have the letter C at the starting point, letter V indicating a vertical line and the letter C at the end point when you complete the vertical line.

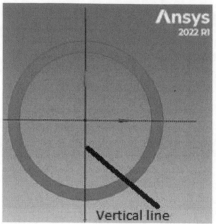

Figure 19.2d) Vertical line through tube

Select Tools>>Face Split from the menu. Turn around the extruded tube and select the face that the line goes through, see Figure 19.2e). Apply the selected face as Target Face in Details View. Select the end points of the vertical line and apply it as Tool Geometry in Details View. Click on Generate and select Face ⬚ . The face has now been split and you can select half the face individually, see Figure 19.2f).

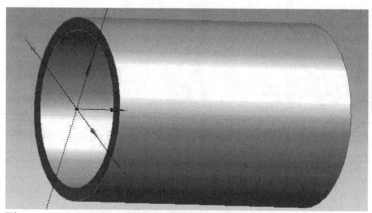

Figure 19.2e) Selected face associated with the line

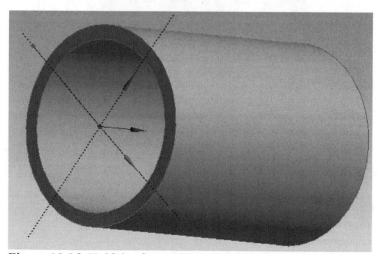

Figure 19.2f) Half the face selected after completed split

Select Tools>>Projection from the menu. Select Edges on Face as Type in Details View. Control select the two radial line segments and apply them as Edges in Details View, see Figure 19.2g). Turn around the extrusion and select the face on the opposite side. Apply this face as Target in Details View, see Figure 19.2g). Click on Generate. You have now created a total of four different flat surfaces, two at each end of the tube.

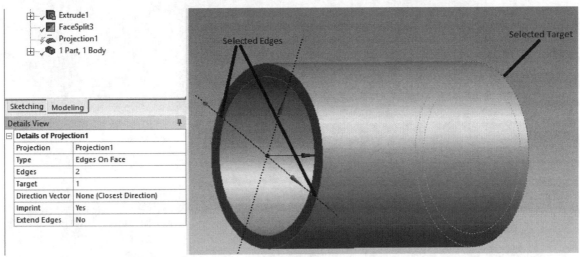

Figure 19.2g) Selected edges and target for projection

Turn the tube around to the other end and select Tools>> Face Split once again from the menu. Apply the inner cylindrical face as Target Face in Details View. Apply the point and edge as Tool Geometry as shown in Figure 19.2h) to split the inner cylindrical face. Click on Generate.

Repeat the face split once again but on the opposite side of the inner cylindrical face, see Figure 19.2i). The inner cylindrical face has now been split and you can select half the face individually, see Figure 19.2j). Close the DesignModeler window.

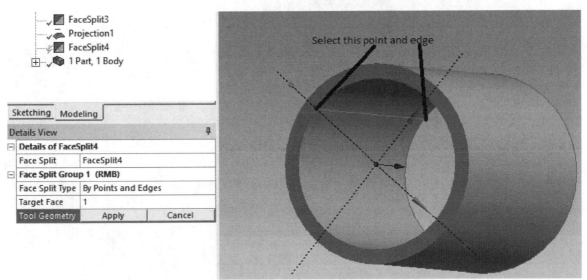

Figure 19.2h) Selected point and edge for face split on one side of inner cylindrical face

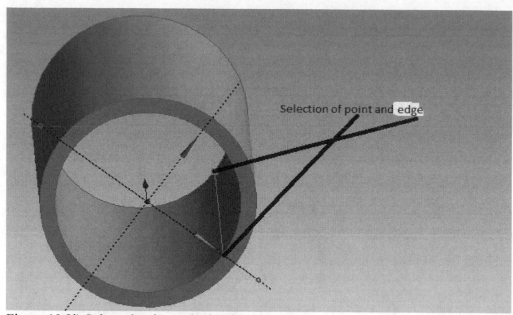

Figure 19.2i) Selected point and edge for split on opposite side of inner cylindrical face

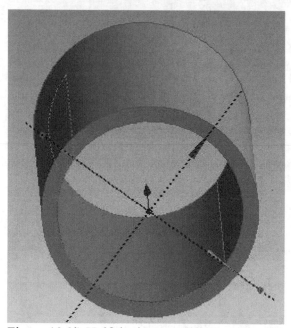

Figure 19.2j) Half the inner cylindrical face selected after completed split

E. Launching ANSYS Meshing

3. Double click on Mesh under Project Schematic in Ansys Workbench. Click on Mesh in the Outline under Project. Select Home>>Tools>>Units>>Metric (mm, kg, N, …) from the menu. Select the Mesh under Project in Outline.

    Open Quality in Details of "Mesh" and set the Smoothing to High, see Figure 19.3a). Generate the Mesh. A tetrahedral mesh is generated.

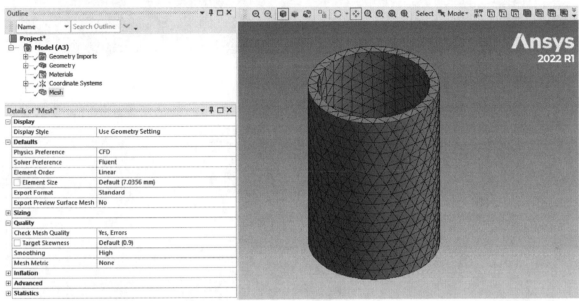

Figure 19.3a) Tetrahedral mesh

Select Mesh>>Controls>>Sizing from the menu. Select Edge , select the radial edge at the top of the extrusion and the corresponding radial line at the bottom of the tube, see Figure 19.3b). Apply the two edges as Geometry in Details of "Edge Sizing" – Sizing. Select Number of Divisions as Type under Definition. Set the Number of Divisions to 32 and Behavior to Hard. Select the third Bias Type - — —— — - and set the Bias Factor to 20, see Figure 19.3c). Repeat this sizing for the other side of the extrusion.

Figure 19.3b) Radial edges on extrusion

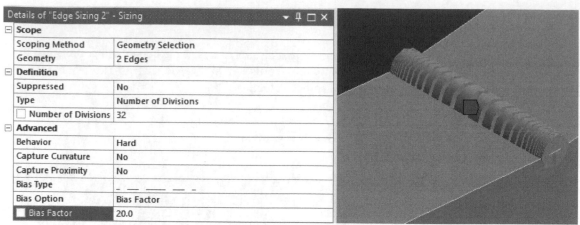

Figure 19.3c) Radial sizing with bias

Right click on Mesh in Outline and select Insert>>Method. Select the extrusion body and Apply it as Geometry under Scope in Details of "Automatic Method" – Method. Select MultiZone as Method under Definition, see Figure 19.3d).

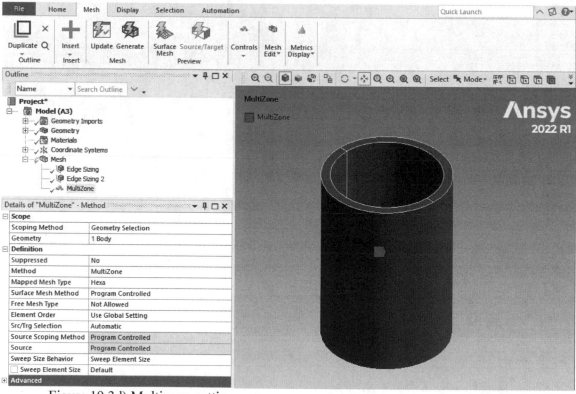

Figure 19.3d) Multizone settings

Select Mesh>>Controls>>Sizing from the menu. Select Edge , control select the eight half-circles at the top and the bottom of the extrusion and Apply the eight Edges as Geometry under Scope in Details of "Sizing" – Sizing, see Figure 19.3e). Select Number of Divisions as Type under Definition. Set the Number of Divisions to 64 and Behavior to Hard. Right click on Mesh in Outline and select Mesh>>Update from the menu.

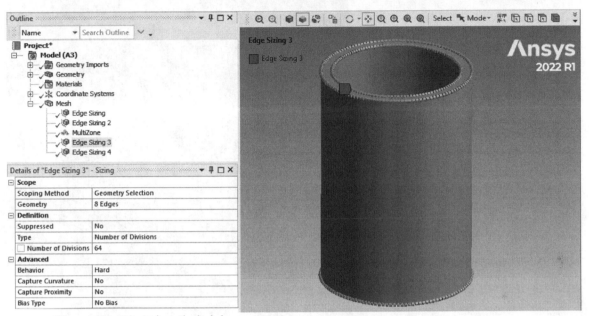

Figure 19.3e) Azimuthal sizing

Select Mesh>>Controls>>Sizing from the menu. Select Edge , control select the two inner vertical lines of the extrusion and Apply the two Edges as Geometry under Scope in Details of "Sizing" – Sizing, see Figure 19.3f). Select Number of Divisions as Type under Definition. Set the Number of Divisions to 256 and Behavior to Hard. Right click on Mesh in Outline and select Mesh>>Update from the menu.

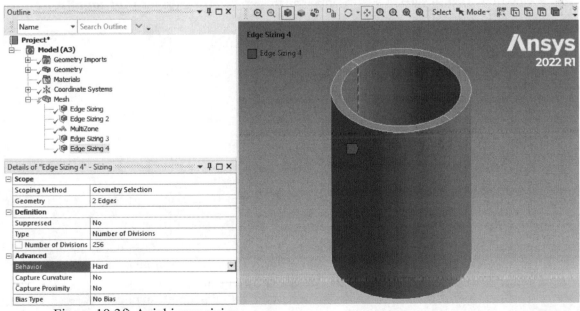

Figure 19.3f) Axial inner sizing

433

Open Statistics in Details of "Mesh" to see the total number of Nodes 1,085568 and total number of Elements 1,048,576.

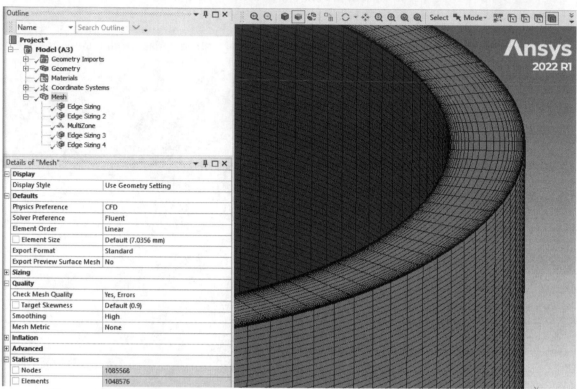

Figure 19.3g) Finished mesh with 1,048,576 elements

Right click in the graphics window and select Cursor Mode>>Face. Select the outer cylindrical face, right click and select Create Named Selection. Enter the name *wall_outer* and click on OK to close the Selection Name window. Repeat this for the two inner cylindrical faces (control-select the two faces) with the name *wall_inner*, the bottom two faces named *wall_bottom* and the top two faces named *wall_top*.

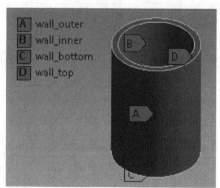

Figure 19.3b) Details for named selections

Select File>>Export...>>Mesh>>FLUENT Input File>>Export from the menu and save the mesh in the *Working Directory* with the name *cylinder-mesh-taylor-couette-flow.msh*. Select File>>Save Project from the menu and save the project with the name *Taylor-Couette.wbpj*. Close the Meshing window and right-click on Mesh under Project Schematic in Ansys Workbench and select Update.

F. Launching ANSYS Fluent

4. Double click on Setup under Project Schematic in Ansys Workbench. Check the Options box Double Precision. Increase the Number of Solver Processes to the same number as the number of computer cores. Click on Start.

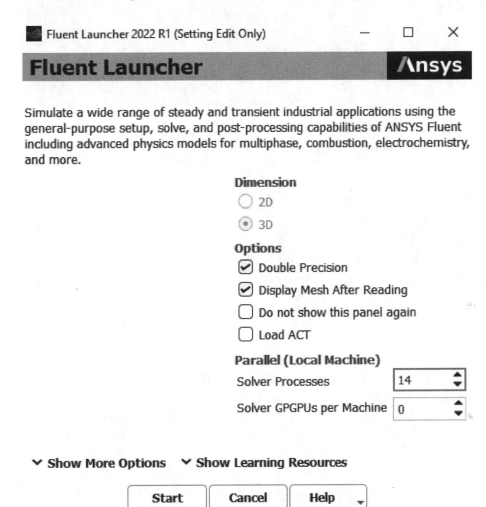

Figure 19.4a) Launching Fluent

Click on Check and Scale… under Mesh in General on the Task Page to verify the Domain Extents. Close the Scale Mesh window. Select Transient as Time under General on the Task Page. Select Relative as Velocity Formulation. Double click on Models and Viscous (SST k-omega) under Setup in the Outline View. Select Laminar as Viscous Model and click OK to close the window.

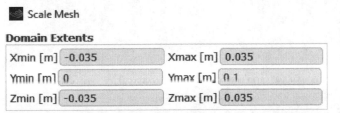

Figure 19.4b) Domain extents for Taylor-Couette flow mesh

Task Page <

General ⑦

Mesh

| Scale... | Check | Report Quality |
| Display... | Units... |

Solver

Type
- ● Pressure-Based
- ○ Density-Based

Velocity Formulation
- ○ Absolute
- ● Relative

Time
- ○ Steady
- ● Transient

☐ Gravity

Figure 19.4c) General settings

Double click on Materials under Setup in the Outline View. Click on Create/Edit for Fluid under Materials on the Task Page. Click on Fluent Database in the Create/Edit Materials window. Scroll down and select *water-liquid (h2o<l>)* as the Fluent Fluid Material. Click on Copy and Close the Fluent Database Materials window. Click on Change/Create and Close the Create/Edit Materials window.

Fluent Database Materials ✕

Fluent Fluid Materials [1/568]

Material Type
fluid ▼

Order Materials by
- ● Name
- ○ Chemical Formula

vinyl-trichlorosilane (sicl3ch2ch)
vinylidene-chloride (ch2ccl2)
water-liquid (h2o<l>)
water-vapor (h2o)
wood-volatiles (wood_vol)

Figure 19.4d) Selection of water-liquid as fluid

5. Open Cell Zone Conditions and Fluid under Setup in the Outline View. Double click on *solid (fluid, id=2)* under Cell Zone Conditions. Select *water-liquid* as Material Name. Check the box for Frame Motion and enter Speed (rad/s) -1.5 as Rotational Velocity. Set the Rotation-Axis Direction as X 0, Y 1, Z 0. Click Apply and Close the Fluid window.

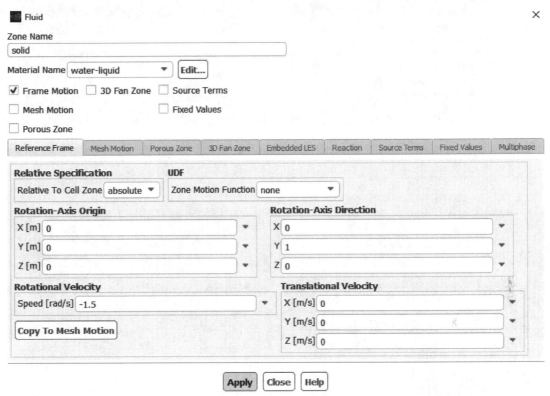

Figure 19.5a) Selection of water-liquid as fluid and setting for rotational velocity

Double click on Boundary Conditions. Select *wall_inner* under Zone in Boundary Conditions on the Task Page. Click on the Edit… button for wall Type. Select *Moving Wall* under Wall Motion. Check Absolute and Rotational under Motion and enter 1.5 as Speed (rad/s). Set the Rotation-Axis Direction as X 0, Y 1, Z 0, see Figure 19.5b). Click Apply and Close the window.

Repeat this step for the *wall_outer* Zone, use Absolute and Rotational Motion, enter 0 as Speed (rad/s) and set the Rotation-Axis Direction as X 0, Y 1, Z 0, see Figure 19.5c). Do the same for the top and bottom walls, see Figure 19.5d) and Figure 19.5e).

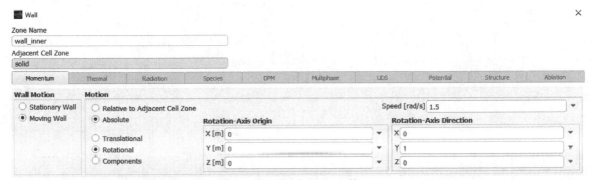

Figure 19.5b) Details of wall motion for inner wall.

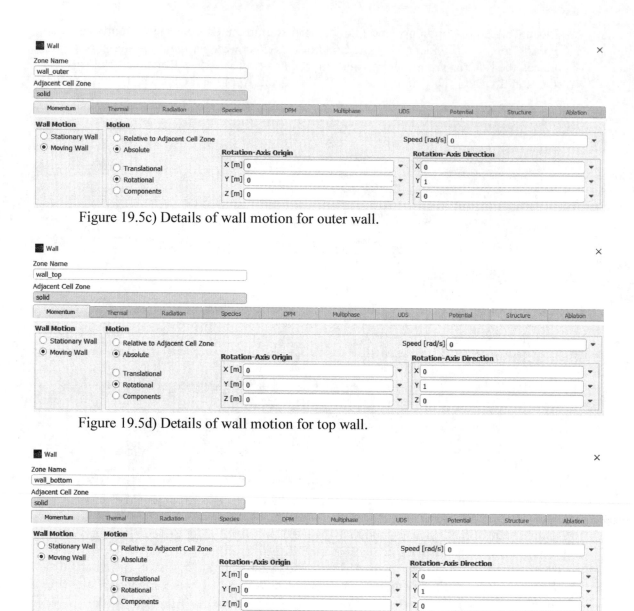

Figure 19.5c) Details of wall motion for outer wall.

Figure 19.5d) Details of wall motion for top wall.

Figure 19.5e) Details of wall motion for bottom wall.

6. Double click on Methods under Solution in the Outline View. Set the Pressure-Velocity Coupling Scheme to PISO and set Momentum under Spatial Discretization to Third-Order MUSCL, see Figure 19.6a).

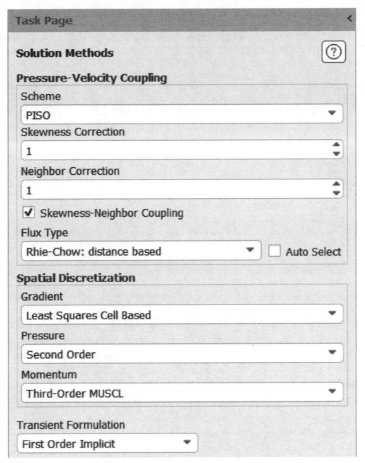

Figure 19.6a) Solution methods

Double click on Monitors and Residual under Solution in the Outline View. Set the Absolute Criterial for all Residuals to 0.0001. Select OK to close the Residual Monitors window.

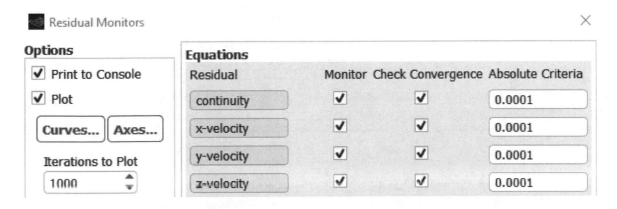

Figure 19.6b) Residual monitors

Double click on Initialization under Solution in the Outline View and select Hybrid Initialization. Click on Initialize. Select File>>Save Project from the menu. Select File>>Export>>Case & Data… from the menu. Save the file as Files of type: CFF Case/Data Files (*.cas.h5 *.dat.h5) with the name *taylor-couette-flow.cas.h5*. Double click on Run Calculation under Solution in the Outline View. Enter 250 for the Number of Time Steps, 0.1 for Time Step Size [s] and 80 for Max Iterations/Time Step. Click on Calculate.

Figure 19.6c) Solution initialization

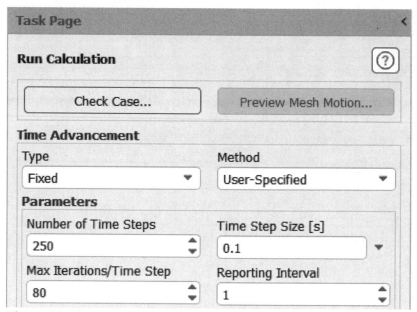

Figure 19.6d) Run calculations

G. Post-Processing

7. The residuals are shown in Figure 19.7a). Double-click on Contours in Graphics under Results in the Outline View. Enter *axial-velocity* as Contour Name in the Contours window. Select Contours of Velocity… and Axial Velocity. Select New Surface>>Plane…. Select YZ Plane as Method and select Create. Check the box Filled under Options. Uncheck Node Values, Boundary Values and Global Range under Options. Select Colormap Options… and select Fixed as Font Behavior under Font. Set the Font Size to 12 and select Apply. Close the Colormap window. Select *wall_outer* under Surfaces and click on Save/Display. Select the red X coordinate axis in the graphics window. Deselect *wall_outer* and select *plane-4* and click on Save/Display. Close the Contours window. Repeat this step for radial and tangential velocities.

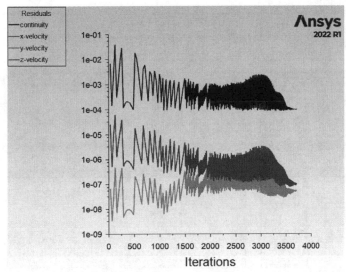

Figure 19.7a) Residuals from calculations

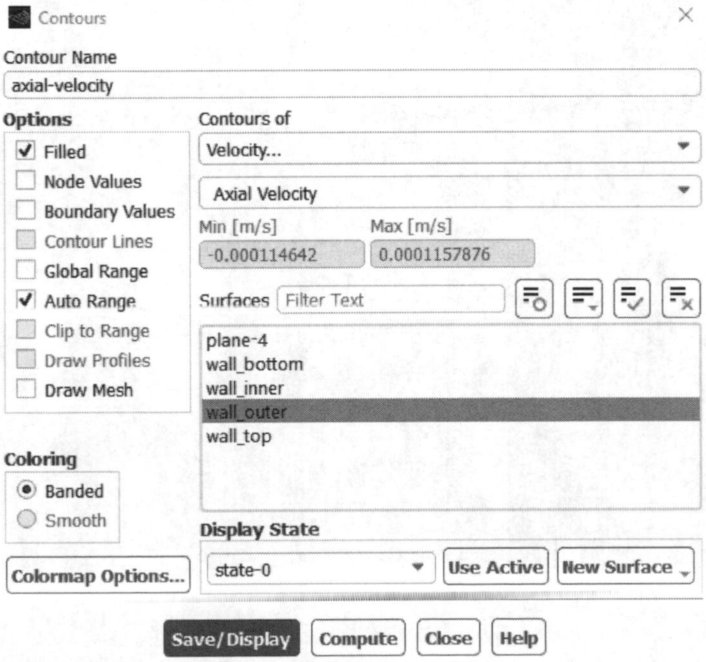

Figure 19.7b) Contours settings

441

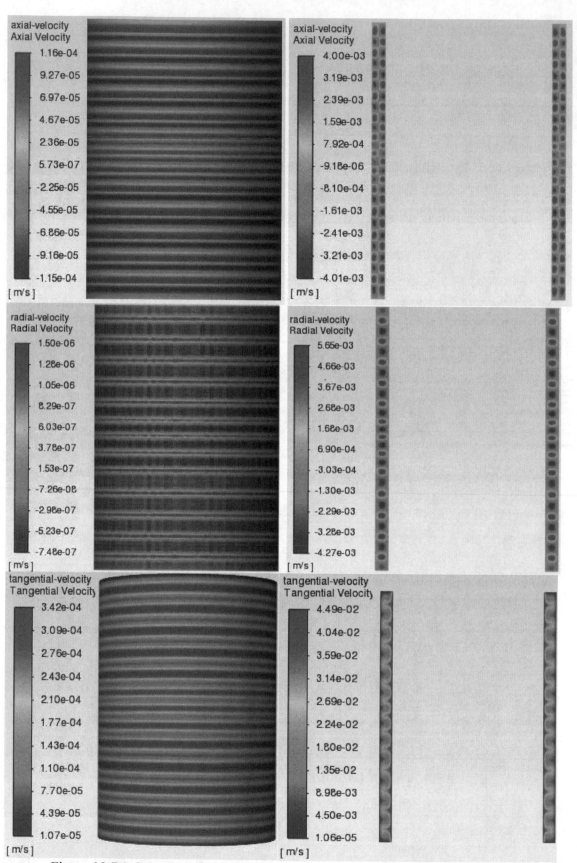

Figure 19.7c) Contours of axial, radial & tangential velocities in front & YZ plane views.

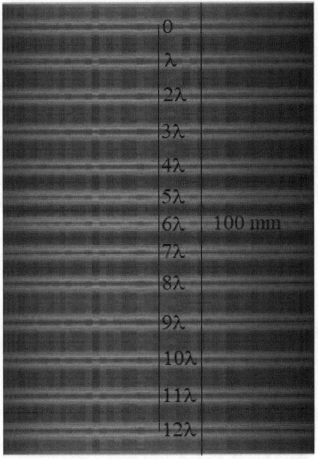

Figure 19.7d) Determining average spanwise wave length using radial velocity contours.

H. Theory

8. The instability of the flow between two vertical rotating cylinders is governed by the so-called Taylor number Ta.

$$Ta = \frac{4\Omega_i^2 d^4}{\nu^2}$$ (19.1)

where Ω_i is the rotation rate of the inner cylinder, d is the distance between the cylinders and ν is the kinematic viscosity of the fluid. Below the critical $Ta_{crit} = 3430$ for a non-rotating outer cylinder, the flow is stable but instabilities will develop above this Taylor number. The non-dimensional wave number α of the centrifugal instability is determined by

$$\alpha = \frac{2\pi d}{\lambda}$$ (19.2)

where λ is the spanwise wave length of the instability. The critical wave numbers $\alpha_{crit} = 3.12$ in the narrow gap limit: $\eta \to 1$. The radius ratio is defined as $\eta = r_i/r_o$ where r_i and r_o is the radius of the inner and outer cylinders, respectively. From figure 19.7c), the wave number can be determined to be

$$\alpha = \frac{2\pi \cdot 0.005}{0.007545} = 4.164 \tag{19.3}$$

In order to determine the wave-length that is used to determine the wave number in equation (19.3), you can measure the total length in the vertical spanwise direction for twelve red or twelve blue successive bands as shown in figure 19.7d). Next, you take this vertical length and divide by twelve to get the average wave-length in the vertical spanwise direction. To get the scale for figure 19.7d), use the total height of the cylinder in the vertical direction in figure 19.7d) that corresponds to the 100 mm height of the cylinder.

The Taylor number in the ANSYS Fluent simulations is

$$Ta = \frac{4 \cdot 1.5^2 \cdot 0.005^4}{(1.004 \cdot 10^{-6})^2} = 5580 \tag{19.4}$$

The neutral stability curve can be given to the first approximation, see figure 19.8.

$$Ta = \frac{2\left(\pi^2 + \alpha^2\right)^3}{(1+\mu)\alpha^2 \left\{ 1 - \frac{16\alpha\pi^2 \cosh^2\left(\frac{\alpha}{2}\right)}{\left[\left(\pi^2 + \alpha^2\right)^2 (\sinh \alpha + \alpha)\right]} \right\}} \tag{19.5}$$

where $\mu = \Omega_o / \Omega_i$ and Ω_o is the rotation rate of the outer cylinder.

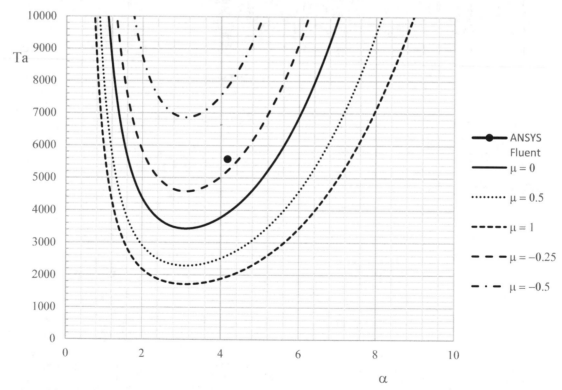

Figure 19.8 Neutral stability curves for Taylor-Couette flow. The filled circle represents result from ANSYS Fluent. The radius ratio is $\eta = \frac{6}{7}$ and $\mu = 0$ in ANSYS Fluent simulations and $\eta = 1$ for the stability curves.

CHAPTER 19. TAYLOR COUETTE FLOW

I. References

1. Chandrasekhar, S., Hydrodynamic and Hydromagnetic Stability, Dover, 1981.
2. Koschmieder, E.L., Benard Cells and Taylor Vortices, Cambridge, 1993.
3. Matsson, J.E, On the Integration of Fluid Flows and Fabrication, *Proceedings of the 2002 American Society for Engineering Education Annual Conference & Exposition.*, Montreal, Canada, (2002).
4. Taylor, G.I., Stability of a Viscous Liquid Contained between Two Rotating Cylinders, *Phil. Trans. Roy. Soc.* London Ser. A, Vol. **223**, (1923).

J. Exercises

19.1 Run the calculations for the flow in the Taylor-Couette apparatus with only the inner cylinder rotating $\mu = 0$ for Taylor numbers $Ta = 10000$, 20000 and 30000 and compare the spanwise wave numbers with the one determined in this chapter for $Ta = 5580$. Include your results in figure 19.8 for comparison with the neutral stability curve.

19.2 Run the calculations for the flow in the Taylor-Couette apparatus with both cylinders rotating $\mu = 1$ and for Taylor numbers $Ta = 3000$, 5000 and 10000 and determine the spanwise wave numbers. Include your results in a graph and compare with the neutral stability curve corresponding to $\mu = -1/2$, see equation (19.5).

K. Notes

CHAPTER 20. DEAN FLOW IN A CURVED CHANNEL

A. Objectives

- Using ANSYS Meshing to Create the Mesh for 3D Dean Flow in a Curved Channel
- Inserting Wall Boundary Conditions
- Running Laminar Steady 3D ANSYS Fluent Simulations
- Using Contour Plots for Visualizations
- Compare Results with Neutral Stability Theory

B. Problem Description

Dean flow appears in a curved tube and can also be shown in a narrow gap curved channel. We will therefore study the flow between two curved walls where the inner wall has a radius of 395 mm and the radius of the outer wall is 405 mm. The height of the curved channel is 100 mm. The centrifugal instability will cause the appearance of counter-rotating vortices at low flow rates.

C. Launching ANSYS Workbench and Selecting Fluent

1. Start by launching ANSYS Workbench. Double click on Fluid Flow (Fluent) under Analysis Systems in the Toolbox. Select Units>>Metric (tonne, mm, s, …) from the menu.

Figure 20.1 Selecting Fluent

D. Launching ANSYS DesignModeler

2. Right click on Geometry in the Project Schematic window and select New DesignModeler Geometry. Select millimeter as the desired length unit from the menu (Units>>Millimeter) in DesignModeler. Select the ZX Plane in the Tree Outline and select Look at Face/Plane/Sketch .

    Select the Sketching tab and click on Arc by Center. Draw two 180 degrees half-circle arcs centered at the origin. Make sure you have the letter *P* when you move the cursor to the origin to define the center of the arc and to start drawing the half circle from the negative horizontal axis with letter *C* showing up to the positive horizontal axis with the letter *C* to finish the arc.

    Click on the Dimensions tab and select Radius. Click on the two arcs in the graphics window and enter 395 mm and 405 mm for the radii of the half-circles in Details View. Select the Line sketch tool and draw two horizontal lines that connects the ends of the arcs.

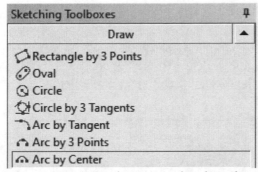

Figure 20.2a) Arc by center sketch tool

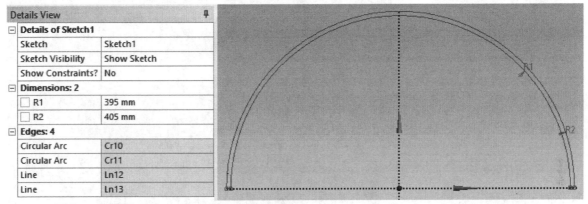

Figure 20.2b) Dimensions for half-circles

Select 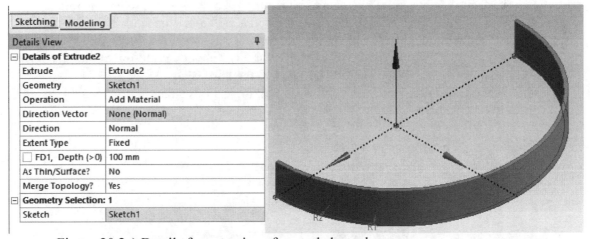 Extrude and apply Sketch1 under ZXPlane in the Tree Outline as Geometry in Details View. Enter 100 mm as the FD1, Depth (>0) of the extrusion. Click on Generate and select isometric view in the graphics window. Close the DesignModeler.

Figure 20.2c) Details for extrusion of curved channel

E. Launching ANSYS Meshing

3. Double click on Mesh under Project Schematic in Ansys Workbench. Select Mesh in the Outline under Project in the Meshing window. Select Home>>Tools>>Units>>Metric (mm, kg, N, ...) from the menu. Select the Mesh under Project in Outline.

   Open Quality in Details of "Mesh" and set the Smoothing to High, see Figure 20.3a). Generate the Mesh.

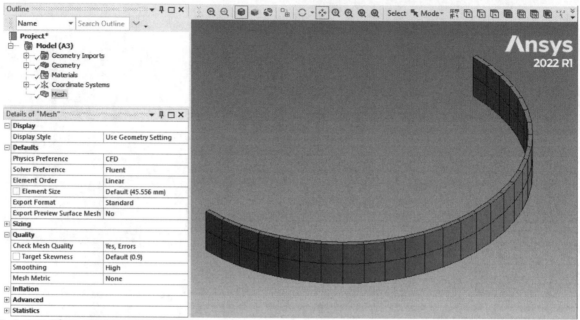

Figure 20.3a) Coarse mesh

Select Mesh>>Controls>>Sizing from the menu. Select Edge 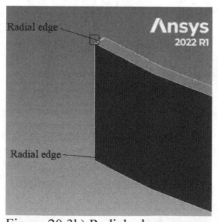, select the radial edge at the top of the curved channel and the corresponding radial line at the bottom of the same channel, see Figure 20.3b). Apply the two edges as Geometry in Details of "Edge Sizing" – Sizing. Select Number of Divisions as Type under Definition. Set the Number of Divisions to 32 and Behavior to Hard. Select the third Bias Type - — —— — - and set the Bias Factor to 20, see Figure 20.3c). Repeat this sizing for the other side of the curved channel.

Figure 20.3b) Radial edges on curved channel

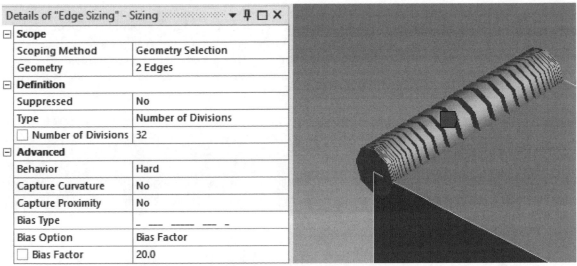

Figure 20.3c) Radial sizing with bias

Right click on Mesh in Outline and select Insert>>Method. Select the curved channel and Apply it as Geometry under Scope in Details of "Automatic Method" – Method. Select MultiZone as Method under Definition, see Figure 20.3d).

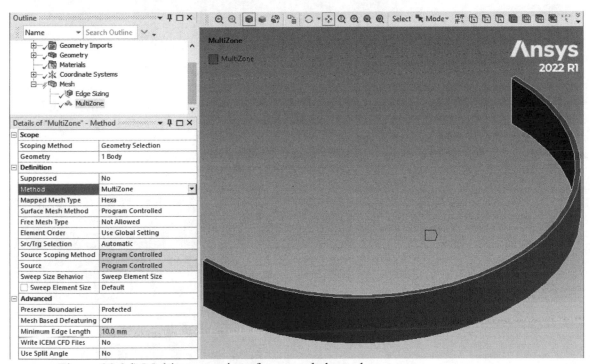

Figure 20.3d) Multizone settings for curved channel

Select Mesh>>Controls>>Sizing from the menu. Select Edge 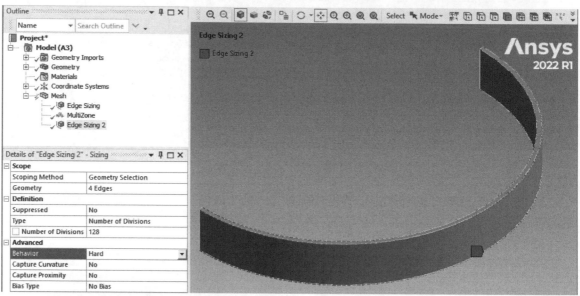, control select the four half-circles at the top and the bottom of the curved channel and Apply the four Edges as Geometry under Scope in Details of "Sizing" – Sizing, see Figure 20.3e). Select Number of Divisions as Type under Definition. Set the Number of Divisions to 128 and Behavior to Hard. Right click on Mesh in Outline and select Mesh>>Update from the menu.

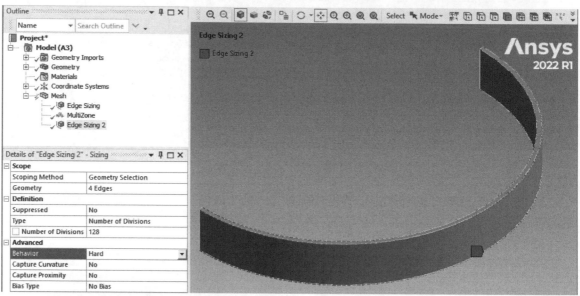

Figure 20.3e) Azimuthal sizing

Select Mesh>>Controls>>Sizing from the menu. Select Edge 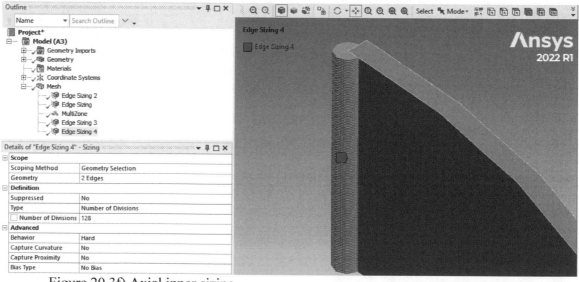, control select the two inner vertical lines of the curved channel and Apply the two Edges as Geometry under Scope in Details of "Sizing" – Sizing, see Figure 20.3f). Select Number of Divisions as Type under Definition. Set the Number of Divisions to 128 and Behavior to Hard. Right click on Mesh in Outline and select Mesh>>Update from the menu.

Figure 20.3f) Axial inner sizing

Open Statistics in Details of "Mesh" to see the total number of Nodes 549,153 and total number of Elements 524,288.

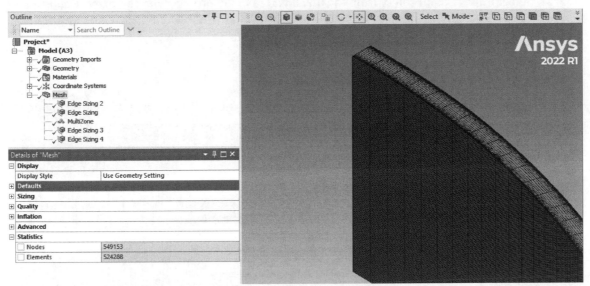

Figure 20.3g) Finished mesh with 524,288 elements

Right click in the graphics window and select Cursor Mode>>Face. Select the outer cylindrical face, right click and select Create Named Selection. Enter the name *wall_outer* and click on OK to close the Selection Name window. Do the same for the inner cylindrical wall and name this *wall_inner*. Furthermore, name the top curved flat surface as *wall_top* and the bottom curved flat surface as *wall_bottom*. Select the flat inlet face and name this *inlet*. Select the flat outlet face and name this *outlet*, see Figure 20.3h).

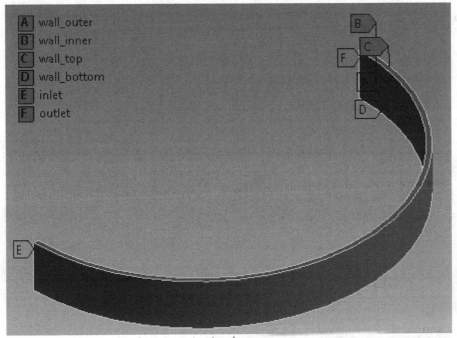

Figure 20.3h) Details for named selections

Select File>>Export…>>Mesh>>FLUENT Input File>>Export from the menu and save the mesh in the *Working Directory* with the name *curved-channel-mesh-dean-flow.msh*. Select File>>Save Project from the menu and save the project with the name *Dean Flow in Curved Narrow Channel.wbpj*. Close the Meshing window and right-click on Mesh under Project Schematic in Ansys Workbench and select Update.

F. Launching ANSYS Fluent

4. Double click on Setup under Project Schematic in Ansys Workbench. Check the Options box Double Precision. Set the Number of Solver Processes to the same number as computer cores. Click on Start.

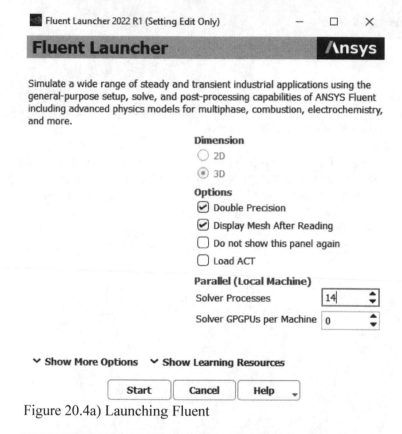

Figure 20.4a) Launching Fluent

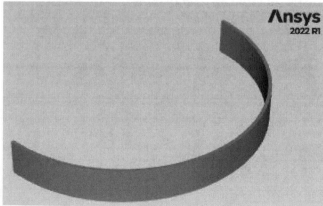

Figure 20.4b) Geometry in Fluent

Click on Check and Scale... under Mesh in General on the Task Page to verify the Domain Extents. Close the Scale Mesh window. Set Time as Transient. Check the box for Gravity. Enter -9.81 as Gravitational Acceleration in the Y direction. Double click on Models and Viscous (SST k-omega) under Setup in the Outline View. Select Laminar as Viscous Model and click OK to close the window.

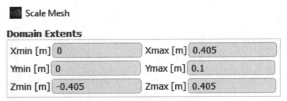

Figure 20.4c) Domain extents for curved channel flow mesh

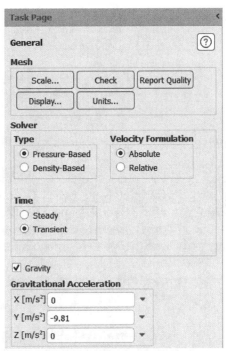

Figure 20.4d) General settings including transient time and gravity

Double click on Materials under Setup in the Outline View. Click on Create/Edit for Fluid under Materials on the Task Page. Click on Fluent Database in the Create/Edit Materials window. Scroll down and select *water-liquid (h2o<l>)* as the Fluent Fluid Material. Click on Copy and Close the Fluent Database Materials window. Click on Change/Create and Close the Create/Edit Materials window.

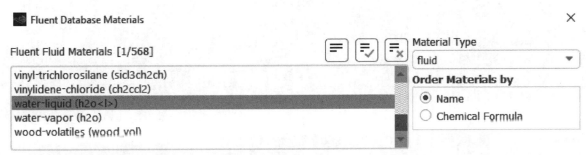

Figure 20.4e) Selection of water-liquid as fluid

5. Open Cell Zone Conditions and Fluid under Setup in the Outline View. Double click on *solid (fluid, id=2)* under Cell Zone Conditions. Select *water-liquid* as Material Name and enter *water-liquid* as the Zone Name. Select Apply and Close the Fluid window.

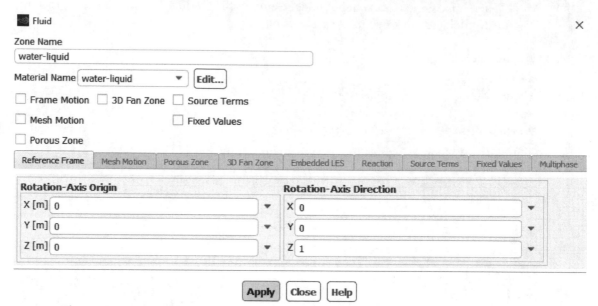

Figure 20.5a) Selection of water-liquid as fluid

Double click on Boundary Conditions. Select *inlet* under Zone in Boundary Conditions on the Task Page. Click on the Edit… button for velocity-inlet Type. Select *Components* as Velocity Specification Method. Select Local Cylindrical (Radial, Tangential, Axial) as Coordinate System. Enter 0.1 as Tangential-Velocity [m/s]. Enter X 0, Y 1 and Z 0 as Axis Direction, see figure 20.5b). Select Apply and Close the Velocity Inlet window.

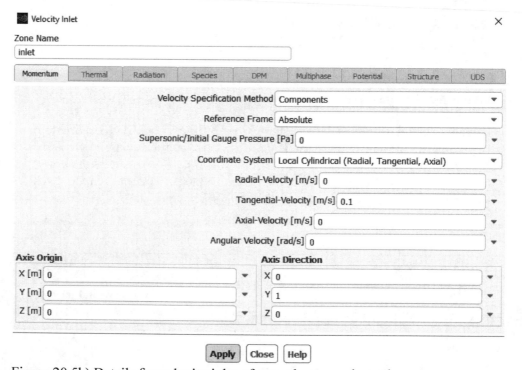

Figure 20.5b) Details for velocity inlet of curved narrow channel

456

6. Double click on Methods under Solution in the Outline View. Use the settings as shown in Figure 20.6a).

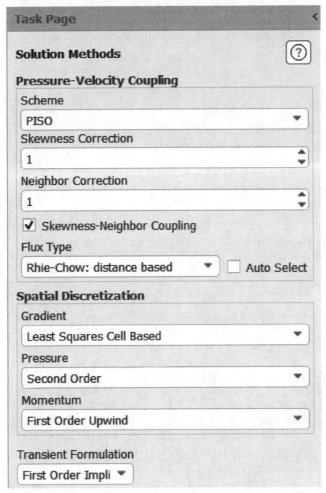

Figure 20.6a) Solution methods

Double click on Monitors and Residual under Solution in the Outline View. Set the Absolute Criteria for all Residuals to 0.0001, see Figure 20.6b). Select OK to close the Residual Monitors window.

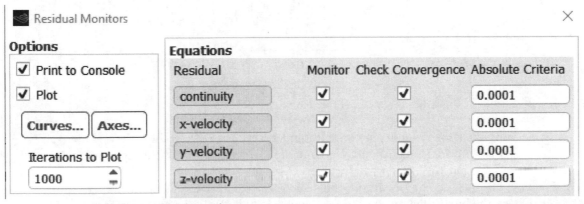

Figure 20.6b) Residual monitors

Double click on Initialization under Solution in the Outline View and select Hybrid Initialization. Click on Initialize. Select File>>Save Project from the menu. Select File>>Export>>Case & Data… from the menu. Save the file as Files of type: CFF Case/Data Files (*.cas.h5 *.dat.h5) with the name *dean-flow.cas.h5*. Double click on Run Calculation under Solution in the Outline View. Enter 700 for the Number of Iterations. Click on Calculate.

Figure 20.6c) Solution initialization

G. Post-Processing

7. The residuals are shown in Figure 20.7a).

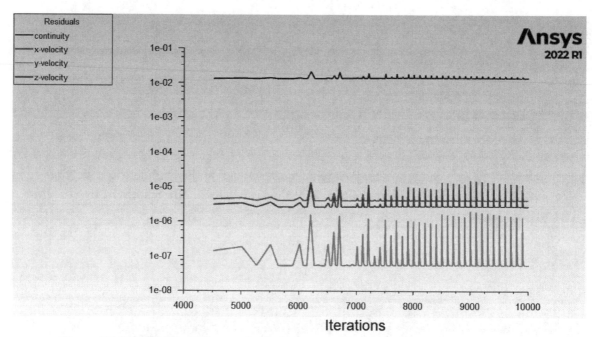

Figure 20.7a) Residuals from calculations

Double-click on Contours in Graphics under Results in the Outline View. Enter *spanwise-y-velocity* as Contour Name in the Contours window. Select Contours of Velocity… and Y Velocity. Select New Surface>>Plane…. Select *XY Plane* as Method, *xy-plane* as New Surface Name and select Create. Close the Plane Surface window. Check the box Filled under Options. Uncheck Node Values, Boundary Values and Global

Range under Options. Select Colormap Options… and select Fixed as Font Behavior under Font. Set the Font Size to 10 and select Apply. Close the Colormap window. Select *wall_outer* under Surfaces and click on Save/Display, see Figure 20.7b). Close the Contours window.

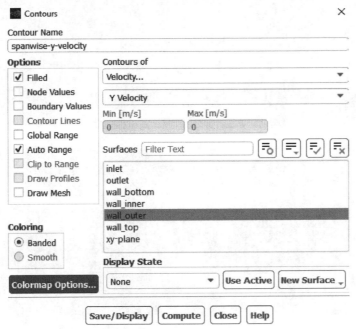

Figure 20.7b) Contours settings

Select Set View From -X Direction, see Figure 20.7c). Manipulate the light properties by unchecking the box for Lighting under Display from View in the menu, see Figure 20.7d).

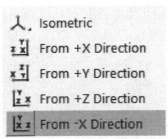

Figure 20.7c) Selecting view

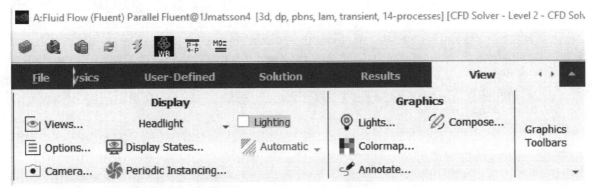

Figure 20.7d) Display settings

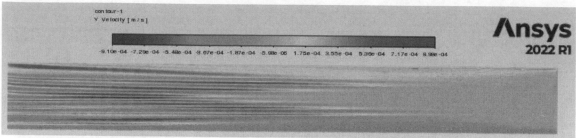

Figure 20.7e) Contours of Y Velocity from the inside of the curved channel

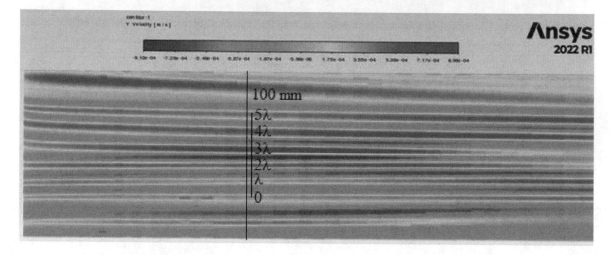

Figure 20.7f) Determination of spanwise wave length for curved channel flow

In Figure 20.7f), you can see six distinct streaks in the middle of the channel. The streaks indicate the presence of streamwise counter rotating vortices with a wavelength in the spanwise Y direction that can be determined from the figure.

H. Theory

8. The centrifugal instability for the flow in a curved channel is determined by the Dean number De.

$$De = Re\sqrt{\gamma} = \frac{U_b d}{v} \sqrt{\frac{d}{R}} \tag{20.1}$$

, where U_b is the bulk flow velocity in the curved channel, d is the channel width, R its radius of curvature at the channel centerline and v is the kinematic viscosity of the fluid. Below the critical $De_{crit} = 36$, the flow is stable but instabilities will develop above this Dean number[1,2]. The non-dimensional wave number α of the centrifugal instability is determined by

$$\alpha = \frac{2\pi d}{\lambda} \tag{20.2}$$

where λ is the spanwise wave length of the instability. The critical wave numbers $\alpha_{crit} = 3.96$. From figure 20.7c), the wave number can be determined to be

$$\alpha = \frac{2\pi \cdot 0.010}{0.00985} = 6.38 \tag{20.3}$$

In order to determine the wave length that is used to determine the wave number in equation (20.3), you can measure the total length in the vertical spanwise direction for three red or three blue successive bands in figure 20.7d). Next, you take this vertical length and divide by three to get the average wave length in the vertical spanwise direction. To get the scale for figure 20.7d), use the total height of the curved channel in the vertical direction in figure 20.7d) that corresponds to 100 mm.

The Dean number in the ANSYS Fluent simulations is

$$De = Re\sqrt{\gamma} = \frac{0.1*0.01}{10^{-6}}\sqrt{\frac{0.01}{0.4}} = 158 \tag{20.4}$$

The neutral stability curve to the first approximation[4], see Figure 20.8.

$$De = \frac{1}{6\alpha}\left[\frac{(\pi^2+\alpha^2)(4\pi^2+\alpha^2)}{(A-B-C)(D+E-F)}\right]^{1/4} \tag{20.5}$$

where

$$A = \frac{2(640-48\pi^2)}{27\pi^4(\alpha^2+\pi^2)^2} + \frac{64(9\pi^2-56)}{27\pi^2(\alpha^2+\pi^2)^3} - \frac{256(\alpha^2-5\pi^2)}{9\pi^2(\alpha^2+\pi^2)^4} \tag{20.6}$$

$$B = \frac{64\pi^2[16\alpha\pi^6+24\alpha\pi^4(\alpha^2+3)+\alpha^3\pi^2(9\alpha^2-38)+\alpha^5(\alpha^2-2)]}{(\alpha^2+4\pi^2)^3(\alpha^2+\pi^2)^4(\sinh\alpha+\alpha)} \tag{20.7}$$

$$C = \frac{64\pi^2\{(24\pi^4-34\alpha^2\pi^2-22\alpha^4)\alpha\cosh\alpha+[16\pi^4(\pi^2+3)+4\alpha^2\pi^2(8\pi^2-21)+\alpha^4(19\pi^2+36)+3\alpha^6]\sinh\alpha\}}{(\alpha^2+4\pi^2)^3(\alpha^2+\pi^2)^4(\sinh\alpha+\alpha)} \tag{20.8}$$

$$D = \frac{2(640-48\pi^2)}{27\pi^4(\alpha^2+4\pi^2)^2} - \frac{64(9\pi^2-104)}{27\pi^2(\alpha^2+4\pi^2)^3} - \frac{256(\alpha^2-20\pi^2)}{9\pi^2(\alpha^2+4\pi^2)^4} \tag{20.9}$$

$$E = \frac{64\pi^2[4\alpha\pi^6+9\alpha\pi^4(\alpha^2+8)+2\alpha^3\pi^2(3\alpha^2+59)+\alpha^5(\alpha^2-2)]}{(\alpha^2+4\pi^2)^4(\alpha^2+\pi^2)^3(\sinh\alpha-\alpha)} \tag{20.10}$$

$$F = \frac{64\alpha\pi^2\{(24\pi^4-46\alpha^2\pi^2-22\alpha^4)\alpha\cosh\alpha+[4\pi^4(\pi^2+12)+\alpha^2\pi^2(17\pi^2-156)+4\alpha^4(4\pi^2+9)+3\alpha^6]\sinh\alpha\}}{(\alpha^2+4\pi^2)^4(\alpha^2+\pi^2)^3(\sinh\alpha-\alpha)} \tag{20.11}$$

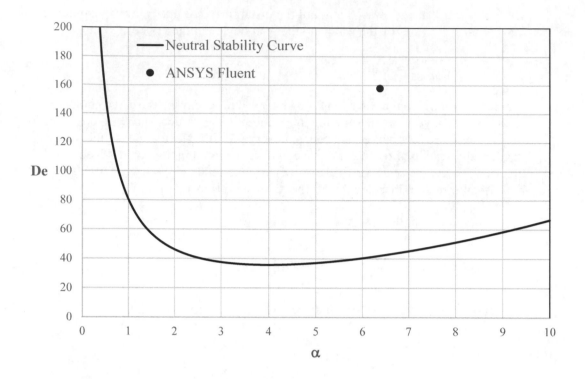

Figure 20.8 Neutral stability curve for Dean flow in a curved channel. The filled circle represents result from ANSYS Fluent.

I. References

1. Chandrasekhar, S., Hydrodynamic and Hydromagnetic Stability, Dover, 1981.
2. Dean, W.R., Fluid motion in a curved channel. *Proc. Roy. Soc.* A**121**, 402, 1928.
3. Matsson, J.E. and Alfredsson, P.H., Curvature- and rotation induced instabilities in channel flow. *J. Fluid Mech.* **210**, 537, (1990).
4. Reid, W.H., On the stability of viscous flow in a curved channel. Proc. Roy. Soc. A, **244**, 1237, (1958).

J. Exercises

20.1 Run the calculations for the flow in the curved channel but use the following tangential velocities: 0.03, 0.06, 0.09 and 0.12 m/s and determine the Dean number and the spanwise wave number for each tangential velocity.

20.2 Run the calculations for the flow in the curved channel as shown in this chapter for different mesh sizes using Element Size 1.6, 1.8, 2.0 and 2.2 mm. Determine the spanwise wave number for each case and compare with the critical value.

CHAPTER 21. ROTATING CHANNEL FLOW

A. Objectives

- Using ANSYS Meshing to Create the Mesh for 3D Plane Channel Flow
- Inserting Boundary Conditions
- Running Laminar Steady 3D ANSYS Fluent Simulations
- Using Contour Plots for Visualizations
- Compare Results with Neutral Stability Theory

B. Problem Description

In this chapter we will study the appearance of streamwise vortices for the flow in a narrow gap rotating plane channel with rectangular cross section. The height of the rectangular cross section is 100 mm and the channel width is 10 mm. The length of the channel is 1000 mm. The Coriolis force due to spanwise system rotation will cause the appearance of counter-rotating vortices and transition to turbulent flow at much lower Reynolds numbers as compared with the non-rotating plane channel flow case.

C. Launching ANSYS Workbench and Selecting Fluent

1. Start by launching ANSYS Workbench. Double click on Fluid Flow (Fluent) under Analysis Systems in the Toolbox. Select Units>>Metric (tonne, mm, s, …) from the menu.

Figure 21.1 Selecting Fluent

D. Launching ANSYS DesignModeler

2. Right click on Geometry in the Project Schematic window and select New DesignModeler Geometry. Select millimeter as the desired length unit from the menu (Units>>Millimeter) in DesignModeler. Select the YZPlane in the Tree Outline and select Look at Face/Plane/Sketch ![icon]. Select the Sketching tab and click on Rectangle. Draw a rectangle from the origin. Make sure you have the letter *P* before you start drawing the rectangle from the origin. Click on the Dimensions tab. Click on the vertical and horizontal sides of the rectangle and enter 10 mm and 100 mm for the vertical and horizontal dimensions in Details View.

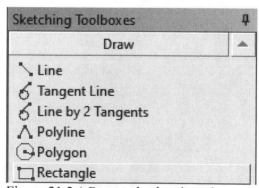

Figure 21.2a) Rectangle sketch tool

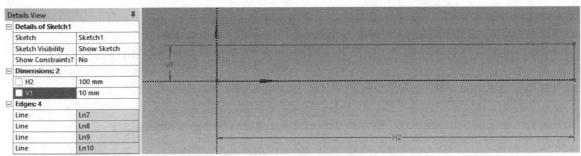

Figure 21.2b) Dimensions for rectangle

Select **Extrude** and apply Sketch1 under YZPlane in the Tree Outline as Geometry in Details View. Enter 1000 mm as the FD1, Depth (>0) of the extrusion. Click on Generate and select isometric view in the graphics window. Close the DesignModeler.

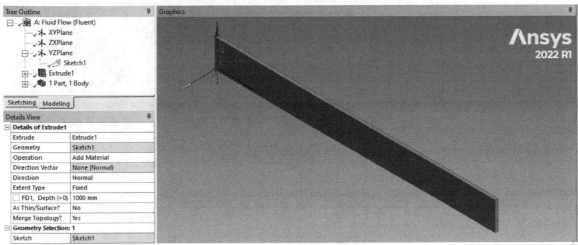

Figure 21.2c) Details for extrusion of straight channel

E. Launching ANSYS Meshing

3. Double click on Mesh under Project Schematic in Ansys Workbench. Select Mesh in the Outline under Project. Select Home>>Tools>>Units>>Metric (mm, kg, N, …) from the menu. Open Quality in Details of "Mesh" and set the Smoothing to High, see Figure 21.3a). Generate the Mesh.

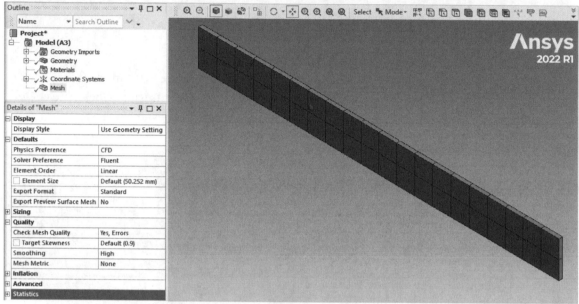

Figure 21.3a) Coarse mesh for straight channel

Select Mesh>>Controls>>Sizing from the menu. Select Edge 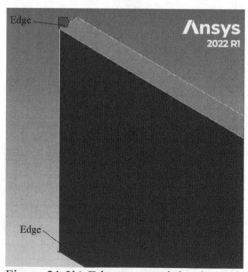, select the edge at the top of the straight channel and the corresponding line at the bottom of the same channel, see Figure 21.3b). Apply the two edges as Geometry in Details of "Edge Sizing" – Sizing. Select Number of Divisions as Type under Definition. Set the Number of Divisions to 32 and Behavior to Hard. Select the third Bias Type - — —— — - and set the Bias Factor to 20, see Figure 21.3c). Repeat this sizing for the other side of the straight channel.

Figure 21.3b) Edges on straight channel

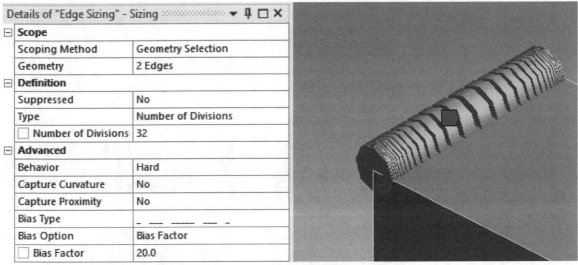

Figure 21.3c) Sizing with bias for straight channel

Right click on Mesh in Outline and select Insert>>Method. Select the straight channel and Apply it as Geometry under Scope in Details of "Automatic Method" – Method. Select MultiZone as Method under Definition, see Figure 21.3d).

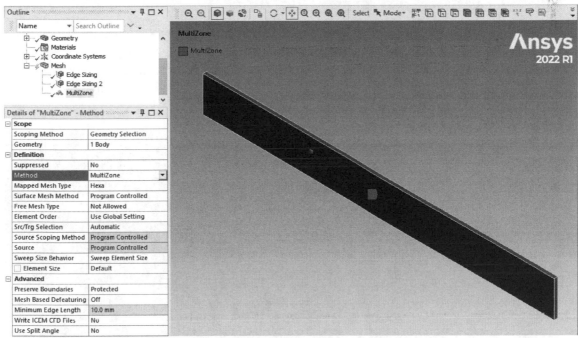

Figure 21.3d) Multizone settings for straight channel

Select Mesh>>Controls>>Sizing from the menu. Select Edge 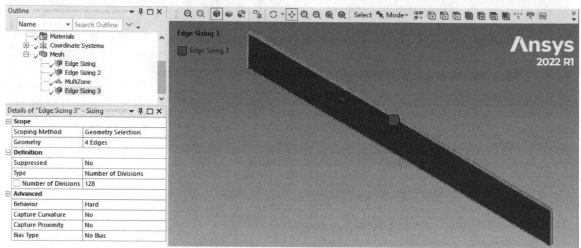, control select the four long straight edges at the top and the bottom of the channel and Apply the four Edges as Geometry under Scope in Details of "Sizing" – Sizing, see Figure 20.3e). Select Number of Divisions as Type under Definition. Set the Number of Divisions to 128 and Behavior to Hard. Right click on Mesh in Outline and select Mesh>>Update from the menu.

Figure 21.3e) Streamwise sizing for straight channel

Select Mesh>>Controls>>Sizing from the menu. Select Edge 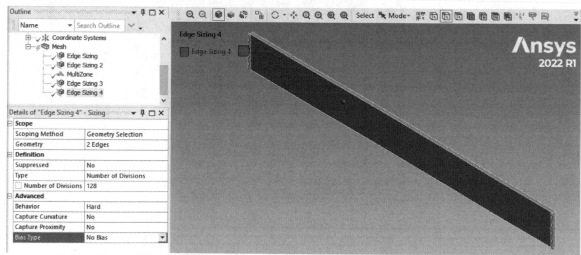, control select the two vertical lines on the straight channel and Apply the two Edges as Geometry under Scope in Details of "Sizing" – Sizing, see Figure 21.3f). Select Number of Divisions as Type under Definition. Set the Number of Divisions to 128 and Behavior to Hard. Right click on Mesh in Outline and select Mesh>>Update from the menu.

Figure 21.3f) Spanwise sizing for straight channel

Open Statistics in Details of "Mesh" to see the total number of Nodes 549,153 and total number of Elements 524,288.

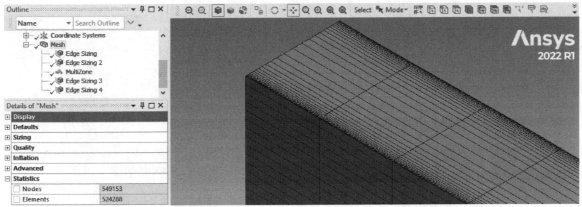

Figure 21.3g) Finished mesh with 524,288 elements for straight channel

Right click in the graphics window and select Cursor Mode>>Face. Select the larger face, right click and select Create Named Selection. Enter the name *wall_right* and click on OK to close the Selection Name window. Do the same for the other larger wall and name this *wall_left*. Furthermore, name the top face as *wall_top* and the bottom face as *wall_bottom*. Select the inlet face and name this *inlet*. Select the outlet face and name this *outlet*, see Figure 21.3h).

Select File>>Export...>>Mesh>>FLUENT Input File>>Export from the menu and save the mesh in the *Working Directory* with the name *straight-channel-mesh.msh*. Select File>>Save Project from the menu and save the project with the name *Straight Rotating Channel Flow.wbpj*. Close the Meshing window and right-click on Mesh under Project Schematic in Ansys Workbench and select Update.

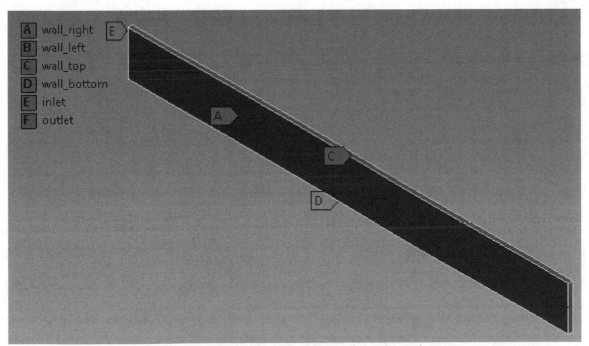

Figure 21.3h) Details for named selections for straight channel

F. Launching ANSYS Fluent

4. Double click on Setup under Project Schematic in Ansys Workbench. Check the Options box for Double Precision. Set the Number of Parallel Solver Processes to the same number as computer cores. Click on Start.

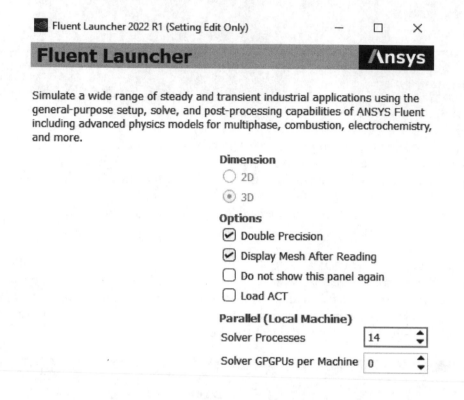

Figure 21.4a) Launching Fluent

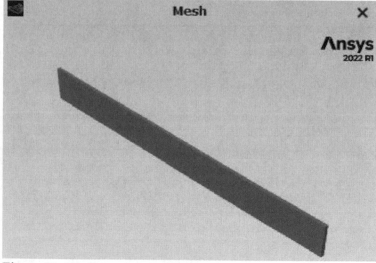

Figure 21.4b) Geometry in Fluent

Click on Check and Scale… under Mesh in General on the Task Page to verify the Domain Extents. Close the Scale Mesh window. Select Time Transient under General on the Task Page. Check the box for Gravity. Enter -9.81 as Gravitational Acceleration in the Y direction.

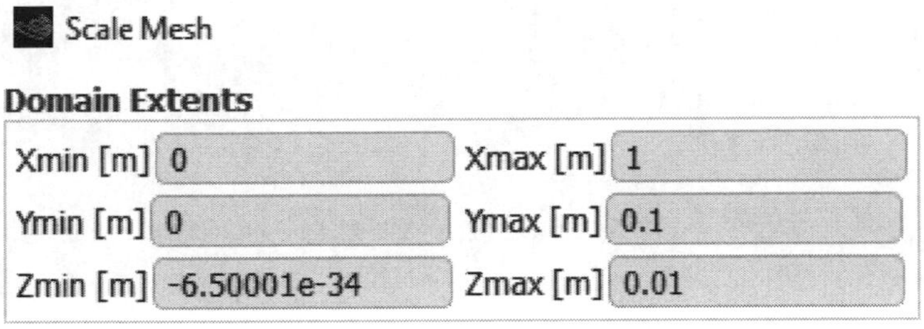

Scale Mesh

Domain Extents

Xmin [m]	0	Xmax [m]	1
Ymin [m]	0	Ymax [m]	0.1
Zmin [m]	-6.50001e-34	Zmax [m]	0.01

Figure 21.4c) Domain extents for rotating channel flow mesh

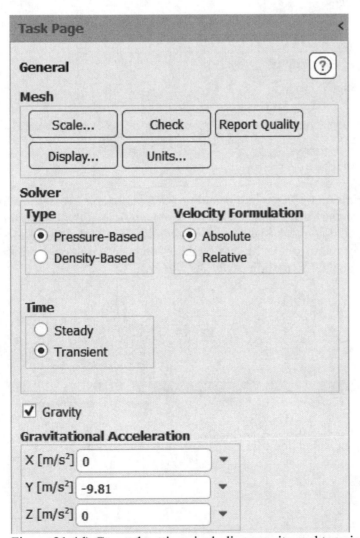

Figure 21.4d) General settings including gravity and transient time

471

Double click on Models and Viscous under Setup in the Outline View. Set the Viscous Model to Laminar and click OK to close the window.

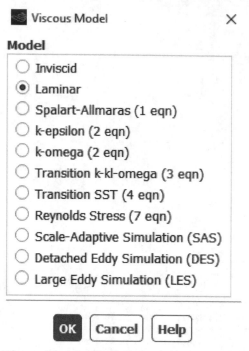

Figure 21.4d) Viscous model

Double click on Materials under Setup in the Outline View. Click on Create/Edit for Fluid under Materials on the Task Page. Click on Fluent Database in the Create/Edit Materials window. Scroll down and select *water-liquid (h2o<l>)* as the Fluent Fluid Material. Click on Copy and Close the Fluent Database Materials window. Click on Change/Create and Close the Create/Edit Materials window.

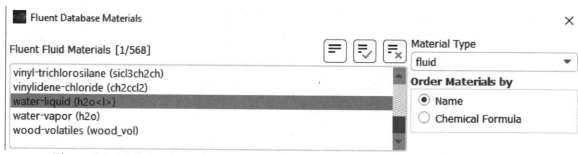

Figure 21.4e) Selection of water-liquid as fluid

5. Open Cell Zone Conditions and Fluid under Setup in the Outline View. Double click on *solid (fluid, id=2)* under Cell Zone Conditions. Select *water-liquid* as Material Name and enter *water-liquid* as the Zone Name. Check the box for Frame Motion. Set the Rotation-Axis Origin to X [m] 0, Y [m] 0 and Z [m] 0.005. Set the Rotation-Axis Direction to X 0, Y -1, and Z 0. Set the Rotational Velocity Speed [rad/s] to 0.5. Select Apply and Close the Fluid window.

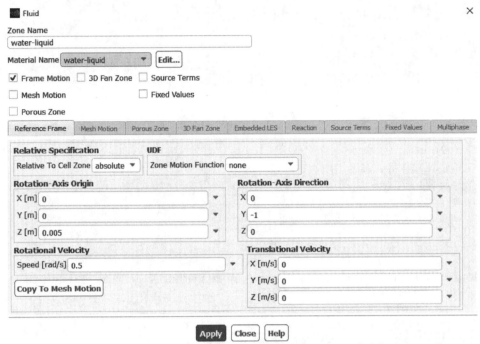

Figure 21.5a) Selection of water-liquid as fluid and settings for frame motion

Double click on Boundary Conditions. Select *inlet* under Zone in Boundary Conditions on the Task Page. Click on the Edit… button for velocity-inlet Type. Select *Components* as Velocity Specification Method. Enter 0.05 as X-Velocity [m/s]. Select Apply and Close the Velocity Inlet window.

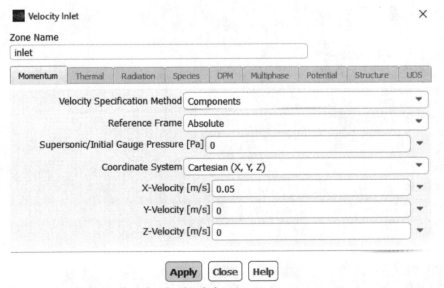

Figure 21.5b) Details of velocity inlet

Select *outlet* under Zone in Boundary Conditions on the Task Page. Click on the Edit…
button for *pressure-outlet* Type. Select *From Neighboring Cell* as Backflow Direction
Specification Method. Select Apply and Close the Pressure Outlet window.

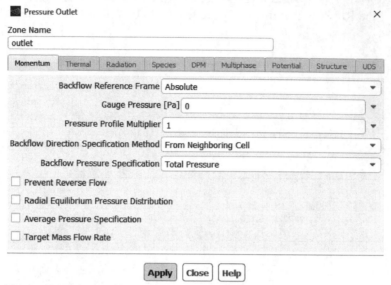

Figure 21.5c) Details of pressure outlet

Double click on Methods under Solution in the Outline View. Use the settings as shown
in Figure 21.5d).

Figure 21.5d) Settings for solution methods

Double click on Monitors and Residual under Solution in the Outline View. Set the
Absolute Criteria to 0.0001 for all equations and click OK to close the window.

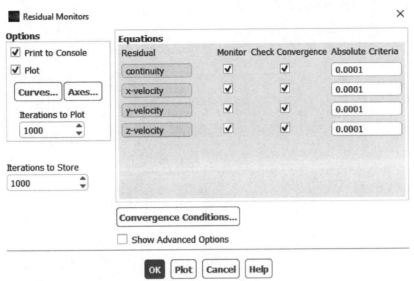

Figure 21.5e) Settings for residual monitors

6. Double click on Initialization under Solution in the Outline View and select Hybrid
    Initialization. Click on Initialize.

    Select File>>Save Project from the menu. Select File>>Export>>Case & Data… from
    the menu. Save the file as Files of type: CFF Case/Data Files (*.cas.h5 *.dat.h5) with the
    name *rotating-channel-flow*.

    Double click on Run Calculation under Solution in the Outline View. Enter the settings
    as shown in Figure 21.6b) and click on Calculate.

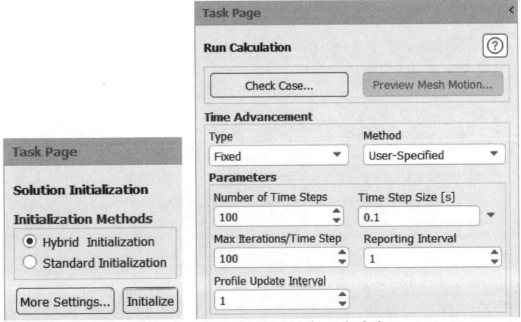

Figure 21.6a) Initialization   Figure 21.6b) Running calculations

G. Post-Processing

7.   The residuals are shown in Figure 21.7a).

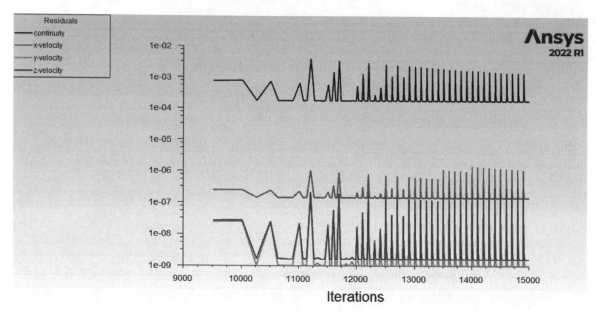

Figure 21.7a) Residuals from calculations

Double-click on Contours in Graphics under Results in the Outline View. Enter *spanwise-y-velocity* as Contour Name in the Contours window. Select Contours of Velocity… and Y Velocity. Check the box Filled under Options. Uncheck Node Values, Boundary Values and Global Range under Options. Select Colormap Options… and select Top as Colormap Alignment. Select Fixed as Font Behavior under Font. Set the Font Size to 10 and select Apply. Close the Colormap window. Select *wall_left* under Surfaces and click on Save/Display, see Figure 21.7b). Close the Contours window.

Figure 21.7b) Contours settings for Y Velocity

Select Set View From +Z Direction, see Figure 21.7c). Manipulate the light properties by unchecking the box for Lighting under Display from View in the menu, see Figure 21.7d).

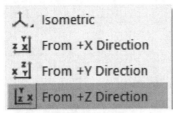

Figure 21.7c) Selecting view

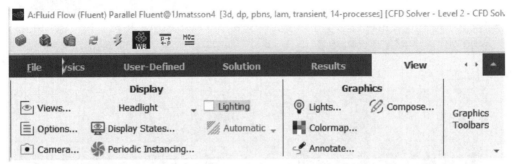

Figure 21.7d) Display settings

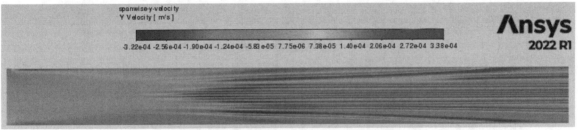

Figure 21.7e) Contours of Y Velocity from the inside of the curved channel

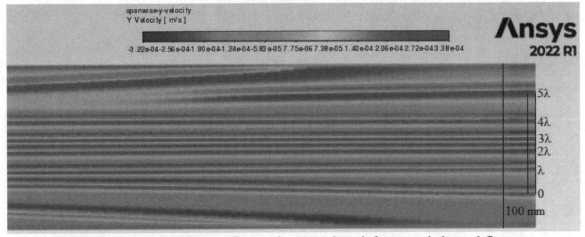

Figure 21.7f) Determination of spanwise wave length for curved channel flow

In Figure 21.7f), you can see six distinct streaks in the middle of the channel. The streaks indicate the presence of streamwise counter rotating vortices with a wavelength in the spanwise Y direction that can be determined from the figure.

Double-click on Contours in Graphics under Results in the Outline View. Enter *streamwise-x-velocity* as Contour Name in the Contours window. Select Contours of Velocity… and X Velocity. Check the box Filled under Options. Uncheck Node Values, Boundary Values and Global Range under Options. Select Colormap Options… and select Fixed as Font Behavior under Font. Set the Font Size to 12 and select Apply. Close the Colormap window.

Select New Surface>>Plane…. Select *YZ Plane* as Method, *yz-plane-x=0.25m* as New Surface Name and select Create. Close the Plane Surface window. Repeat this step and create *yz-plane-x=0.5m* and *yz-plane-x=0.75m*.

Select *inlet, yz-plane-x=0.25m, yz-plane-x=0.5m, yz-plane-x=0.75m* and *outlet* under Surfaces and click on Save/Display, see Figure 21.7g). Close the Contours window. Select Set View From +X Direction. Repeat this step but instead select Contours of Velocity… and Y Velocity.

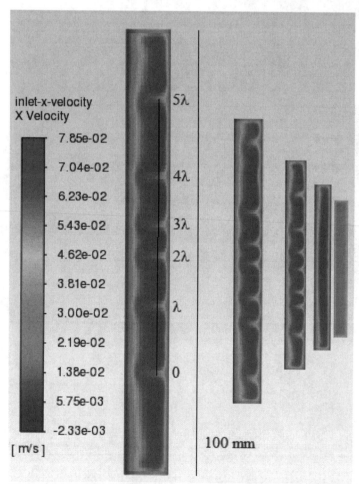

Figure 21.7g) Contours of X Velocity in YZ Plane at inlet, $x = 0.25$ m, $x = 0.5$ m, $x = 0.75$m and at the outlet of rotating plane channel.

H. Theory

8. The Coriolis instability for the flow in a plane channel with system rotation is governed by the Reynolds number Re and the rotation number Ro.

$$Re = \frac{U_b d}{v} \tag{21.1}$$

$$Ro = \frac{\Omega d}{U_b} \tag{21.2}$$

where U_b is the bulk flow velocity in the channel, d is the channel width, Ω is the rotation rate and v is the kinematic viscosity of the fluid. The critical Reynolds number Re_{crit} is 5770 caused by Tollmien-Schlichting waves without rotation. Roll cell instabilities will develop below this Reynolds number due to the Coriolis force. The non-dimensional wave number α of the Coriolis instability is determined by

$$\alpha = \frac{2\pi d}{\lambda} \tag{21.3}$$

where λ is the spanwise wave-length of the instability. From Figure 21.7d), the wave number can be determined to be

$$\alpha = \frac{2\pi \cdot 0.01}{0.012417} = 5.06 \tag{21.4}$$

In order to determine the wave-length that is used to determine the wave number in equation (21.3), you can measure the total length in the vertical spanwise direction for six successive streaks in Figures 21.7e) – g). Next, you take this vertical length and divide by five to get the average wave-length in the vertical spanwise direction. To get the scale for figures 21.7e) – g), use the total height of the channel in the vertical direction in Figures 21.7e) – g) that corresponds to 100 mm.

The Reynolds and rotation numbers in the Ansys Fluent simulations is

$$Re = \frac{U_b d}{v} = \frac{0.05*0.01}{10^{-6}} = 1000 \tag{21.5}$$

$$Ro = \frac{\Omega d}{U_b} = \frac{0.5*0.01}{0.05} = 0.1 \tag{21.6}$$

I. References

1. Alfredsson, P.H. and Persson, H., Instabilities in channel flow with system rotation. *J. Fluid Mech.*, **202**, 543, (1989).
2. Chandrasekhar, S., Hydrodynamic and Hydromagnetic Stability, Dover, 1981.
3. Hart, J.E., Instability and secondary motion in a rotating channel flow. *J. Fluid Mech.*, **45**, 341, (1971).
4. Lezius, D.K. and Johnston, J.P., The structure and stability of turbulent boundary layers in rotating channel flow, *J. Fluid Mech.*, **77**, 153, (1976).
5. Speziale, C.G. and Thangam, S., Numerical study of secondary flows and roll-cell instabilities in rotating channel flow. *J. Fluid Mech.*, **130**, 377, (1983).

J. Exercises

1. Run the calculations for the flow in the rotating channel in this chapter but use the following bulk flow velocities: 0.03, 0.06, 0.09 and 0.12 m/s and determine the Reynolds number, rotation number and the spanwise wave number for each velocity.

2. Run the calculations for the flow in the rotating channel as shown in this chapter at $Re = 500$ for different rotation numbers $Ro = 0.2, 0.3, 0.4$ and 0.5. Determine the spanwise wave number for each case and compare with the critical value.

CHAPTER 22. COMPRESSIBLE FLOW PAST A BULLET

A. Objectives

- Using Mesh Created in ICEM CFD for 2D Axisymmetric Flow Past a Bullet
- Inserting Boundary Conditions
- Running Turbulent Steady 2D Axisymmetric ANSYS Fluent Simulations
- Using Contour Plots for Visualizations
- Compare Drag Coefficient Results with Experiments

B. Problem Description

In this chapter we will study the flow around the Berger 7 mm 180 grain VLD bullet. We will be using a parabolic mesh for the compressible airflow at different Mach number past the bullet. Results from simulation drag coefficients will be compared with experiments for the bullet. ICEM CFD is not available in Ansys Student. You will be reading the mesh in Ansys Workbench.

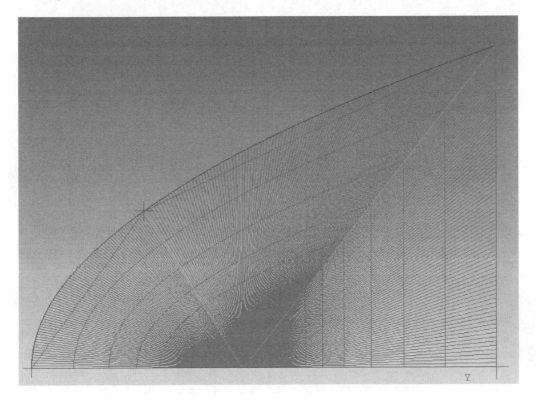

C. Starting Ansys Workbench and Setting Up ANSYS Fluent

1. Start by launching Ansys Workbench. Double click on Fluent under Toolbox>>Component Systems. Analysis Systems in the Toolbox. Select Units>>Metric (kg, m, s, …) from the menu. Right click on Mesh in the Project Schematic window and select Import Mesh File…>>Browse…. Select *BergerBullet7mm180grainVLD.msh* and click on Open. File>>Save Project from the menu and save the project with the name *Berger Bullet.wbpj*. Double click on Setup in Project Schematic. Select Double Precision under Options and set the number of Solver Processes under Parallel (Local Machine) to the same number as the number of computer cores. Click on Start.

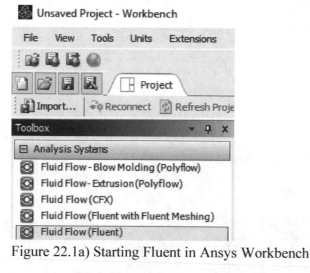

Figure 22.1a) Starting Fluent in Ansys Workbench

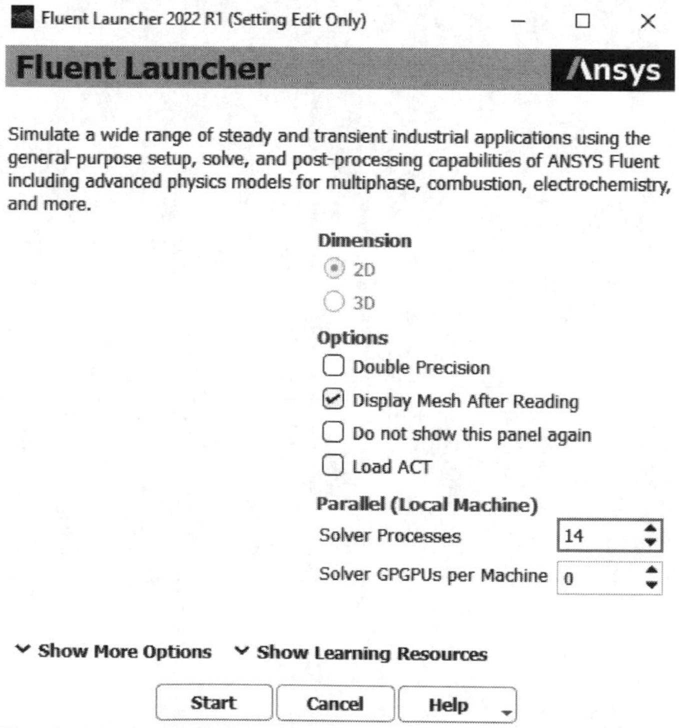

Figure 22.1b) Fluent launcher

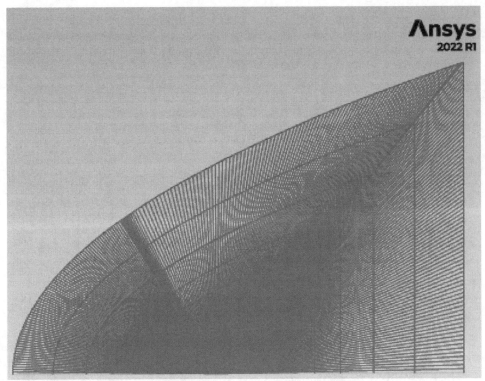

Figure 22.1c) Mesh in Ansys Fluent

2. Double click on General under Setup in Outline View. Set 2D Space to *Axisymmetric*. Double click on Models and Viscous (SST k-omega) under Setup in Outline View. Set Model to *Spalart-Allmaras (1 eqn)*. Select *Strain/Vorticity-Based* under Spalart-Allmaras Production. Click OK to close the Viscous Model window.

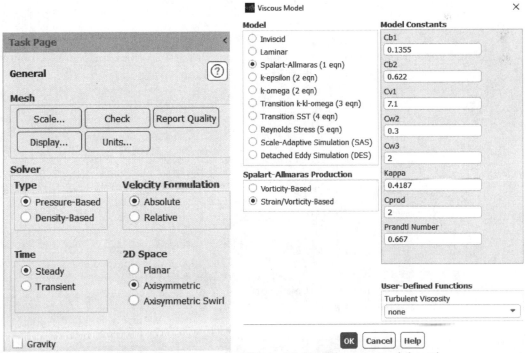

Figure 22.2a) General settings   Figure 22.2b) Viscous model settings

3. Double click on Materials, Fluid and air under Setup in the Outline View. Select *ideal-gas* for Density and select *sutherland* for Viscosity. Select OK to close the Sutherland Law window. Click on Change/Create and Close the Create/Edit Materials window.

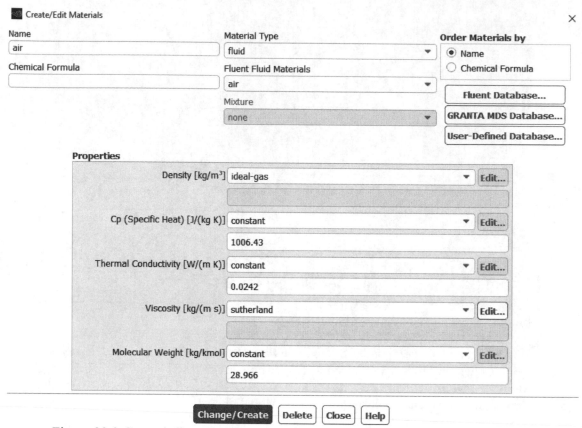

Figure 22.3 Create/edit materials

4. Double click on Boundary Conditions under Setup in the Outline View. Select *farfield* as Zone under Boundary Conditions on the Task Page. Click on Edit. Set Mach number to *2.68*. Select Apply and Close the Pressure Far-Field window.

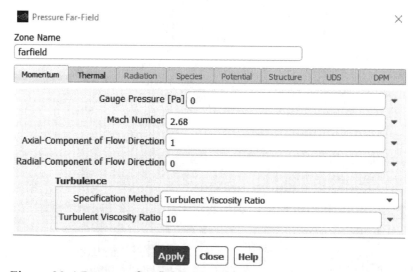

Figure 22.4 Pressure far-field

5. Double click on Methods under Solution in the Outline View. Select *Second Order Upwind* for Modified Turbulent Viscosity under Spatial Discretization. Double click on Controls under Solution in the Outline View. Set *Density* to 0.5 and *Modified Turbulent Viscosity* to 0.9 under Pseudo Time Explicit Relaxation Factors.

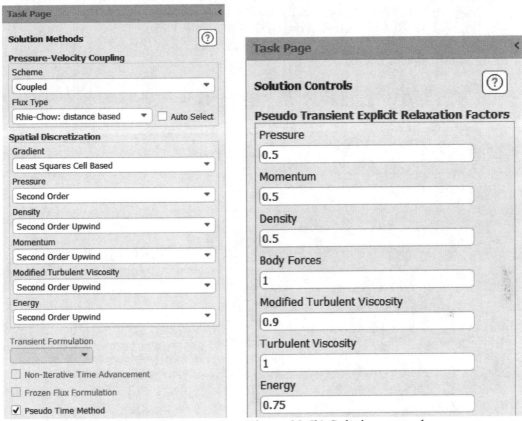

Figure 22.5a) Solution methods   Figure 22.5b) Solution controls

6. Double click on Monitors and Residual under Solution in the Outline View. Set all Residuals to 1e-06. Click OK to close the Residual Monitors window.

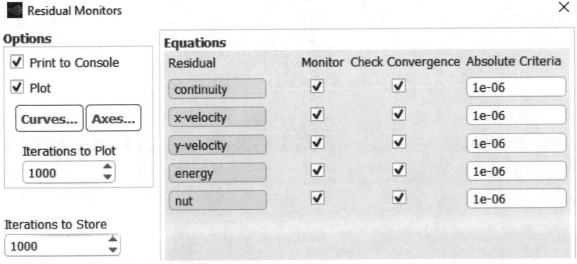

Figure 22.6 Residual monitors

7. Double click on Reference Values under Setup in the Outline View. Select Compute from *farfield* under Reference Values on the Task Page. Set the Area (m2) to 0.0000408894 and set Length (m) to 0.038735.

Double click on Report Definitions under Solution in the Outline View. Select New>>Force Report>>Drag in the Report Definitions window. Check the boxes for Report File, Report Plot and Print to Console under Create. Select *bullet* under Wall Zones. Click OK to close the Drag Report Definition window. Close the Report Definitions window.

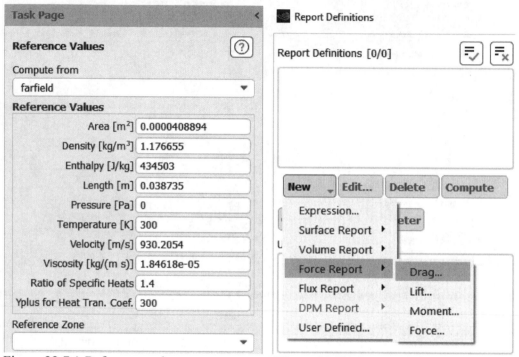

Figure 22.7a) Reference values   Figure 22.7b) Report definitions

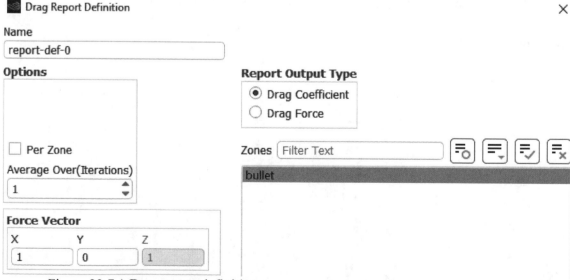

Figure 22.7c) Drag report definitions

8. Double click on Initialization under Solution in the Outline View. Select Hybrid Initialization and click on Initialize. Move the cursor to the Console and press *enter* on the keyboard to get the command prompt.

   a) Enter *solve/initialize/set-fmg-initialization* and press enter 3 times.
   b) Enter *100* as number of cycles on level 1 and press enter 2 times.
   c) Enter *100* as number of cycles on level 2 and press enter 8 times.
   d) Answer *yes* to the question *enable FMG verbose?* and press enter.
   e) Enter *solve/initialize/fmg-initialization* and press enter.
   f) Answer *yes* to the question *Enable FMG initialization?*

```
> solve/initialize/set-fmg-initialization


Customize your FMG initialization:
  set the number of multigrid levels [5]

  set FMG parameters on levels ..

   residual reduction on level 1 is:  [0.001]
   number of cycles on level 1 is:  [10] 100

   residual reduction on level 2 is:  [0.001]
   number of cycles on level 2 is:  [50] 100

   residual reduction on level 3 is:  [0.001]
   number of cycles on level 3 is:  [100]

   residual reduction on level 4 is:  [0.001]
   number of cycles on level 4 is:  [500]

   residual reduction on level 5 [coarsest grid] is:  [0.001]
   number of cycles on level 5 is:  [500]


 Number of FMG (and FAS  geometric multigrid) levels: 5
* FMG customization summary:
*   residual reduction on level 0 [finest grid] is: 0.001
*   number of cycles on level 0 is: 1
*   residual reduction on level 1 is: 0.001
*   number of cycles on level 1 is: 100
*   residual reduction on level 2 is: 0.001
*   number of cycles on level 2 is: 100
*   residual reduction on level 3 is: 0.001
*   number of cycles on level 3 is: 100
*   residual reduction on level 4 is: 0.001
*   number of cycles on level 4 is: 500
*   residual reduction on level 5 [coarsest grid] is: 0.001
*   number of cycles on level 5 is: 500
* FMG customization complete


  set FMG courant-number [0.75]

  enable FMG verbose? [no] yes

> solve/initialize/fmg-initialization
Enable FMG initialization? [no] yes
```
Figure 22.8 Full multigrid initialization

9. Double click on Run Calculation under Solution in the Outline View. Enter 400 as Number of Iterations. Click on Calculate.

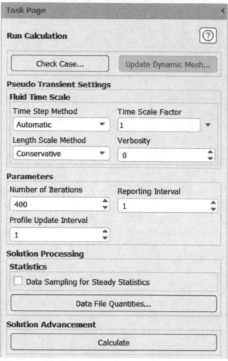

Figure 22.9a) Running calculations

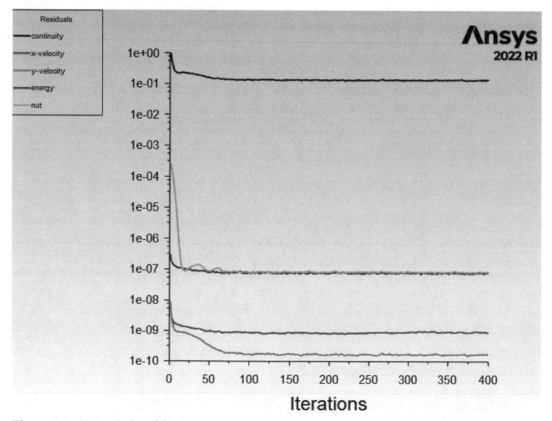

Figure 22.9b) Scaled residuals

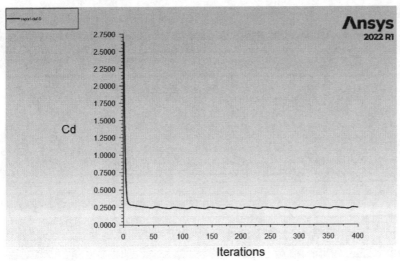

Figure 22.9c) Drag coefficient

10. Double click on Graphics and Contours under Results in the Outline View. Select Contours of Pressure and Static Pressure. Deselect all surfaces and use the settings and check boxes as shown in Figure 22.10a). Select Colormap Options…. Set Font Behavior to Fixed and Font Size to 11. Choose Top Colormap Alignment. Select Apply and Close the Colormap window. Enter the name *static-pressure-field* as Contour Name. Click on Save/Display. Zoom in on the bullet.

Create another contour plot and use the same setting but change to Contours of Velocity and Mach Number. Select Colormap Options…. Set Type to float under Number Format. Enter the name *mach-number-field* as Contour Name. Click on Save/Display. Zoom in on the bullet.

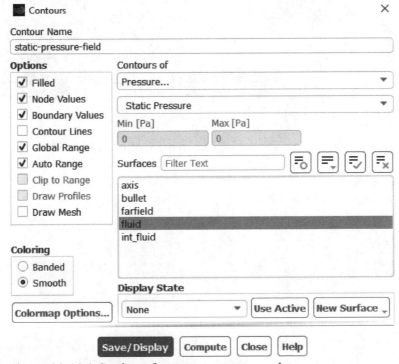

Figure 22.10a) Settings for pressure contour plot

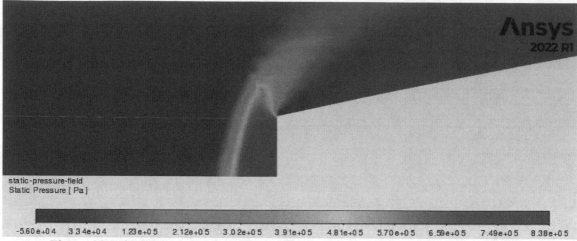

Figure 22.10b) Pressure at the tip of bullet

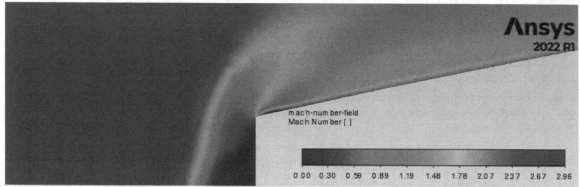

Figure 22.10c) Mach number around tip of bullet

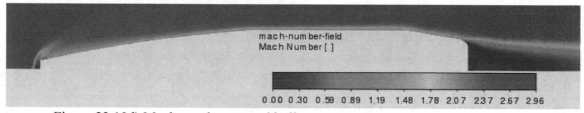

Figure 22.10d) Mach number around bullet

Double click on Reports and Forces under Results in the Outline View. Select *bullet* under Wall Zones and click on Print in the force Reports window. The value for the drag coefficient 0.242 at Ma = 2.68 has a 1.1 % difference in comparison with the experimental drag coefficient value 0.239.

```
Forces - Direction Vector (1 0 0)
                       Forces [N]                             Coefficients
Zone                   Pressure      Viscous       Total      Pressure      Viscous       Total
bullet                 4.2212706     0.80677021    5.0280408  0.20279417    0.038758069   0.24155224
-----------------      ------------  ------------  ---------  ------------  ------------  ------------
Net                    4.2212706     0.80677021    5.0280408  0.20279417    0.038758069   0.24155224
```

Figure 22.10e) Forces and coefficients at Mach number 2.68

11. You will need to create the case and scheme files to complete a batch job. Select File>>Write>>Case & Data... from the menu. Save the case and data files in the *Working Directory* folder with the name ***BergerBullet7mm180grainVLD.cas.h5***. Open Notebook and create and save the scheme file with the name ***berger_bulletIII.scm*** for the batch job. The scheme file is shown in Figure 22.11a) and is available for download from *sdcpublications.com*. Select File>>Read>>Scheme... from the menu in Ansys Fluent. Open the scheme file to start the batch job. Tables 22.1, 22.2 and Figures 22.11b), 22.11c) are showing the results from the batch job in comparison with experiments, G7 drag profile and G7 ballistic coefficient.

```
(Do ((x 0.700 (+ x 0.025))) ((> x 1.150))
(begin
    (Ti-menu-load-string (format #f "file read-case BergerBullet7mm180grainVLD.cas.h5"))
    (Ti-menu-load-string (format #f "define boundary-conditions pressure-far-field farfield n 0 n ~a n 300 n 1 n 0 n n y n 10" x))
    (Ti-menu-load-string (format #f "report reference-values compute pressure-far-field farfield"))
    (Ti-menu-load-string (format #f "solve initialize hyb-initialization yes"))
    (Ti-menu-load-string (format #f "solve/initialize/set-fmg-initialization 5 0.001 100 0.001 100 0.001 100 0.001 500 0.001 500 0.75 y"))
    (Ti-menu-load-string (format #f "solve/initialize/fmg-initialization y"))
    (Ti-menu-load-string (format #f "solve iterate 400"))
    (Ti-menu-load-string (format #f "report forces wall-forces y 1 0 y Berger_bullet_drag_coefficient_data_Ma=~a" x))
    (Ti-menu-load-string (format #f "display objects display static-pressure-field"))
    (Ti-menu-load-string (format #f "display save-picture Berger_bullet_static_pressure_contour_plot_Ma=~a.jpg yes" x))
    (Ti-menu-load-string (format #f "display objects display mach-number-field"))
    (Ti-menu-load-string (format #f "display save-picture Berger_bullet_Mach_number_contour_plot_Ma=~a.jpg yes" x))
    (Ti-menu-load-string (format #f "plot residuals y y y y y"))
    (Ti-menu-load-string (format #f "display save-picture Berger_bullet_residuals_plot_Ma=~a.jpg yes" x))
    (Ti-menu-load-string (format #f "plot file report-def-0-rfile.out"))
    (Ti-menu-load-string (format #f "display save-picture Berger_bullet_drag_coefficient_plot_Ma=~a.jpg yes" x))
))
(Do ((x 1.150 (+ x 0.100))) ((> x 2.000))
(begin
    (Ti-menu-load-string (format #f "file read-case BergerBullet7mm180grainVLD.cas.h5"))
    (Ti-menu-load-string (format #f "define boundary-conditions pressure-far-field farfield n 0 n ~a n 300 n 1 n 0 n n y n 10" x))
    (Ti-menu-load-string (format #f "report reference-values compute pressure-far-field farfield"))
    (Ti-menu-load-string (format #f "solve initialize hyb-initialization yes"))
    (Ti-menu-load-string (format #f "solve/initialize/set-fmg-initialization 5 0.001 100 0.001 100 0.001 100 0.001 500 0.001 500 0.75 y"))
    (Ti-menu-load-string (format #f "solve/initialize/fmg-initialization y"))
    (Ti-menu-load-string (format #f "solve iterate 400"))
    (Ti-menu-load-string (format #f "report forces wall-forces y 1 0 y Berger_bullet_drag_coefficient_data_Ma=~a" x))
    (Ti-menu-load-string (format #f "display objects display static-pressure-field"))
    (Ti-menu-load-string (format #f "display save-picture Berger_bullet_static_pressure_contour_plot_Ma=~a.jpg yes" x))
    (Ti-menu-load-string (format #f "display objects display mach-number-field"))
    (Ti-menu-load-string (format #f "display save-picture Berger_bullet_Mach_number_contour_plot_Ma=~a.jpg yes" x))
    (Ti-menu-load-string (format #f "plot residuals y y y y y"))
    (Ti-menu-load-string (format #f "display save-picture Berger_bullet_residuals_plot_Ma=~a.jpg yes" x))
    (Ti-menu-load-string (format #f "plot file report-def-0-rfile.out"))
    (Ti-menu-load-string (format #f "display save-picture Berger_bullet_drag_coefficient_plot_Ma=~a.jpg yes" x))
))
```

Figure 22.11a) Scheme file for batch job

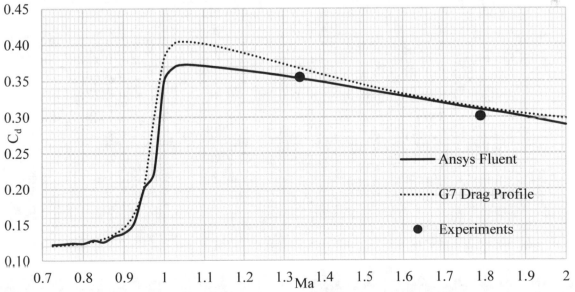

Figure 22.11b) Comparison between ANSYS Fluent, experiments[4] and G7 Drag Profile

491

Ma	V (m/s)	$C_{d,Fluent}$	$C_{d,G7}$	$i_{G7,Fluent}$	$BC_{G7,Fluent}$ (lb/in^2)
0.725	251.9535	0.1222	0.1207	1.0124	0.3156
0.75	260.6416	0.1230	0.1215	1.0121	0.3157
0.775	269.3296	0.1242	0.1226	1.0133	0.3154
0.8	278.0177	0.1237	0.1242	0.9958	0.3209
0.825	286.7057	0.1283	0.1266	1.0138	0.3152
0.85	295.3938	0.1257	0.1306	0.9624	0.3320
0.875	304.0818	0.1338	0.1368	0.9782	0.3267
0.9	312.7699	0.1384	0.1464	0.9454	0.3380
0.925	321.4579	0.1529	0.166	0.9213	0.3468
0.95	330.146	0.2016	0.2054	0.9816	0.3255
0.975	338.834	0.2237	0.2993	0.7473	0.4276
1	347.5221	0.3494	0.3803	0.9187	0.3478
1.025	356.2101	0.3688	0.4015	0.9185	0.3479
1.05	364.8982	0.3725	0.4043	0.9213	0.3468
1.075	373.5862	0.3721	0.4034	0.9223	0.3465
1.1	382.2743	0.3708	0.4014	0.9238	0.3459
1.125	390.9623	0.3692	0.3987	0.9261	0.3450
1.15	399.6504	0.3678	0.3955	0.9299	0.3436
1.2	417.0265	0.3644	0.3884	0.9381	0.3406
1.25	434.4026	0.3609	0.381	0.9471	0.3374
1.3	451.7787	0.3569	0.3732	0.9564	0.3341
1.35	469.1548	0.3523	0.3657	0.9633	0.3317
1.4	486.5309	0.3480	0.358	0.9720	0.3287
1.5	521.2831	0.3378	0.344	0.9820	0.3254
1.55	538.6592	0.3332	0.3376	0.9869	0.3238
1.6	556.0353	0.3286	0.3315	0.9913	0.3224
1.65	573.4114	0.3239	0.326	0.9935	0.3216
1.7	590.7875	0.3187	0.3209	0.9932	0.3217
1.75	608.1636	0.3137	0.316	0.9927	0.3219
1.8	625.5398	0.3093	0.3117	0.9924	0.3220
1.85	642.9159	0.3052	0.3078	0.9917	0.3222
1.9	660.292	0.3002	0.3042	0.9870	0.3238
1.95	677.6681	0.2945	0.301	0.9784	0.3266
2	695.0442	0.2889	0.298	0.9696	0.3296

Table 22.1 Drag coefficient, form factor and ballistic coefficient from Ansys Fluent in comparison with G7 model.

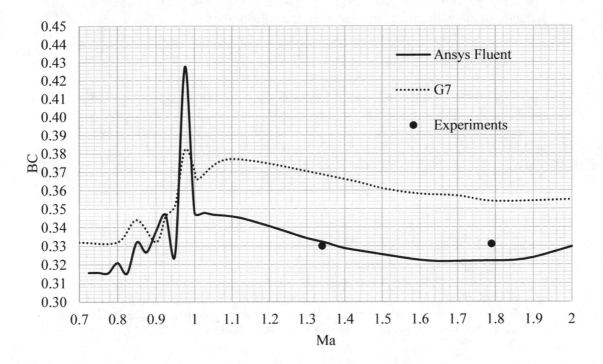

Figure 22.11c) Comparison between Ansys Fluent, experiments and G7 ballistic coefficients BC

D. Theory

12. The drag coefficient for a bullet is defined as

$$C_d = \frac{F_d}{\frac{1}{2}\rho V^2 A} \tag{22.1}$$

where F_d (N) is the drag force on the bullet, $\rho = 1.1767$ (kg/m^3) is air density, V (m/s) is the velocity of the bullet and A is frontal area of the bullet. The Mach number Ma is defined as the velocity of the bullet divided by the speed of sound $a = 347.5221$ (m/s) in the air

$$Ma = \frac{V}{a} \tag{22.2}$$

where the speed of sound depends on the absolute temperature of the air $T = 300$ (K), the gas constant $R = 287.5514$ (J/kg-K) and γ is the adiabatic index.

$$a = \sqrt{\gamma RT} \tag{22.3}$$

The ballistic coefficient BC (lb/m^2) for the bullet is determined from

$$BC = \frac{m}{id^2} \tag{22.4}$$

493

where i is the form factor, $m = 0.02571$(lb) is the bullet mass and $d = 0.2837$ (in) is the diameter of the bullet. The form factor is determined from

$$i = \frac{c_d}{c_{d,GX}}$$ (22.4)

where $c_{d,GX}$ is the drag coefficient from the GX drag profile as shown in Figure 22.11b) and Table 22.1 for drag profile G7.

E. References

1. ANSYS Fluent Tutorial Guide, Release 18.0, January 2017.
2. Beech, D., Aerodynamic Drag Measurement and Modeling for Small Arms – Improving on Ballistic Coefficients, Applied Ballistics, 2019.
3. Litz, B., Ballistic Coefficient Testing of the Sierra .308 155 grain Matchking PALMA Bullet., Applied Ballistics, ABDOC107.2, (2007).
4. Litz, B., Berger's 7 mm VLD Bullets Part 1: Properties and Test Results, Applied Ballistics, ABDOC109.1, (2008).
5. Muruganantham, V.R. and Babin, T., Numerical Investigation of hybrid blend design target bullets, MATEC Web of Conferences **172**, 01006 (2018).

F. Exercises

22.1 Run the simulations in this chapter for the Berger 7mm 168 VLD bullet. Bullet dimensions in inches are given in Reference 4. Use the file "BergerBullet7mm168grainVLD.dat" to read the geometry for the bullet. This file is available for download at sdcpublications.com. Fill out the missing information in Table 22.3. Draw the G7 drag profile and include comparison with experiments and ANSYS Fluent results in the same graph. The G7 drag profile is also available for download at sdcpublications.com.

Ma	V (m/s)	C_d, Exp.	C_d, Fluent	Percent Diff.
0.89	308.9115	------		-----
1.12	388.7425	------		-----
1.34	465.1027	0.335		
1.79	621.2939	0.297		
2.23	774.0142	0.270		
2.69	930.2054	0.249		

Table 22.2 Drag coefficient from Ansys Fluent in comparison with experiments for Berger 7mm 168 grain VLD

22.2 Run the simulations in this chapter for the Sierra .30 cal 155 grain PALMA MatchKing. Bullet dimensions in inches are given in Reference 3. Use the file "SierraBullet.30cal155grainPALMAMatchKing.dat" to read the geometry for the bullet. This file is available for download at sdcpublications.com. Fill out the missing information in Table 22.4. Draw the G7 drag profile and include comparison with experiments and ANSYS Fluent results in the same graph. The G7 drag profile is also available for download at sdcpublications.com.

Ma	V (m/s)	C_d, Exp.	C_d, Fluent	Percent Diff.
0.89	308.9115	------		-----
1.12	388.7425	------		-----
1.34	465.1027	0.406		
1.79	621.2939	0.348		
2.23	774.0142	0.309		
2.69	930.2054	0.280		

Table 22.3 Drag coefficient from Ansys Fluent in comparison with experiments for Sierra .30 cal 155 grain PALMA MatchKing

G. Notes

CHAPTER 23. VERTICAL AXIS WIND TURBINE FLOW

A. Objectives

- Using Dynamic Mesh and 6DOF Solver
- Inserting Rotating Mesh
- Running Turbulent Transient 3D ANSYS Fluent Simulations for VAWT
- Using Contour Plots for Visualizations
- Comparison of Rotor RPM with Experiments

B. Problem Description

In this chapter we will study the airflow that causes the drag based helical Savonius rotor VAWT to spin. We will be using a combination of fixed and rotating mesh domains for the incompressible airflow. Results for rotor RPM will be compared with experiments. We will create a mesh with less than 512 000 elements so that the ANSYS Fluent student version can be used. The performance of this rotor has been studied by Zakaria and Ibrahim[1-2] using the commercial code AcuSolve and compared with experiments. The helical rotor has180 degree sweep angle, height over diameter ratio of 2, an overlap ratio of 0.242, and end plates with a diameter that is 10% larger than the rotor diameter of 0.5 m. The helical rotor was modeled in SOLIDWORKS.

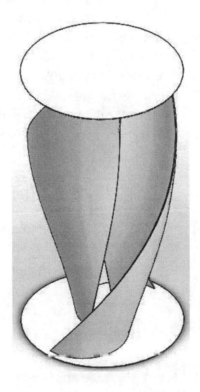

C. Launching ANSYS Workbench and Selecting Fluent

1. Start by launching ANSYS Workbench. Double click on Fluid Flow (Fluent) under Analysis Systems in the Toolbox. Select Units>>Metric (tonne, mm, s, …) from the menu.

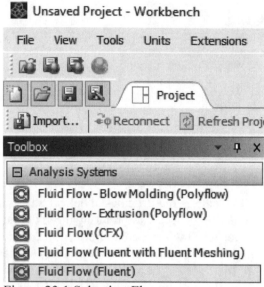

Figure 23.1 Selecting Fluent

D. Launching ANSYS DesignModeler

2. Right click on Geometry in the Project Schematic window and select New DesignModeler Geometry. Select millimeter as the desired length unit from the menu (Units>>Millimeter) in DesignModeler. Select File>>Import External Geometry File … from the menu. Select and Open the file *Helical Savonius Turbine.IGS*. This and other files can be downloaded from *sdcpublications.com*. Click on Generate. Figure 23.2b) shows the imported rotor placed inside the rotating domain. Set Simplify Topology in Details View to Yes. Click on Generate once again. The rotating domain has a diameter of 605 mm and a height of 1100 mm.

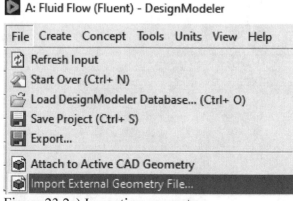

Figure 23.2a) Importing geometry

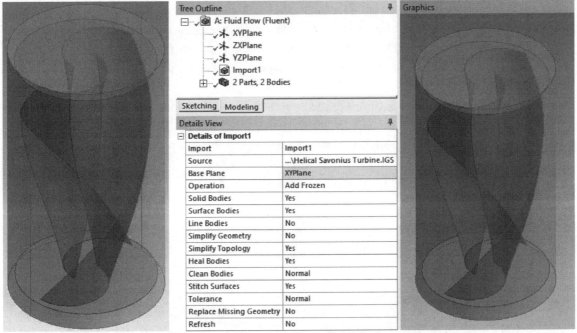

Figure 23.2b) Imported rotor before and after simplified topology.

Select Create>>Boolean from the menu. Select Subtract as the Operation in Details View. Apply the cylindrical rotating domain as the Target Body and Apply the rotor as the Tool Body. To select the rotor, select the second Solid under 2 Parts, 2 Bodies in the Tree Outline. Click on Generate. You will now have 1 Part, 1 Body in the Tree Outline.

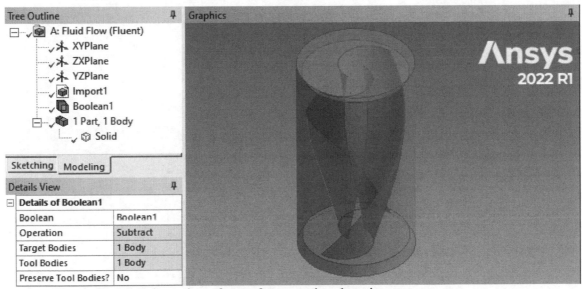

Figure 23.2c) Subtraction of rotor from rotating domain

3. Select File>>Import External Geometry File … from the menu. Select and Open the file *Fixed Domain for HST.IGS*. Click on Generate. Figure 23.3a) shows the imported fixed domain with the cylinder in the middle of the fixed domain. The fixed domain has a length of 15,000 mm, a square inlet and a square outlet with dimension 8,000 mm.

Select Create>>Boolean from the menu. Select Subtract as the Operation in Details View. Apply the fixed domain as the Target Body and Apply the cylinder as the Tool Body. To select the cylinder, you will need to select the green rectangle in the lower left corner of the graphics window, see Figure 23.3a). Click on Generate. You will now have 2 Parts, 2 Bodies in the Tree Outline. Close DesignModeler.

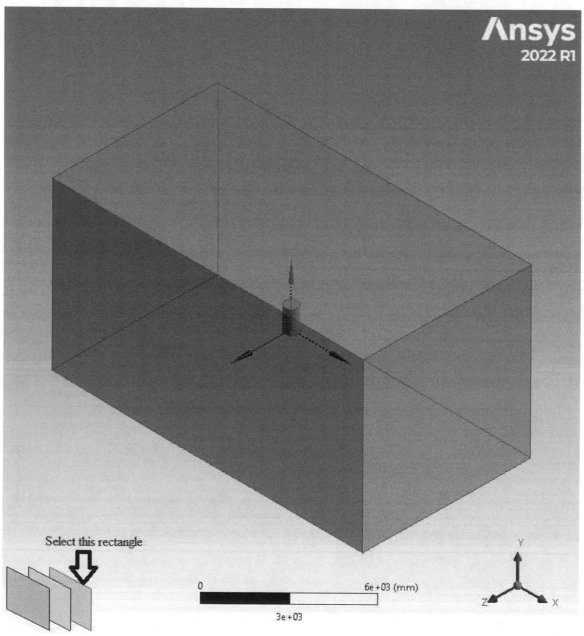

Figure 23.3a) Fixed outer domain, rotating domain and turbine

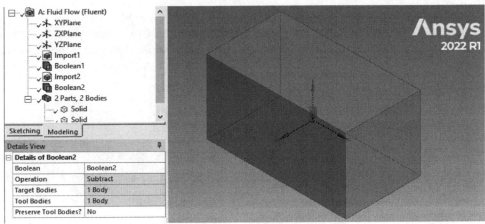

Figure 23.3b) Completed subtraction of rotating domain from fixed domain

E. Launching ANSYS Meshing

4. Double click on Mesh under Project Schematic in ANSYS Workbench. Select Mesh in the Outline under Project. Select Home>>Tools>>Units>>Metric (mm, kg, N, ...) from the menu. Right click on Mesh under Project in Outline and select Generate Mesh. A coarse mesh is generated.

   Right click on Coordinate System under Project and Model in the Outline. Select Insert>>Coordinate System. Select Define By Global Coordinates under Origin in Details of "Coordinate System". Set Origin Y to 500 mm. For Axis Y under Orientation About Principal Axis, select Define By Global Z Axis, see Figure 23.4a).

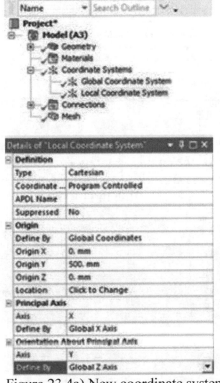

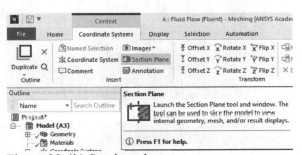

Figure 23.4a) New coordinate system   Figure 23.4b) Section plane

Right click on the new Coordinate System and select Rename. Rename the coordinate system as Local Coordinate System. Right click on the Local Coordinate System and select Create Section Plane, see Figure 23.4b). Click on Section Plane in the menu, see Figure 23.4c).

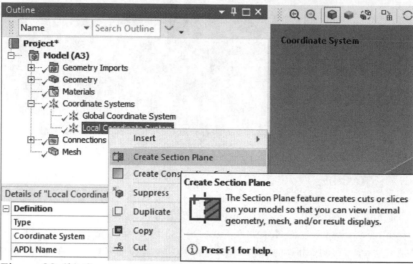

Figure 23.4b) Create section plane

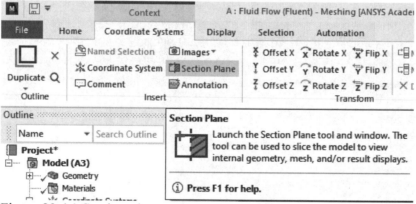

Figure 23.4c) Section plane

Right click in the graphics window and select View>>Top. Select Mesh in the Outline and use your mouse wheel to zoom in and out on the circular rotating domain, see Figure 23.4d). Right click in the graphics window, select Isometric View and zoom out using the mouse wheel. Uncheck the box for Section Plane 1 to show the full mesh.

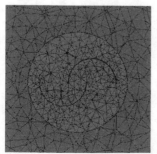 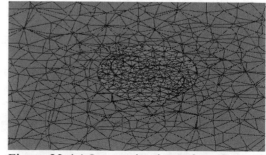

Figure 23.4d) Coarse mesh for rotor   Figure 23.4e) Isometric view of mesh

5. Select Mesh in the Outline. Select Mesh>>Controls>>Sizing from the menu. Right click in the graphics window and select Cursor Mode>>Body. Select the larger outer fixed domain in the graphics window and Apply as Geometry under Scope in Details of "Body Sizing" - Sizing. Set the Element Size to 900.0 mm and set Capture Curvature and Capture Proximity to Yes. Right click on Body Sizing under Mesh in the Outline and select Rename. Rename the body sizing to *Body Sizing Fixed Domain*.

Open Geometry under Project and Model in the Outline. Right click on the first Solid and select Rename. Rename this Solid to *Rotating Domain*. Rename the second Solid to *Fixed Domain*. Right click on Fixed Domain under Geometry and select Hide Body.

Select Mesh in the Outline and create another body sizing but this time select the cylindrical rotating domain in the graphics window. Set the Element Size to 60.0 mm and set Capture Curvature and Capture Proximity to Yes. Rename the body sizing to Body Sizing Rotating Domain. Right click on Fixed Domain under Geometry and select Hide Body.

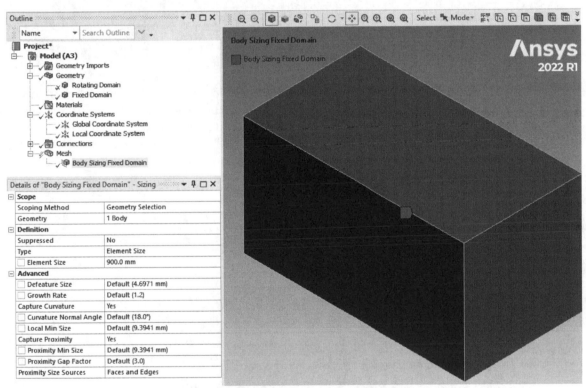

Figure 23.5a) Body sizing for fixed mesh

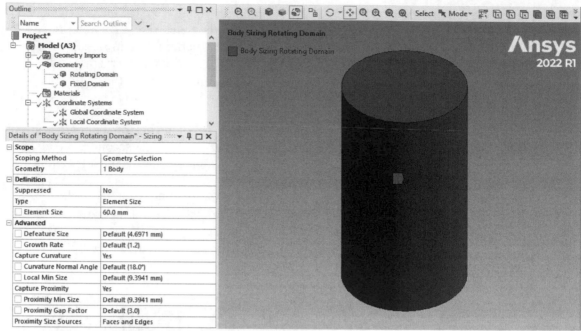

Figure 23.5b) Body sizing for rotating mesh

6. Right click on Fixed Domain under Geometry and select Hide Body. Right click in the graphics window and select Cursor Mode>>Face. Right click in the graphics window and select View>>Left. Zoom out and select the square inlet face, right click and select Create Named Selection…. Enter the name *inlet* as Selection Name and click OK to close the window. Repeat this step but select Right view and name the face *outlet*. Select

    Isometric View in the graphics window and zoom out. Select Wireframe .

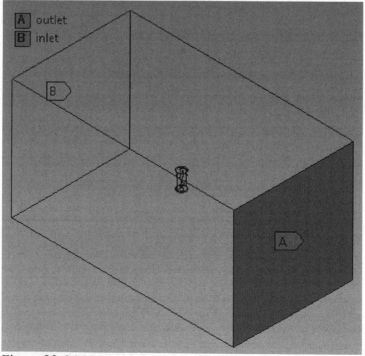

Figure 23.6a) Named selections for inlet and outlet

Right click in the graphics window and select Cursor Mode>>Body. Select Fixed Domain under Geometry in the Outline. Select the outer fixed domain, right click and select Create Named Selection…. Enter the name *fixed_domain* as Selection Name and click OK to close the window.

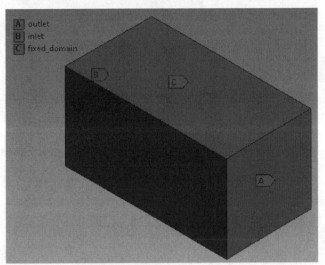

Figure 23.6b) Named selections for inlet, outlet and fixed domain

7. Right click on Fixed Domain under Geometry in the Outline and select Hide Body. Right click in the graphics window and select Cursor Mode>>Box Zoom. Zoom in on the rotating domain, right click in the graphics window and select Cursor Mode>>Face.

Right click in the graphics window and select Select All. Control deselect the cylindrical outer face of the rotating domain, control deselect the top circular face of the rotating domain, and finally turn the model around by holding the mouse wheel and dragging the mouse followed by control deselection of the bottom circular face of the rotating domain. Right click in the graphics window and select Create Named Selection…. Enter the name *rotor* as Selection Name and click OK to close the window.

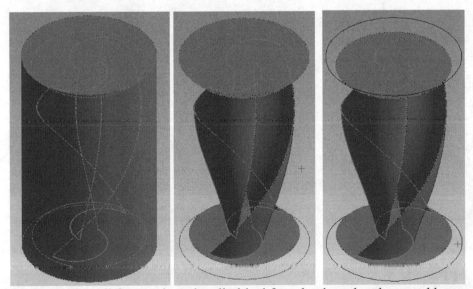

Figure 23.7a) All faces selected, cylindrical face deselected and top and bottom circular faces deselected

Select *rotor* under Named Selections in the Outline. You will have 14 Faces selected as Geometry for the rotor, see Figure 23.7b).

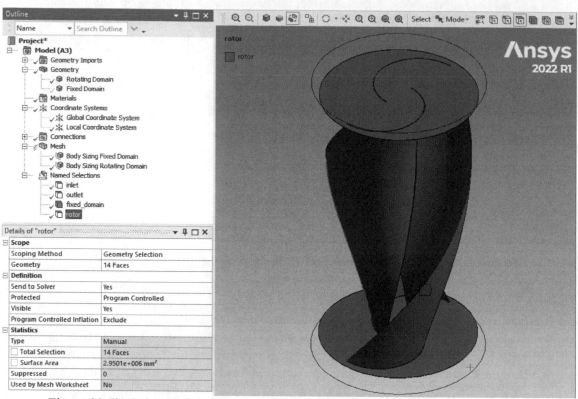

Figure 23.7b) Selected faces for helical Savonius rotor

Right click in the graphics window and select Cursor Mode>>Body. Select the cylindrical rotating domain in the graphics window, right click and select Create Named Selection…. Enter the name *rotating_domain* as Selection Name and click OK to close the window.

Figure 23.7c) Selected body for rotating domain

8. Right click in the graphics window and select Cursor Mode>>Face. Select Mesh in the Outline. Select Mesh>>Controls>>Sizing from the menu. Select Named Selection as Scoping Method under Scope in Details of "Sizing" - Sizing. Select *rotor* as Named Selection. Set the Element Size to **30.0 mm** and set **Capture Curvature** and **Capture Proximity** to **Yes**. Right click on Face Sizing under Mesh in the Outline and select Rename. Rename the face sizing to Face Sizing Rotor.

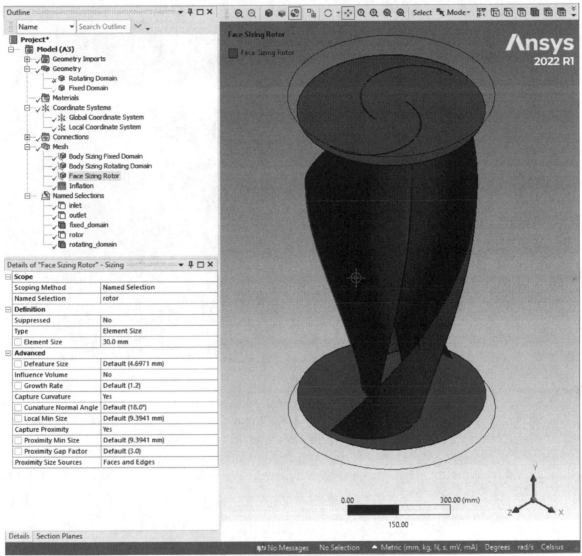

Figure 23.8a) Face sizing of helical Savonius rotor

Select Mesh>>Controls>>Inflation from the menu. Select Named Selection as Scoping Method under Scope in Details of "Inflation" - Inflation. Select *rotating_domain* as Named Selection under Scope. Select Named Selections as Boundary Scoping Method under Definition in Details of "Inflation" - Inflation. Select *rotor* as Boundary under Definition. Select First Layer Thickness as Inflation Option and set First Layer Height to **1.0 mm**. Set Maximum Layers to **3** and growth rate to **1.2**, see Figure 23.8b).

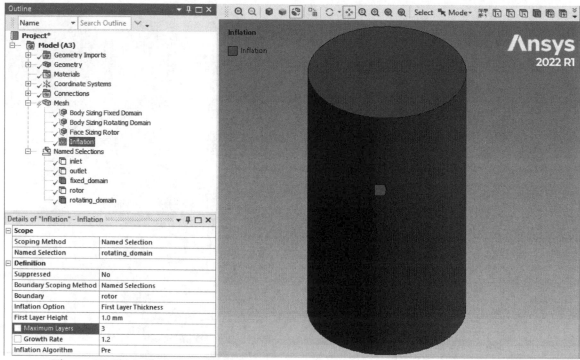

Figure 23.8b) Inflation settings

9. Right click on Mesh in the Outline and select Generate Mesh. Open Quality in Details of Mesh and select Orthogonal Quality as Mesh Metric. The Min value is 0.10131 which is larger than the minimum recommended value of 0.1, see Figure 23.9a).

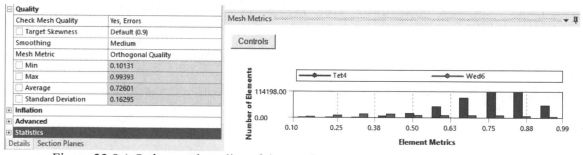

Figure 23.9a) Orthogonal quality of the mesh

Select Mesh in the Outline. Select the Section Planes tab in the lower left corner and check the box for Section Plane1. Select Section Plane1 and click on Edit Section Plane. Select Section Plane1 once again next to the check box and drag the mesh up and down in graphics window until you get Figure 23.9b). Close the Section Plane.

Open Statistics in Details of "Mesh". The number of Nodes is 120506 and the number of Elements 506532. Select File>>Export>>Mesh>>Fluent Input File>>Export from the menu and save the mesh file with the name *helical-savonius-turbine.msh*. Close Ansys Meshing. Right click on Mesh under Project Schematic in Ansys Workbench and select Update. Select File>>Save As from the menu in Ansys Workbench. Save the project with the name *Helical Savonius Turbine Flow.wbpj*.

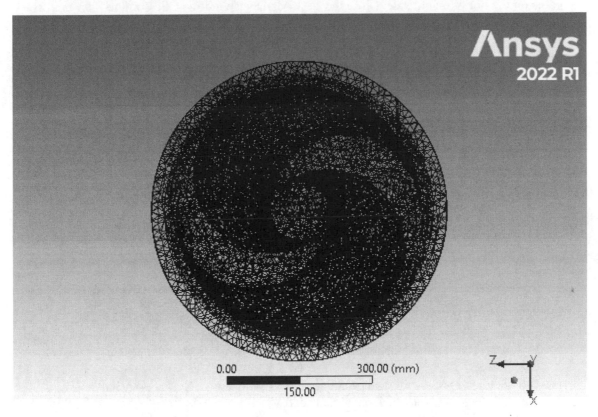

Figure 23.9b) Finished rotor mesh

10. Double-click on Setup under Project Schematic in Ansys Workbench. Check the box for Double Precision under Options in the Ansys Fluent Launcher. Click on Show More Options and take a note of the location of the *Working Directory*. Set the number of processes equal to the number of processor cores for your computer. To check the number of physical cores, press the Ctrl + Shift + Esc keys simultaneously to open the Task Manager. Go to the Performance tab and select CPU from the left column. You will see the number of physical cores on the bottom-right side. Close the Task Manager window. Click on the Start button in the Fluent Launcher window.

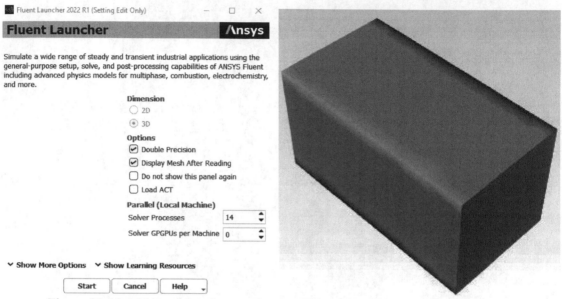

Figure 23.10a) Fluent launcher   Figure 23.10b) Mesh in Fluent

11. Double click on General under Setup in the Outline View. Select Transient as Time on the Task Page. Double click on Models and Viscous (SST k-omega). Select the Spalart-Allmaras (1 eqn) Viscous Model and click on OK to close the window.

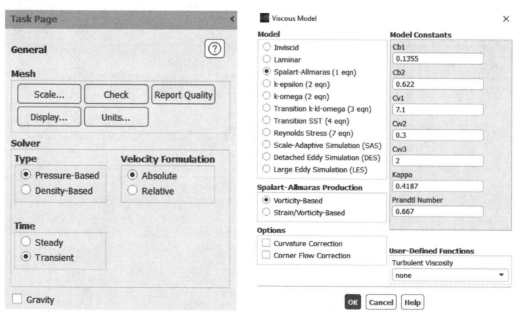

Figure 23.11a) Transient time setting   Figure 23.11b) Viscous model selection

12. Open Boundary Conditions and Inlet under Setup in the Outline View. Double click on inlet_(velocity-inlet, id = 8) , select Components as Velocity Specification Method and set X-Velocity [m/s] to 5. Click Apply and Close the Velocity Inlet window. Open Wall under Boundary Conditions. Double click on wall-fixed_domain (wall, id=7) and select Specified Shear as Shear Condition. Click Apply and Close the Wall window.

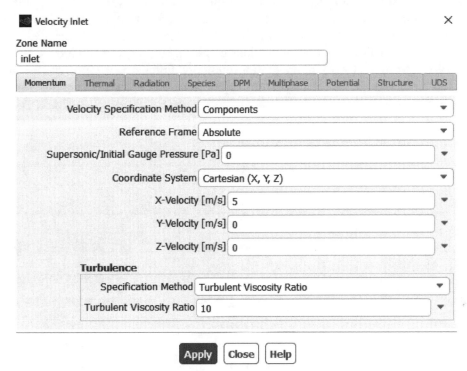

Figure 23.12a) Velocity inlet settings

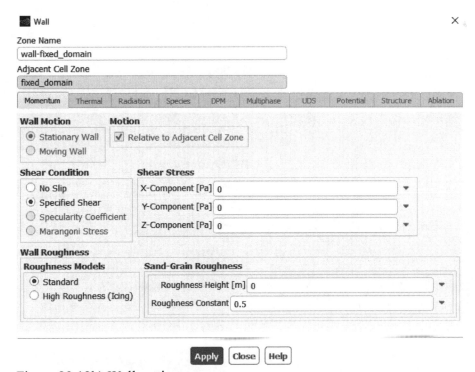

Figure 23.12b) Wall settings

13. Double click on Dynamic Mesh under Setup in the Outline View. Check the box for Dynamic Mesh, uncheck Smoothing under Mesh Methods and check the box for Six DOF under Options. Click on Settings… under Options. Select Create/Edit … in the Options window.

Check the box for One DOF Rotation and set the Mass (kg) to 4.01319583. Set the Axis under One DOF to X 0, Y 1, Z 0. Set the Center of Rotation to X (m) 0, Y (m) 0.50000169, Z (m) 0. Set Moment of Inertia (kg/m2) to 0.13735168. Enter *one-dof-rotation* as Name. Click on Create and Close the Six DOF Properties window. Check the box for Write Motion History in the Options window. Enter *HST Motion History* as File Name and direct the File Name to your *Working Directory*. Select OK to close the Options window.

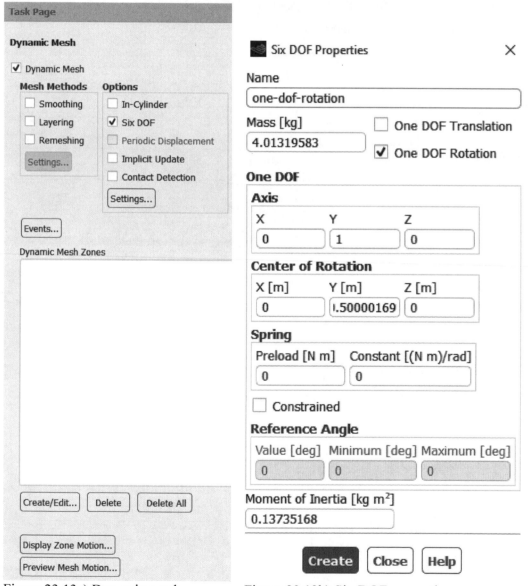

Figure 23.13a) Dynamic mesh   Figure 23.13b) Six DOF properties

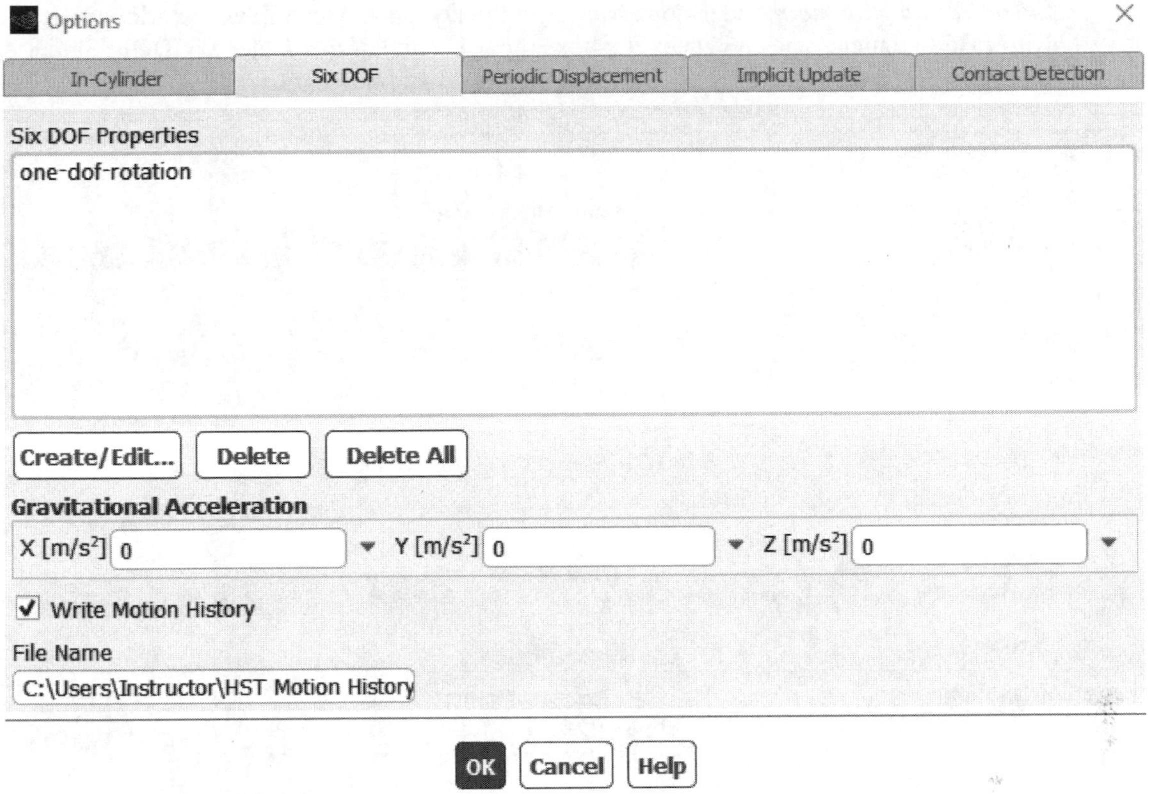

Options ✕

| In-Cylinder | Six DOF | Periodic Displacement | Implicit Update | Contact Detection |

Six DOF Properties

one-dof-rotation

Create/Edit... Delete Delete All

Gravitational Acceleration

X [m/s²] 0 ▼ Y [m/s²] 0 ▼ Z [m/s²] 0 ▼

☑ Write Motion History

File Name

C:\Users\Instructor\HST Motion History

OK Cancel Help

Figure 23.13c) Options settings

Select Create/Edit… under Dynamic Mesh Zones on the Task Page. Select *fixed_domain* under Zone Names in the Dynamic Mesh Zones window. Select Stationary under Type and click on Create.

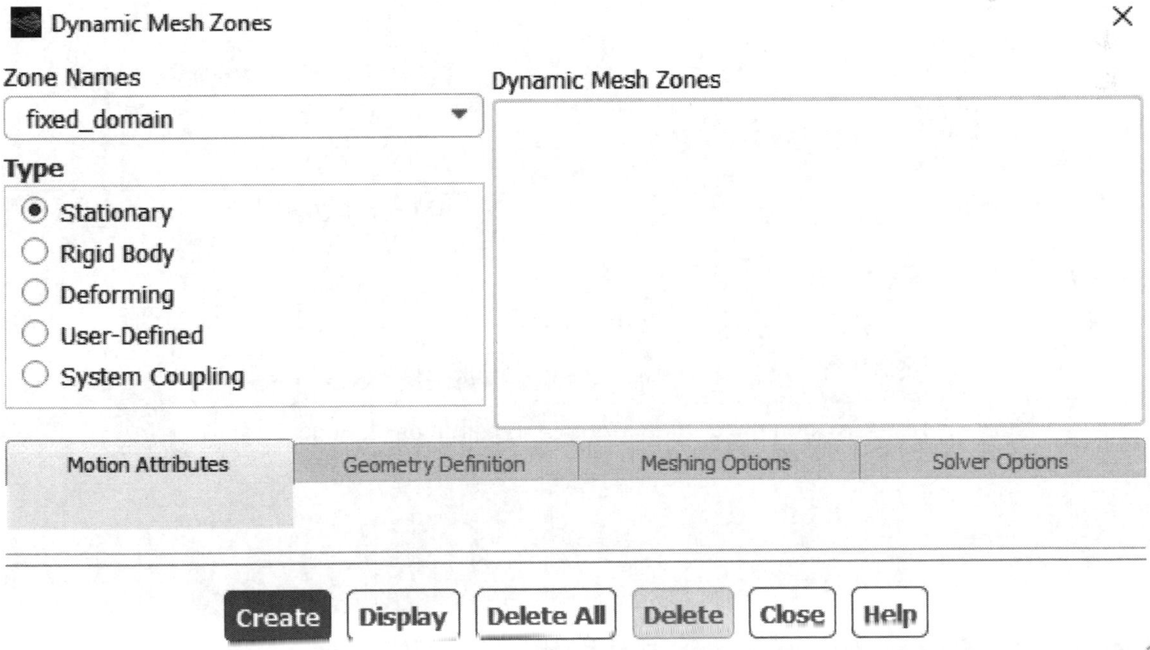

Dynamic Mesh Zones ✕

Zone Names

fixed_domain ▼

Dynamic Mesh Zones

Type

◉ Stationary
○ Rigid Body
○ Deforming
○ User-Defined
○ System Coupling

| Motion Attributes | Geometry Definition | Meshing Options | Solver Options |

Create Display Delete All Delete Close Help

Figure 23.13d) Creation of *fixed_domain* dynamic mesh zone

Select *rotating_domain* under Zone Names in the Dynamic Mesh Zones window. Select Rigid Body under Type and check the boxes for On and Passive under Six DOF. Select *one-dof-rotation* under Six DOF UDF/Properties. Click on Create.

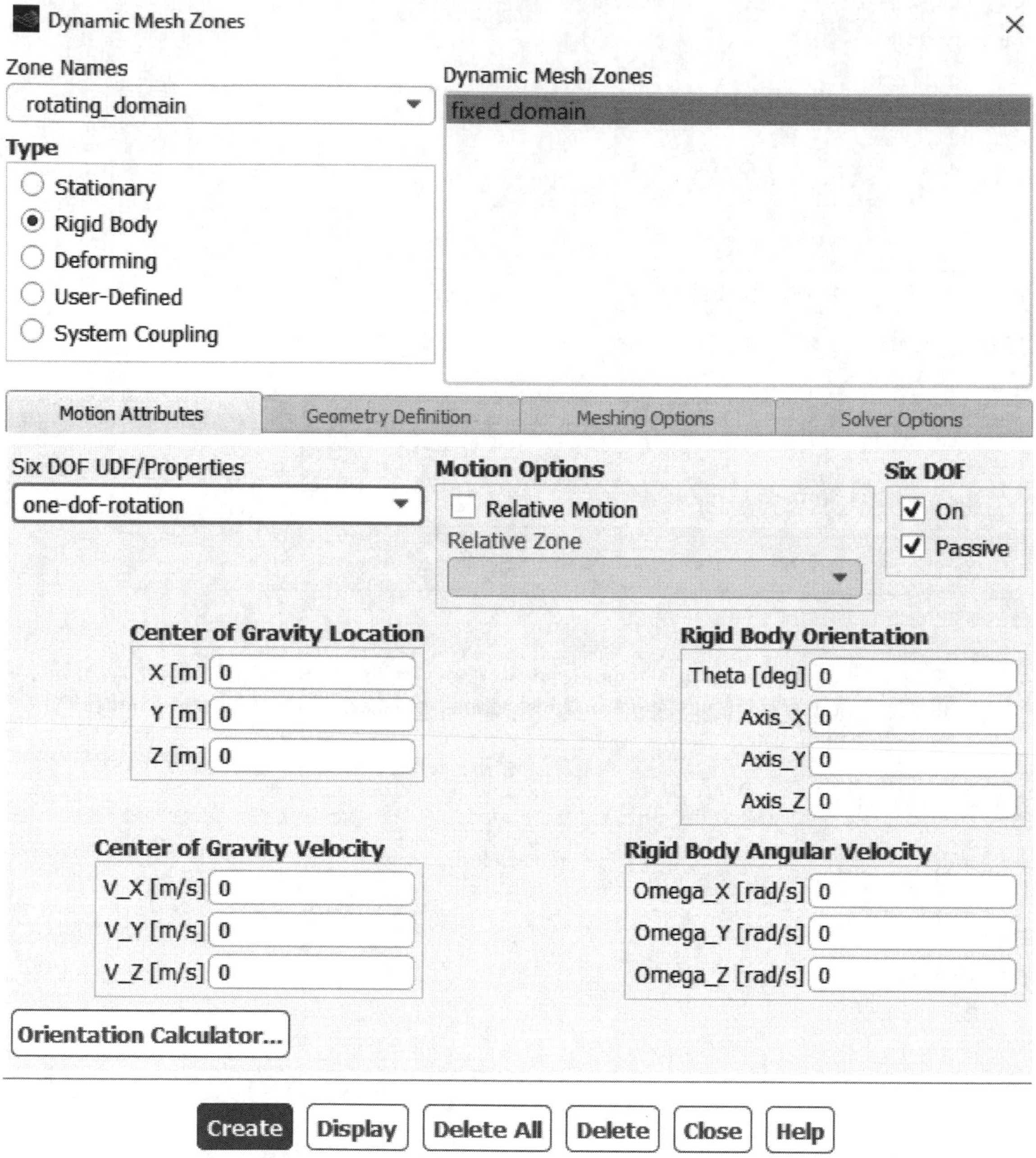

Figure 23.13e) Creation of *rotating_domain* dynamic mesh zone

Select *rotor* under Zone Names in the Dynamic Mesh Zones window. Select Rigid Body under Type and uncheck the box for Passive under Six DOF. Select *one-dof-rotation* under Six DOF UDF/Properties. Click on Create. Close the Dynamic Mesh Zones window.

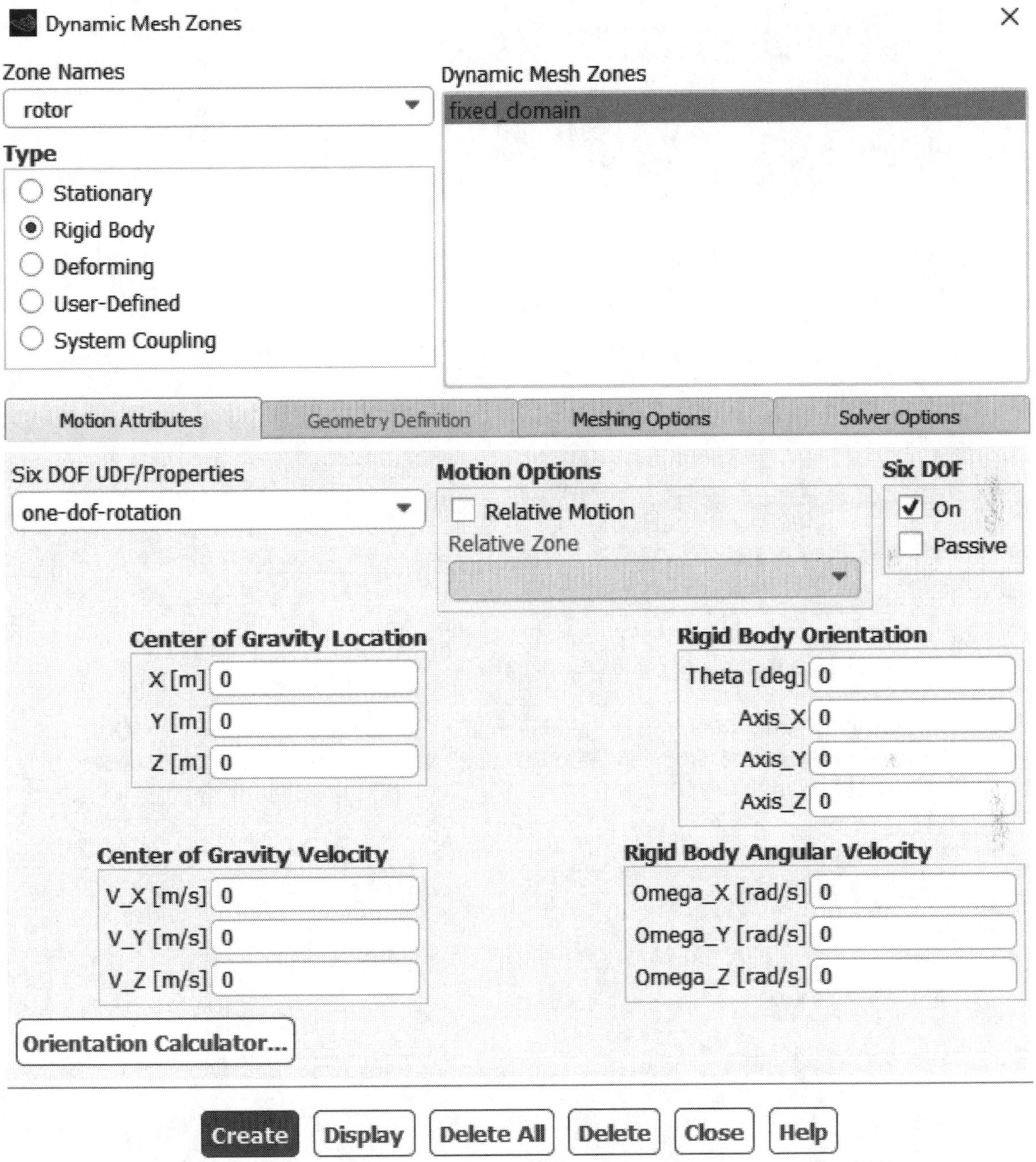

Figure 23.13f) Creation of *rotor* dynamic mesh zone

Double click on Reference Values under Setup in the Outline View. Select Compute from *inlet* under Reference Values on the Task Page. Set the Area [m²] to 0.5 which is the swept area of the turbine and set Length [m] to 0.25. The reference area is used to calculate the force and moment coefficients. The reference length is used to compute the moment coefficient.

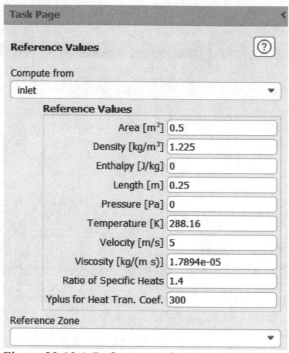

Figure 23.13g) Reference values

14. Double click on Methods under Solution in the Outline View. Select Second Order Upwind for Modified Turbulent Viscosity and select Second Order Implicit under Transient Formulation.

Figure 23.14a) Solution methods

Open Monitors under Solution in the Outline View and double click on Residual. Set the Absolute Criteria for all Residuals to 1e-06. Click on OK to close the Residual Monitors window.

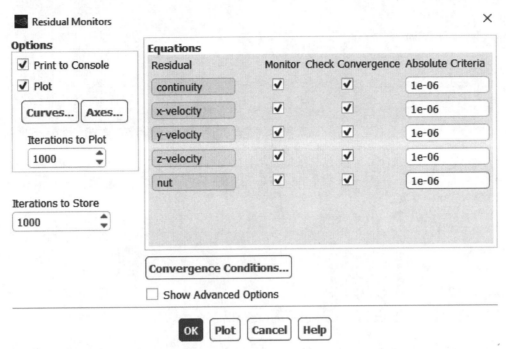

Figure 23.14b) Residual monitors

15. Open Monitors under Solution in the Outline View and double click on Report Files. Select New… in the Report Files Definition window. Select New>>Force Report>>Force… in the New Report File window. Select the Force Vector as X 1, Y 0 and Z 0. Select *rotor* under Zones. Check the boxes for Report File, Report Plot and Print to Console under Create. Enter *force-x* as Name. Select OK to close the Force Report Definition window.

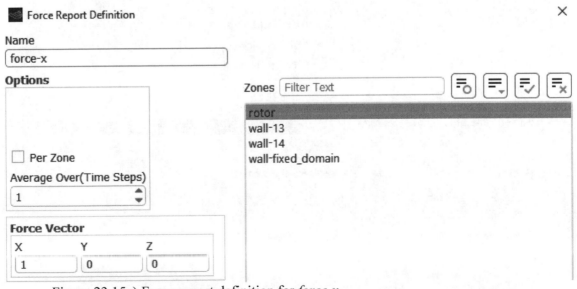

Figure 23.15a) Force report definition for *force-x*

Select New>>Force Report>>Force… in the New Report File window. Select the Force Vector as X 0, Y 0 and Z 1. Select *rotor* under Zones. Check the boxes for Report File, Report Plot and Print to Console under Create. Enter *force-z* as Name. Select OK to close the window.

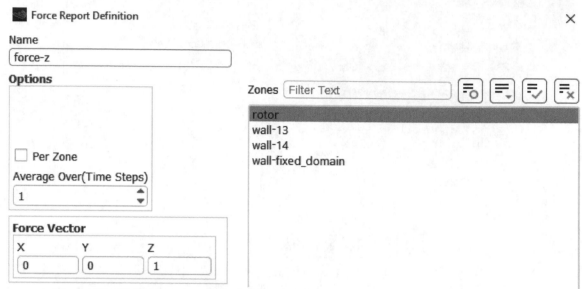

Figure 23.15b) Force report definition for *force-z*

Select New>>Force Report>>Drag… in the New Report File window. Select *Drag Coefficient* as Report Output Type. Select the Force Vector as X 1, Y 0 and Z 0. Select *rotor* under Zones. Check the boxes for Report File, Report Plot and Print to Console under Create. Enter *downwind-drag-force-coefficient* as Name. Select OK to close the window.

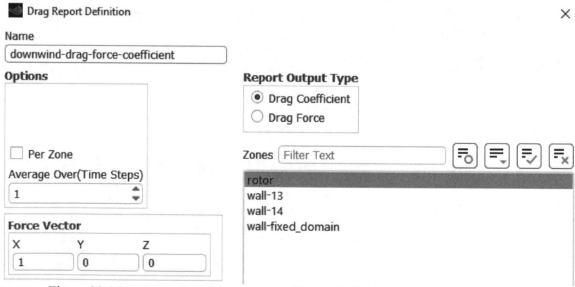

Figure 23.15c) Downwind drag force coefficient definition

Select New>>Force Report>>Lift... in the New Report File window. Select Lift Coefficient as Report Output Type. Select the Force Vector as X 0, Y 0 and Z 1. Select *rotor* under Zones. Check the boxes for Report File, Report Plot and Print to Console under Create. Enter *side-force-coefficient* as Name. Select OK to close the window.

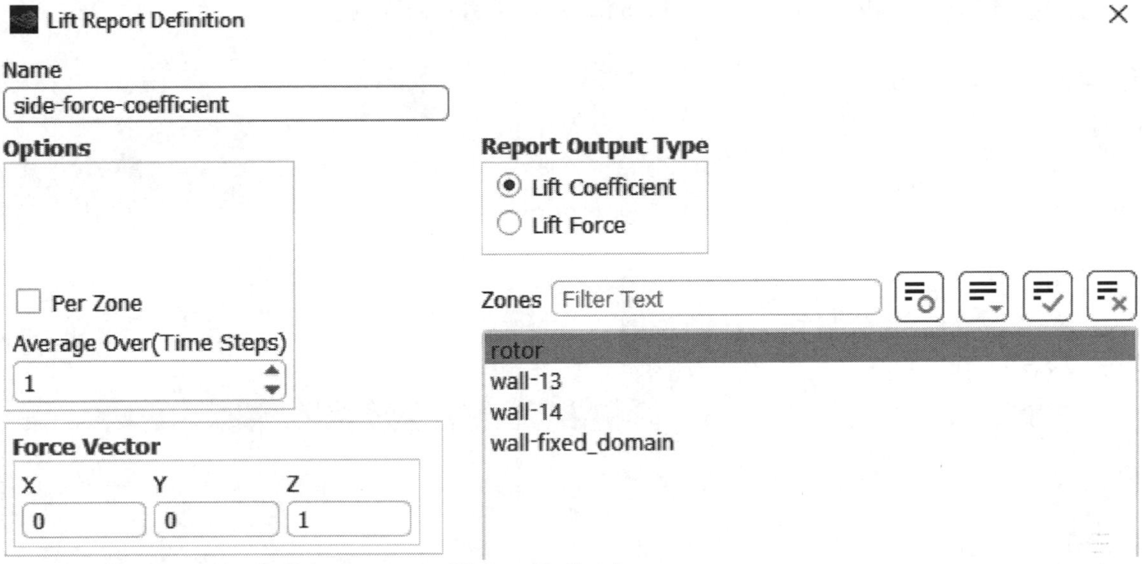

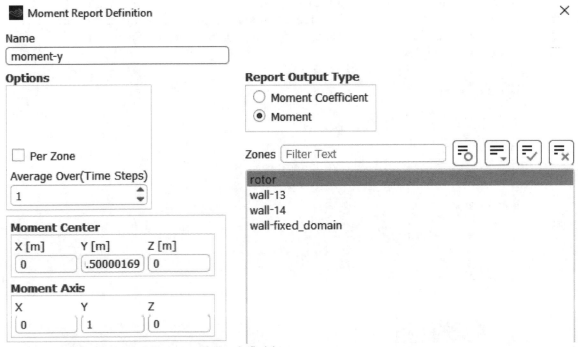

Figure 23.15d) Side force coefficient definition

Select New>>Force Report>>Moment... in the New Report File window. Select Moment as Report Output Type. Select Moment Center as X (m) 0, Y (m) 0.50000169 and Z (m) 0. Select Moment Axis as X 0, Y 1 and Z 0. Select *rotor* under Zones. Check the boxes for Report File, Report Plot and Print to Console under Create. Enter *moment-y* as Name. Select OK to close the window.

Figure 23.15e) Moment report definition

Select New>>Force Report>>Moment… in the New Report File window. Select Moment Coefficient as Report Output Type. Select Moment Center as X (m) 0, Y (m) 0.50000169 and Z (m) 0. Select Moment Axis as X 0, Y 1 and Z 0. Select *rotor* under Zones. Check the boxes for Report File, Report Plot and Print to Console under Create. Enter *torque-coefficient* as Name. Select OK to close the window. Close the Report File Definitions window and select OK to close the New Report File window.

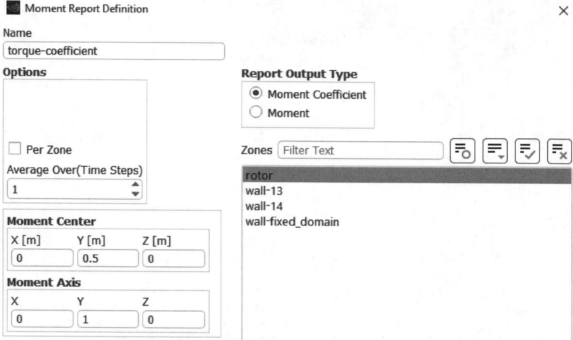

Figure 23.15f) Torque coefficient definition

16. We are going to define a point at the outer edge of the rotor in the rotating domain. Select the Results tab from the menu and Create Point under Surface. Set the Coordinates to x [m] 0, y [m] 0 and z [m] 0.275. Enter *end-plate-edge-point* as the Name. Click on Create and Close the window.

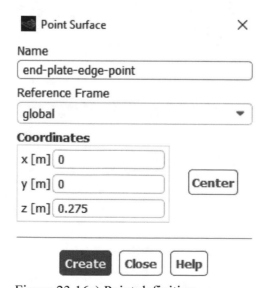

Figure 23.16a) Point definition

Double click on Report Files under Solution and Monitors in the Outline View. Select New... in the Report File Definitions window. Select New>>Surface Report>>Vertex Average... in the New Report File window. Select Velocity and Velocity Magnitude as Field Variable in the Surface Report Definition window. Select *end-plate-edge-point* under Surfaces. Check the boxes for Report File, Report Plot and Print to Console under Create. Enter *end-plate-tip-velocity-magnitude* as Name. Select OK to close the window.

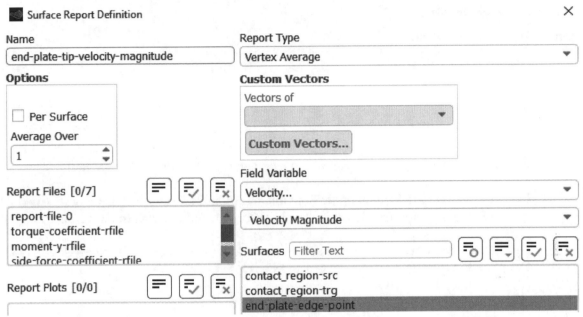

Figure 23.16b) End plate tip velocity-magnitude definition

Select New>>Expression in the New Report File window. Select *end-plate-tip-velocity-magnitude* from Report Definitions. Divide by the radius of the rotor 0.275. Multiply by 30 and divide by PI to get rotor-rpm, see Figure 23.16c). Enter *end-plate-tip-rpm* as the name. Check the boxes for Report File, Report Plot and Print to Console under Create. Select OK to close the window.

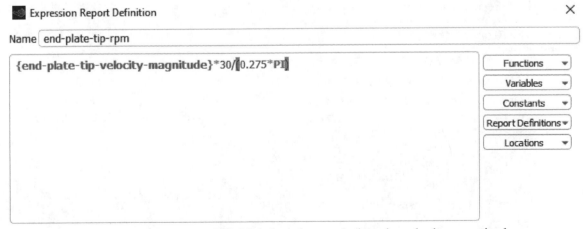

Figure 23.16c) Rotor rpm definition based on end plate tip velocity magnitude

Select New>>Expression in the New Report File window. Select *end-plate-tip-velocity-magnitude* from Report Definitions. Divide by the **incoming wind velocity 5** to get tip speed ratio. Enter *end-plate-tip-velocity-ratio* as the name. Check the boxes for Report File, Report Plot and Print to Console under Create. Select OK to close the window. Close the Report File Definitions window. Select OK to close the New Report File window.

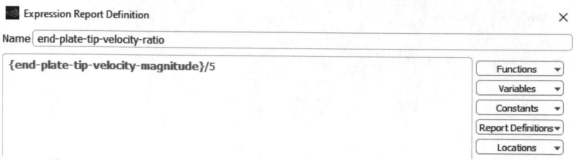

Figure 23.16d) Tip speed ratio definition based on end-plate tip velocity magnitude

17. Double click on Initialization under Solution in the Outline View. Set **X Velocity (m/s)** to **5** and click on Initialize. Select Hybrid Initialization and click on Initialize once again.

Open Calculation Activities and double click on Autosave (Every Time Steps). Set Save Data File Every to 25 Time Steps. Click OK to close the window.

Figure 23.17a) Autosave settings

Double click on Solution Animations under Calculation Activities. Select New Object>>Contours…. Select Contours of Velocity and Velocity Magnitude. Check the box for Filled under Options. Select *rotor* under Surfaces. Enter *velocity-magnitude* as Contour Name and select Save/Display.

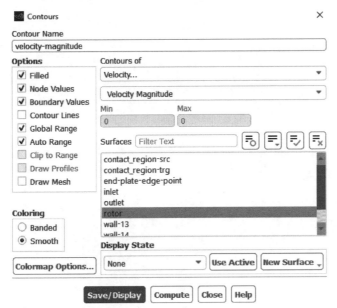

Figure 23.17b) Contours settings

Select *velocity-magnitude* as Animation Object. Name the animation as *velocity-magnitude-animation*. Select OK to close the Animation Definition window. Close the Contours window.

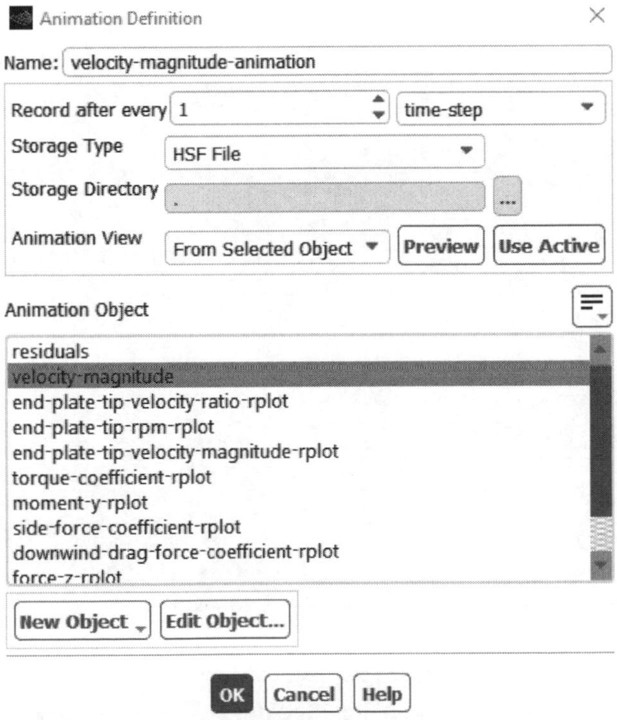

Figure 23.17c) Animation definition

18. Select File>>Save Project from the menu. Select File>>Export>>Case & Data ... from the menu. Save the file in your *Working Directory* with the name *Helical Savonius Turbine Flow.cas*.h5. Double click on Run Calculation. Set the Time Step Size (s) to 0.01 and set the Number of Time Steps to 2000. Set Max Iterations/Time Step to 50. Click on Calculate.

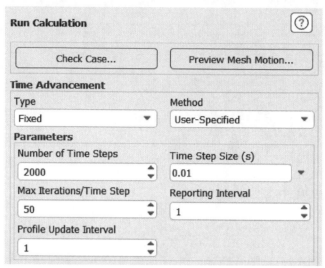

Figure 23.18a) Run calculation settings

Figure 23.18b) shows the velocity magnitude at the edge of the rotor end plate during the first 20 seconds. The velocity magnitude levels off at around 7.1386 m/s corresponding to rotor tip speed $v = 6.4896$ m/s and a rotor rpm of 248. This value is 19 % higher than the corresponding value of 208 rpm as listed by Zakaria and Ibrahim[1] in their experiments. The rotor tip speed $v = 6.4896$ m/s translates to a tip speed ratio of 1.30 since the free stream velocity is 5 m/s.

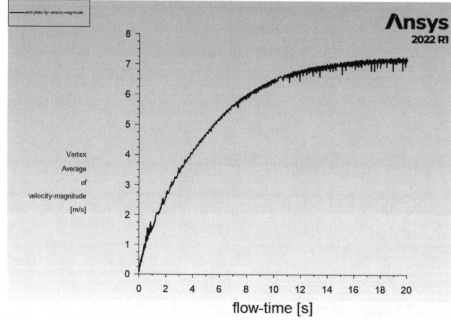

Figure 23.18b) Velocity magnitude at rotor end plate edge during the first 20 seconds

When calculations are complete, click on Copy Screenshot of Active Window to Clipboard in the graphics window, see Figure 23.18c). The velocity magnitude at the edge of the rotor end plate plot can be pasted into a Word document.

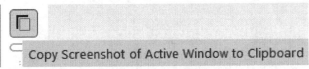

Figure 23.18c) Copying screenshot

Start Excel and open the file *end-plate-tip-velocity-magnitude-rfile.out*. Select Next> in Step 1 of the Text Import Wizard. Check the box for Space under Delimiters and select Next> in Step 2. Select Finish in Step 3. Copy the last 101 values from columns *B* and *C* and paste the data in columns *E* and *F* with the heading *t (s)* for column *E* and heading *Velocity (m/s)* for column *F*. Determine the average value for the velocities in column *E*. The average value is 7.1386 m/s.

F	G	H	I
t (s)	*Velocity (m/s)*		
19	7.13554287	Average (m/s) =	7.1386
19.01	7.149167538	d (m) =	0.5
19.02	7.212070465	D (m) =	0.55
19.03	7.10932827	*v* (m/s) = Average*d/D =	6.4896
19.04	7.136461258	N (rpm) = 60**v*/(p**d*) =	247.8845
19.05	7.19170475	N -exp. (rpm) =	208
19.06	7.022022724	% diff. =	19
19.07	7.103362083	V (m/s) =	5
19.08	7.177595139	TSR = *v*/V =	1.2979
19.09	7.054239273		
19.1	7.098936558		
19.11	7.166029453		
19.12	7.137376308		
19.13	7.129078865		
19.14	7.19861412		
19.15	7.041828156		
19.16	7.165311337		
19.17	7.175068378		
19.18	7.007740974		
19.19	7.111739159		
19.2	7.177062988		

Table 23.1 Evaluation of data for *end-plate-tip-velocity-magnitude-rfile.out*

Open Graphics and Contours under Results in the Outline View. Double click on Velocity Magnitude under Contours. Uncheck Auto Range under Options in the Contours window. Set Max to 8 and select Colormap Options…. Set Font Behavior to Fixed and Font Size to 12 under Font in the Colormap window. Set Type to general and Precision to 2 under Number Format. Select Apply and Close the Colormap window. Select

Save/Display and Close the Contours window. Figure 23.18d) shows the contours of velocity magnitude for the helical Savonius rotor after 20 seconds. It is seen that the velocity magnitude increases as expected further away from the rotation axis up to a maximum value of around 7.2 m/s at the edge of the circular end plate.

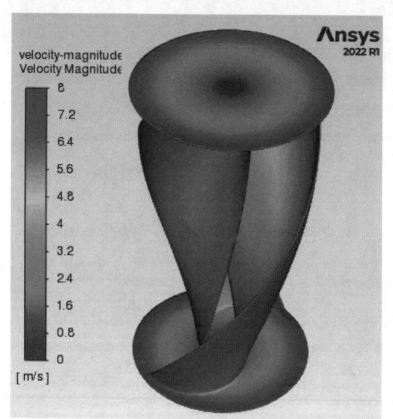

Figure 23.18d) Contours of velocity magnitude for the rotor after 20 seconds

In Figure 23.18e) we see the time signal of the drag and lift coefficients on the rotor during one second. The average value of the corresponding lift force 10.64 N is 14% higher than the average value of the drag force at 9.32 N. The corresponding average values for the drag and lift coefficients are 1.22 and 1.39, correspondingly.

Start Excel and open the files *downwind-drag-force-coefficient-rfile.out* and *side-force-coefficient-rfile.out*. Select Next> in Step 1 of the Text Import Wizard. Check the box for Space under Delimiters and select Next> in Step 2. Select Finish in Step 3.

Copy the last 101 values from columns *B* and *C* from both files and paste the data in columns *A to D* in a new Excel file with heading *t (s)* for column *A*, heading C_{Fx} for column *B*, *t (s)* for column *C* and heading C_{Fz} for column *D*. The period *T* = 0.12 s in Figure 23.18e) corresponds to a 180 degrees rotation of the rotor. The average value for the drag coefficient is 1.2176 and the average value for the lift coefficient is 1.3901.

A	B	C	D	E	F
t (s)	C_{Fx}	t (s)	C_{Fz}	ρ (kg/m^3) =	1.225
19	1.348684	19	1.362314	V (m/s) =	5
19.01	1.312634	19.01	1.303586	A_s =	0.5
19.02	1.251845	19.02	1.271013	C_{Fx-ave} =	1.2176
19.03	1.184117	19.03	1.281653	F_{x-ave} (N) =	9.3221
19.04	1.127277	19.04	1.323998	C_{Fz-ave} =	1.3901
19.05	1.10004	19.05	1.380236	F_{z-ave} (N) =	10.6432
19.06	1.10494	19.06	1.435386		
19.07	1.136749	19.07	1.478236		
19.08	1.187358	19.08	1.503679		
19.09	1.249462	19.09	1.501547		
19.1	1.305366	19.1	1.473558		
19.11	1.340649	19.11	1.419547		
19.12	1.347277	19.12	1.354569		
19.13	1.30364	19.13	1.298327		
19.14	1.240377	19.14	1.269213		
19.15	1.170378	19.15	1.284042		
19.16	1.117255	19.16	1.329208		
19.17	1.094841	19.17	1.386537		
19.18	1.104925	19.18	1.438804		
19.19	1.138854	19.19	1.480706		
19.2	1.188813	19.2	1.502208		

Table 23.2 Evaluation of data for drag and lift coefficients

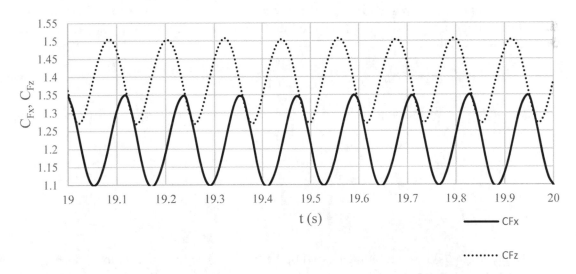

Figure 23.18e) Drag and lift coefficients on rotor during one second

F. Theory

19. In defining the geometry we follow the approach by Rogowski and Maronski[3]. The cross sectional geometry of the rotor is shown in Figure 23.19.

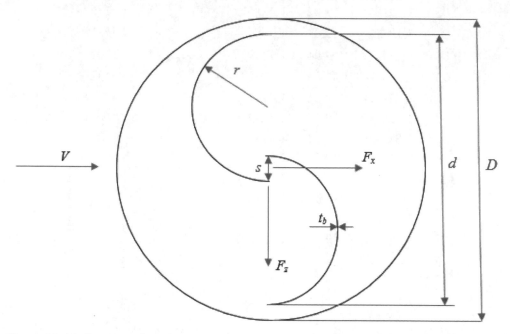

Figure 23.19 Cross sectional geometry for the helical Savonius rotor

The bucket gap width s is 0.121 m and the bucket thickness t_b is 3 mm. The bucket radius r is 0.15525 m and the rotor diameter $d = 4r − s = 0.5$ m. The overlap ratio is $s/d =$ 0.242. The twist angle of the helical rotor is 180 degrees. The end plate diameter D is 0.55 m and the end plate thickness t_e is 2 mm. The free stream velocity is V (m/s). The downwind drag force is F_x (N) and the side force or lift force is F_z (N). The rotor swept area is defined as

$$A_s = (4r − s)h \tag{23.1}$$

where $h = 1$ m is the height of the rotor. We define the rotor tip speed v (m/s) and the tip speed ratio TSR as

$$v = \omega d/2 \tag{23.2}$$

$$TSR = \frac{\omega d}{2V} = \frac{v}{V} \tag{23.3}$$

where ω (rad/s) is the rotational speed of the turbine. The turbine rotational speed N (rpm) is then given by

$$N = \frac{30\omega}{\pi} = \frac{60v}{\pi d} \tag{23.4}$$

The drag and lift coefficients can be defined as

$$C_{F_x} = \frac{F_x}{\frac{1}{2}\rho V^2 A_s} \qquad (23.5)$$

$$C_{F_z} = \frac{F_z}{\frac{1}{2}\rho V^2 A_s} \qquad (23.6)$$

where ρ (kg/m^3) is air density. We can also define the torque coefficient

$$C_T = \frac{T}{\frac{1}{2}\rho V^2 A_s \frac{d}{2}} \qquad (23.7)$$

where T (Nm) is the turbine torque. Finally, we define the power coefficient as

$$C_P = C_T * TSR \qquad (23.8)$$

G. References

1. Zakaria, A. and Ibrahim, M.S.N., "Analysis of Savonius Rotor Performance Operating at Low Wind Speeds Using Numerical Study.", International Journal of Engineering & Technology, **7**, 1549-1552, 2018.
2. Zakaria, A. and Ibrahim, M.S.N., "Numerical Performance Evaluation of Savonius Rotors by Flow-Driven and Sliding-Mesh Approaches.", International Journal of Advanced Trends in Computer Science and Engineering, **8**, No. 1, 2019.
3. Rogowski, K and Maronski, R, "CFD Computation of the Savonius rotor.", Journal of Theoretical and Applied Mechanics, **53**, 37-45, 2015.
4. Blackwell, B.F., Sheldahl, R.E. and Feltz, L.V. "Wind Tunnel Performance Data for Two- and Three-Bucket Savonius Rotors", Sandia Laboratories, Report SAND76-0131, 1977.

H. Exercises

23.1 Run the simulations in this chapter for the Helical Savonius Turbine at the different free stream velocities as shown in Table 23.3. Fill out the table and plot drag and lift coefficients versus tip speed ratio. Also, plot N (rpm) from simulations and experiments versus free stream velocity V.

V (m/s)	N (rpm)	TSR	F_x (N)	C_{F_x}	F_z (N)	C_{F_z}	N (rpm)-Exp.	% diff.	Δt (s)	N_t
2							83		0.01	2000
3							124		0.01	2000
4							170		0.01	2000
5	248	1.298	9.322	1.218	10.643	1.390	208	19	0.01	2000
6							230		0.01	2000

Table 23.3 Turbine rotational speed from ANSYS Fluent in comparison with experiments[1].

23.2 Run the simulations in this chapter for the Helical Savonius Turbine at the different aspect ratios h/d as shown in Table 23.4. Fill out the table and plot drag and lift coefficients versus tip speed ratio. Also, plot N (rpm) from simulations versus free aspect ratio h/d.

h/d	V (m/s)	N (rpm)	TSR	F_x (N)	C_{F_x}	F_z (N)	C_{F_z}	Δt (s)	N_t
2	5	248	1.298	9.322	1.218	10.643	1.390	0.01	2000
8	5							0.01	2000
32	5							0.01	2000

Table 23.4 Input data and results from ANSYS Fluent simulations

h/d	A_s (m^2)	h (m)	d (m)	D (m)	end-plate edge-point (m)	s (m)	Reference Length (m)
2	0.5	1	0.5	0.55	0.275	0.121	0.25
8	0.5	2	0.25	0.275	0.1375	0.0605	0.125
32	0.5	4	0.125	0.1375	0.06875	0.03025	0.0625

Table 23.5 Input data for ANSYS Fluent simulations

h/d	No. Nodes	No. Elements	Orthogonal Quality Min	Growth Rate	Rotor Mesh Size (mm)
2	120506	506532	0.10131	1.2	30
8					
32					

Table 23.6 Input data and results from ANSYS Meshing

h/d	Center of Mass Y (m)	Moment of Inertia (kg-m2)	Mass (kg)
2	0.50000169	0.13735168	4.01319583
8	1.00000073	0.02699234	3.22839079
32			

Table 23.6 Input data for ANSYS Fluent simulations

CHAPTER 24. CIRCULAR HYDRAULIC JUMP

A. Objectives

- Using ANSYS Fluent to Study the Flow in a Circular Hydraulic Jump
- Inserting Boundary Conditions
- Using Volume of Fluid Model for Multiphase Flow with Surface Tension
- Running Laminar 2D Axisymmetric ANSYS Fluent Simulations
- Using Contour Plots for Visualizations of Volume Fraction
- Using Excel for Free Surface Plots

B. Problem Description

We will study the circular hydraulic jump for laminar flow on a flat plate. The water exits the circular tube nozzle with radius R_n and velocity U_0 and spreads out over the horizontal flat plate. The flow enters the hydraulic jump at radius R_j with a water height H after the jump.

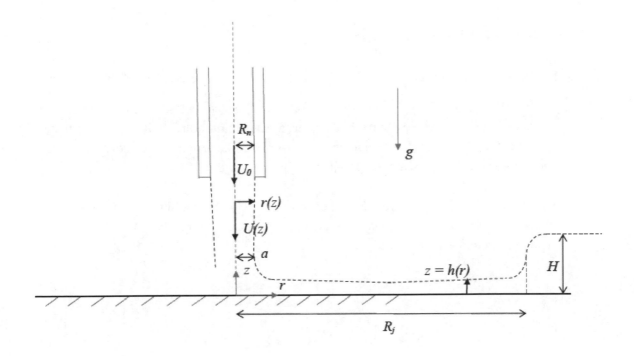

C. Launching ANSYS Workbench and Selecting Fluent

1. Start by launching ANSYS Workbench. Launch Fluid Flow (Fluent) that is available under Analysis Systems in ANSYS Workbench. Select Geometry under Project Schematic, right click and select Properties. Select 2D Analysis Type under Advanced Geometry Options. Right click on Geometry in Project Schematic and select New DesignModeler Geometry to start DesignModeler.

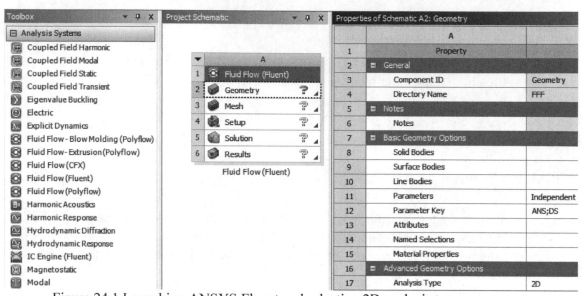

Figure 24.1 Launching ANSYS Fluent and selecting 2D analysis type

D. Launching ANSYS DesignModeler

2. Select **Units>>Millimeter** from the menu in DesignModeler. Select the XYPlane in the Tree Outline. Select Look At Face/Plane/Sketch ⬚. Select the Sketching tab and Polyline. Draw the shape with the dimensions as shown in Figure 24.2. Make sure that the letter P is showing when you move the cursor over the origin of the coordinate system and before you start drawing the polyline.

Select Concept>>Surfaces from Sketches from the menu. Select Sketch 1 under XY Plane in the Tree Outline. Apply the sketch as a Base Object in Details View. Click on Generate and close DesignModeler.

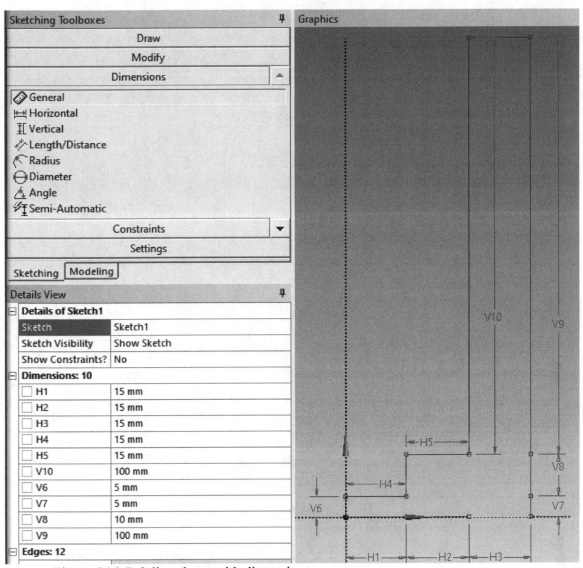

Figure 24.2 Polyline shape with dimensions

E. Launching ANSYS Meshing

3. Double click on Mesh under Project Schematic in ANSYS Workbench. Right-click on Mesh under Project and Model (A3) in the Meshing window and select Update. Select Mesh>>Controls>Face Meshing from the menu. Click on the polyline shape in the graphics window and Apply it as Geometry in Details of Face Meshing.

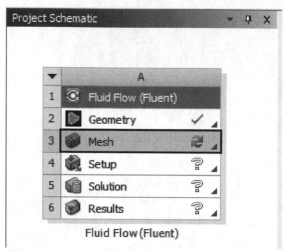

Figure 24.3a) Launching meshing window

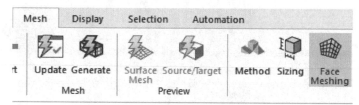

Figure 24.3b) Selecting face meshing

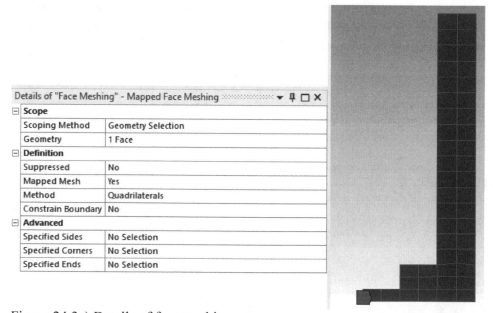

Figure 24.3c) Details of face meshing

Select Mesh>>Controls>>Sizing from the menu. Select the Edge tool 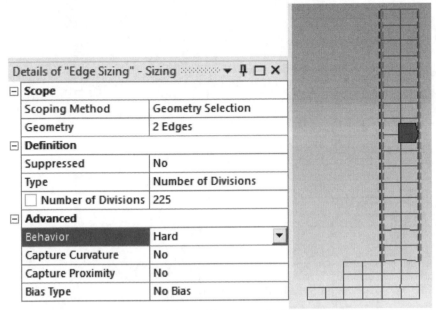, control click on the two long vertical edges and Apply them as Geometry in Details of Sizing. Select Number of Divisions as Type and enter 225. Select Hard as Behavior, see Figure 24.3d).

Details of "Edge Sizing" - Sizing	▼ ⏚ □ ✕
Scope	
Scoping Method	Geometry Selection
Geometry	2 Edges
Definition	
Suppressed	No
Type	Number of Divisions
☐ Number of Divisions	225
Advanced	
Behavior	Hard
Capture Curvature	No
Capture Proximity	No
Bias Type	No Bias

Figure 24.3d) Details for sizing of long vertical edges

Repeat this step but control select the two right most horizontal edges (one located at the top and the other at the bottom), enter 70 as the Number of Divisions and select Hard as Behavior, see Figure 24.3e). Select the first Bias Type and set the Bias Factor to 40. Select the horizontal line at the top and apply the line next to Reverse Bias.

Details of "Edge Sizing 2" - Sizing	▼ ⏚ □ ✕
Scope	
Scoping Method	Geometry Selection
Geometry	2 Edges
Definition	
Suppressed	No
Type	Number of Divisions
☐ Number of Divisions	70
Advanced	
Behavior	Hard
Capture Curvature	No
Capture Proximity	No
Bias Type	____ ___ _ _
Bias Option	Bias Factor
☐ Bias Factor	40.0
Reverse Bias	1 Edge

Figure 24.3e) Details for sizing of right horizontal edges

Repeat this step but control select the two middle horizontal edges, enter 60 as the Number of Divisions and select Hard as Behavior, see Figure 24.3f).

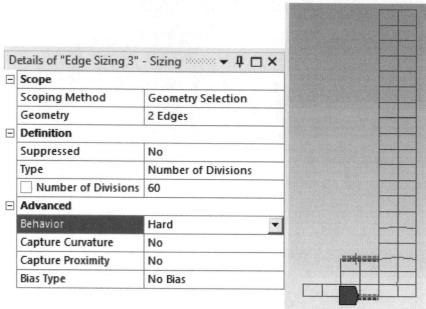

Figure 24.3f) Details for sizing of middle horizontal edges

Repeat this step but control select the two left horizontal edges, enter 25 as the Number of Divisions and select Hard as Behavior, see Figure 24.3g).

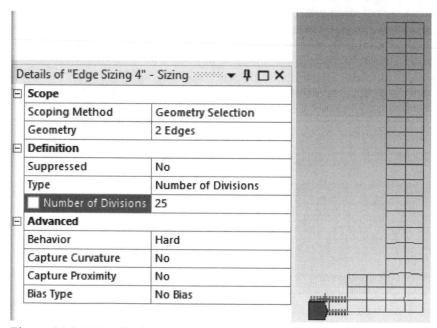

Figure 24.3g) Details for sizing of left horizontal edges

Repeat this step but control select the two bottom vertical edges, enter 40 as the Number of Divisions and select Hard as Behavior, see Figure 24.3h). Select the first Bias Type and set the Bias Factor to 5. Select the vertical line to the left and apply the line next to Reverse Bias.

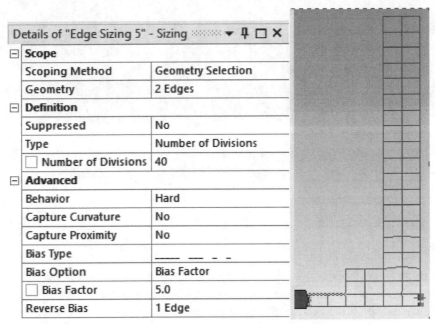

Details of "Edge Sizing 5" - Sizing ▼ 🔲 ❌

⊟ **Scope**	
Scoping Method	Geometry Selection
Geometry	2 Edges
⊟ **Definition**	
Suppressed	No
Type	Number of Divisions
☐ Number of Divisions	40
⊟ **Advanced**	
Behavior	Hard
Capture Curvature	No
Capture Proximity	No
Bias Type	___ __ _ _
Bias Option	Bias Factor
☐ Bias Factor	5.0
Reverse Bias	1 Edge

Figure 24.3h) Details for sizing of bottom vertical edges

Repeat this step but control select the two middle vertical edges, enter 40 as the Number of Divisions and select Hard as Behavior, see Figure 24.3i). Select the second Bias Type and set the Bias Factor to 2. Right-click on Mesh under Project and Model (A3) in the Meshing window and select Update. The finished mesh is shown in Figure 24.3j). The mesh has 27611 Nodes and 27150 Elements.

Details of "Edge Sizing 6" - Sizing ▼ 🔲 ❌

⊟ **Scope**	
Scoping Method	Geometry Selection
Geometry	2 Edges
⊟ **Definition**	
Suppressed	No
Type	Number of Divisions
☐ Number of Divisions	40
⊟ **Advanced**	
Behavior	Hard
Capture Curvature	No
Capture Proximity	No
Bias Type	_ _ __ ___
Bias Option	Bias Factor
☐ Bias Factor	2.0
Reverse Bias	No Selection

Figure 24.3i) Details for sizing of middle vertical edges

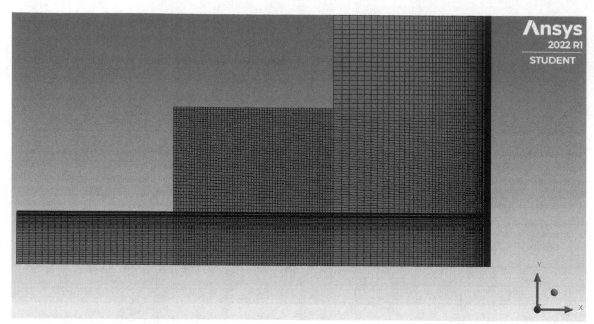

Figure 24.3j) Finished mesh

Select the three lower horizontal edges, right click and select Create Named Selection. Enter the name *axis* and click OK to close the window. Control-select the three right most vertical edges and the left most horizontal edge that was not selected as an axis edge, right click and select Create Named Selection. Enter the name *wall*. Name the left most vertical edge as *inlet*. Finally, name the remaining four edges as *outlet*.

Figure 24.3k) Named selections

Select File>>Export...>>Mesh>>FLUENT Input File>>Export from the menu. Save the mesh with the name *circular-hydraulic-jump.msh*. Select File>>Save Project from the menu and save the project with the name *Circular Hydraulic Jump.wbpj*. Close the meshing window. Right click on Mesh in the Project Schematic and select Update.

F. Launching ANSYS Fluent

4. Double click on Setup under Project Schematic in ANSYS Workbench. Check the box for Double Precision. Click on Show More Options and write down the location of your *Working Directory*. You will need this information later. Set the number of parallel processes equal to the number of processor cores for your computer. To check the number of physical cores, press the Ctrl + Shift + Esc keys simultaneously to open the Task Manager. Go to the Performance tab and select CPU from the left column. You'll see the number of physical cores on the bottom-right side. Close the Task Manager window. Click on the Start button in the Fluent Launcher window. Click OK when you get the Key Behavioral Changes window.

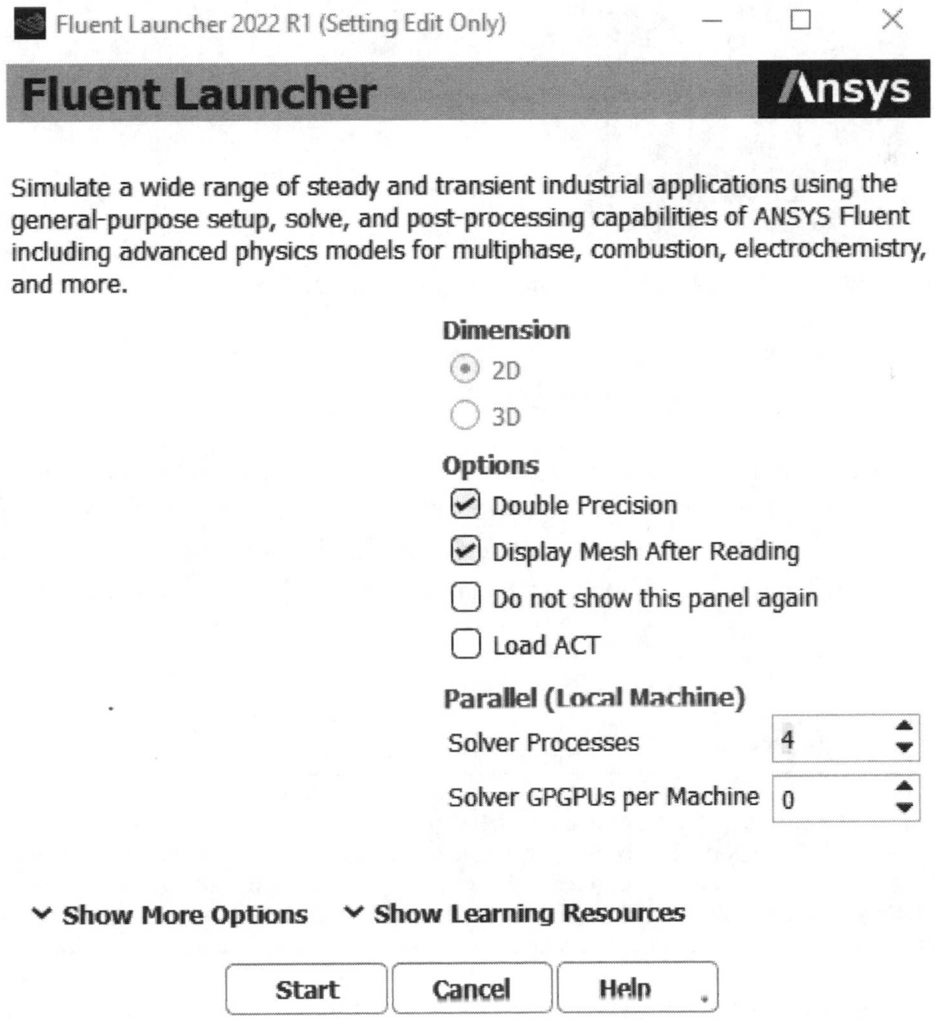

Figure 24.4a) Fluent launcher

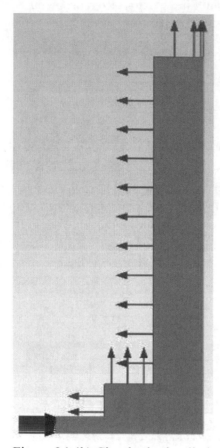

Figure 24.4b) Circular hydraulic jump with mesh in ANSYS Fluent

5. Select Axisymmetric for 2D Space in the General Solver settings on the Task Page in Ansys Fluent. Select Transient for Time in the General Solver settings.

    Open Models under Setup in the Outline View and double click on Viscous (SST k-omega). Select Laminar Model and click on OK to close the Viscous Model window. Double click on Multiphase. Select the Volume of Fluid Model and check the box for Implicit Body Force under Body Force Formulation. Click on Apply and Close the Multiphase Model window.

Why did we select Axisymmetric and Transient for Time? This problem is axisymmetric. We choose Transient since we are interested in studying the shape of the free surface, hydraulic jump and how it deforms over time.

Why did we select Volume of Fluid as Multiphase model and Implicit Body Force? The Volume of Fluid model is used when we have two or more immiscible fluids where the position of the interface between the fluids is of interest. The applications of the VOF Model includes free-surface flows, stratified flows and the tracking of liquid-gas interfaces in general. The VOF model is limited to the pressure based solver and doesn't allow for void regions without a fluid. Implicit Body Force is used when we have large body forces such as is the case when we have gravity and surface tension in multiphase flows. Checking the box for Implicit Body Force will make the solution more robust with better convergence.

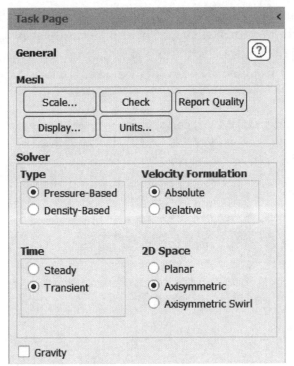

Figure 24.5a) General settings

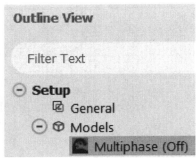

Figure 24.5b) Multiphase model

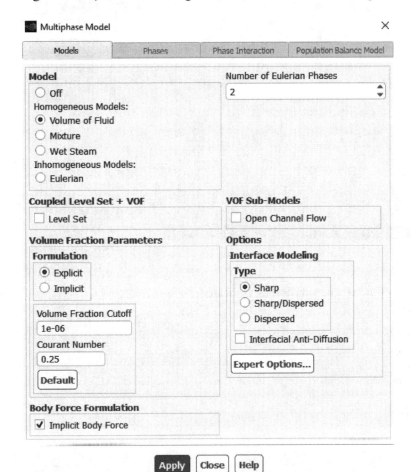

Figure 24.5c) Selecting volume of fluid model

6. Next, we double click on Materials under Setup in the Outline View. Select Fluid under Materials on the Task Page and click on Create/Edit…. Click on Fluent Database… in the Create/Edit Materials window. Scroll down in the Fluent Fluid Material section and select *water-liquid (h2o<l>)*. Click Copy at the bottom of the Fluent Database Materials window. Close the two windows.

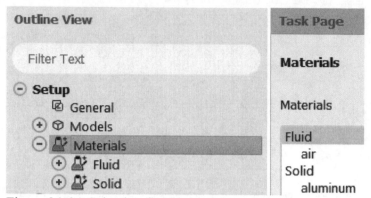

Figure 24.6a) Selecting fluid materials

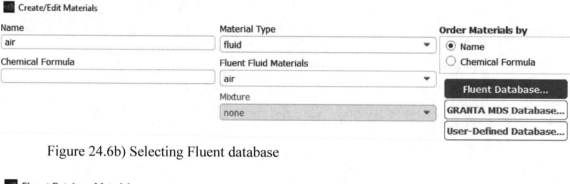

Figure 24.6b) Selecting Fluent database

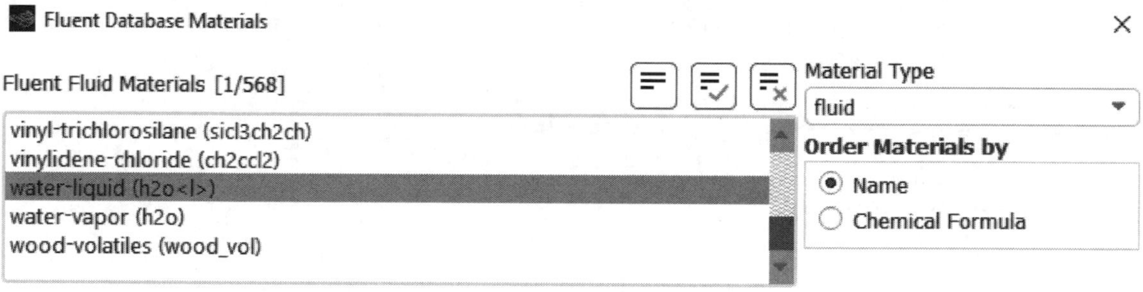

Figure 24.6c) Selection of water-liquid as fluid material

7. Select the Physics tab in the menu and select Models>>Multiphase Model. Select the Phases tab at the top of the Multiphase Model window. Select phase-1-Primary Phase and enter *air* as name for the Primary Phase. Select phase-2-Secondary Phase, select *water-liquid* as the Phase Material and enter *water* as the Name for the Secondary Phase. Click on Apply in the Multiphase Model window. Click on the Phase Interaction tab at the top of the Multiphase Model window. Select constant for Surface Tension Coefficients [N/m] and enter 0.07286 as the value. Click on Apply and Close the Multiphase Model window.

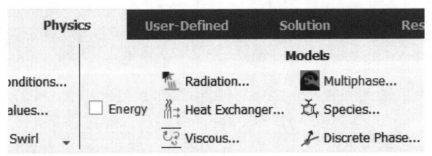
Figure 24.7a) Setting up the physics

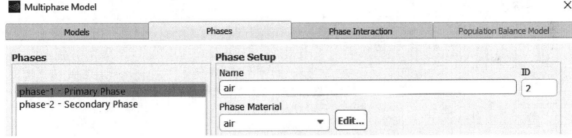
Figure 24.7b) Editing phase-1-Primary Phase

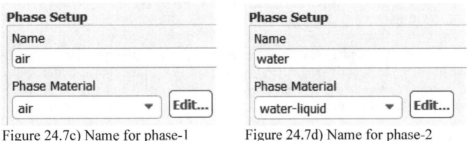

Figure 24.7c) Name for phase-1   Figure 24.7d) Name for phase-2

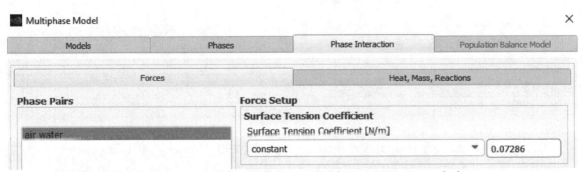
Figure 24.7e) Entering value for surface tension between water and air

543

8. Select the Physics tab in the menu and select Operating Conditions… under Solver. Check the box for Gravity and enter 9.81 m/s² for Gravitational Acceleration in the positive X direction. Select user-input as Operating Density Method. Enter 1.225 as Operating Density [kg/m³]. Set the Reference Pressure Location for X [m] to 0.0375 and for Y [m] to 0.115. Click on the OK button to close the window.

Figure 24.8a) Selecting operating conditions

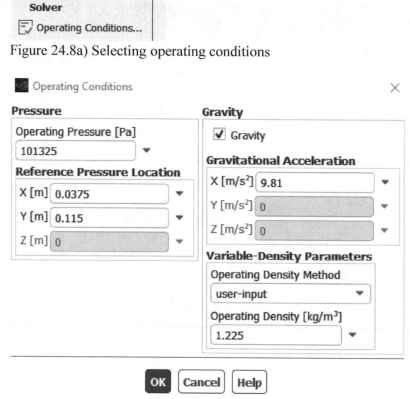

Figure 24.8b) Including gravity and specified operating density

9. Double click on Boundary Conditions under Setup in the Outline View. Select *inlet* under Zone in Boundary Conditions on the Task Page. Click on the Edit… button. Select Components as Velocity Specification Method. Set Axial-Velocity [m/s] to 0.4. Select Apply and Close the window.

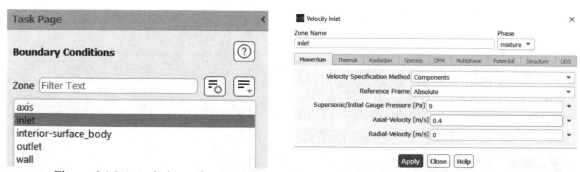

Figure 24.9a) Axis boundary condition   Figure 24.9b) Velocity-inlet boundary

10. For *inlet*, select water from the Phase drop-down menu under Boundary Conditions on the Task Page. Click on the Edit... button, make sure that the value for the Volume Fraction is 1 and click Apply and Close the window.

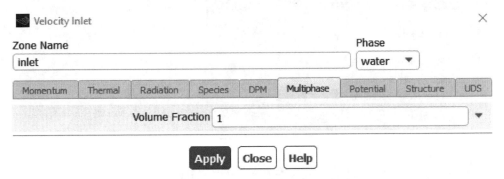

Figure 24.10 Details for pressure-inlet boundary condition

11. Double click on Methods under Solution in the Outline View and choose Body Force Weighted for Pressure and First Order Upwind for Momentum under Spatial Discretization on the Task Page.

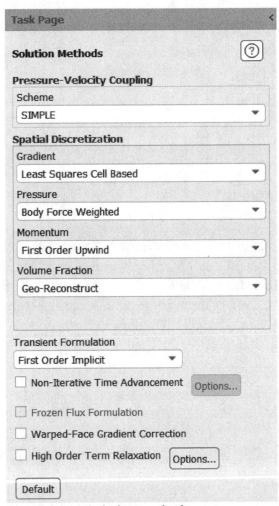

Figure 24.11 Solution methods

12. Double click on Initialization under Solution in the Tree and select Compute from *inlet* on the Task Page. Set water Volume Fraction to 0. Click on Initialize.

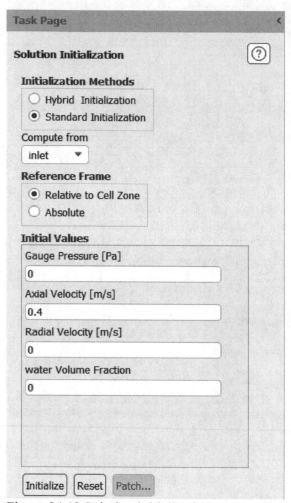

Figure 24.12 Solution initialization

13. Select the *Domain* tab in the menu and select Adapt>>Manual…. Select Cell Registers>>New>>Region from the Manual Mesh Adaption window. Set X Max [m] to 0.015 and Y Max (m) to 0.005. Click on Display Options…. Select red as Color and no Symbol. Check the box for Filled and uncheck the boxes for Wireframe and Marker under Options. Click on OK to close the Cell Register Display Options window. Click on Save/Display in the Region Register window. Close the windows.

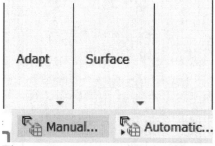

Figure 24.13a) Domain adaption

Manual Mesh Adaption ✕

Refinement Criterion [_____] ▾   **Predefined Criteria** ▾

Coarsening Criterion [_____] ▾   **Cell Registers** ▾

		New ▸	Region...
		Manage	Boundary...
		~~Displ~~	Field Variable...
		General A	Limiter...
		Copy to Au	Residuals...
			Volume...
			Yplus/Ystar...

Adapt Display Close Help

Figure 24.13b) Adaption controls

Region Register   ✕

Name [region_0]

Options   **Input Coordinates**

◉ Inside   X Min [m]   X Max [m]
○ Outside   [0]   [0.015]

              Y Min [m]   Y Max [m]
Shapes   [0]   [0.005]

◉ Quad   Z Min [m]   Z Max [m]
○ Circle   [0]   [0]
○ Cylinder
              Radius [m]
              [0]

              [Select Points with Mouse]

[Save/Display] [Save] [**Display Options...**] [Close] [Help]

Figure 24.13c) Settings for region register

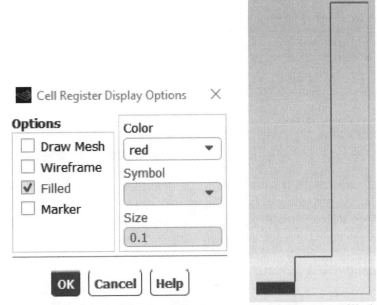

Cell Register Display Options ✕

Options   Color

☐ Draw Mesh   [red]   ▾
☐ Wireframe   Symbol
☑ Filled   [_____] ▾
☐ Marker   Size
               [0.1]

OK [Cancel] [Help]

Figure 24.13d) Cell register display   Figure 24.13e) Filled inlet section

Click on Copy Screenshot of Active Window to Clipboard, see Figure 24.13f). The contour plot can be pasted into a Word document.

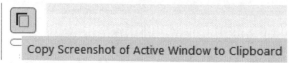

Figure 24.13f) Copying screenshot

14. Click on the Patch... button on the Solution Initialization Task Page. Select *water* as Phase in the Patch window and select Volume Fraction as Variable. Select region_0 as Registers to Patch and set the Value to 1. Click on the Patch button. Select Mixture as Phase and choose Radial Velocity as Variable. Set the Value [m/s] to 0 and click on the Patch button. Close the Patch window.

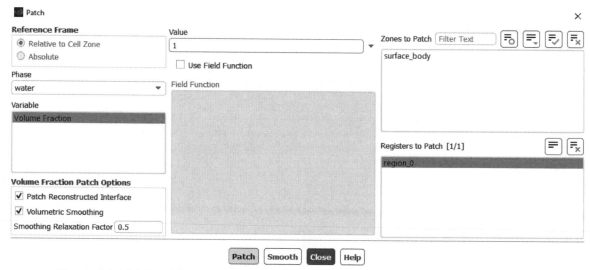

Figure 24.14a) Patching settings for water

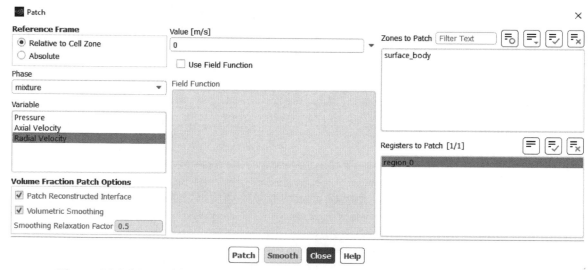

Figure 24.14b) Patching settings for mixture

15. Double click on Graphics and Contours under Results in the Outline View. Select Contours of Phases… and Volume fraction. Select *water* as the Phase. Deselect all Surfaces. Click on Colormap Options. Select general as Type and 0 for Precision under Number Format in the Colormap window. Set Colormap Alignment to Top. Select Apply and Close the Colormap window. Click on Save/Display and Close the Contours window.

Contours   ✕

Contour Name

> contour-1

Options

Contours of

- ☑ Filled
- ☑ Node Values
- ☑ Boundary Values
- ☐ Contour Lines
- ☑ Global Range
- ☑ Auto Range
- ☐ Clip to Range
- ☐ Draw Profiles
- ☐ Draw Mesh

> Phases... ▾

> Volume fraction ▾

Phase

> water ▾

Min   Max

> 0   0

Surfaces   Filter Text

> axis
> inlet
> interior-surface_body
> outlet
> surface_body
> wall

Coloring

- ○ Banded
- ● Smooth

Colormap Options...

Display State

> None ▾   **Use Active**   **New Surface** ▾

Save/Display   **Compute**   **Close**   **Help**

Figure 24.15a) Contours window

Select the View tab in the menu and click on Views…. Select *axis* as Mirror Plane and click on Apply. Zoom out and translate the view if needed so that the entire model is visible in the graphics window. Click on the Camera… button in the Views window. Use your left mouse button to rotate the dial clockwise until the model rotates 90 degrees clockwise and appears upright. Close the Camera Parameters window. Click on the Save button under Actions in the Views window and Close the Views window. Click on Copy Screenshot of Active Window to Clipboard, see Figure 8.13f). The water volume fraction contour plot can be pasted into a Word document.

| File | Domain | Physics | User-Defined | Solution | Results | View | |

| Display | | Graphics | | Graphics Toolbars |
| ◉ Views...   Headlight ▾   ☐ Lighting | ⊙ Lights...   ◿ Compose... | ☑ Mesh Display ☑ View | ☑ Copy |

Figure 24.15b) Displaying views

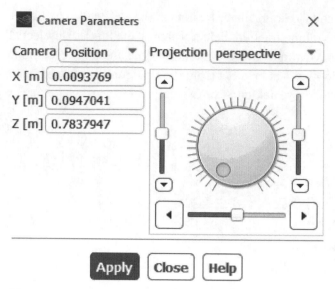

Figure 24.15c) Camera parameters window

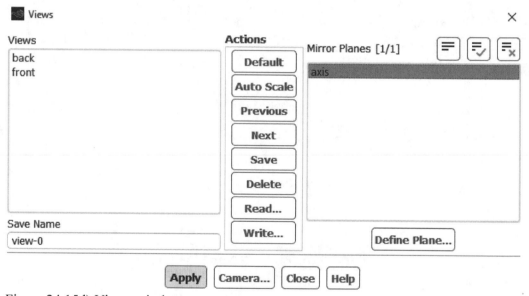

Figure 24.15d) Views window

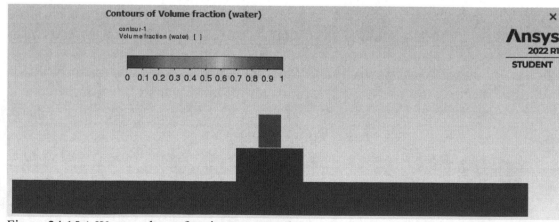

Figure 24.15e) Water volume fraction contour plot

16. Select File>>Export>>Case and Data… from the menu. Save the Case and Data File with the name Circular Hydraulic Jump.cas.h5. Double click on Run Calculation under Solution in the Outline View and set the Time Step Size to 0.0001 s. Set the Number of Time Steps to 20000. Click on the Calculate button on the Task Page. Click OK in the Information window when the calculation is complete. Click on Copy Screenshot of Active Window to Clipboard, see Figure 24.13f). The Scaled Residuals plot can be pasted into a Word document.

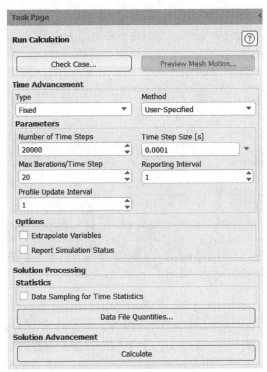

Figure 24.16a) Calculation settings

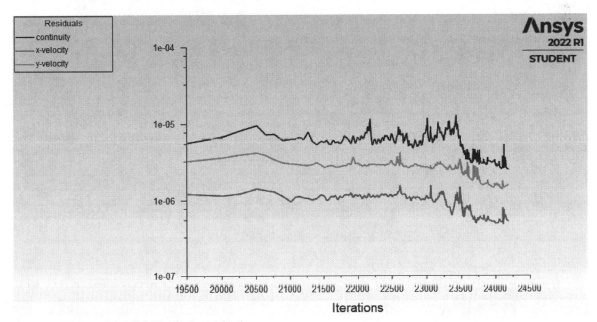

Figure 24.16b) Scaled residuals

G. Post-Processing

Double click on Graphics and Contours under Results in the Outline View. Select Contours of Phases... and Volume Fraction. Select *water* as the Phase. Deselect all Surfaces. Click on Save/Display. Select the View tab in the menu and click on Views.... Select *axis* as Mirror Plane, select view-0 under Views and click on Apply. Zoom out and ⊕ Pan the view if needed so that the entire model is visible in the graphics window. Close the Views window. Click on Copy Screenshot of Active Window to Clipboard, see Figure 8.13f). The water volume fraction contour plot can be pasted into a Word document.

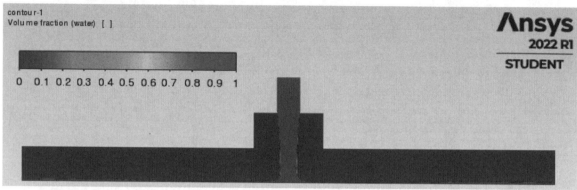

Figure 24.16c) Contours of water volume fraction at $t = 2$ s

Double click on Graphics and Contours under Results in the Outline View. Select Contours of Velocity... and Velocity Magnitude. Select *mixture* as the Phase. Deselect all Surfaces. Click on Colormap Options. Select general as Type and 0 for Precision under Number Format in the Colormap window. Set Colormap Alignment to Top. Select Apply and Close the Colormap window. Click on Save/Display and Close the Contours window. Repeat this step for Axial Velocity and Radial Velocity for the mixture.

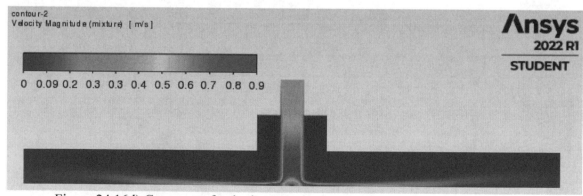

Figure 24.16d) Contours of velocity magnitude at $t = 2$ s

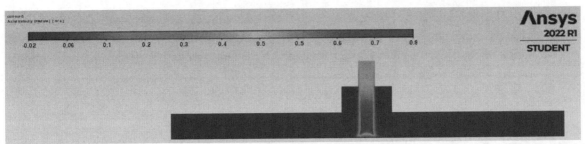

Figure 24.16e) Contours of axial velocity at $t = 2$ s

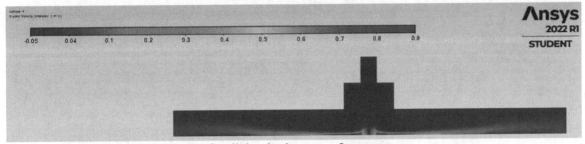

Figure 24.16f) Contours of radial velocity at $t = 2$ s

Double click on Plots and XY Plot under Results in the Outline View. Select Velocity and Velocity Magnitude for Y Axis Function and Direction Vector for X Axis Function.

Select New Surface >> Line/Rake from the drop–down menu. Enter the following end points: x0 [m] = 0, y0 [m] = 0 and x1 [m] = 0.045, y1 [m] = 0. Select Create and Close the window.

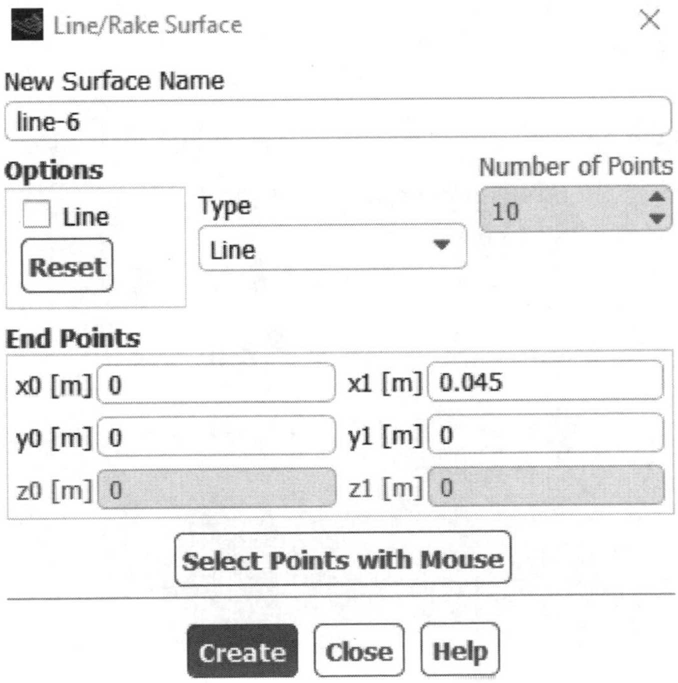

Figure 24.16g) Line/rake surface settings

553

Select *line-6* under the Surfaces section in the Solution XY Plot window. Click on Curves..., select Curve #0 and select the first available pattern with no (blank) symbol. Click on Apply and select Curve # 1, select the second available pattern and no symbol as shown in Figure 24.16h).

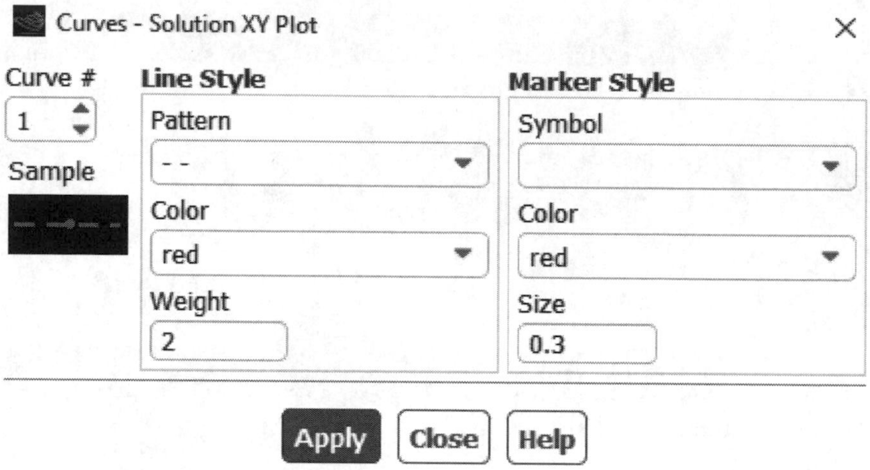

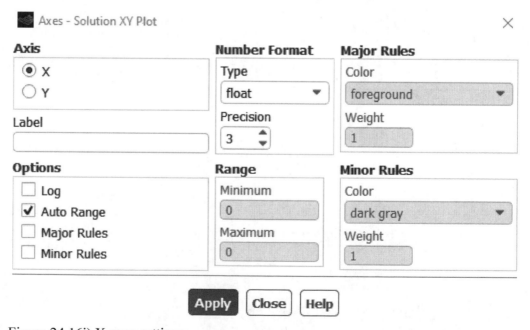

Figure 24.16h) Curve settings

Click on Apply and Close the Curves-Solution XY Plot window. Click on Axes... in the Solution XY Plot window. Select X Axis, set Type to float, set Precision under Number Format to 3 and click on Apply. Select Y Axis, set Type to general, set Precision under Number Format to 2 and click on Apply. Close the Axes – Solution XY Plot window. Click on Save/Plot in the Solution XY Plot window. Click on Copy Screenshot of Active Window to Clipboard in the graphics window, see Figure 24.13f). The XY plot can be pasted into a Word document.

Figure 24.16i) X-axes settings

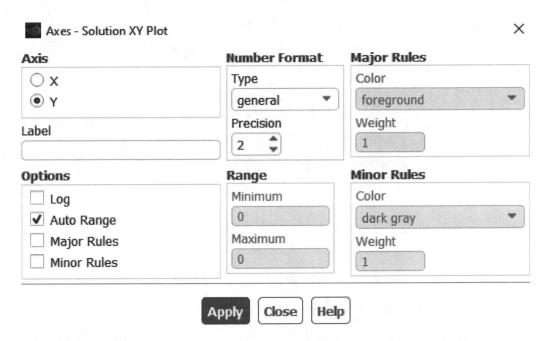

Figure 24.16j) Y-axes settings

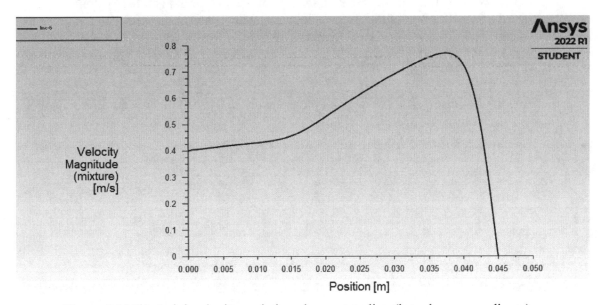

Figure 24.16k) Axial velocity variation along centerline (based on *x* coordinate)

Select File>>Export>>Solution Data... from the menu. Select ASCII as File Type, Node under Location and Space as Delimiter. Select *surface_body* under Cell Zones and *line-6* under Surfaces. Select Axial Velocity under Quantities. Click on Write and save the ASCII File in the Working Directory with the name *centerline axial velocity data*. Close the Export window.

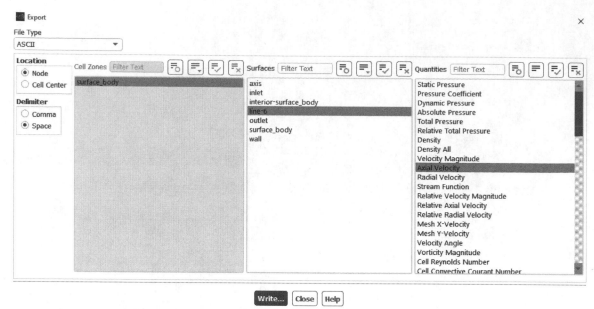

Figure 24.16l) Export settings for axial velocity data

Open the saved file in Excel. Click on Next in the Text Import Wizard – Step 1 of 3. Click on Next in Text Import Wizard – Step 2 of 3. Click on Finish in Text Import Wizard – Step 3 of 3. Plot the axial velocity along the centerline. The definition of the two coordinate systems (x, y) and (r, z) is

$$z = Z - x = 0.045 - x, \ r = y \tag{24.1}$$

Label the D column as U (m/s). Enter the equation =0.045-B2 in the E column and label this column z (m).

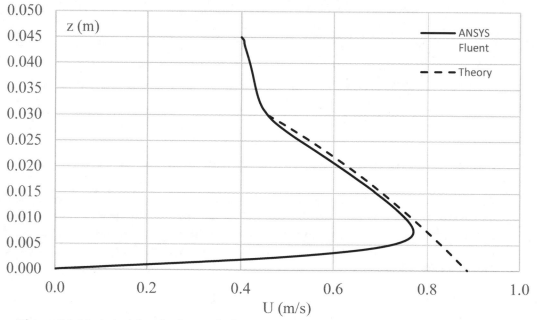

Figure 24.16m) Axial velocity variation along centerline (based on z coordinate)

x-coordinate	U (m/s)	z (m)	x-coordinate	U (m/s)	z (m)
0.00E+00	**0.400000**	**4.500E-02**	2.10E-02	0.546686	2.400E-02
6.00E-04	0.404913	4.440E-02	2.13E-02	0.551003	2.375E-02
1.20E-03	0.405686	4.380E-02	2.15E-02	0.555311	2.350E-02
1.80E-03	0.407256	4.320E-02	2.18E-02	0.559608	2.325E-02
2.40E-03	0.409259	4.260E-02	2.20E-02	0.563890	2.300E-02
3.00E-03	0.411339	4.200E-02	2.23E-02	0.568157	2.275E-02
3.60E-03	0.413381	4.140E-02	2.25E-02	0.572407	2.250E-02
4.20E-03	0.415350	4.080E-02	2.28E-02	0.576639	2.225E-02
4.80E-03	0.417236	4.020E-02	2.30E-02	0.580851	2.200E-02
5.40E-03	0.419032	3.960E-02	2.33E-02	0.585042	2.175E-02
6.00E-03	0.420740	3.900E-02	2.35E-02	0.589213	2.150E-02
6.60E-03	0.422370	3.840E-02	2.38E-02	0.593362	2.125E-02
7.20E-03	0.423934	3.780E-02	2.40E-02	0.597489	2.100E-02
7.80E-03	0.425449	3.720E-02	2.43E-02	0.601595	2.075E-02
8.40E-03	0.426937	3.660E-02	2.45E-02	0.605679	2.050E-02
9.00E-03	0.428422	3.600E-02	2.48E-02	0.609742	2.025E-02
9.60E-03	0.429934	3.540E-02	2.50E-02	0.613783	2.000E-02
1.02E-02	0.431515	3.480E-02	2.53E-02	0.617804	1.975E-02
1.08E-02	0.433215	3.420E-02	2.55E-02	0.621804	1.950E-02
1.14E-02	0.435103	3.360E-02	2.58E-02	0.625785	1.925E-02
1.20E-02	0.437266	3.300E-02	2.60E-02	0.629747	1.900E-02
1.26E-02	0.439814	3.240E-02	2.63E-02	0.633691	1.875E-02
1.32E-02	0.442876	3.180E-02	2.65E-02	0.637617	1.850E-02
1.38E-02	0.446590	3.120E-02	2.68E-02	0.641527	1.825E-02
1.44E-02	0.451524	3.060E-02	2.70E-02	0.645422	1.800E-02
1.50E-02	**0.456553**	**3.000E-02**	2.73E-02	0.649303	1.775E-02
1.53E-02	0.458883	2.975E-02	2.75E-02	0.653170	1.750E-02
1.55E-02	0.461676	2.950E-02	2.78E-02	0.657025	1.725E-02
1.58E-02	0.464425	2.925E-02	2.80E-02	0.660867	1.700E-02
1.60E-02	0.467319	2.900E-02	2.83E-02	0.664697	1.675E-02
1.63E-02	0.470406	2.875E-02	2.85E-02	0.668514	1.650E-02
1.65E-02	0.473662	2.850E-02	2.88E-02	0.672318	1.625E-02
1.68E-02	0.477062	2.825E-02	2.90E-02	0.676107	1.600E-02
1.70E-02	0.480592	2.800E-02	2.93E-02	0.679877	1.575E-02
1.73E-02	0.484243	2.775E-02	2.95E-02	0.683626	1.550E-02
1.75E-02	0.488006	2.750E-02	2.98E-02	0.686908	1.525E-02
1.78E-02	0.491869	2.725E-02	3.00E-02	0.690300	1.500E-02
1.80E-02	0.495820	2.700E-02	3.08E-02	0.702161	1.420E-02
1.83E-02	0.499849	2.675E-02	3.16E-02	0.712892	1.344E-02
1.85E-02	0.503943	2.650E-02	3.23E-02	0.722963	1.272E-02
1.88E-02	0.508095	2.625E-02	3.30E-02	0.731860	1.204E-02
1.90E-02	0.512293	2.600E-02	3.36E-02	0.739783	1.140E-02
1.93E-02	0.516529	2.575E-02	3.42E-02	0.746913	1.078E-02
1.95E-02	0.520797	2.550E-02	3.48E-02	0.753259	1.020E-02
1.98E-02	0.525087	2.525E-02	3.53E-02	0.758747	9.653E-03
2.00E-02	0.529394	2.500E-02	3.59E-02	0.763289	9.132E-03
2.03E-02	0.533713	2.475E-02	3.64E-02	0.766789	8.638E-03
2.05E-02	0.538037	2.450E-02	3.68E-02	0.769135	8.169E-03
2.08E-02	0.542363	2.425E-02	**3.73E-02**	**0.770204**	**7.725E-03**

Figure 24.16n) Excel data for axial velocity along centerline

We find from the data that the axial velocity increases in the 15 mm long pipe section from $U_{in} = 0.4$ m/s at $z = 0.045$ m to $U_0 = 0.456553$ m/s at $z = Z = 0.03$ m. The axial velocity increases to a maximum velocity 0.770204 m/s at $z = 0.007725$ m.

17. Select File>>Export>>Data… from the menu. Save the data file in the *Working Directory* folder with the name *CHJ-t=2s.dat.h5*. Select the Results tab from the menu and select Surface>>Create>>Iso-Surface…. Select Surface of Constant Phases… and Volume fraction. Select *air* as the Phase and set Iso-Values to 0.5. Select *surface_body* under From Zones and *surface_body* under From Surface. Click on Create.

Figure 24.17a) Iso-surface settings for air phase

Select *water* as the Phase and set Iso-Values to 0.5. Select *surface_body* under From Zones and *surface_body* under From Surface. Enter *chj-free-surface-t-2s* as New Surface Name and click on Create. Close the window.

Select File>>Export>>Solution Data… from the menu. Select ASCII as File Type, Node under Location and Space as Delimiter. Select *surface_body* under Cell Zones and *chj-free-surface-t-2s* under Surfaces. Select Volume fraction (water) under Quantities. Click on Write… and save the ASCII File in the Working Directory with the name *chj-free-surface-coordinates-t-2s*. Close the Export window.

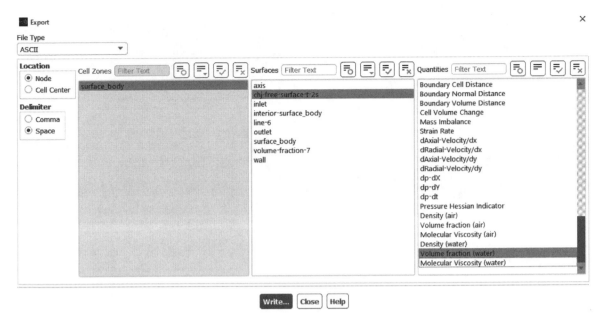

Figure 24.17b) Export settings for free surface data

Open the saved file in Excel. Click on Next in the Text Import Wizard – Step 1 of 3. Click on Next in Text Import Wizard – Step 2 of 3. Click on Finish in Text Import Wizard – Step 3 of 3. Plot the free surface. The two coordinate systems (x, y) and (r, z) are related as shown in Equation (24.1). Copy the y-coordinate column in the Excel file and paste it to column F. Label this column r (m). Enter the equation =0.045-B2 in the G column and label this column z (m).

	A	B	C	D	E	F	G
1	nodenumber	x-coordinate	y-coordinate	axial-velocity	water-vof	r (m)	z (m)
2	1	3.85E-02	3.58E-03	0.742913	5.00E-01	0.003582	0.006526
3	2	3.88E-02	3.58E-03	0.753174	5.00E-01	0.003583	0.006167
4	3	3.81E-02	3.58E-03	0.746378	5.00E-01	0.003584	0.006905
5	4	3.92E-02	3.59E-03	0.766766	5.00E-01	0.003586	0.005827
6	5	3.79E-02	3.59E-03	0.746822	5.00E-01	0.003586	0.007145
7	6	3.92E-02	3.59E-03	0.766899	5.00E-01	0.003586	0.005819
8	7	3.77E-02	3.59E-03	0.746513	5.00E-01	0.003588	0.007304
9	8	3.95E-02	3.59E-03	0.769151	5.00E-01	0.003593	0.005505
10	9	3.73E-02	3.59E-03	0.738585	5.00E-01	0.003595	0.007725
11	10	3.98E-02	3.60E-03	0.768775	5.00E-01	0.003602	0.005200

Figure 24.17c) Excel data of dimensional free surface elevation

ΔH

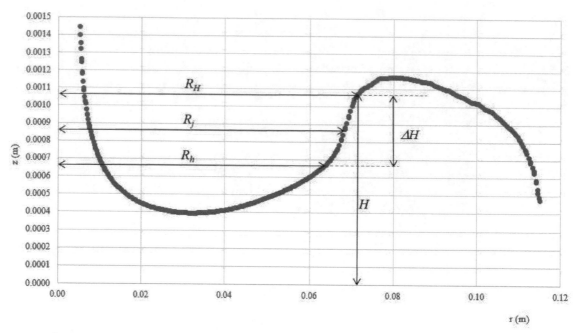

Figure 24.17d) Free surface elevation on horizontal plate versus radial distance from axis

Copy the *r* (m) column *F* in the Excel file and paste it to column *H*. Label this column *r (m)*. Enter the equation =(G4-G2)/(H4-H2) in the *I* column and label this column *dz/dr*.

	A	B	C	D	E	F	G	H	I	J
1	nodenumber	x-coordinate	y-coordinate	axial-velocity	water-vof	r (m)	z (m)	r (m)	dz/dr	
2	1	3.85E-02	3.58E-03	0.742913	5.00E-01	0.003582	0.006526	0.003582		
3	2	3.88E-02	3.58E-03	0.753174	5.00E-01	0.003583	0.006167	0.003583	=(G4-G2)/(H4-H2)	
4	3	3.81E-02	3.58E-03	0.746378	5.00E-01	0.003584	0.006905	0.003584	-9.75E+01	
5	4	3.92E-02	3.59E-03	0.766766	5.00E-01	0.003586	0.005827	0.003586	9.15E+01	
6	5	3.79E-02	3.59E-03	0.746822	5.00E-01	0.003586	0.007145	0.003586	-4.92E+01	
7	6	3.92E-02	3.59E-03	0.766899	5.00E-01	0.003586	0.005819	0.003586	9.21E+01	
8	7	3.77E-02	3.59E-03	0.746513	5.00E-01	0.003588	0.007304	0.003588	-5.04E+01	
9	8	3.95E-02	3.59E-03	0.769151	5.00E-01	0.003593	0.005505	0.003593	6.43E+01	
10	9	3.73E-02	3.59E-03	0.738585	5.00E-01	0.003595	0.007725	0.003595	-3.14E+01	
11	10	3.98E-02	3.60E-03	0.768775	5.00E-01	0.003602	0.005200	0.003602	5.22E+01	

Figure 24.17e) Excel data for gradient *dz/dr*

We define the location of the jump *Rj* based on the radial location where the gradient *dz/dr* (as determined using the central finite difference formula in the Excel file for the derivative) has a maximum as seen in Figure 24.17f). The location of the hydraulic jump is determined to be at *Rj* = 0.068443 m.

r (m)	z (m)	dz/dr
0.063200	**0.000657**	**0.0200**
0.063444	0.000662	0.0211
0.063889	0.000672	0.0237
0.064333	0.000683	0.0265
0.064778	0.000696	0.0280
0.064806	0.000697	0.0293
0.065222	0.000709	0.0308
0.065667	0.000723	0.0362
0.066111	0.000741	0.0423
0.066416	0.000755	0.0466
0.066556	0.000762	0.0501
0.067000	0.000784	0.0549
0.067444	0.000810	0.0609
0.067529	0.000816	0.0678
0.067889	0.000841	0.0695
0.068333	0.000872	0.0731
0.068443	**0.000881**	**0.0793**
0.068778	0.000907	0.0777
0.069222	0.000942	0.0773
0.069321	0.000950	0.0774
0.069667	0.000976	0.0725
0.070111	0.001007	0.0667
0.070351	0.001022	0.0627
0.070556	0.001035	0.0526
0.071000	0.001056	0.0431
0.071444	0.001073	0.0342
0.071889	0.001086	0.0261
0.072333	0.001096	0.0214
0.072360	**0.001097**	**0.0200**

Figure 24.17f) Excel data for gradient dz/dr

The two different radii corresponding to before the jump $R_h = 0.0632$ mm and after the jump $R_H = 0.072360$ mm are shown in Figures 24.17d) and 24.17g). R_h and R_H were both chosen to correspond to $dz/dr = 0.02$. From R_H and R_h it is possible to find $H = 0.001097$ m and $\Delta H = 0.00044$ m.

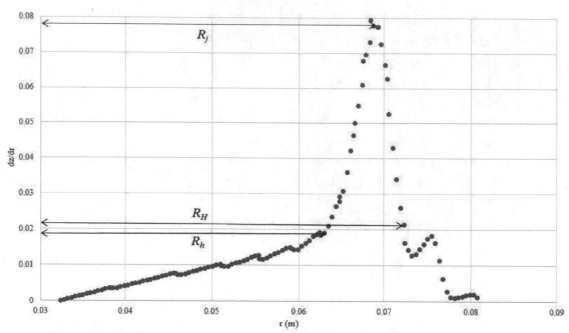

Figure 24.17g) Surface slope versus radial distance from axis

We can find the minimum radius $a = 0.003582$ m for the vertical jet from the data as shown in Figures 24.17c) and 24.17h). The corresponding vertical location is $z_a = 0.006526$ m with an axial velocity $U = 0.742913$ m/s.

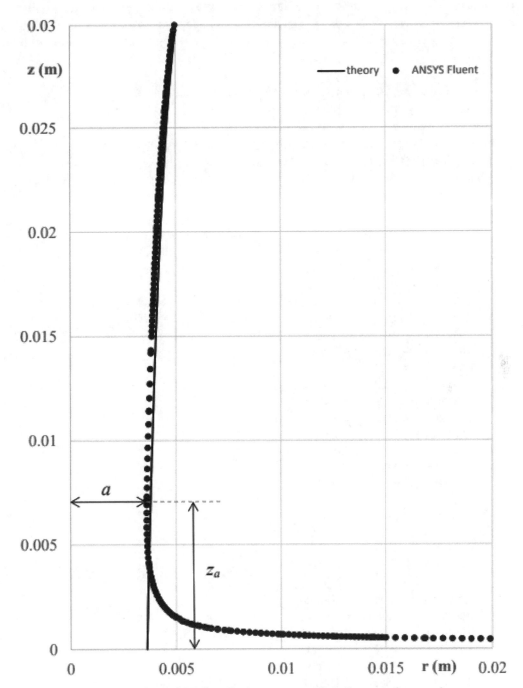

Figure 24.17h) Free surface elevation for jet versus radial distance from axis

In Figures 24.17i) and 24.17j) is the Mathematica code and the theoretical data points shown.

```
RN = 0.005; g = 9.81; ρ = 998.2; σ = 0.07286; Z = 0.030; μ = 0.001003; v = μ / ρ; U0 = 0.456554;
H = 0.0011; dH = 0.0004; Fr = U0^2 / (g * RN); We = ρ * U0^2 * RN / σ; i = 0; Clear[r]; Clear[U];
Do[{eq1 = NSolve[a / RN - (1 + 2 * (Z - z) / (Fr * RN) + 2 * (1 - (RN / a)) / We)^(-1/4) == 0, a];
    i = i + 1;
    zz[i] = z;
    r[i] = a //. eq1; U[i] = U0 * (RN / a)^2 //. eq1;
    s = NDSolve[{y'[x] == y[x] Cos[x + y[x]], y[0] == 1}, y[x], {x, 0, 30}];
    f[x_] := Evaluate[y[x] /. s];
    Print[i, r[i][[1]], zz[i], U[i][[1]]]}, {z, 0.03, 0, -0.001}];
m = Table[{r[i][[1]], zz[i], U[i][[1]]}, {i, 1, 31, 1}];
m = Prepend[m, {"r (m)", "z (m)", "U (m/s)"}];
m // TableForm
Export["mfile.xls", m]
```

Figure 24.17i) Mathematica code used to generate the data in Figure 24.17j)

r (m)	z (m)	U (m/s)
0.005	0.03	0.456554
0.0048923	0.029	0.4768842
0.004795	0.028	0.4964207
0.0047066	0.027	0.515249
0.0046257	0.026	0.5334405
0.0045511	0.025	0.5510553
0.0044822	0.024	0.5681446
0.004418	0.023	0.5847527
0.0043582	0.022	0.6009177
0.0043022	0.021	0.6166732
0.0042495	0.02	0.6320486
0.0041999	0.019	0.64707
0.004153	0.018	0.6617607
0.0041086	0.017	0.6761415
0.0040665	0.016	0.6902312
0.0040264	0.015	0.7040466
0.0039882	0.014	0.7176031
0.0039517	0.013	0.7309147
0.0039168	0.012	0.7439941
0.0038834	0.011	0.756853
0.0038513	0.01	0.7695022
0.0038206	0.009	0.7819516
0.003791	0.008	0.7942102
0.0037625	0.007	0.8062867
0.003735	0.006	0.8181888
0.0037085	0.005	0.8299239
0.0036829	0.004	0.8414987
0.0036582	0.003	0.8529197
0.0036342	0.002	0.8641928
0.003611	0.001	0.8753235
0.0035886	0	0.8863172

Figure 24.17j) Data values for theoretical curves

H. Theory

18. Following Bush and Aristoff (2003), we define the jet Reynolds number for this flow

$$Re = \frac{Q}{a\nu} \qquad (24.2)$$

where Q (m^3/s) is volume flow rate, a (m) is the minimum radius of the jet and ν (m^2/s) is kinematic viscosity of the liquid.

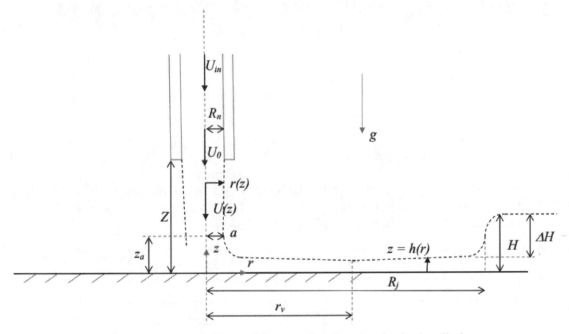

Figure 24.18 Geometry and shape of free surface for circular hydraulic jump

The minimum jet radius a (m) and corresponding maximum axial jet speed U_m (m/s) can be determined from

$$a = \frac{R_n}{\left\{1+\frac{2}{U_0^2}[gZ+\frac{\sigma}{\rho R_n a}(a-R_n)]\right\}^{1/4}} \qquad (24.3)$$

$$U_m = U_0\left(\frac{R_n}{a}\right)^2 \qquad (24.4)$$

where U_0 (m/s) is the jet velocity corresponding to nozzle radius R_n (m), g (m/s^2) is acceleration due to gravity, Z (m) is the height of the nozzle exit above the horizontal plate, σ (N/m) is surface tension and ρ (kg/m^3) is liquid density.

The free jet radius and axial velocity at any vertical location can be determined from

$$r(z) = \frac{R_n}{\left[1+\frac{2(Z-z)}{FrR_n}+\frac{2}{We}\left(1-\frac{R_n}{r(z)}\right)\right]^{1/4}} \qquad (24.5)$$

$$U(z) = U_0\left(\frac{R_n}{r(z)}\right)^2 \qquad (24.6)$$

565

$$Fr = \frac{U_0^2}{gR_n} \tag{24.7}$$

$$We = \frac{\rho U_0^2 R_n}{\sigma} \tag{24.8}$$

where Fr is the Froude number and We is the Weber number. The jump radius R_j (m) is given by the following equations

$$R_j g \left(\frac{aH}{Q}\right)^2 \left(1 + \frac{2}{Bo}\right) + \frac{1}{2R_j H}\left(\frac{a}{\pi}\right)^2 = 0.10132 - 0.1297\frac{1}{Re^{1/2}}\left(\frac{R_j}{a}\right)^{3/2} ; r(z) < r_v \tag{24.9}$$

$$R_j g \left(\frac{aH}{Q}\right)^2 \left(1 + \frac{2}{Bo}\right) + \frac{1}{2R_j H}\left(\frac{a}{\pi}\right)^2 = \frac{0.01676}{0.1826 + \frac{1}{Re}\left(\frac{R_j}{a}\right)^3} ; r(z) > r_v \tag{24.10}$$

$$Bo = \frac{\rho g R_j \Delta H}{\sigma} \tag{24.11}$$

$$r_v = 0.315 a Re^{1/3} \tag{24.12}$$

where Bo is Bond number, H (m) is the height of the liquid after the jump and ΔH (m) is the increase in liquid height through the jump. Table 24.1 shows a summary of numerical values used for the different variables in this chapter.

	Definition	Theory	ANSYS Fluent	Percent Difference
	Coordinates and Axial Velocity			
z	vertical coordinate defined from wall			
r	radial coordinate defined from centerline			
$r(z)$	free jet radius			
$U(z)$	axial velocity along centerline			
	Nozzle Geometry and Nozzle Velocity			
R_n (m)	nozzle radius	0.005	0.005	
Z (m)	distance from nozzle exit to wall	0.03	0.03	
U_{in} (m/s)	nozzle inlet axial velocity	0.4	0.4	
U_0 (m/s)	jet axial velocity @ nozzle exit	0.456553	0.456553	
	Jet Geometry, Max. Axial Velocity & Flow Rate			
a (m)	minimum radius of jet	0.003749	0.003582	4
z_a (m)	distance from minimum jet radius to wall	0.006526	0.006526	
U_m (m/s)	maximum axial jet velocity at minimum radius of jet	0.811928	0.742913	9
Q (m³/s)	volume flow rate	3.5851E-05	2.9946E-05	16

Non-Dimensional Parameters and Flow Rate				
Bo	Bond number	3.83	4.04	6
Fr	Froude number	4.25	4.25	
Re	Reynolds number	9517	8320	13
We	Weber number	14.3	14.3	
Hydraulic Jump Location and Thickness				
r_v (m)	radial distance to minimum liquid thickness	0.0250261	0.0228637	9
R_h (m)	radial location before hydraulic jump	0.0632	0.0632	
R_j (m)	radial location of hydraulic jump	0.0647	0.0684	6
R_j / a	non-dimensional radial location of hydraulic jump	17.3	19.1	11
R_H (m)	radial location after hydraulic jump	0.0724	0.0724	
H	height of liquid after hydraulic jump	0.001097	0.001097	
ΔH	increase in liquid height through hydraulic jump	0.00044	0.00044	
Fluid Properties				
σ (N/m)	surface tension	0.07286	0.07286	
ρ (kg/m^3)	liquid density	998.2	998.2	
μ (kg/m-s)	liquid dynamic viscosity	0.001003	0.001003	
ν (m^2/s)	kinematic viscosity of liquid	1.0048E-06	1.00481E-06	

Table 24.1 Numerical values from theory and simulations

I. References

1. Bush, J.W. M and Aristoff, J.M., "The influence of surface tension on the circular hydraulic jump", *Journal of Fluid Mechanics,* **489**, 229-238, (2003).
2. Bush, J.W. M, Aristoff, J.M. and Hosoi, A.E., "An experimental investigation of the stability of the circular hydraulic jump", *Journal of Fluid Mechanics,* **558**, 33-52, (2006).
3. Lokman Hosain, M.D., "CFD Simulation of Jet Cooling and Implementation of Flow Solvers in GPU",*Master of Science Thesis*, Stockholm, Sweden, (2013).
4. Passandideh-Fard, M., Teymourtash, A.R. and Khavari, M., "Numerical Study of Circular Hydraulic Jump Using Volume-of-Fluid Method", *J. Fluids Engineering*, **133**, (2011).
5. Raghav, G.H, "Numerical analysis of hydraulic jump by an impinging jet.", *Master of Technology Thesis*, Rourkela, (2014).

J. Exercises

24.1 Use ANSYS Fluent to run the simulation for different inlet velocities U_{in} = 0.2, 0.6 and 0.8 m/s and compare results in Table 24.2. Plot a graph with Reynolds number Re on the horizontal axis and R_j/a on the vertical axis based on your data in Table 24.2.

	ANSYS Fluent	ANSYS Fluent	ANSYS Fluent	ANSYS Fluent
a (m)		0.003582		
Bo		3.75		
Q (m³/s)		0.0000308173		
Re		8562		
R_j (m)		0.0697		
R_j/a		19.5		
r_v (m)		0.0230833		
Um (m/s)		0.764528		
Fr		4.25		
H		0.0011		
ΔH		0.0004		
R_n (m)	0.005	0.005	0.005	0.005
U_{in} (m/s)	0.2	0.4	0.6	0.8
U_0 (m/s)		0.456554		
We		14.3		
Z (m)	0.030	0.030	0.030	0.030
σ (N/m)	0.07286	0.07286	0.07286	0.07286
ρ (kg/m³)	998.2	998.2	998.2	998.2
ν (m²/s)	0.00000100481	0.00000100481	0.00000100481	0.00000100481

Table 24.2 Numerical values for Exercise 24.1

CHAPTER 25. OPTIMIZED ELBOW

A. Objectives

- Using Ansys Fluent to Study the Flow in an Optimized Elbow
- Inserting Boundary Conditions
- Running Ansys Fluent Simulations
- Using Contour Plots for Visualizations of Velocity and Pressure Fields

B. Problem Description

We will study the pressure drop in a 90° elbow for an optimized design and compare the pressure drop with a standard design.

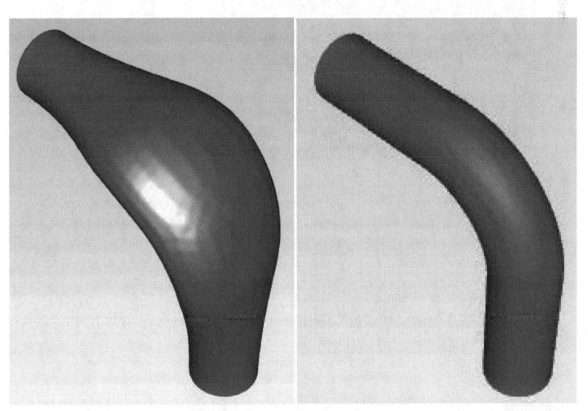

C. Launching Ansys Fluent

1. Start by launching Ansys Fluent in standalone mode.

Figure 25.1 Launching ANSYS Fluent

2. Check the box for Double Precision and Display Mesh After Reading. Select Meshing in the upper left portion of the Fluent Launcher window. Click on Show More Options and write down the location of your *Working Directory*. You will need this information later. Select Parallel Processing Options and set the number of Meshing Processes equal to the number of processor cores for your computer. The number of Meshing Processes is limited to 4 if you are using Ansys Student. To check the number of physical cores, press the Ctrl + Shift + Esc keys simultaneously to open the Task Manager. Go to the Performance tab and select CPU from the left column. You'll see the number of physical cores on the bottom-right side. Close the Task Manager window.

   Click on the Start button in the Fluent Launcher window. Click OK when you get the Key Behavioral Changes window.

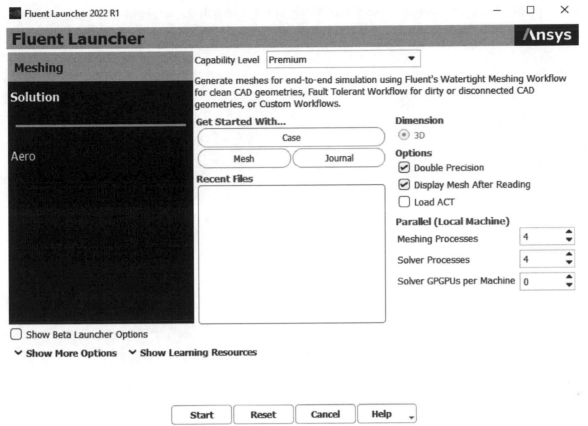

Figure 25.2 Fluent launcher window

D. Ansys Fluent Meshing

3. Select Workflow Type as Watertight Geometry.

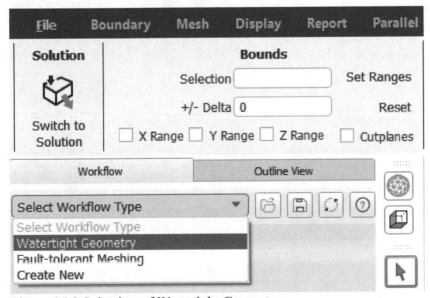

Figure 25.3 Selection of Watertight Geometry

4. Select Import Geometry and select **Mesh** as File Format. Select *in* as File Units and *cm* as Units. Browse for the File Name and open the file ***optimized_elbow.msh.h5***. Select Import Geometry.

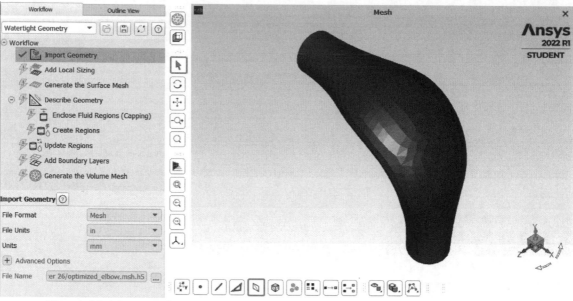

Figure 25.4 Imported geometry

5. Right click on Add Local Sizing and select Update. Right click on Generate the Surface Mesh and select Update.

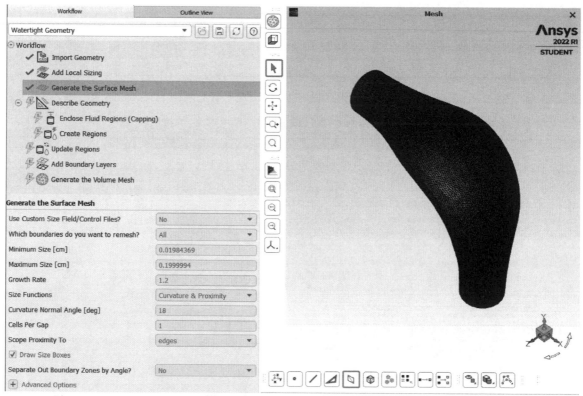

Figure 25.5 Generated surface mesh

6. Select Describe Geometry. As Geometry Type, select "The geometry consists of only fluid regions with no voids". Right click on Describe Geometry and select Update. Select Update Boundaries, right click and select update. Select Update Regions, right click and select update.

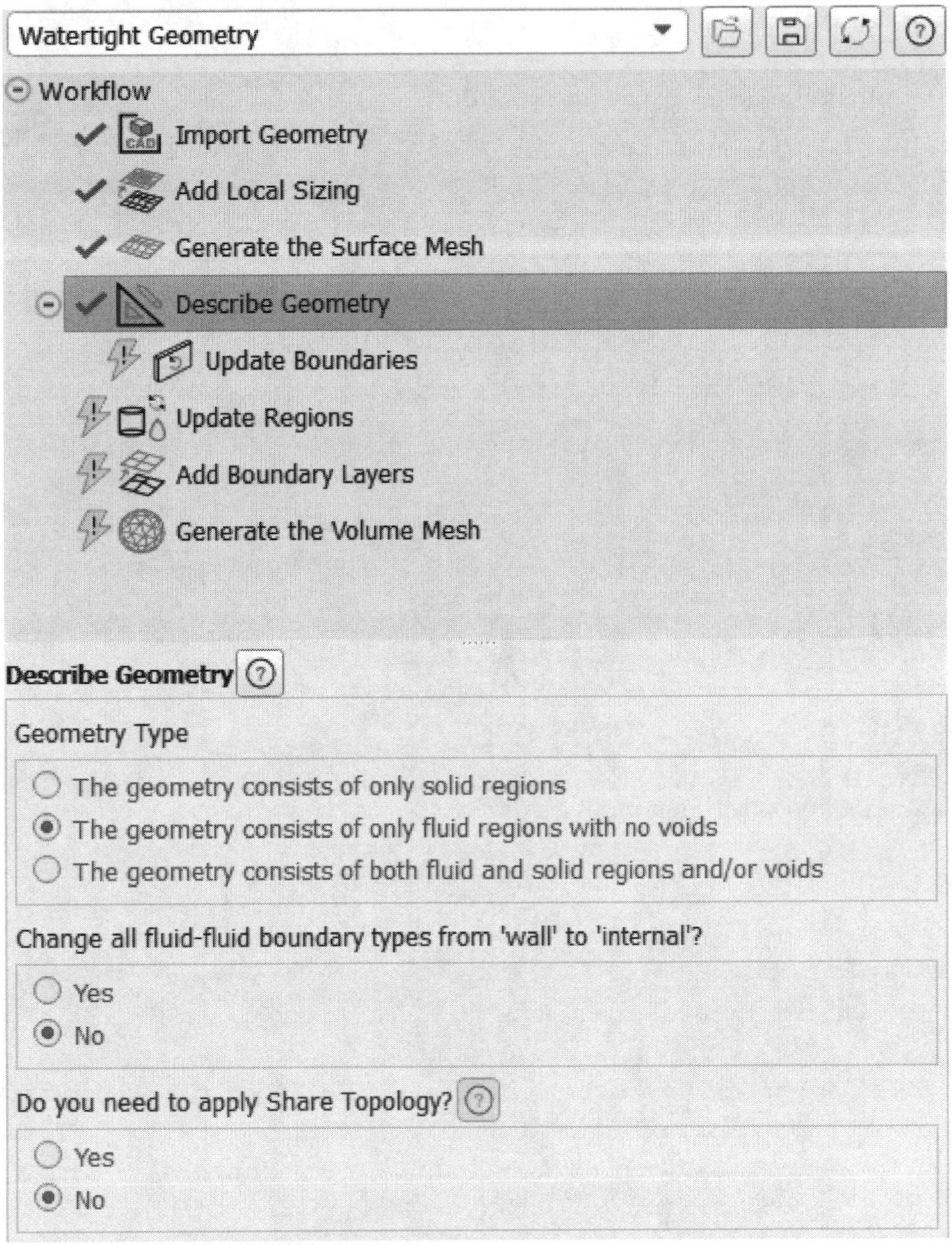

Figure 25.6a) Describe geometry

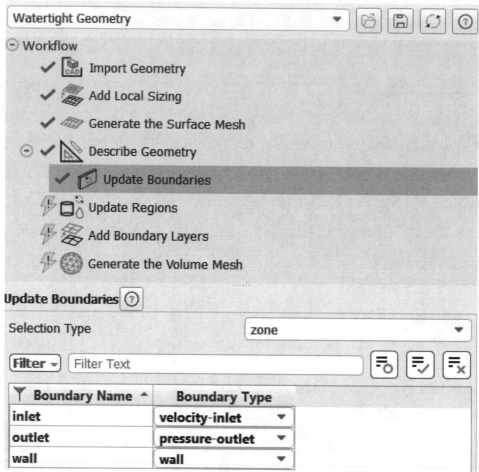

Figure 25.6b) Update boundaries

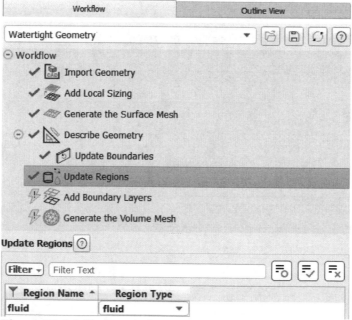

Figure 25.6c) Update regions

7. Select Add Boundary Layers. Set the Offset Method Type to aspect-ratio, Number of Layers to 6 and First Aspect Ratio to 12. Select Add Boundary Layers.

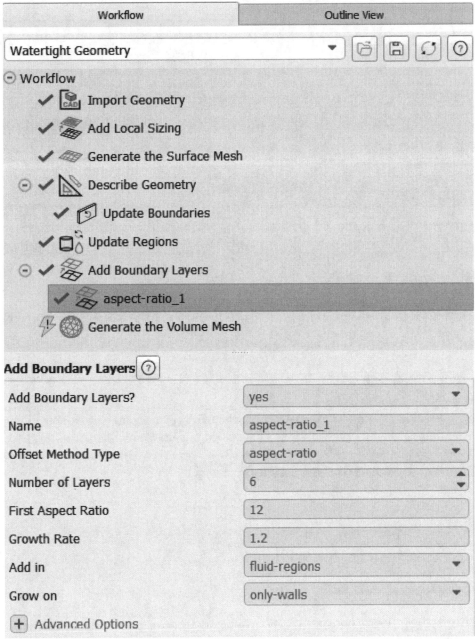

Figure 25.7 Adding boundary layers

8. Select Generate the Volume Mesh. Right click on Generate the Volume Mesh and select Update.

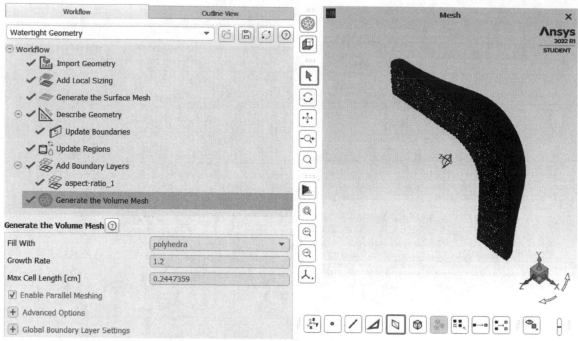

Figure 25.8 Generated volume mesh

E. Ansys Fluent Solution

9. Select Switch to Solution. Answer Yes to the question that you get. Double click on General under Setup in the Outline View. Click on Scale under Mesh on the Task Page.

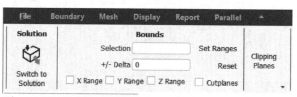

Figure 25.9a) Switching to solution

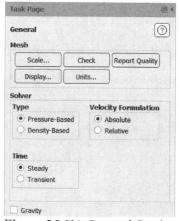

Figure 25.9b) General Settings

576

Select Specify Scaling Factors under Scaling. Set the Scaling Factors in the X, Y, Z directions to **0.875**. Click on Scale. The Domain Extent will change. Close the Scale Mesh window.

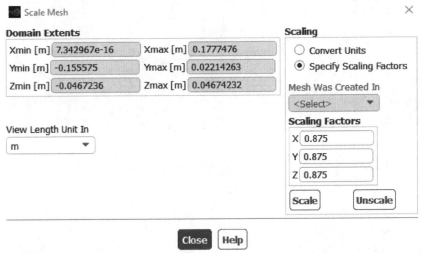

Figure 25.9c) Scale Mesh

Check the Gravity box under General on the Task Page. Set the Y [m/s^2] value to **-9.81**.

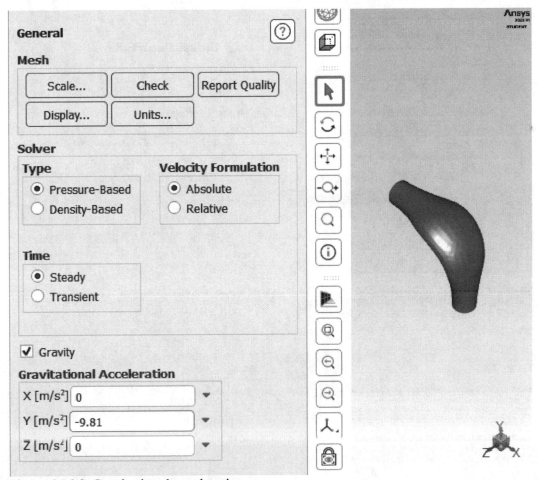

Figure 25.9d) Gravitational acceleration

10. Open Models under Setup in the Outline View and double click on Viscous (SST k-omega). Select k-epsilon (2 eqn), Realizable k-epsilon Model and Standard Wall Functions as Near-Wall Treatment and set the Model Constants and User-Defined Functions as listed in Figure 25.10a). Select OK to close the Viscous Model window.

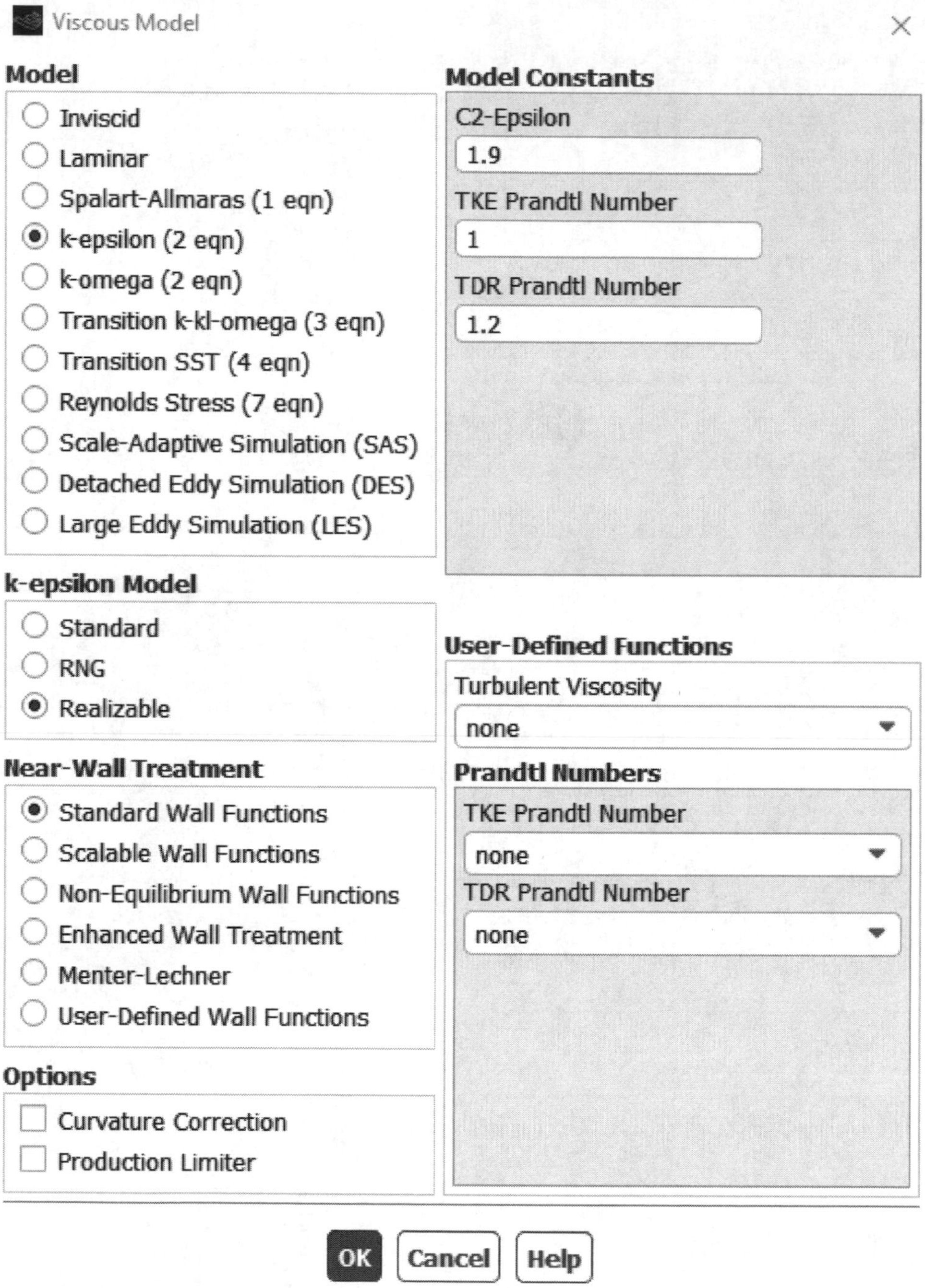

Figure 25.10a) Viscous model

Open Materials and Fluid under Setup in the Outline View and double click on *air*. Set Density [kg/m³] to **1.190** and Viscosity [kg/(m s)] to **1.8286e-05** to correspond to values in experiments. Select Change/Create and Close the Create/Edit Materials window.

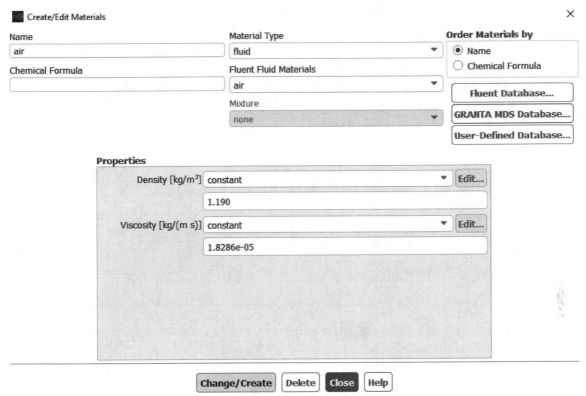

Figure 25.10b) Create/Edit Materials

11. Open Boundary Conditions under Setup in the Outline View and double click on Inlet. Set the Velocity Magnitude [m/s] to **25.99**. Select Apply and Close the Velocity Inlet window.

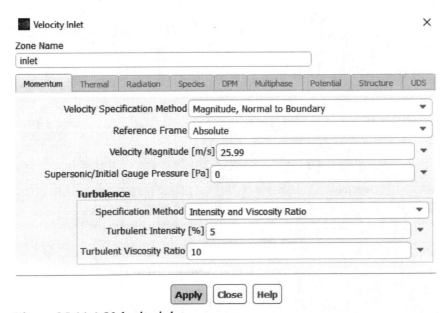

Figure 25.11a) Velocity inlet

Open Boundary Conditions under Setup in the Outline View and double click on Wall. Set the Roughness Height [m] under Sand-Grain Roughness to **0.00023**. Set the Roughness Constant to **0.5**. Select Apply and Close the Velocity Inlet window.

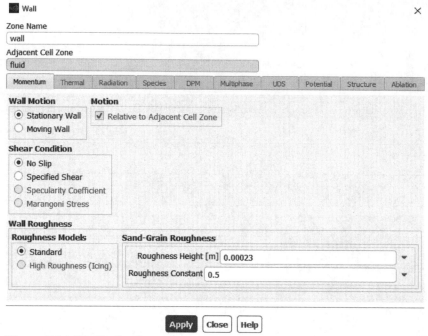

Figure 25.11b) Wall

Double click on Reference Values under Setup in the Outline View. Select Compute from inlet under Reference Values on the Task Page. Set the Area [m²] to **0.001552** and the Length [m] to **0.04445**. This is the inlet diameter. Select fluid as Reference Zone.

Figure 25.11b) Velocity inlet

12. Double click on Methods under Solution in the Outline View. Set Turbulent Kinetic Energy and Turbulent Dissipation Rate to Second Order Upwind. Set Gradient under Spatial Discretization to Least Squares Cell Based.

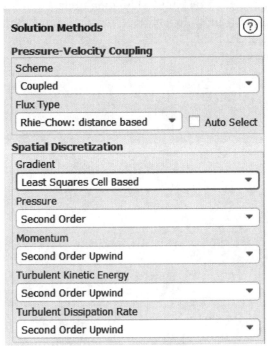

Figure 25.12a) Solution methods

Open Monitors under Solution in the Outline View and double click on Residual under Monitors. Set the Absolute Criteria for all Equations to **1e-04**. Select OK to close the Residual Monitors window.

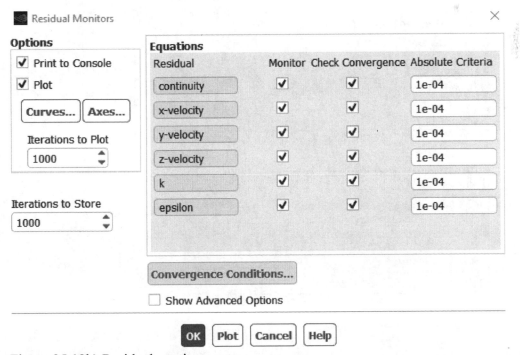

Figure 25.12b) Residual monitors

F. Initialization and Calculations

13. Double click on Initialization under Solution in the Outline View. Select Hybrid Initialization and select Initialize. Double click on Run Calculation under Solution in the Outline View. Select Check Case… under Run Calculation on the Task Page. Select OK in the Information window that appears. Set the Number of Iterations to 500 under Parameter. Select Calculate to start the calculations.

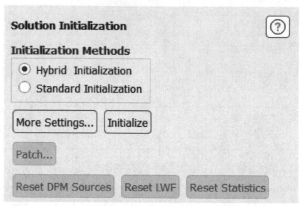

Figure 26.13a) Solution initialization

Figure 25.13b) Run calculations

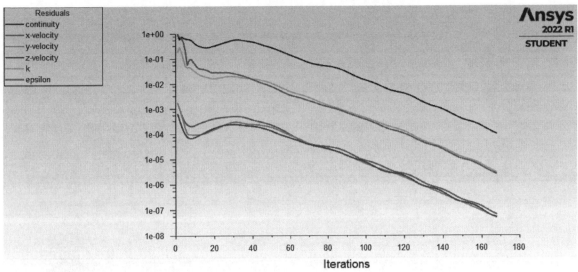

Figure 25.13c) Residuals

G. Post-processing

14. Double click on Graphics under Results in the Outline View. Right click on Contours and select New…. Select Contours of Pressure and Static Pressure. Select *wall* under Surfaces. Select Colormap Options… and set Font Size under Font to 10. Select Apply and Close the Colormap window. Click on Save/Display and Close the Contours window.

Figure 25.14a) Contours window

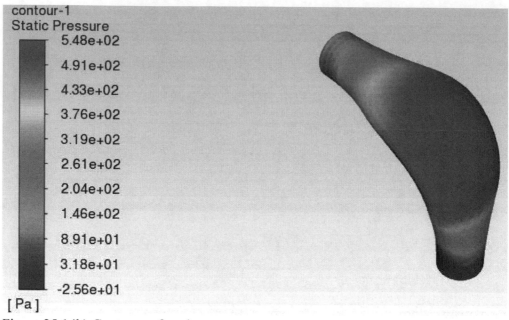

Figure 25.14b) Contours of static pressure

15. Right click on Surfaces under Results in the Outline View. Select New Plane.... Select XY Plane as Method. Select Create and Close the Plane Surface window.

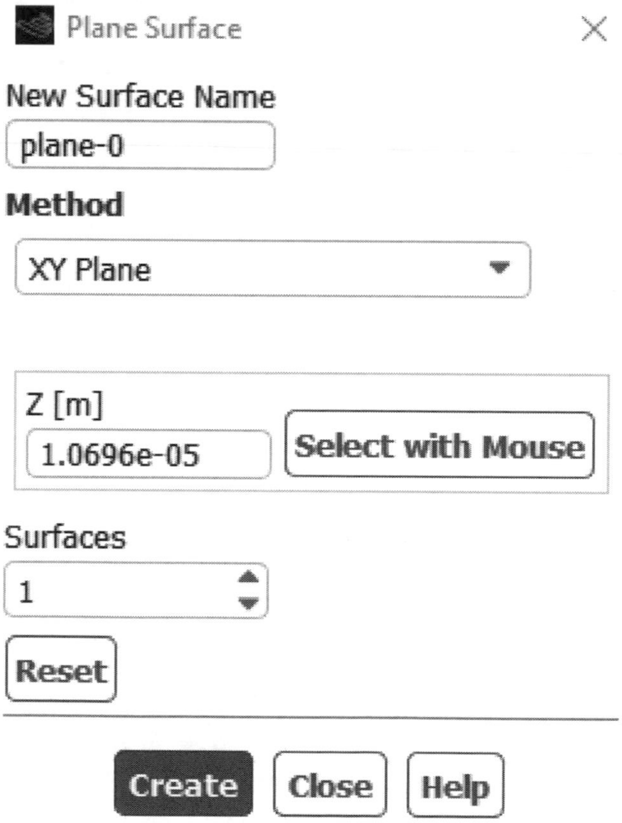

Figure 25.15 Plane surface

16. Double click on Graphics under Results in the Outline View. Right click on Contours and select New…. Select Contours of Velocity and Velocity Magnitude. Select *plane-0* under Surfaces. Select Colormap Options… and set Font Size under Font to 10. Select Apply and Close the Colormap window. Click on Save/Display and Close the Contours window. Select the blue Z axis for the coordinate system in the graphics window to view the contour plot in the *XY* plane.

Figure 25.16a) Contours window

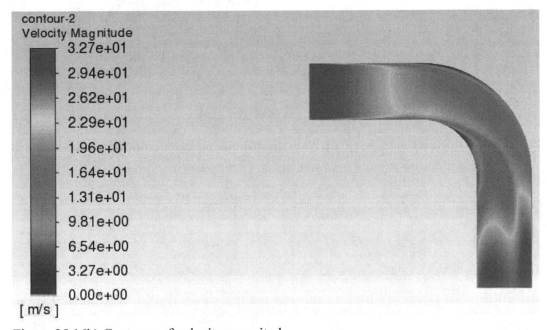

Figure 25.16b) Contours of velocity magnitude

17. Double click on Reports and Surface Integrals under Results in the Outline View. Select Area-Weighted Average as Report Type. Select Pressure… and Static Pressure as Field Variable. Select *inlet* and *outlet* as Surfaces. Select Compute and Close the Surface Integrals window. Write down the inlet static pressure from the console.

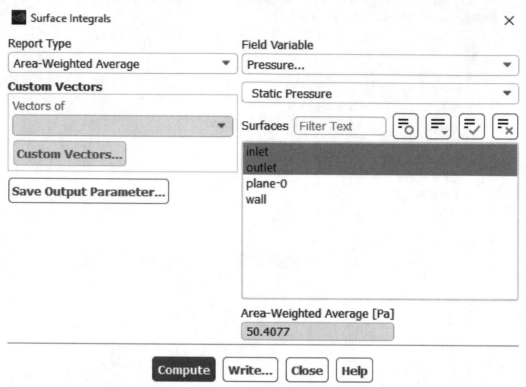

Figure 25.17a) Surface integrals for static pressure

"Surface Integral Report"

Area-Weighted Average Static Pressure	[Pa]
inlet	91.139689
outlet	0
Net	45.569834

Figure 25.17b) Results from console for static pressure

18. Select Area-Weighted Average as Report Type. Select Pressure... and Pressure Coefficient as Field Variable. Select *inlet* and *outlet* as Surfaces. Select Compute and Close the Surface Integrals window. Write down the inlet pressure coefficient from the console.

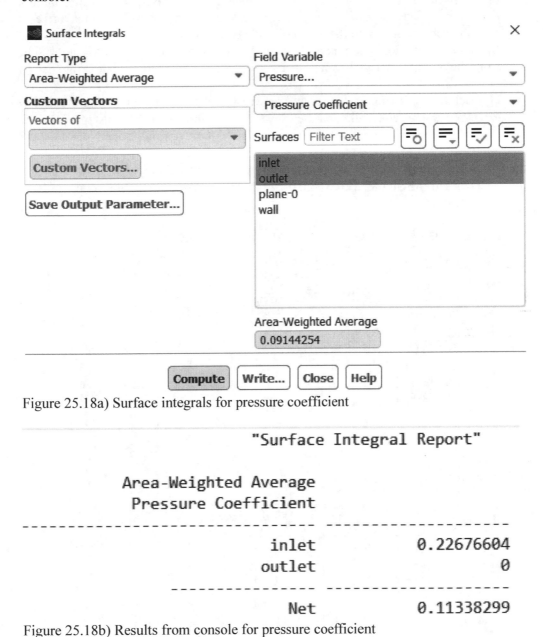

Figure 25.18a) Surface integrals for pressure coefficient

"Surface Integral Report"

Area-Weighted Average
Pressure Coefficient

inlet	0.22676604
outlet	0
Net	0.11338299

Figure 25.18b) Results from console for pressure coefficient

H. Batch Jobs

19. You will need to create the case and scheme files to complete a batch job. Select File>>Write>>Case & Data… from the menu. Save the case and data files in the *Working Directory* folder with the name ***optimized_elbow.cas.h5***. Open Notebook and create and save the scheme file with the name ***elbow_optimized.scm*** for the batch job. The scheme file is shown in the Appendix at the end of this chapter. Select File>>Read>>Scheme… from the menu in Ansys Fluent. Open the scheme file to start the batch job.

    Go back to step 1 in this chapter and repeat all the steps but instead use the *standard elbow*. In step 4, you will open the file ***standard_elbow.msh.h5***. In step 19, you will save the case and data files with the name ***standard_elbow.cas.h5***. Open Notebook and create and save the scheme file with the name ***elbow_standard.scm*** for the batch job. The scheme file is shown in the Appendix at the end of this chapter. Select File>>Read>>Scheme… from the menu in Ansys Fluent. Open the scheme file to start the batch job.

U (m/s)	Re	Optimized $\Delta p_{Exp.}$ (Pa)	Optimized $K_{Exp.}$	U (m/s)	Re	Optimized Δp_{Fluent}(Pa)	Standard Δp_{Fluent}(Pa)	Optimized K_{Fluent}	Standard K_{Fluent}	Ratio K
38.47	111305	199.3	0.2263	40	115719	227.8925	380.3952	0.2394	0.3995	0.5991
38.10	110226	195.9	0.2268	39	112826	216.0891	360.7739	0.2387	0.3986	0.5990
37.57	108685	191.0	0.2274	38	109933	204.6047	341.6763	0.2381	0.3976	0.5988
36.86	106642	184.1	0.2277	37	107040	193.4377	323.1022	0.2375	0.3966	0.5987
35.69	103252	172.9	0.2281	36	104147	182.5981	305.0535	0.2368	0.3956	0.5986
33.88	98018	155.3	0.2273	35	101254	172.0706	287.5376	0.2361	0.3945	0.5984
31.65	91562	135.1	0.2266	34	98361	161.8450	270.5427	0.2353	0.3933	0.5982
30.19	87329	123.5	0.2278	33	95468	151.9230	254.0773	0.2344	0.3921	0.5979
28.62	82785	110.5	0.2268	32	92575	142.1980	238.1450	0.2334	0.3908	0.5971
25.99	75200	91.2	0.2268	31	89682	132.6719	222.7478	0.2320	0.3895	0.5956
24.55	71024	81.6	0.2275	30	86789	123.6848	207.8861	0.2309	0.3882	0.5950
23.77	68763	76.6	0.2278	29	83896	115.0491	193.5434	0.2299	0.3867	0.5944
21.78	63017	64.5	0.2284	28	81003	106.7218	179.7254	0.2288	0.3852	0.5938
20.62	59652	57.3	0.2265	27	78110	98.7269	166.4504	0.2276	0.3837	0.5931
19.10	55255	49.4	0.2276	26	75217	91.1397	153.7200	0.2266	0.3821	0.5929
18.37	53154	45.6	0.2270	25	72324	83.8336	141.5045	0.2254	0.3805	0.5924
16.65	48167	37.6	0.2279	24	69431	76.8600	129.8240	0.2242	0.3788	0.5920
15.59	45115	33.3	0.2301	23	66538	70.0422	118.4808	0.2225	0.3764	0.5912
14.32	41414	27.9	0.2288	22	63645	63.6494	107.7822	0.2210	0.3742	0.5905
13.43	38843	24.5	0.2284	21	60752	57.6031	97.4351	0.2195	0.3713	0.5912
12.41	35891	21.4	0.2337	20	57859	51.8692	87.5658	0.2179	0.3679	0.5923
10.89	31497	16.4	0.2325	19	54966	46.3277	78.2623	0.2157	0.3643	0.5920
9.69	28030	12.5	0.2238	18	52073	41.1383	69.5926	0.2134	0.3610	0.5911
8.94	25865	9.9	0.2081	17	49180	36.1460	61.4542	0.2102	0.3573	0.5882
8.25	23861	7.9	0.1952	16	46287	31.3724	53.4743	0.2059	0.3510	0.5867
6.73	19458	5.2	0.1932	15	43395	26.3380	45.7071	0.1967	0.3414	0.5762
5.35	15464	3.3	0.1941	14	40502	22.9196	36.6386	0.1965	0.3141	0.6256
4.59	13280	2.6	0.2074	13	37609	20.1377	31.2254	0.2002	0.3105	0.6449
3.51	10151	1.5	0.2075	12	34716	17.2320	26.0039	0.2011	0.3035	0.6627
				11	31823	15.2604	22.8686	0.2119	0.3176	0.6673
				10	28930	13.4700	19.9241	0.2264	0.3348	0.6761
				9	26037	11.2566	16.7440	0.2335	0.3474	0.6723
				8	23144	9.2187	13.7980	0.2421	0.3623	0.6681
				7	20251	7.3591	11.0906	0.2524	0.3804	0.6635
				6	17358	5.6826	8.6279	0.2653	0.4028	0.6586
				5	14465	4.1927	6.4182	0.2818	0.4314	0.6532
				4	11572	2.8944	4.4723	0.3040	0.4697	0.6472
				3	8678.9	1.7994	2.8072	0.3360	0.5242	0.6410
				2	5786	0.9316	1.4584	0.3914	0.6127	0.6388
				1	2893	0.3211	0.4978	0.5396	0.8365	0.6450

Table 25.1 Experimental results and Ansys Fluent data for $T = 21.67$ C, $p_{atm} = 100,700$ Pa, $\rho = 1.19$ kg/m^3, $\mu = 1.8286 \cdot 10^{-5}$ kg/(m s), $D = 0.04445$ m, sand-grain roughness height 0.00023 m and roughness constant 0.5.

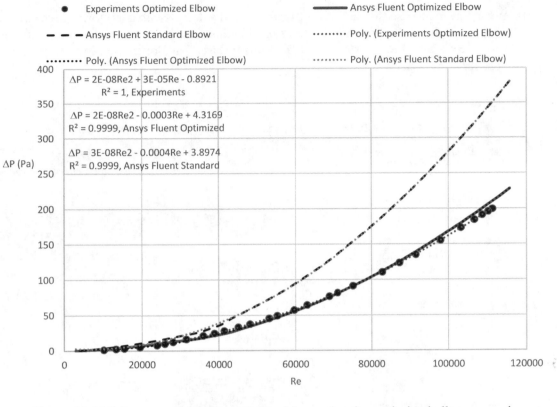

Figure 25.19a) Pressure drop versus Reynolds number for optimized elbow experiments

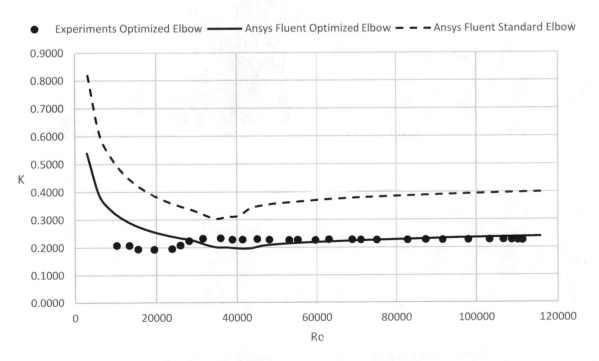

Figure 25.19b) Pressure coefficient versus *Re* for standard and optimized elbows

I. Sculptor

20. We are going to use the Sculptor software to find a design with even lower pressure drop than the optimized design that we have used in this chapter. Open the Sculptor software and select File>>Import Mesh>>CFD>>Fluent Case… and open the file *optimal_elbow.cas.*

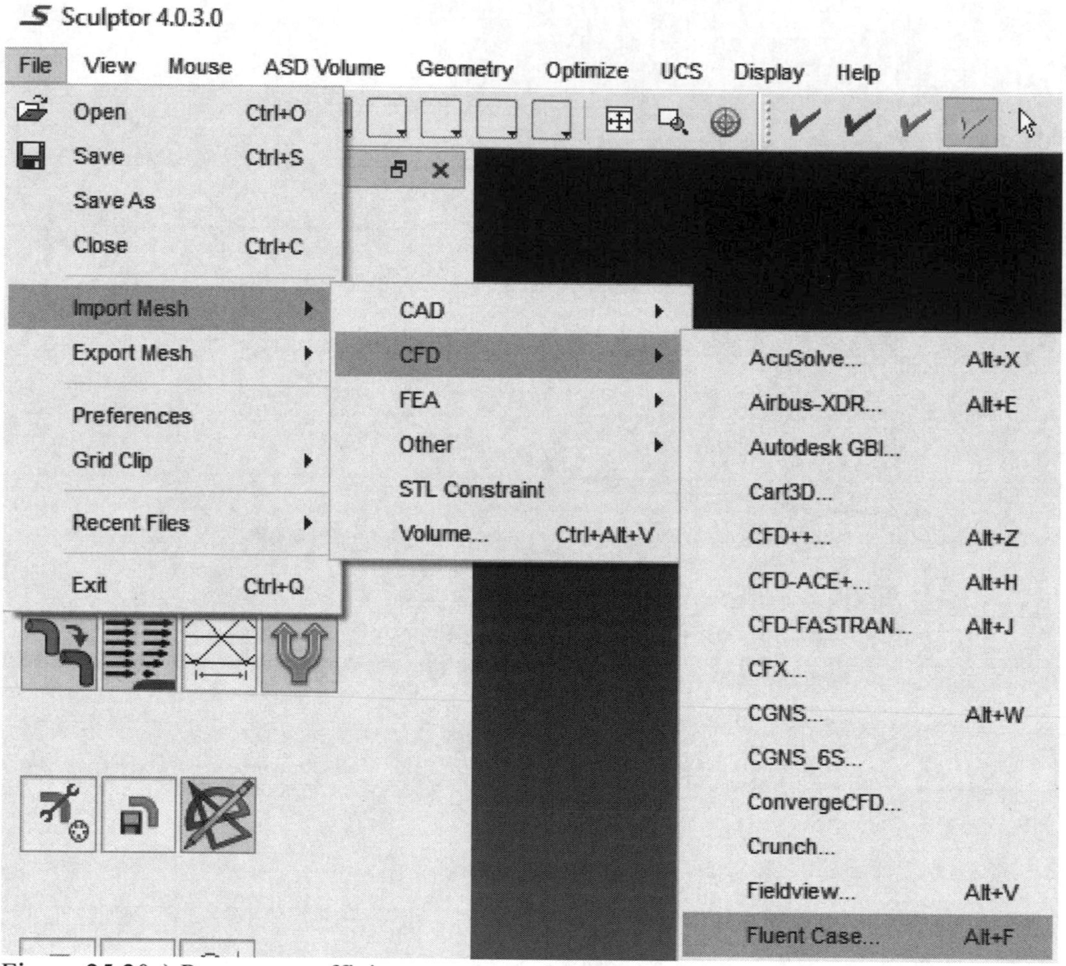

Figure 25.20a) Pressure coefficient versus *Re* for standard and optimized elbows

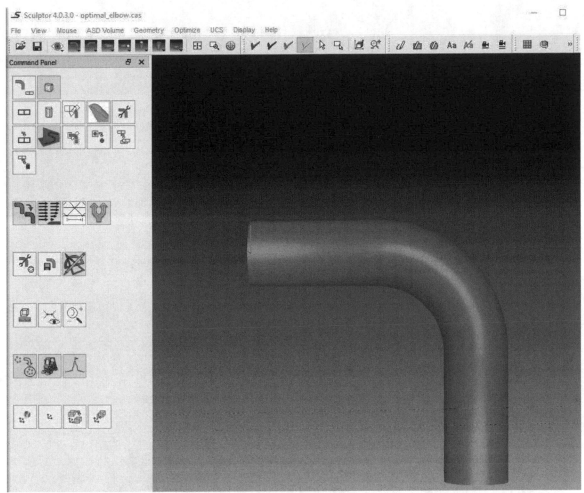

Figure 25.20b) Optimal elbow opened in Sculptor

Select Create Volume 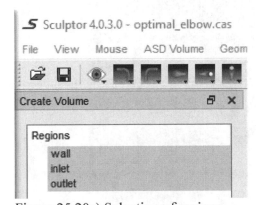 from the Command Panel. Select Regions wall, inlet and outlet. Set the Dimensions to Length 8, Height 2.2, Width 4.3 and set the Center of the volume to be X 4, Y 0, Z 0. Select Apply and OK.

Figure 25.20c) Selection of regions

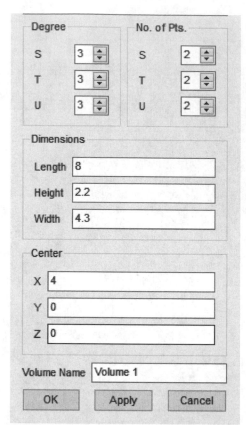

Figure 25.20d) Selection of dimensions and center for volume

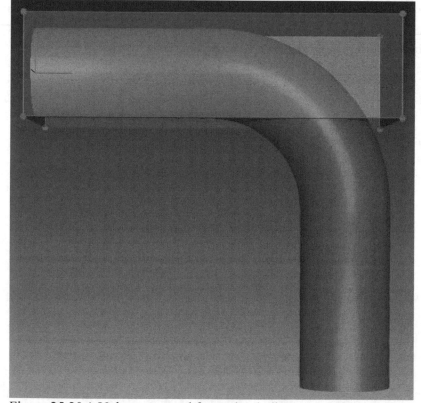

Figure 25.20e) Volume created for optimal elbow

21. Select Modify Volume 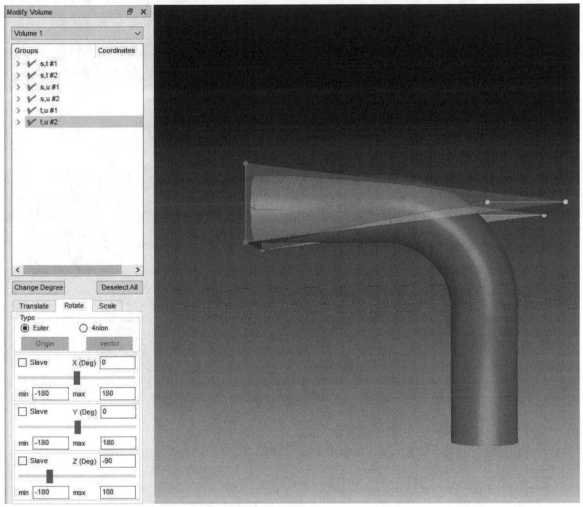 from the Main Menu. Select plane t,u #2 from Groups under Volume 1. Select the Rotate tab. Move the Z (deg) slider to the left so that you have -90. Select the Translate tab. Change the values of max to 7 and min to -7 for X, Y and Z coordinates. Move the Y slider to the left so that you have -7. Move the X slider to the left and enter the value so that you have -1. Select OK to exit.

Figure 25.21a) Plane t,u #2 rotated

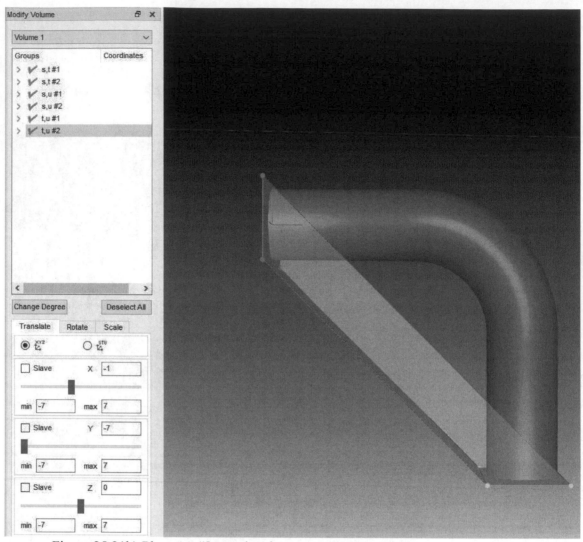

Figure 25.21b) Plane t,u #2 translated

22. Select Insert/Remove Control Points 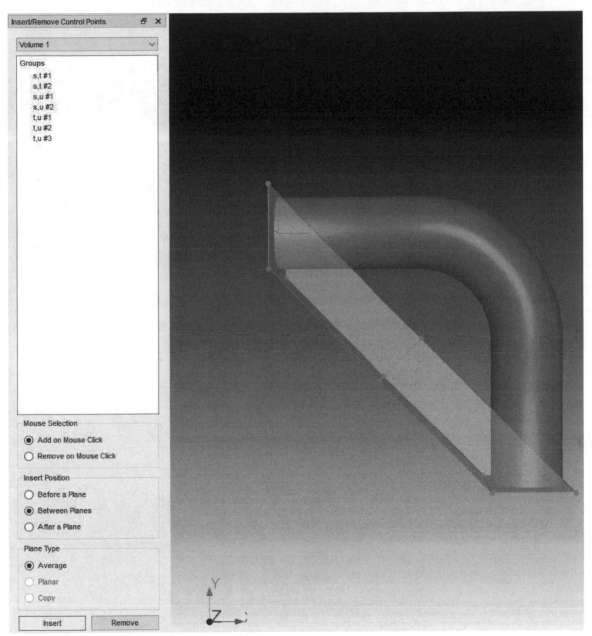 from the Main Menu. Select plane t,u #1 and t,u #2 from Groups. Select Between Planes as Insert Position. Select Insert and Done.

Figure 25.22 Inserting a new plane

23. Select Modify Volume 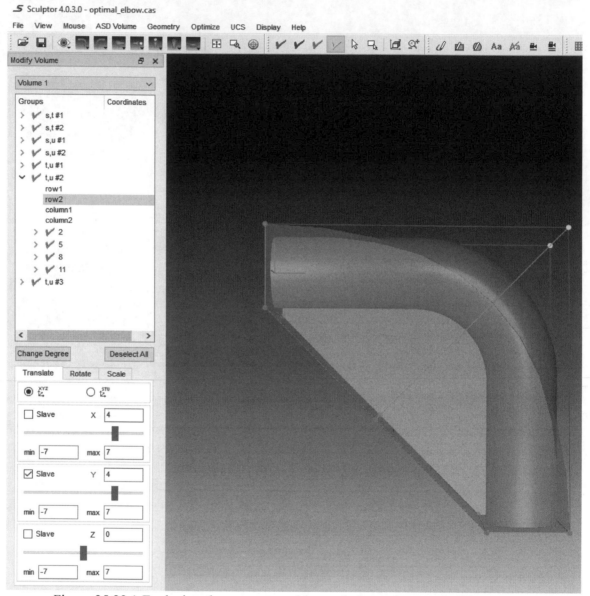 from the Main Menu. Open t,u #2 under Groups and select row2. Check the Slave box for the Y coordinate. Set X to 4 and select Enter on the keyboard. Deselect row2 and select row1. Set X to 2.4 and Enter. Select OK.

Figure 25.23a) Enclosing the geometry with translation using row2 under t,u#2

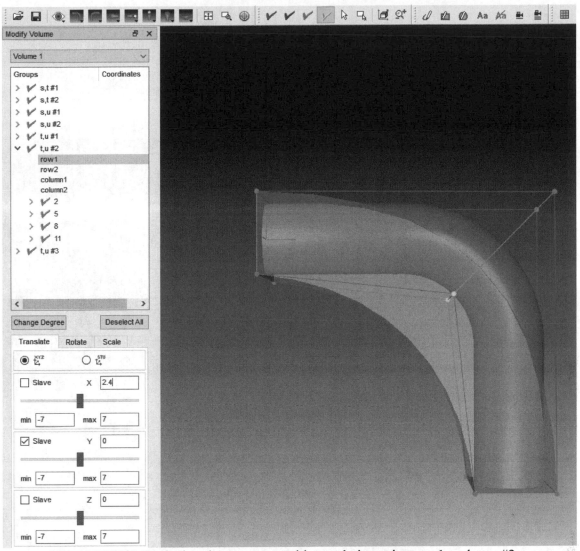

Figure 25.23b) Enclosing the geometry with translation using row1 under t,u#2

24. Select Insert/Remove Control Points 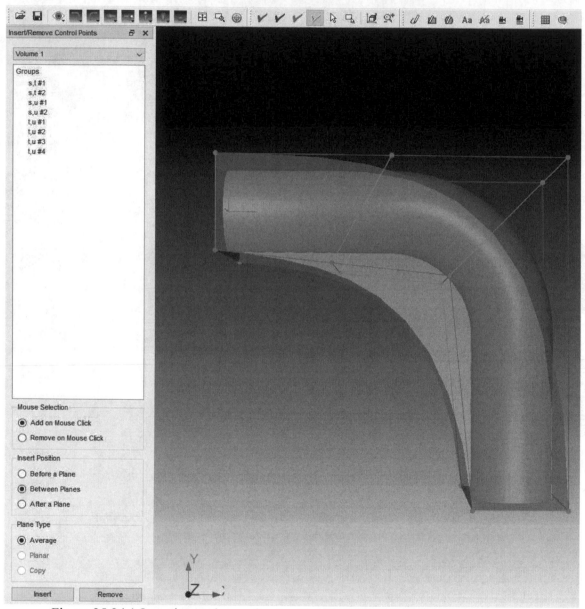 from the Main Menu. Select plane t,u #1 and t,u #2 from Groups. Select Between Planes as Insert Position. Select Insert. Select plane t,u #3 and t,u #4 from Groups. Select Between Planes as Insert Position. Select Insert and Done.

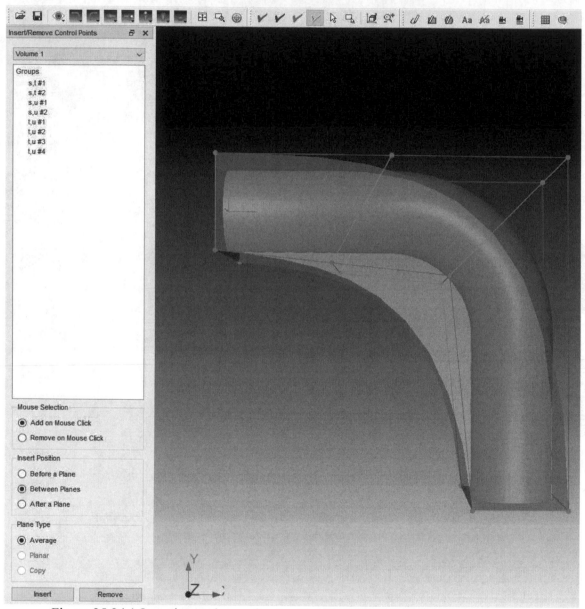

Figure 25.24a) Inserting a plane using t,u #1 and t,u #2

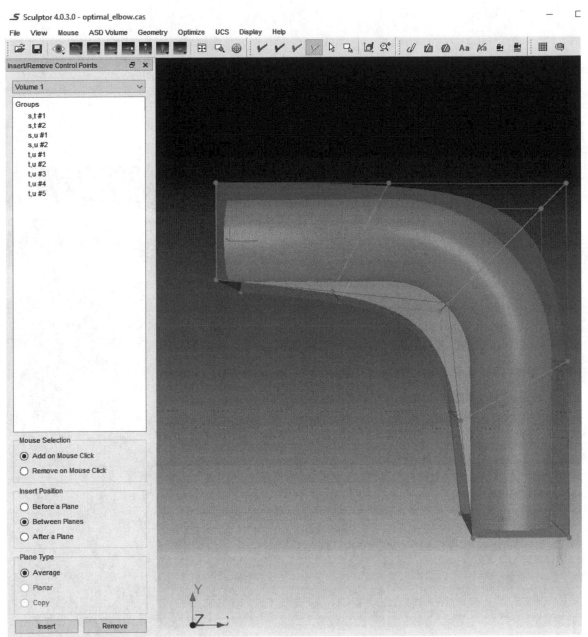

Figure 25.24b) Inserting another plane using t,u #3 and t,u #4

25. Select Modify Volume 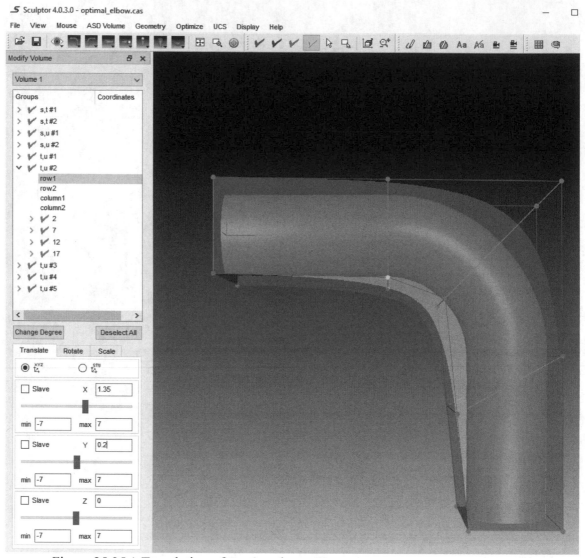 from the Main Menu. Open t,u #2 under Groups and select row1. Uncheck the Slave box for the Y coordinate. Set X to 1.35 and Y to 0.2 and select Enter on the keyboard. Deselect row1 under t,u #2 and select row1 under t,u #4. Set X to 0.2 and Y to 1.35 and Enter. Enter 1.35 next to Y and 0.2 next to X. Select OK.

Figure 25.25a) Translation of row1 under t,u#2

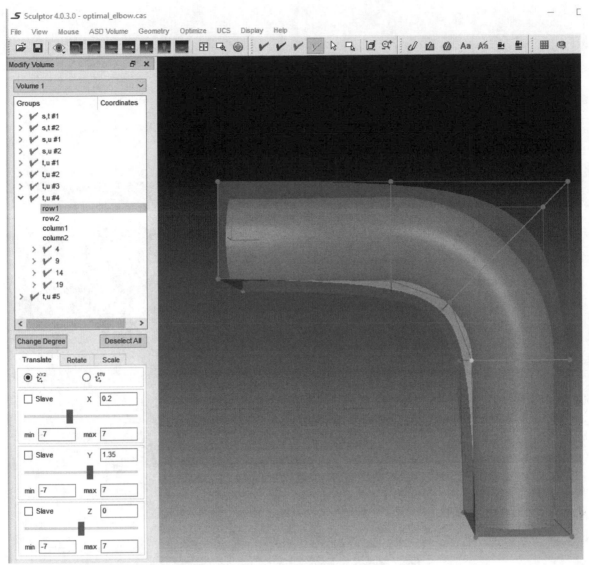

Figure 25.25b) Translation of row1 under t,u#4

26. Select Insert/Remove Control Points 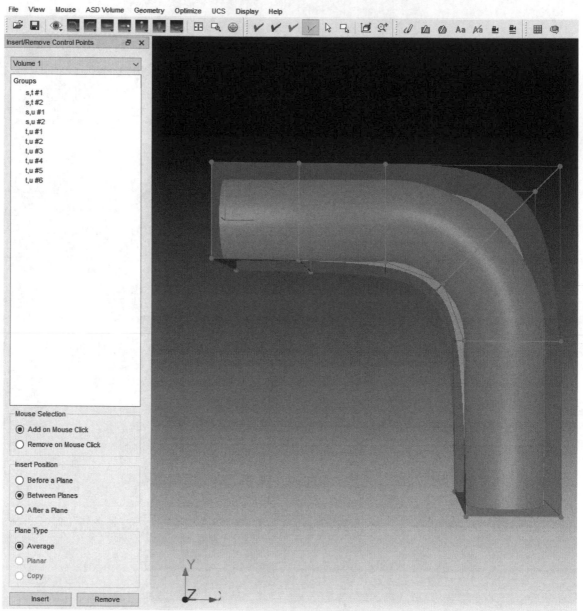 from the Main Menu. Select plane t,u #1 and t,u #2 from Groups. Select Between Planes as Insert Position. Select Insert. Select plane t,u #5 and t,u #6 from Groups. Select Between Planes as Insert Position. Select Insert and Done. Select Freeze Grid 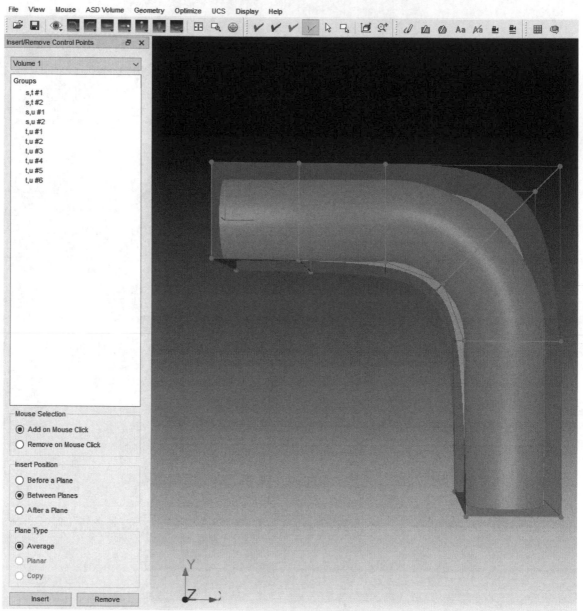. Answer OK to the question.

Figure 25.26a) Inserting another plane using t,u #1 and t,u #2

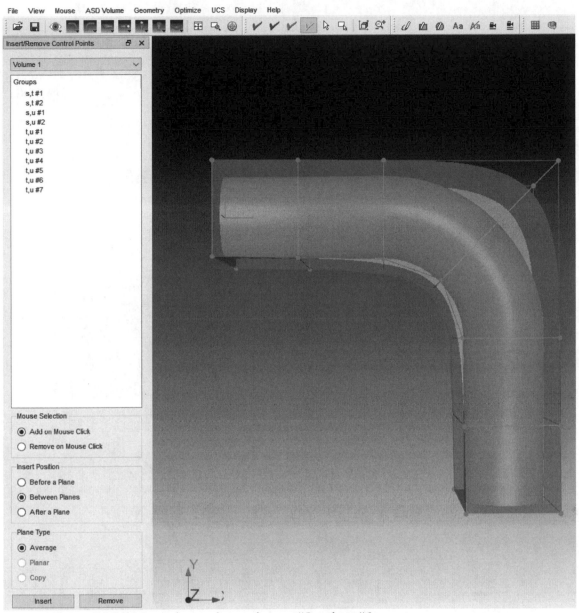

Figure 25.26b) Inserting a plane using t,u #5 and t,u #6

27. Select Insert Control Point Groups 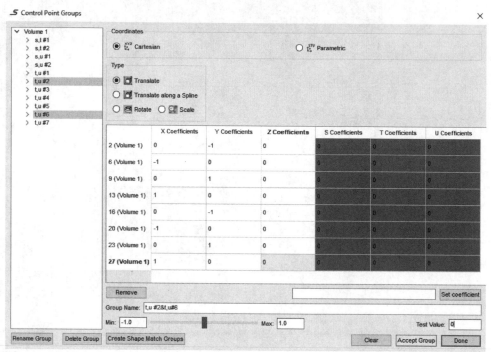 from the Main Menu. Select planes t,u #2 and t,u#6. Enter the values for the X, Y and Z Coefficients as shown in Figure 25.27a). Enter the Group Name: Y-DIR:t,u #2&X-DIR:t,u#6. Accept Group. Select planes t,u #3 and t,u#5. Enter the values for the X, Y and Z Coefficients as shown in Figure 25.27b). Enter the Group Name: Y-DIR:t,u #3&X-DIR:t,u#5. Accept Group.

Figure 25.27a) Control point groups in Y and X directions for planes t,u#2 and t,u#6

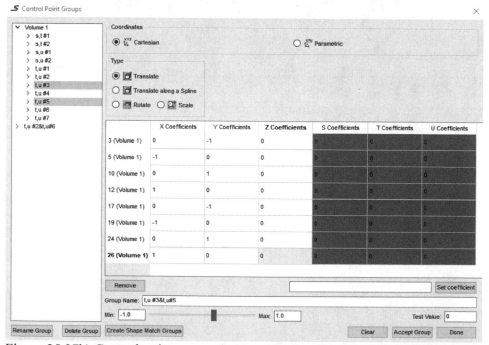

Figure 25.27b) Control point groups in Y and X directions for planes t,u#3 and t,u#5

28. Select planes t,u #2 and t,u#6. Enter the values for the X, Y and Z Coefficients as shown in Figure 25.28a). Enter the Group Name: Z-DIR:t,u #2&Z-DIR:t,u#6. Accept Group. Select planes t,u #3 and t,u#5. Enter the values for the X, Y and Z Coefficients as shown in Figure 25.28b). Enter the Group Name: Z-DIR:t,u #3&Z-DIR:t,u#5. Accept Group.

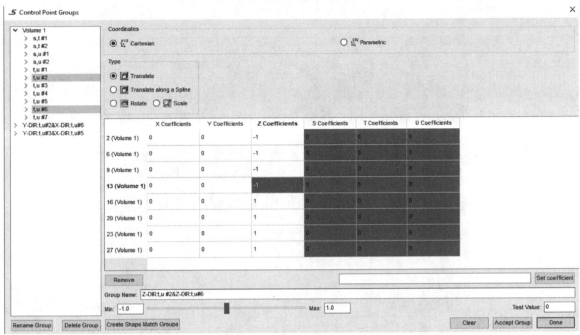

Figure 25.28a) Control point groups in Z direction for planes t,u#2 and t,u#6

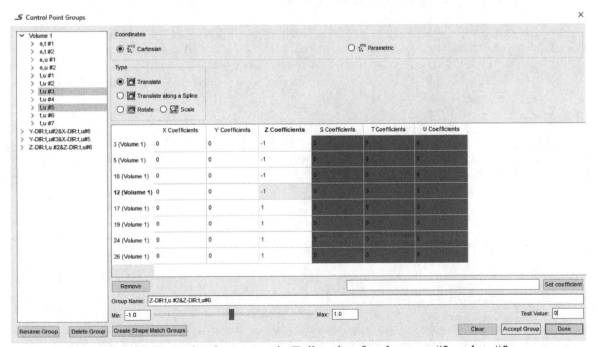

Figure 25.28b) Control point groups in Z direction for planes t,u#3 and t,u#5

29. Select plane t,u#4. Enter the values for the X, Y and Z Coefficients as shown in Figure 25.29a). Enter the Group Name: Z-DIR:t,u #4. Accept Group. Select plane t,u#4. Enter the values for the S, T and U Coefficients as shown in Figure 25.29b). Enter the Group Name: T-DIR:t,u #4. Accept Group. Select Done to exit the Control Point Groups window. Answer Yes to the question that you get.

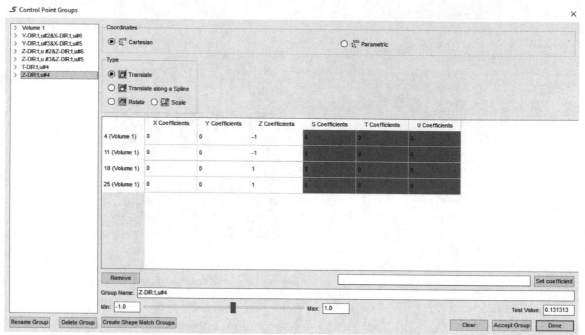

Figure 25.29a) Control point groups in Z direction for plane t,u#4

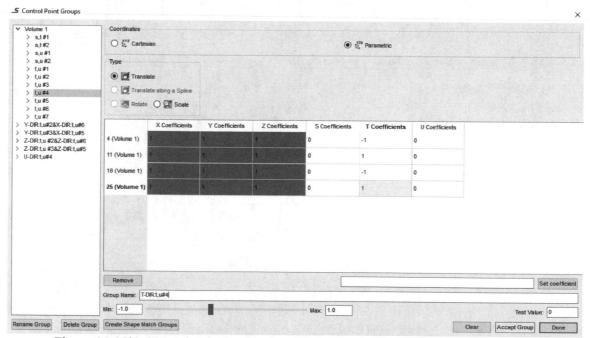

Figure 25.29b) Control point groups in T direction for plane t,u#4

30. Select Optimize>>Optimization from the menu. This will open the Optimization Control Center window. Select the General tab and check all the boxes and use the settings as shown in Figure 25.30b). Select the Analysis Variables tab and check all the boxes and use the settings as shown in Figure 25.30c).

Figure 25.30a) Starting the optimization process

Figure 25.30b) General settings

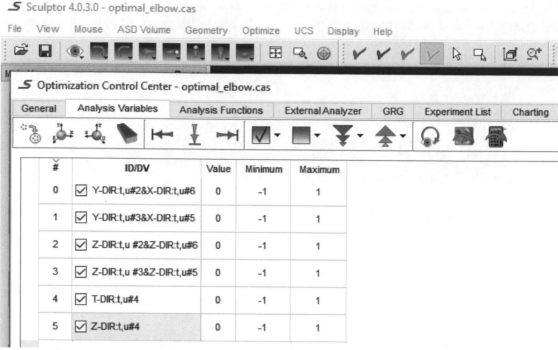

Figure 25.30c) Analysis variables settings

31. Create a desktop folder named OCC-elbow. Go to the Sculptor menu and select File>>Save As and save the file with the name *OCC-elbow.mdf*. Go back to the Optimization Control Center window. Select the External Analyzer tab and enter the Replacement Parameters, Directory, Command, Arguments, Instances and Control Template File code as shown in Figure 25.31b) and 25.31c). You may need to modify the Command and Arguments lines based upon the version of Ansys that you are using and the location of the *fluent.exe* file.

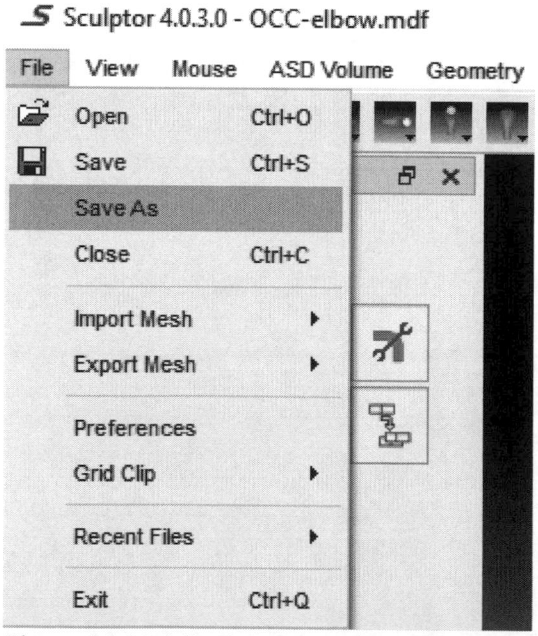

Figure 25.31a) Saving the sculptor file

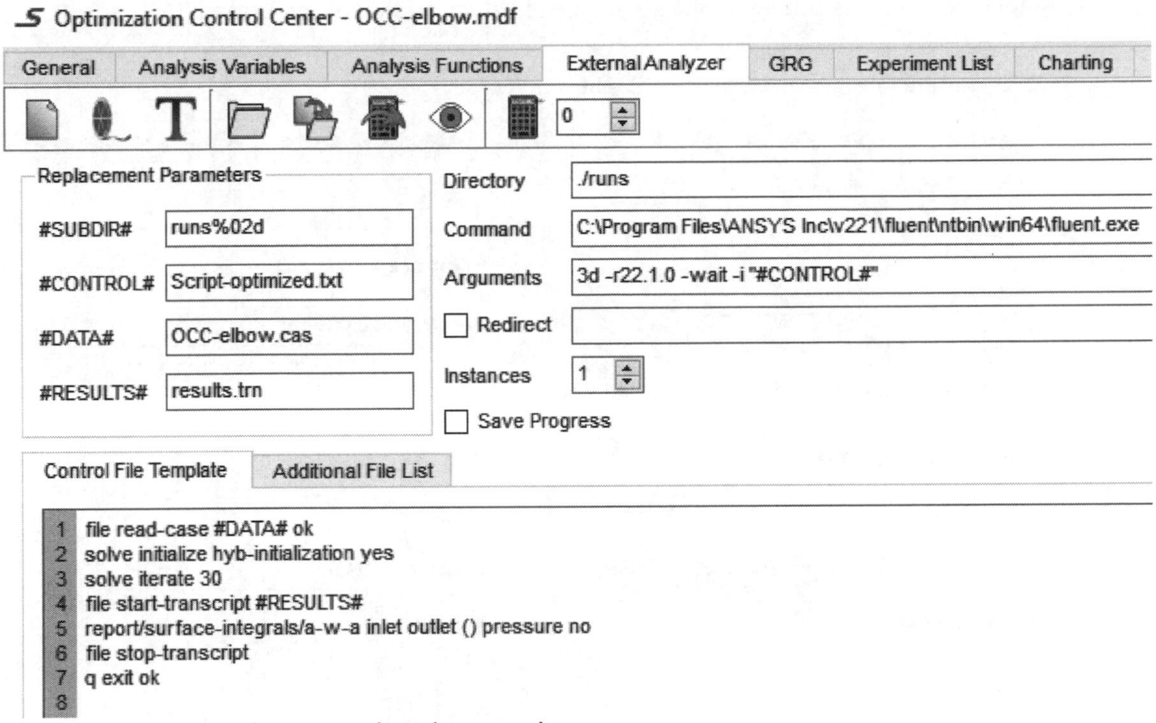

Figure 25.31b) External analyzer settings

Replacement Parameters

#SUBDIR#	runs%02d
#CONTROL#	Script-optimized.txt
#DATA#	OCC-elbow.cas
#RESULTS#	results.trn

Directory   ./runs

Command   C:\Program Files\ANSYS Inc\v221\fluent\ntbin\win64\fluent.exe

Arguments   3d -r22.1.0 -wait -i "#CONTROL#"

Control File Template

```
1       file read-case #DATA# ok
2       solve initialize hyb-initialization yes
3       solve iterate 30
4       file start-transcript #RESULTS#
5       report/surface-integrals/a-w-a inlet outlet () pressure no
6       file stop-transcript
7       q exit ok
8
```

Figure 25.31c) Replacement parameters, directory, command, arguments and control template file.

32. Select the GRG tab and enter the settings and values as shown in Figure 25.32.

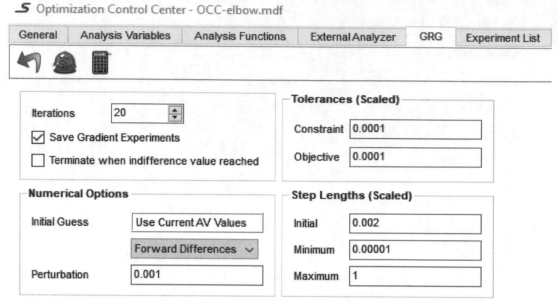

Figure 25.32 GRG settings

33. Select the External Analyzer tab and click on the calculator [calculator icon] to complete a test. Fluent will open and 30 iterations of calculations will be completed. Select the Analysis Functions tab and Open Results File [icon]. Double click on the *runs* folder and the *runs00* folder. Open the *results.trn* file. Create three rows under Variable and name these inlet, outlet and difference, respectively. Click in the window under Expression and to the right of inlet. Highlight the word inlet on line 25 under Area-Weighted Average Static Pressure, right click and select Set lastRegexp.

Do the corresponding under Expression and next to outlet by highlighting outlet on line 26, right click and select Set lastRegexp. Finally, write inlet-outlet in the area for Expression next to difference and choose the red downward arrow as Objective for the difference. Enter 1000 as the value for Allowable difference and 10 as the Indifference value for difference, see Figure 25.33. Delete the runs00 folder. Select the GRG tab and Run Optimization to complete optimization calculations.

∫ Optimization Control Center - OCC-elbow.mdf

| General | Analysis Variables | Analysis Functions | External Analyzer | GRG | Experiment List | Ch |

	Variable	Expression	Value	Objective	Allowable	Indifference
	inlet	lastRegexp("inlet(.*)$")	503.82372			
	outlet	lastRegexp("outlet(.*)$")	0			
	difference	inlet-outlet	inlet-outlet ↓		100	10

Results Selector

```
1   Transcript Start Time: 10:33:23, 30 Apr 2022 Central Daylight Time
2   Current Directory: "C:\Users\Instructor\Desktop\OCC-Elbow\runs\runs00"
3
4   Build Time: Nov 29 2021 12:23:13 EST  Build Id: 10213
5   Executable Path: C:/PROGRA~1/ANSYSI~1/v221/fluent/ntbin/win64/fluent.exe
6
7   _____
8   ID  Hostname  Core O.S.      PID  Vendor
9   _____
10  n0*  1Jmatsson4  1/28  Windows-x64  10756  Intel(R) Xeon(R) W-2275
11  host 1Jmatsson4       Windows-x64  21356  Intel(R) Xeon(R) W-2275
12
13  MPI Option Selected: intel
14  Selected system interconnect: default
15  _____
16
17  Cortex Process ID: 22948
18  License Server Path: 1055@10.200.1.245
19
20  > report/surface-integrals/a-w-a inlet outlet () pressure no
21
22      Area-Weighted Average
23        Static Pressure        [Pa]
24  _____    _____
25        inlet      503.82372
26        outlet         0
27  _____    _____
28         Net       251.91186
29
30  > file stop-transcript
31  Transcript Stop Time: 10:33:23, 30 Apr 2022 Central Daylight Time
32  Total Transcript Time: 0 Minute 0 Second.
```

Figure 25.33a) Analysis functions settings

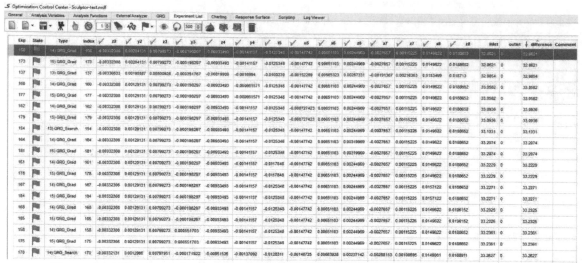

Figure 25.33b) Example out from running optimizations

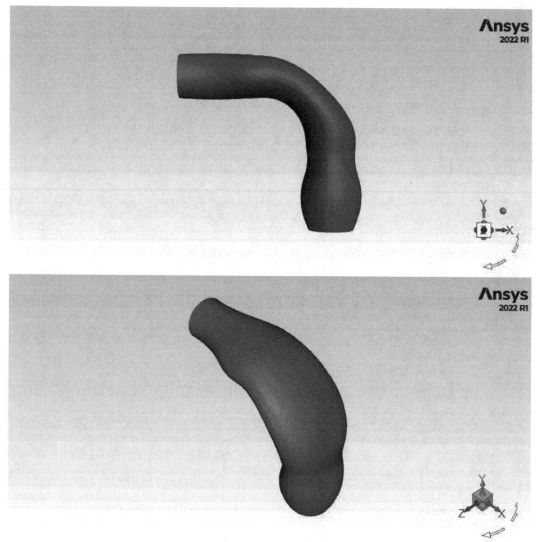

Figure 25.33c) Example design with pressure drop 32.86 Pa. This can be compared with 91.14 Pa for the original optimized design.

J. Theory

The temperature dependence for the dynamic viscosity of air can be described using the following equation

$$\mu = aT^b \tag{26.1}$$

where the constants are $a = 2.791 \cdot 10^{-7}$ and $b = 0.7355$, and T (K) is the temperature of the air. The density of air is given by the ideal gas law expressed as

$$\rho = \frac{p_{atm}}{RT} \tag{26.2}$$

where p_{atm} is the atmospheric pressure and $R = 287$ (J/(kg K)) is the gas constant. The Reynolds number is defined as

$$Re = \frac{UD\rho}{\mu} \tag{26.3}$$

where U (m/s) is the average velocity and $D = 0.04445$ (m) is the inner diameter of the pipe. The pressure coefficient can be expressed as

$$K = \frac{\Delta p}{\frac{1}{2}\rho U^2} \tag{26.4}$$

where Δp (Pa) is the pressure drop from the inlet to the outlet of the elbow. The outlet of the elbow will be at atmospheric pressure and the inlet of the elbow will be at a higher pressure than the atmospheric pressure. The percentage difference between experimental results and Ansys Fluent results for the pressure coefficient can be determined as

$$Percent\ Error = 100 \cdot \frac{|K_{Exp.} - K_{Fluent}|}{K_{Fluent}} \tag{26.5}$$

K. References

1. Baukal, C.E., Gershtein, V.Y., Li, X., "Computational Fluid Dynamics in Industrial Combustion.", CRC Press (2001).

L. Exercises

25.1 Run the calculations shown in this chapter for different surface height roughness and compare with the results in this chapter.

25.2 Run the calculations shown in this chapter for different Absolute Criteria for all Equations and compare with the results in this chapter.

25.3 Run the calculations shown in this chapter for different mesh sizes and compare with the results in this chapter.

25.4 Run the ANSYS Fluent calculations shown in this chapter for different velocities U (m/s) and compare with static pressure and pressure coefficient results as shown in Table 25.2. Determine the percent error between experiments and Ansys Fluent for the pressure coefficient K.

	U (m/s)	Re	$\Delta p_{Exp.}$ (Pa)	$K_{Exp.}$	$K_{Ansys\ Fluent}$	% Error
Student 1	38.47	111305	199.3	0.2263		
Student 2	38.10	110226	195.9	0.2268		
Student 3	37.57	108685	191.0	0.2274		
Student 4	36.86	106642	184.1	0.2277		
Student 5	35.69	103252	172.9	0.2281		
Student 6	33.88	98018	155.3	0.2273		
Student 7	31.65	91562	135.1	0.2266		
Student 8	30.19	87329	123.5	0.2278		
Student 9	28.62	82785	110.5	0.2268		
Student 10	25.99	75200	91.2	0.2268		
Student 11	24.55	71024	81.6	0.2275		
Student 12	23.77	68763	76.6	0.2278		
Student 13	21.78	63017	64.5	0.2284		
Student 14	20.62	59652	57.3	0.2265		
Student 15	19.10	55255	49.4	0.2276		
Student 16	18.37	53154	45.6	0.2270		
Student 17	16.65	48167	37.6	0.2279		
Student 18	15.59	45115	33.3	0.2301		
Student 19	14.32	41414	27.9	0.2288		
Student 20	13.43	38843	24.5	0.2284		

Table 25.2 Data for mean velocity, Reynolds number, pressure drop, pressure coefficient and percentage error

M. Appendix

This is the scheme file for the batch job shown in this chapter.

```
(Do ((y 2 (- y 1))) ((< y 1))
(begin
(Do ((x 40 (- x 1))) ((< x 1))
(begin
(if (> y 1)
        (Ti-menu-load-string (format #f "file read-case optimized_elbow.cas.h5 ok"))
        (Ti-menu-load-string (format #f "file read-case standard_elbow.cas.h5 ok")))
(if (> x 10)
        (begin
                (Ti-menu-load-string (format #f "define models viscous ke-realizable yes"))
                (Ti-menu-load-string (format #f "define boundary-conditions velocity-inlet inlet n n y y n ~a n 0 n n y 5 10" x))
                (Ti-menu-load-string (format #f "define boundary-conditions wall wall n n n 0.00023 n 0.5"))
                (Ti-menu-load-string (format #f "solve set discretization-scheme mom 1"))
                (Ti-menu-load-string (format #f "solve monitors residual convergence-criteria 0.0001 0.0001 0.0001 0.0001")))
        (begin
                (Ti-menu-load-string (format #f "define models viscous laminar yes"))
                (Ti-menu-load-string (format #f "define boundary-conditions velocity-inlet inlet n n y y n ~a n 0" x))
                (Ti-menu-load-string (format #f "solve set discretization-scheme mom 0"))
                (Ti-menu-load-string (format #f "solve monitors residual convergence-criteria 0.00001 0.00001 0.00001 0.00001"))))
(Ti-menu-load-string (format #f "solve initialize hyb-initialization yes"))
(Ti-menu-load-string (format #f "solve iterate 300"))
(if (> y 1)
        (begin
                (Ti-menu-load-string (format #f "report s-i a-w-a inlet outlet () p y optimized_elbow_static_pressure_data_U=~amps no ok" x))
                (Ti-menu-load-string (format #f "report s-i a-w-a inlet outlet () p-c y optimized_elbow_pressure_coeff__data_U=~amps no ok" x)))
        (begin
                (Ti-menu-load-string (format #f "report s-i a-w-a inlet outlet () p y standard_elbow_static_pressure_data_U=~amps no ok" x))
                (Ti-menu-load-string (format #f "report s-i a-w-a inlet outlet () p-c y standard_elbow_pressure_coeff__data_U=~amps no ok" x))))
(Ti-menu-load-string (format #f "display objects display contour-1"))
(Ti-menu-load-string (format #f "display views restore-view isometric"))
(if (> y 1)
        (Ti-menu-load-string (format #f "display save-picture optimized_elbow_static_pressure_contour_plot_U=~amps.jpg yes" x))
        (Ti-menu-load-string (format #f "display save-picture standard_elbow_static_pressure_contour_plot_U=~amps.jpg yes" x)))
(Ti-menu-load-string (format #f "display objects display contour-2"))
(Ti-menu-load-string (format #f "display views restore-view front"))
(if (> y 1)
        (Ti-menu-load-string (format #f "display save-picture optimized_elbow_velocity_magnitude_contour plot U=~amps.jpg yes" x))
        (Ti-menu-load-string (format #f "display save-picture standard_elbow_velocity_magnitude_contour_plot_U=~amps.jpg yes" x)))
(Ti-menu-load-string (format #f "plot residuals y y y y y"))
(if (> y 1)
        (Ti-menu-load-string (format #f "display save-picture optimized_elbow_residuals_plot_U=~amps.jpg yes" x))
        (Ti-menu-load-string (format #f "display save-picture standard_elbow_residuals_plot_U=~amps.jpg yes" x)))))))))
```

Figure 25.19d) Scheme file for standard and optimized elbows

N. Notes

CHAPTER 26. GOLF BALL

A. Objectives

- Using Ansys Fluent to Study the Flow around a Volvik ProBismuth Golf Ball
- Loading the *stl* file for the Volvik ProBismuth Golf Ball
- Using Fault-Tolerant Meshing in Ansys Fluent Meshing
- Inserting Boundary Conditions
- Running Ansys Fluent Simulations
- Using Contour Plots for Visualizations of Velocity and Pressure Fields

B. Problem Description

We will study the flow around a Volvik ProBismuth golf ball. This golf ball has an octahedron pattern of dimples and the number of dimples is 446.

C. Launching Ansys Fluent

1. Start by launching Ansys Fluent in standalone mode.

Figure 26.1 Launching ANSYS Fluent

2. Check the box for Double Precision. Select Meshing in the upper left portion of the Fluent Launcher window. Click on Show More Options and write down the location of your *Working Directory*. You will need this information later. Set the number of Meshing and Solver Processes equal to the number of processor cores for your computer. The number of Meshing Processes is limited to 4 if you are using Ansys Student. To check the number of physical cores, press the Ctrl + Shift + Esc keys simultaneously to open the Task Manager. Go to the Performance tab and select CPU from the left column. You'll see the number of physical cores on the bottom-right side. Close the Task Manager window. Click on the Start button in the Fluent Launcher window.

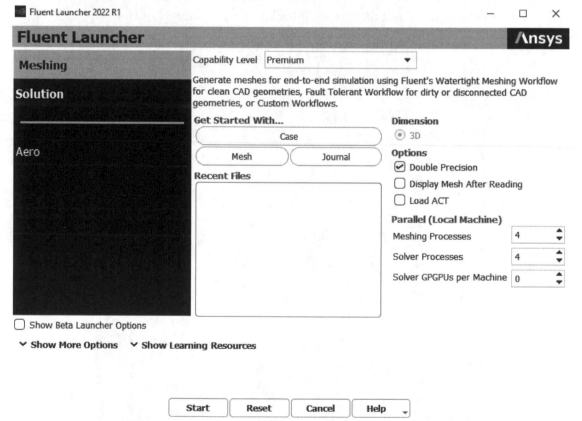

Figure 26.2 Fluent launcher window

D. Ansys Fluent Meshing

3. Select Workflow Type as Fault-Tolerant Meshing.

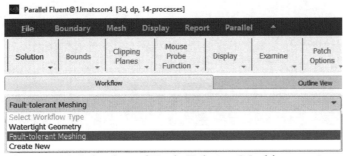

Figure 26.3 Selection of Fault-Tolerant Meshing

4. Select Import CAD and Part Management. Select *mm* as Display Unit. Browse for the CAD File. Select Other CAD Formats (*.jt *.plmxml *.stl, ...) as Files of type. Open the file ***Volvik Pro Bismuth 18 scans.stl***. Select *mm* as File Unit. Select Custom as Create Meshing Objects. Select Load and check the box next to CAD Model. Right-click on Meshing Model and select Create Object. Right-click on Object and select Rename. Change the name to *golf-ball*. Drag and drop the part "Volvik Pro Bismuth 18 scans" from CAD Model to *golf-ball* under Meshing Model. Select Create Meshing Objects.

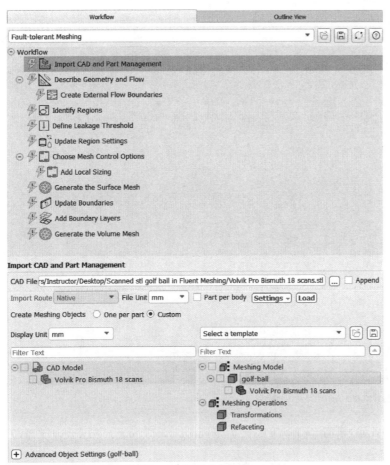

Figure 26.4 Import CAD and Part Management

5. Select External flow around an object as Flow Type under Describe Geometry and Flow. Select Yes for Add an enclosure? and Yes for Add local refinement regions? Open Advanced Options and select Yes for Identify regions? and Yes for Close leakages? Select Describe Geometry and Flow.

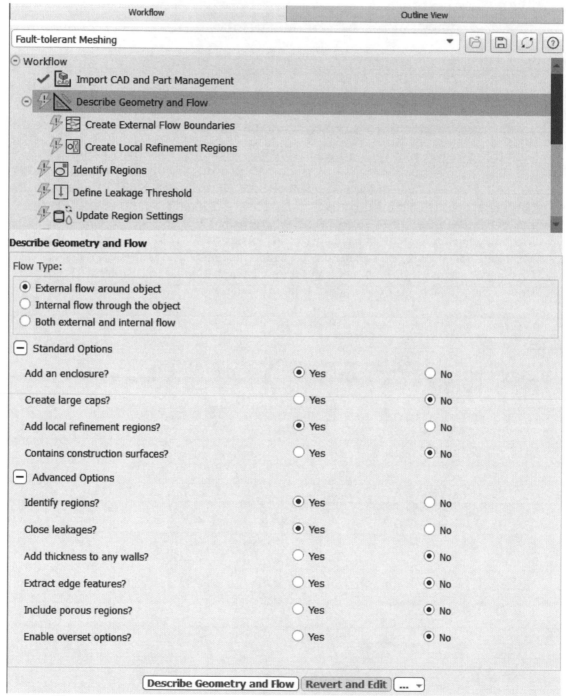

Figure 26.5 Describe geometry and flow

6. Choose the settings as shown in Figure 26.6 including Box Parameters to Create External Flow Boundaries. Set X Min, X Max to 2,4, Y Min, Y Max to 2,2 and Z Min, Z Max to 2,2. Select the *golf_ball* and select Create External Flow Boundaries in the lower left corner.

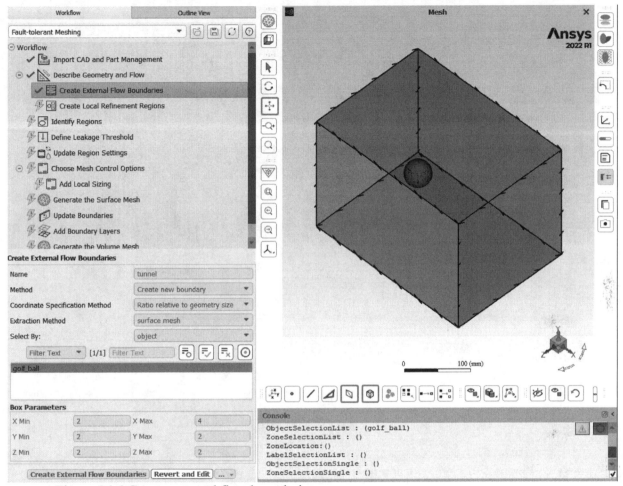

Figure 26.6 Create external flow boundaries

7. Choose the settings as shown in Figure 26.7 to Create Local Refinement Regions. Select Offset Box as Type and set Wake Levels to 2. Select the *golf_ball* and select Create Local Refinement Regions.

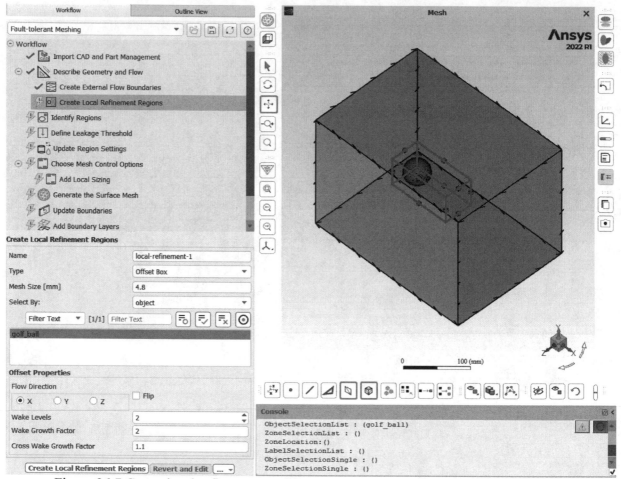

Figure 26.7 Create local refinement regions

8. Select Identify Regions in the Workflow. Choose the settings as shown in Figure 26.8. Select *tunnel* and select Identify Region in the lower left corner.

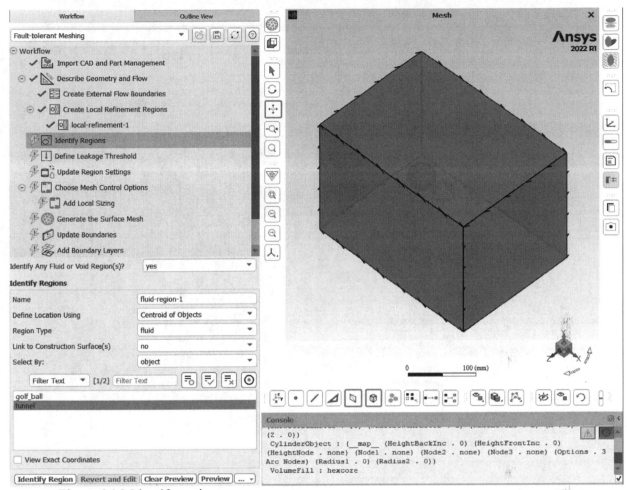

Figure 26.8 Identify regions

9. Select Define Leakage Threshold in the Workflow. Choose the settings as shown in Figure 26.9. Select Yes for Would you like to close any leakages for your region(s)?

Select *fluid-region-1* and select Preview Leakages. Select Set View 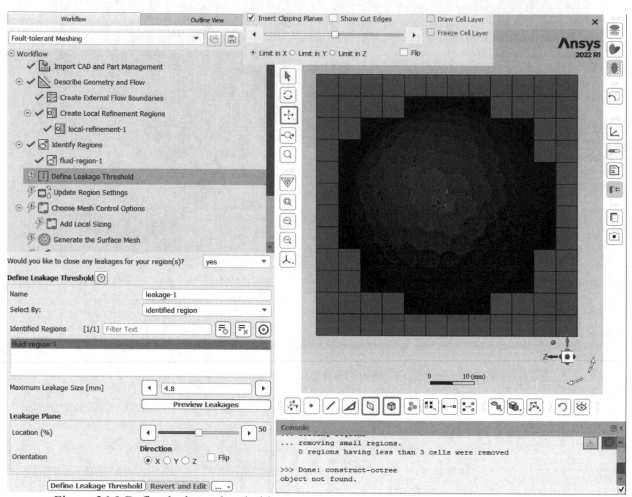 From +X Direction in the graphics window. Select Clipping Planes under the menu and check the box for Insert Clipping Planes. Select Define Leakage Threshold in the lower left corner.

Figure 26.9 Define leakage threshold

10. Select Update Region Settings in the Workflow. Choose the settings as shown in Figure 26.10 to Update Region Settings. Select Update Regions in the lower left corner.

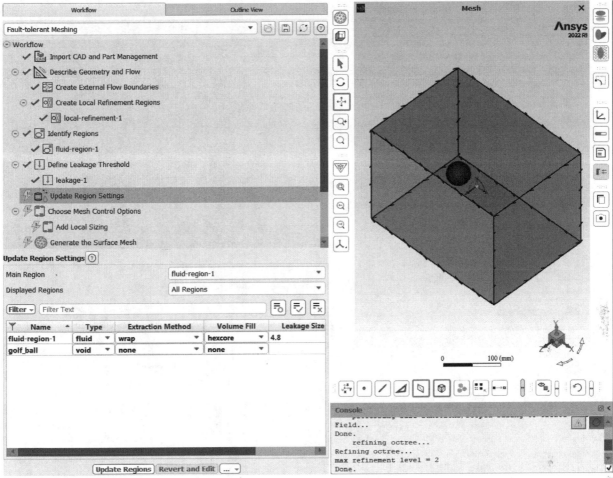

Figure 26.10 Update region settings

11. Choose the settings as shown in Figure 26.11 to Choose Mesh Control Settings. Select Choose Options in the lower left corner.

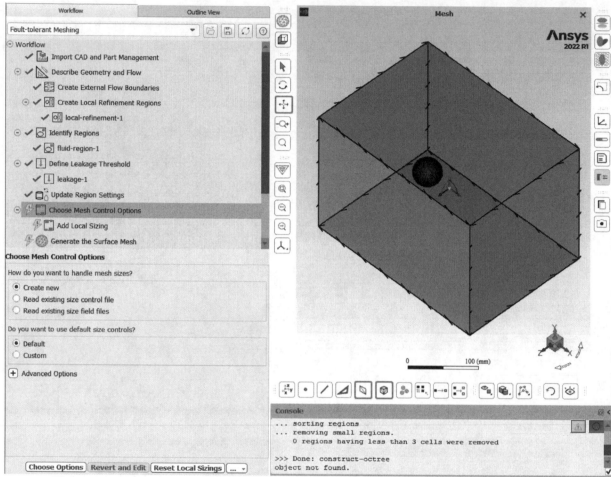

Figure 26.11 Choose mesh control options

12. Choose the settings as shown in Figure 26.12 to Add Local Sizing. Select Add Local Sizing under Choose Mesh Control Options in the Workflow. Select *soft* as Size Functions. Enter 0.4 as Maximum Size [mm]. Select Scope To Faces and select *golf_ball*. Select Add Local Sizing in the lower left corner.

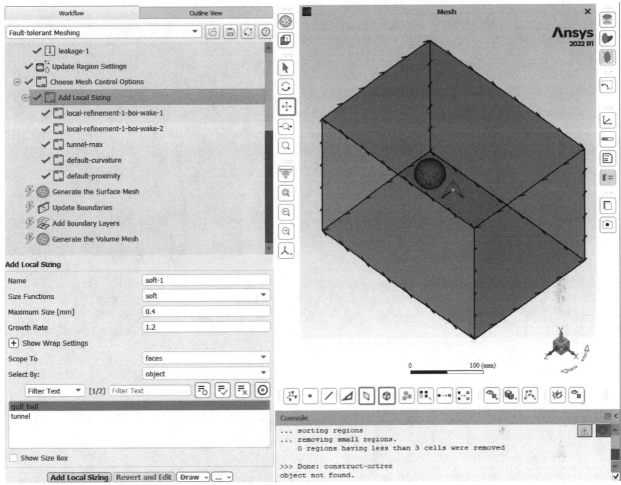

Figure 26.12 Choose mesh control options

13. Select Generate the Surface Mesh in the Workflow. Choose the settings as shown in Figure 26.13 to Generate the Surface Mesh. Open Advanced Options and select Yes for Auto Assign Zone Types? Set the Global Minimum [mm] to 0.2. Select Generate the Surface Mesh in the lower left corner.

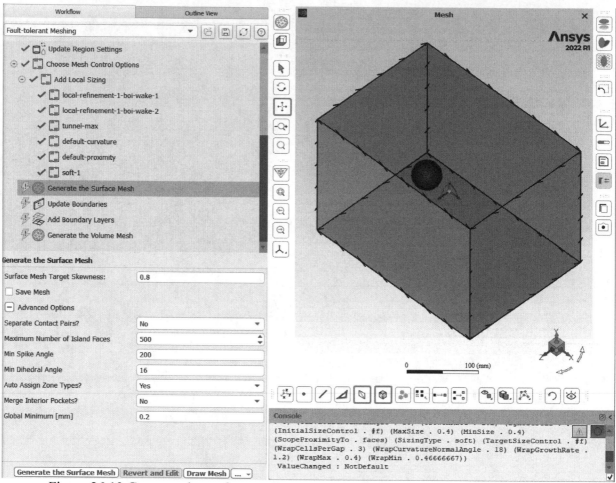

Figure 26.13 Generate the surface mesh

14. Right-click on Generate the Surface Mesh and select Insert Next Task>>Identify
    Deviated Surfaces. Choose the settings as shown in Figure 26.14 and select
    Deviation>>Compute. Select Deviation>>Draw Contour. Select Identify Deviated Faces
    in the lower left corner.

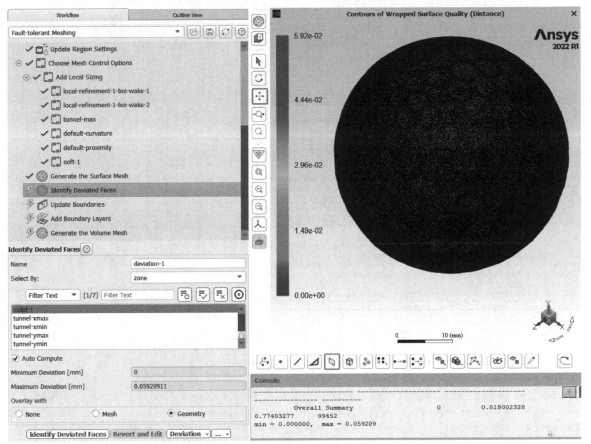

Figure 26.14 Identify Deviated Faces

15. Select Update Boundaries in the workflow. Choose the settings as shown in Figure 26.15. Select Update Boundaries in the lower left corner.

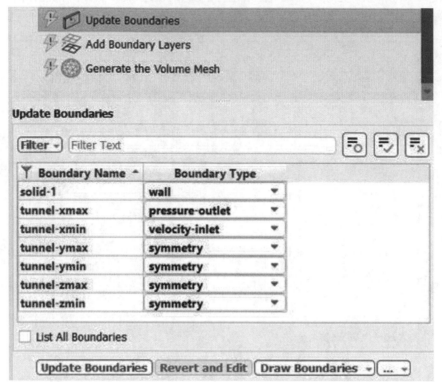

Figure 26.15 Update boundaries

16. Select Add Boundary Layers in the Workflow. Choose the settings as shown in Figure 26.16. Select Add Boundary Layers in the lower left corner.

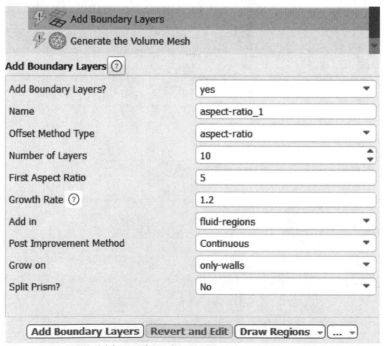

Figure 26.16 Add boundary layers

17. Select Generate the Volume Mesh in the Workflow. Choose the settings as shown in Figure 26.17a). Select Generate Volume Mesh in the lower left corner. Use the Clipping Planes to study the mesh. Select Switch to Solution in the upper left corner. Answer yes to the question that you get.

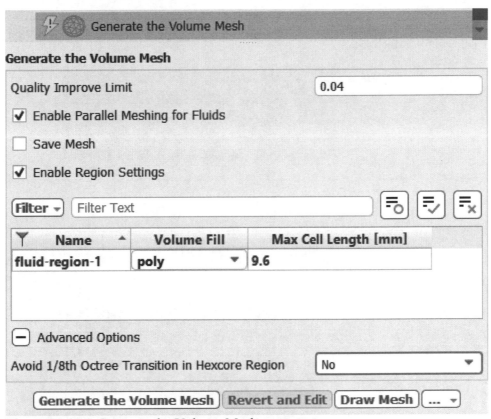

Figure 26.17a) Generate the Volume Mesh

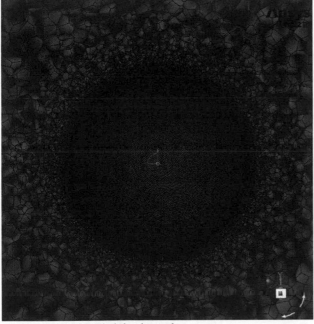

Figure 26.17b) Finished mesh

E. Ansys Fluent

18. Double click on General under Setup in the Outline View. Select Display… under Mesh on the Task Page. Select the settings as shown in Figure 26.18a) and click on Display.

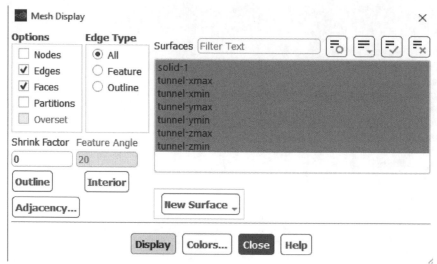

Figure 26.18a) Mesh display settings

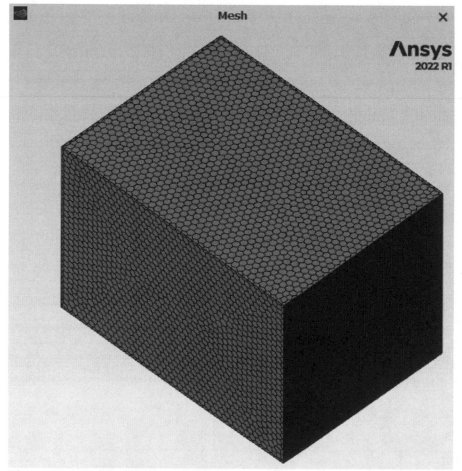

Figure 26.18b) Mesh in Ansys Fluent

19. Double click on Models under Setup in the Outline View. Double click on Viscous (SST k-omega) under Models. Select OK to close the window.

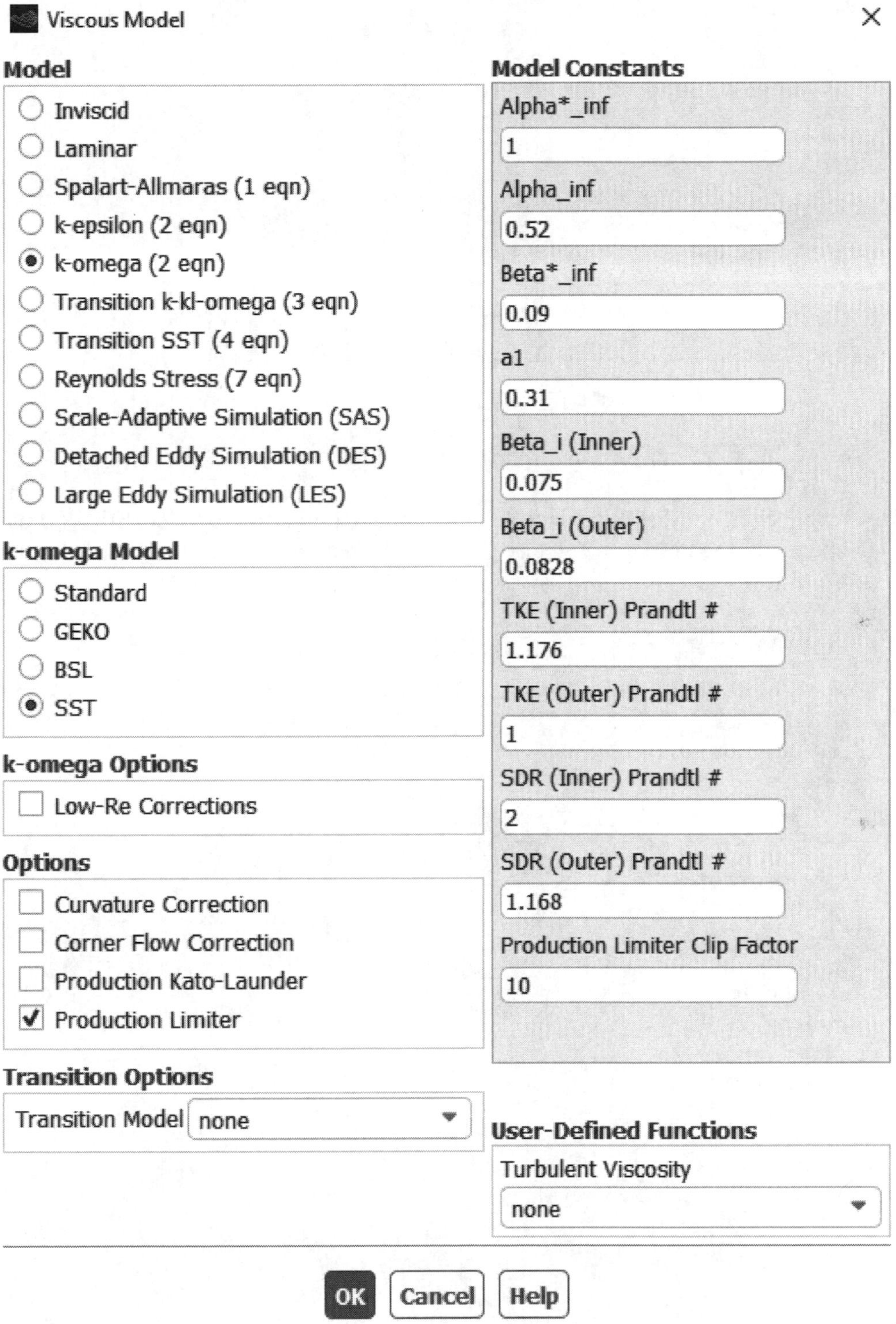

Viscous Model   ✕

Model

- ◯ Inviscid
- ◯ Laminar
- ◯ Spalart-Allmaras (1 eqn)
- ◯ k-epsilon (2 eqn)
- ⦿ k-omega (2 eqn)
- ◯ Transition k-kl-omega (3 eqn)
- ◯ Transition SST (4 eqn)
- ◯ Reynolds Stress (7 eqn)
- ◯ Scale-Adaptive Simulation (SAS)
- ◯ Detached Eddy Simulation (DES)
- ◯ Large Eddy Simulation (LES)

k-omega Model

- ◯ Standard
- ◯ GEKO
- ◯ BSL
- ⦿ SST

k-omega Options

- ☐ Low-Re Corrections

Options

- ☐ Curvature Correction
- ☐ Corner Flow Correction
- ☐ Production Kato-Launder
- ☑ Production Limiter

Transition Options

Transition Model | none ▾

Model Constants

Alpha*_inf
| 1 |

Alpha_inf
| 0.52 |

Beta*_inf
| 0.09 |

a1
| 0.31 |

Beta_i (Inner)
| 0.075 |

Beta_i (Outer)
| 0.0828 |

TKE (Inner) Prandtl #
| 1.176 |

TKE (Outer) Prandtl #
| 1 |

SDR (Inner) Prandtl #
| 2 |

SDR (Outer) Prandtl #
| 1.168 |

Production Limiter Clip Factor
| 10 |

User-Defined Functions

Turbulent Viscosity
| none ▾ |

OK   Cancel   Help

Figure 26.19 Viscous model in Ansys Fluent

20. Double click on Boundary Conditions and Inlet under Setup in the Outline View. Set the Velocity Magnitude [m/s] to 30. Select Apply and Close the Velocity Inlet window.

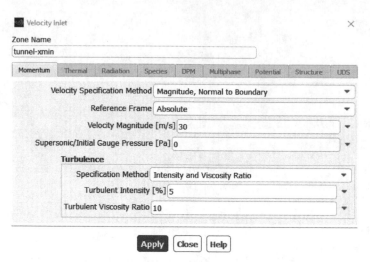

Figure 26.20 Velocity inlet settings in Ansys Fluent

21. Double click on Methods under Solution in the Outline View. Select SIMPLE as the Scheme for the Pressure-Velocity Coupling. Select the other settings as shown in Figure 26.21.

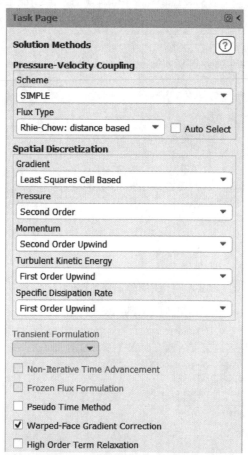

Figure 26.21 Solution methods

22. Double click on Projected Areas under Results and Reports in the Outline View. Select *solid-1* as the Surface and enter 0.000001 as Min Feature Size [m], select X as Projection Direction and click on Compute. The computed frontal Area [m²] is 0.0014424. Close the window.

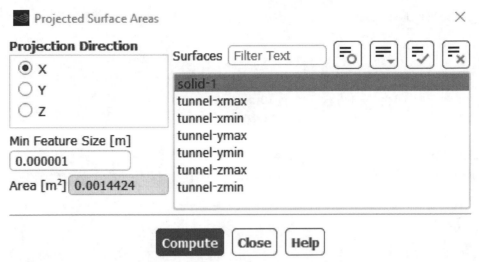

Figure 26.22 Final scaled residuals

23. Select the Physics tab from the menu and select Reference Values under Solver. Select Compute from *solid-1* and enter 0.0014424 as Area [m²]. Enter 30 as Velocity [m/s]. Select fluid-region-1 as Reference Zone.

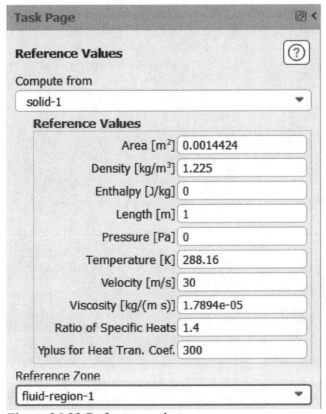

Figure 26.23 Reference values

24. Double-click on Report Definitions under Solution in the Outline View. Select New>>Force Report>>Drag... from the drop-down menu. Select *solid-1* under Zones. Check the box for Print to Console under Create. Select the OK button to close the Drag Report Definition window. Close the Report Definitions window.

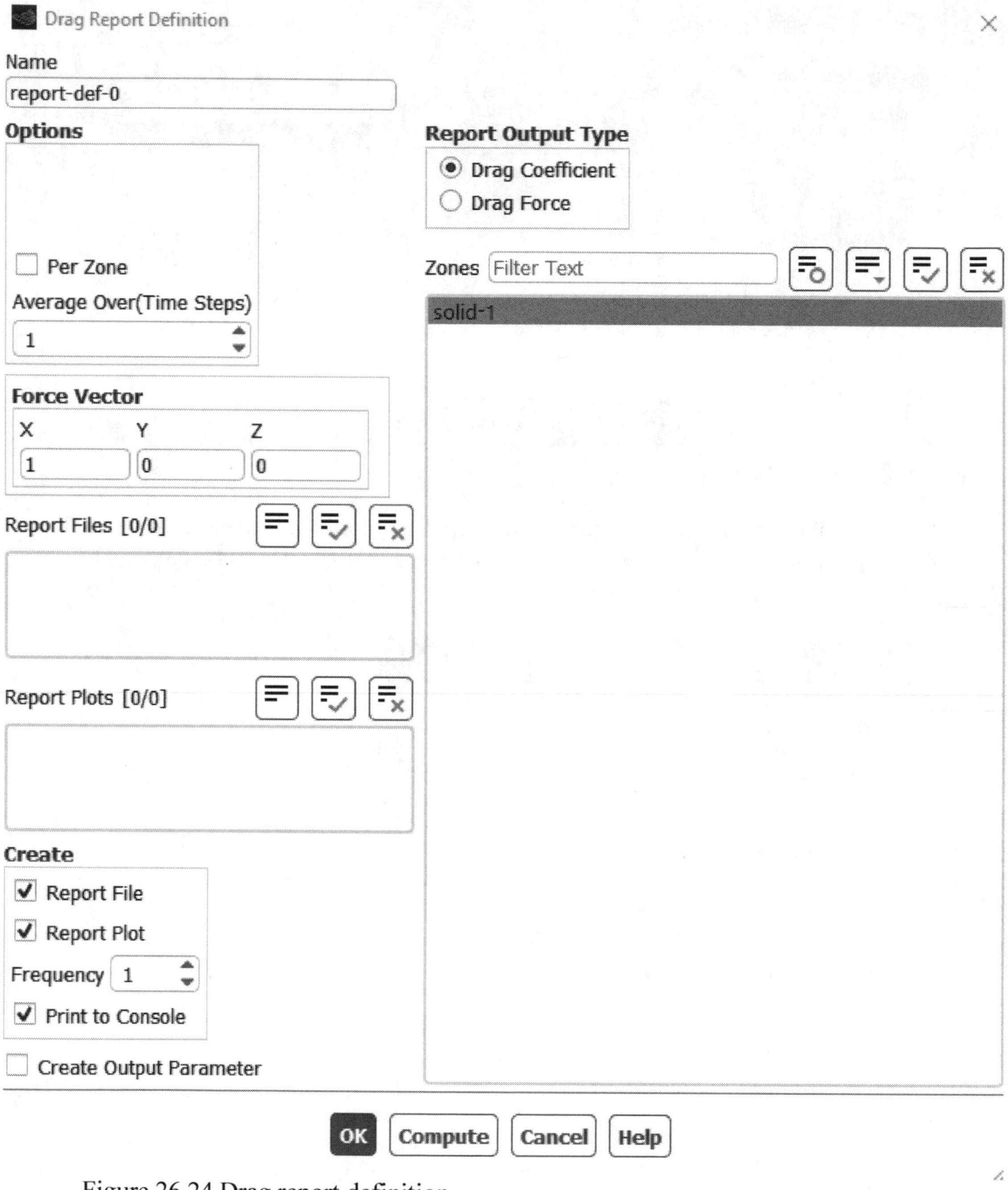

Figure 26.24 Drag report definition

25. Double click on Initialization under Solution in the Outline View. Select Hybrid Initializations and click on Initialize. Select File>>Write>>Case and enter the name *golf_ball_flow.cas.h5*

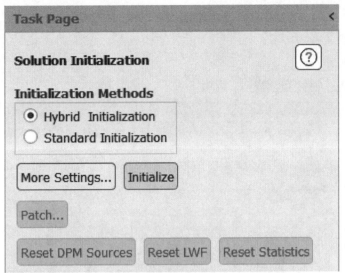

Figure 26.25 Solution initialization in Ansys Fluent

26. Double click on Run Calculation under Solution in the Outline View. Set the Number of Iterations under Parameters to 70.

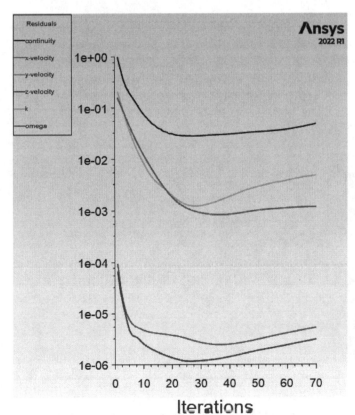

Figure 26.26 Residuals for the first 70 iterations

27. Double click on General under Setup in the Outline View. Select Transient under Time on the Task Page.

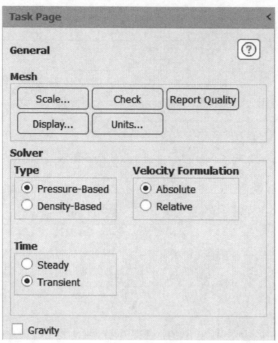

Figure 26.27 Transient simulation setting

28. Double click on Controls under Solution in the Outline View. Set the Under-Relaxation Factor for Momentum to 0.6.

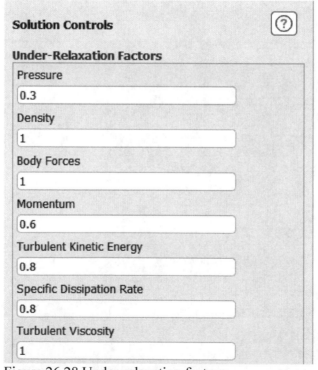

Figure 26.28 Under-relaxation factors

29. Double click on Methods under Solution in the Outline View. Set the Turbulent Kinetic Energy and Specific Dissipation Rate under Spatial Discretization to Second Order Upwind.

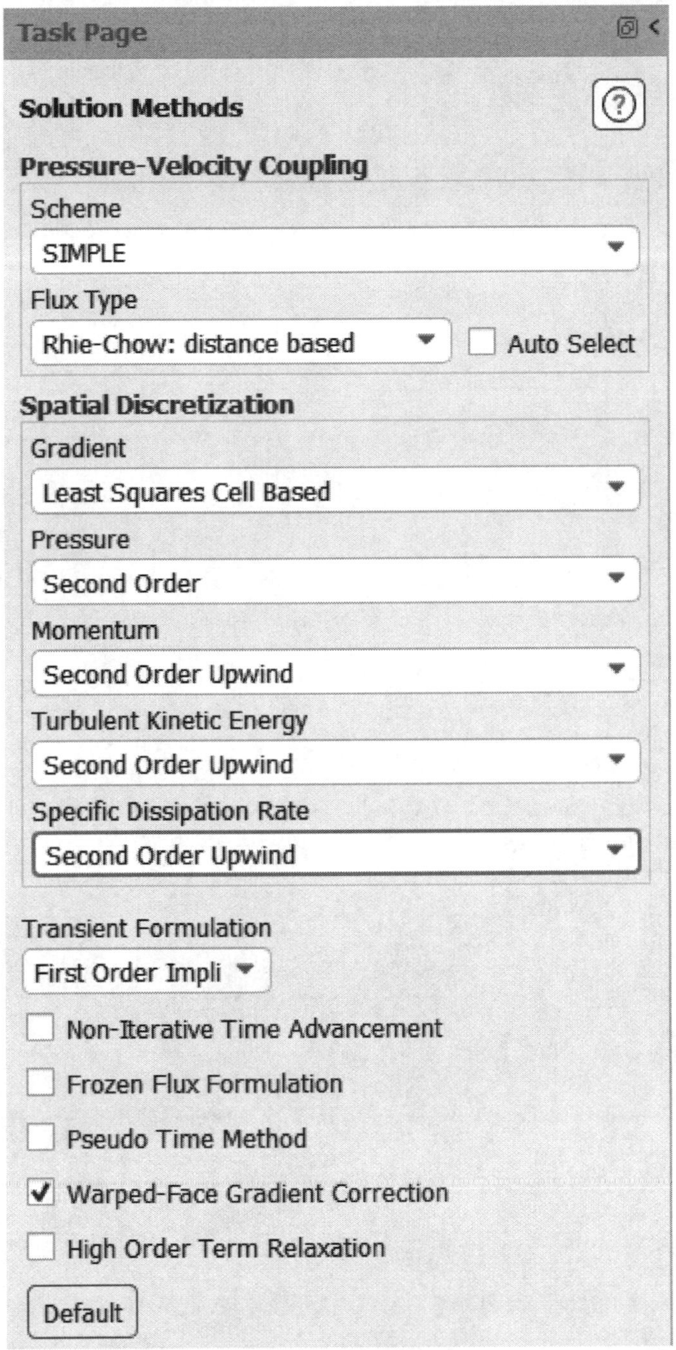

Figure 26.29 Solution method settings

30. Double click on Run Calculation. Set the Time Step Size [s] to 0.001. Set the Number of Time Steps to 200 and the Max Iterations/Time Step to 50. Select Calculate under Solution Advancement.

Figure 26.30a) Run calculation settings

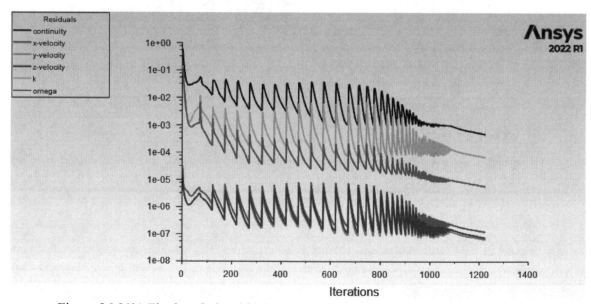

Figure 26.30b) Final scaled residuals

F. Post-Processing

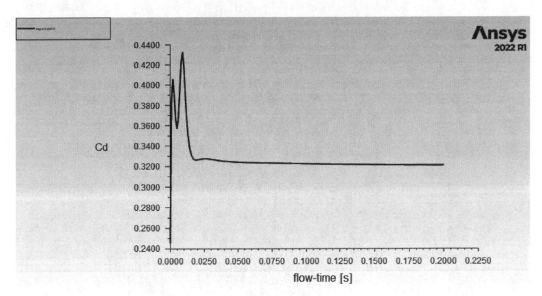

Figure 26.30c) Drag coefficient versus flow time

31. Double click on Reports under Results in the Outline View and double click on Forces under Reports. Select the *solid-1* zone and select the Print button.

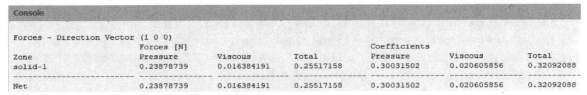

Figure 26.31 Drag coefficient from console

32. Double click on Graphics and Contours under Results in the Outline View. Select the settings as shown in Figure 26.32a) and click on Save/Display.

Figure 26.32a) Contours window settings for static pressure

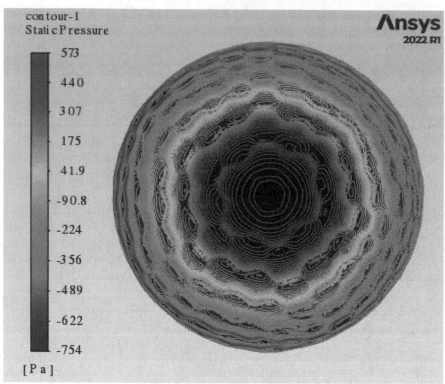

Figure 26.32b) Contours of static pressure for golf ball

33. Select Domain from the menu and select Create>>Plane…. Under Surface. Choose the settings as shown in Figure 26.33 and select Create. Close the Plane Surface window.

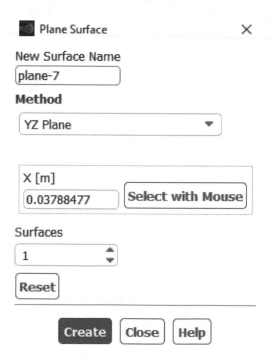

Figure 26.33 Creating a new plane

34. Double-click on Graphics and Pathlines under Results in the Outline View. Select the settings as shown in Figure 26.34a). Select *plane-7* and *solid-1* under Release from Surfaces and select Save/Display.

Figure 26.34a) Pathline settings

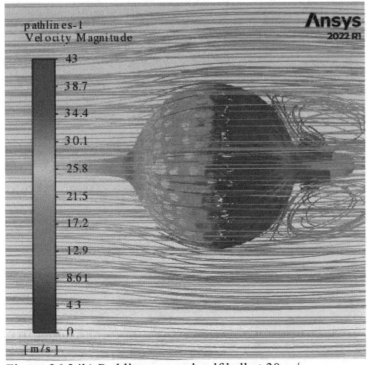

Figure 26.34b) Pathlines around golf ball at 30 m/s.

643

G. Batch Jobs

35. Double click on Monitors and Residual under Solution in the Outline View. Set the Absolute Criteria for all residuals to 0.0001. Select OK to close the Residual Monitors window.

You will need to create the case and scheme files to complete a batch job. Select File>>Write>>Case & Data... from the menu. Save the case and data files in the *Working Directory* folder with the name **volvik_golf_ball_flow.cas.h5**.

Open Notebook and create and save the scheme file with the name **volvik_golf_ball_scheme_file.scm** for the batch job. The scheme file is shown in Figure 26.35a). Select File>>Read>>Scheme... from the menu in Ansys Fluent. Open the scheme file to start the batch job.

Table 26.1 and Figure 26.35b) show the results from the batch job in comparison with the two models for smooth sphere (eq. 26.2) and golf ball (eq. 26.3) as described in the theory section. Figure 26.35b) also include results from experimental testing of different golf balls, see Jenkins *et al.* [1].

```
(Do ((x 100 (- x 5))) ((< x 5))
(begin
  (Ti-menu-load-string (format #f "file read-case volvik_golf_ball_flow.cas.h5"))
  (Ti-menu-load-string (format #f "define boundary-conditions velocity-inlet tunnel-xmin y y n ~a n 0 y n 1 n 0 n 0 n n y 5 10" x))
  (Ti-menu-load-string (format #f "report reference-values velocity ~a" x))
  (Ti-menu-load-string (format #f "solve set discretization-scheme k 0"))
  (Ti-menu-load-string (format #f "solve set discretization-scheme omega 0"))
  (Ti-menu-load-string (format #f "solve set under-relaxation mom 0.7"))
  (Ti-menu-load-string (format #f "solve set time-step 0.001"))
  (Ti-menu-load-string (format #f "solve initialize hyb-initialization yes"))
  (Ti-menu-load-string (format #f "solve dual-time-iterate 1 70"))
  (Ti-menu-load-string (format #f "solve set discretization-scheme k 1"))
  (Ti-menu-load-string (format #f "solve set discretization-scheme omega 1"))
  (Ti-menu-load-string (format #f "solve set under-relaxation mom 0.6"))
  (Ti-menu-load-string (format #f "solve set time-step 0.001"))
  (Ti-menu-load-string (format #f "solve dual-time-iterate 200 50"))
  (Ti-menu-load-string (format #f "report forces wall-forces y 1 0 0 y volvik_golf_ball_drag_coefficient_data_U=~amps" x))
  (Ti-menu-load-string (format #f "display objects display contour-1"))
  (Ti-menu-load-string (format #f "display views restore-view left"))
  (Ti-menu-load-string (format #f "display save-picture volvik_golf_ball_static_pressure_contour_plot_U=~amps.jpg" x))
  (Ti-menu-load-string (format #f "display objects display pathlines-1"))
  (Ti-menu-load-string (format #f "display views restore-view front"))
  (Ti-menu-load-string (format #f "display save-picture volvik_golf_ball_path_lines_U=~amps.jpg" x))
  (Ti-menu-load-string (format #f "plot residuals y y y y y y"))
  (Ti-menu-load-string (format #f "display save-picture volvik_golf_ball_residuals_plot_U=~amps.jpg" x))
  (Ti-menu-load-string (format #f "plot file report-def-0-rfile.out"))
  (Ti-menu-load-string (format #f "display save-picture volvik_golf_ball_drag_coefficient_plot_U=~amps.jpg" x))))
```

Figure 26.35a) Scheme file for golf ball batch job

U (m/s)	Re	Cd, eq. 26.3	Cd, Fluent
5	14669	0.454181	0.43352488
10	29337	0.484592	0.37572094
15	44006	0.424748	0.35089201
20	58675	0.245559	0.3410472
25	73344	0.234798	0.33262889
30	88012	0.233924	0.32674455
35	102681	0.236257	0.32149642
40	117350	0.238722	0.31080765
45	132019	0.241107	0.30696753
50	146687	0.243356	0.30392807
55	161356	0.245461	0.30117214
60	176025	0.247429	0.2984325
65	190694	0.249270	0.29505383
70	205362	0.250997	0.29156131
75	220031	0.252621	0.28907414
80	234700	0.254153	0.2878027
85	249368	0.255602	0.28638794
90	264037	0.256975	0.28588134
95	278706	0.258281	0.28454641
100	293375	0.259525	0.28261509

Table 26.1 Comparison between Ansys Fluent and model for drag crisis (equation 26.2)

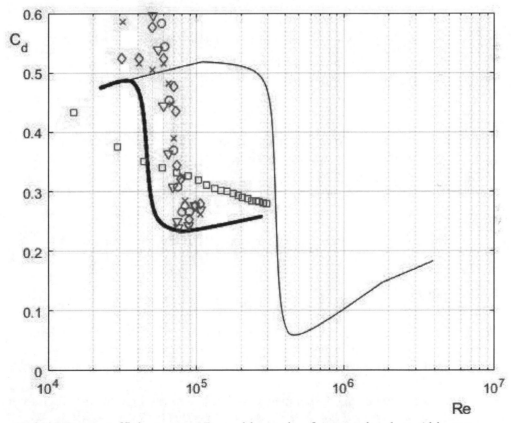

Figure 26.35b) Drag coefficient versus Reynolds number for smooth sphere (thin curve, model eq. 26.2), golf ball (thick curve, model eq. 26.3), □ (Ansys Fluent for Volvik Pro Bismuth), x (Tournament Plus, Exp.), o (Titlest, Exp.), ∇ (Callaway, Exp.), ◊ (TaylorMade, Exp.)

H. Theory

1. The drag coefficient from experiments for a smooth sphere has been found in a large Reynolds number range using the following curve fit-formula from Morrison [1]:

$$C_d = \frac{24}{Re} + \frac{2.6\left(\frac{Re}{5}\right)}{1+\left(\frac{Re}{5}\right)^{1.52}} + \frac{0.411\left(\frac{Re}{2.63\times10^5}\right)^{-7.94}}{1+\left(\frac{Re}{2.63\times10^5}\right)^{-8}} + \frac{0.25\left(\frac{Re}{10^6}\right)}{1+\left(\frac{Re}{10^6}\right)} \qquad 0.1 < Re < 10^6 \qquad (26.1)$$

The model for the drag crisis for a smooth sphere based on Achenbach [3] can be determined as

$$C_d = \frac{280}{10^3} - \frac{159}{10^3} atan\left[\frac{3}{50}\left(\frac{Re}{10^3} - 340\right)\right] + \frac{1}{30} log\left(\frac{Re}{11\cdot10^4}\right)[1 + \tanh(11 \cdot 10^4 - Re)] +$$
$$+ \frac{1}{11} log\left(\frac{Re}{42\cdot10^4}\right)[1 + \tanh(Re - 42 \cdot 10^4)] - \frac{1}{28} log\left(\frac{Re}{18\cdot10^5}\right)[1 + \tanh(Re - 18 \cdot 10^5)]$$
$$2 \cdot 10^4 < Re < 4 \cdot 10^6 \qquad (26.2)$$

The model for the drag crisis for a golf ball based on Bearman and Harvey [2] is

$$C_d = \frac{369}{10^3} - \frac{183}{2\cdot10^3} atan\left[\frac{7}{20}\left(\frac{Re}{10^3} - 46\right)\right] + \frac{1}{34} log\left(\frac{Re}{8\cdot10^4}\right)[1 + \tanh(Re - 8 \cdot 10^4)] +$$
$$+ \frac{1}{16} log\left(\frac{Re}{37\cdot10^3}\right)[1 + \tanh(37 \cdot 10^3 - Re)] \qquad 2 \cdot 10^4 < Re < 3 \cdot 10^5 \qquad (26.3)$$

We define the Reynolds number as

$$Re = \frac{UD\rho}{\mu} \qquad (26.4)$$

where U is free stream velocity, D is sphere or golf ball diameter, ρ is density and μ is dynamic viscosity. The drag coefficient is defined as

$$C_d = \frac{F_d}{\frac{1}{2}\rho U^2 A} \qquad (26.5)$$

where F_d is drag force and A is frontal area of the sphere or golf ball.

I. References

1. P. E. Jenkins, J. Arellano, M. Ross and M. Snell, "Drag Coefficients of Golf Balls", *World Journal of Mechanics*, **8**, 236-241, (2018).
2. F. A. Morrison, "An Introduction to Fluid Mechanics", *Cambridge University Press*, (2013).
3. E. Achenbach, "Experiments on the flow past spheres at very high Reynolds numbers", *J. Fluid Mech.*, **54** (3), 565-575, (1972).
4. P.W. Bearman and J.K. Harvey, Golf Ball Aerodynamics, *Aeronautical Quarterly*, **27**, 112-122, (1976).

J. Exercise

26.1. Run Ansys Fluent calculations shown in this chapter for different velocities U (m/s) and compare your static pressures, pathlines and drag coefficient results as shown in Table 26.2. Determine the percent error between the golf ball model as shown in eq. 26.3 and Ansys Fluent results for the drag coefficient.

Student	U (m/s)	Re	Cd, eq. 26.3	Cd, Fluent	Percent Difference
A	7.5				
B	12.5				
C	17.5				
D	22.5				
E	27.5				
F	32.5				
G	37.5				
H	42.5				
I	52.5				
J	57.5				
K	62.5				
L	67.5				
M	72.5				
N	77.5				
O	82.5				
P	87.5				
Q	92.5				
R	97.5				

Table 26.2 Free stream velocity, Reynolds number, drag coefficients and percentage error for 18 different velocities

26.2. Run Ansys Fluent calculations shown in this chapter for the golf balls as shown in Table 26.3 and 26.4. Compare the static pressure, pathlines and drag coefficient results with this chapter results. Determine the percent error between the golf ball model as shown in eq. 26.3 and Ansys Fluent drag coefficients for the different commercial golf balls at a free stream velocity of $U = 28$ m/s.

Manufacturer	Model	Design	Number of Dimples
Callaway	ERC SOFT	Hex	332
Mizuno	RD 566	D micro dimple	566
Volvik	ProBismuth	Octahedron	446
Wilson Staff	50 ELITE	--------	302

Table 26.3 Model, design and number of dimples for different commercial golf balls

Manufacturer	A (m²)	Re	Cd, eq. 26.3	Cd, Fluent	Percent Difference
Callaway					
Mizuno					
Volvik	0.0014423734				
Wilson Staff					

Table 26.4 Frontal area, Reynolds number, golf ball model drag coefficient, Ansys Fluent drag coefficient and percentage error for different commercial golf balls at $U = 28$ m/s.

K. Notes

CHAPTER 27. MODEL CAR: AUDI R8

A. Objectives

- Using Ansys Fluent to Study the Flow around an Audi R8
- Loading the *stl* file for the Audi R8
- Using Fault-Tolerant Meshing in Ansys Fluent Meshing
- Inserting Boundary Conditions
- Running Ansys Fluent Simulations
- Using Contour Plots for Visualizations of Velocity and Pressure Fields

B. Problem Description

We will study the flow around an Audi R8 model car.

C. Launching Ansys Fluent

1. Start by launching Ansys Fluent in standalone mode.

Figure 27.1 Launching ANSYS Fluent

2. Check the box for Double Precision. Select Meshing in the upper left portion of the Fluent Launcher window. Click on Show More Options and write down the location of your *Working Directory*. You will need this information later. Set the number of Meshing and Solver Processes equal to the number of processor cores for your computer. The number of Meshing Processes is limited to 4 if you are using Ansys Student. To check the number of physical cores, press the Ctrl + Shift + Esc keys simultaneously to open the Task Manager. Go to the Performance tab and select CPU from the left column. You'll see the number of physical cores on the bottom-right side. Close the Task Manager window. Click on the Start button in the Fluent Launcher window.

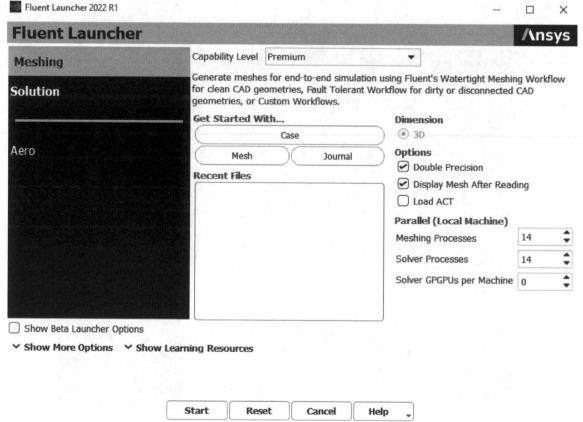

Figure 27.2 Fluent launcher window

D. Ansys Fluent Meshing

3. Select Workflow Type as Fault-Tolerant Meshing.

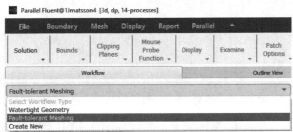

Figure 27.3 Selection of Fault-Tolerant Meshing

4. Select Import CAD and Part Management. Select *mm* as Display Unit. Browse for the CAD File. Select Other CAD Formats (*.jt *.plmxml *.stl) as Files of type. Open the file *Car Scan 2 Rotated Final.stl*. Select *mm* as File Unit. Select Custom as Create Meshing Objects. Select Load and check the box next to CAD Model. Right-click on Meshing Model and select Create Object. Right-click on Object and select Rename. Change the name to *Audi-R8*. Drag and drop the part *Car Scan 2 Rotated Final* from CAD Model to *Audi-R8* under Meshing Model. Select Create Meshing Objects.

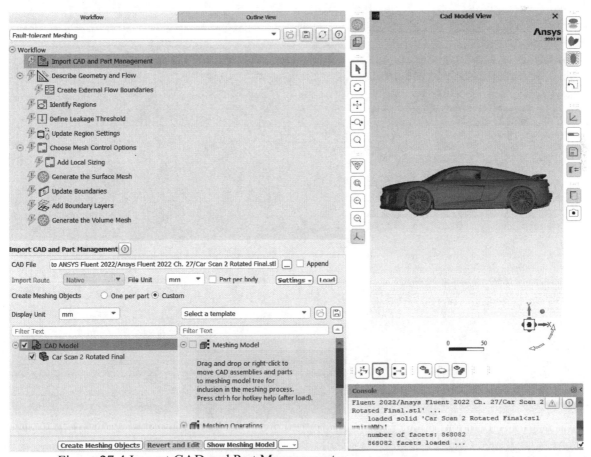

Figure 27.4 Import CAD and Part Management

5. Select External flow around object as Flow Type under Describe Geometry and Flow. Select Yes for Add an enclosure? and Yes for Add local refinement regions? Open Advanced Options and select Yes for Identify regions?, Yes for Close leakages? and Yes for Extract Edge Features. Select Describe Geometry and Flow.

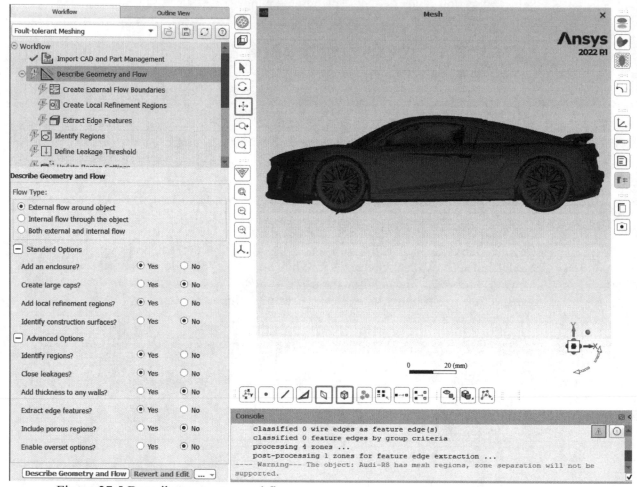

Figure 27.5 Describe geometry and flow

6.   Choose the settings as shown in Figure 27.6a) to Create External Flow Boundaries. Select Create External Flow Boundaries in the lower left corner.

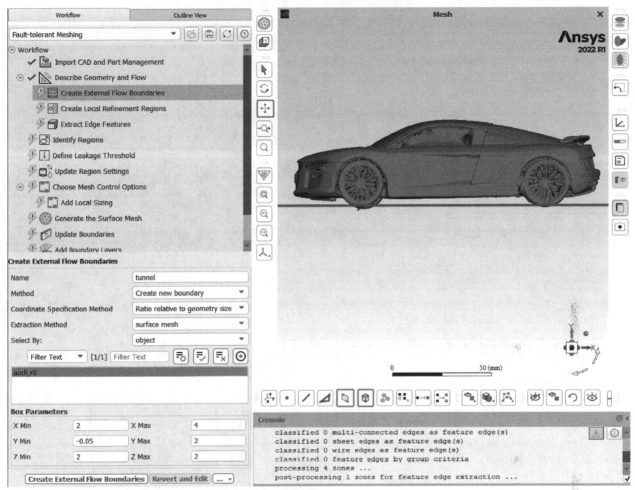

Figure 27.6a) Create external flow boundaries

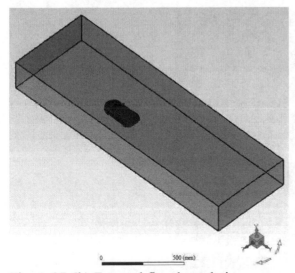

Figure 27.6b) External flow boundaries

7. Choose the settings as shown in Figure 27.7a) to Create Local Refinement Regions. Select Create Local Refinement Regions in the lower left corner.

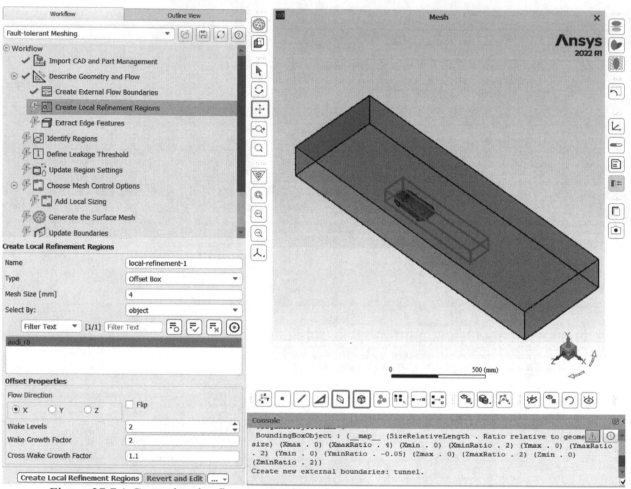

Figure 27.7a) Create local refinement regions

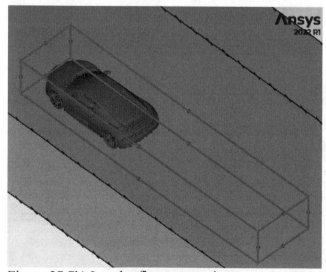

Figure 27.7b) Local refinement region

Select Extract Edge Features in the Workflow. Choose the settings as shown in Figure 26.7c) to Extract Edge Features. Select Extract Edge Features in the lower left corner.

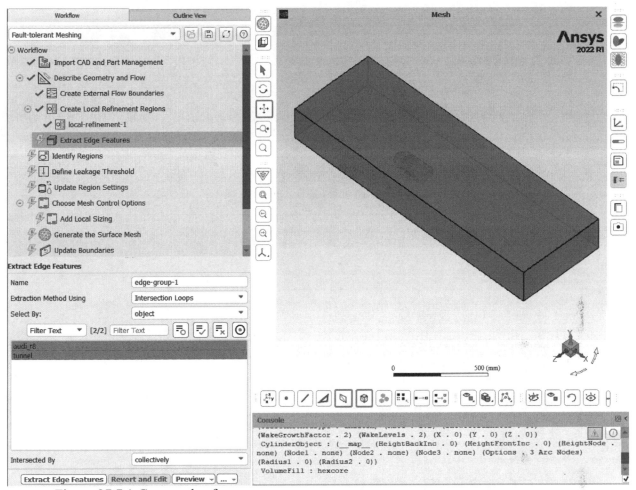

Figure 27.7c) Create edge features

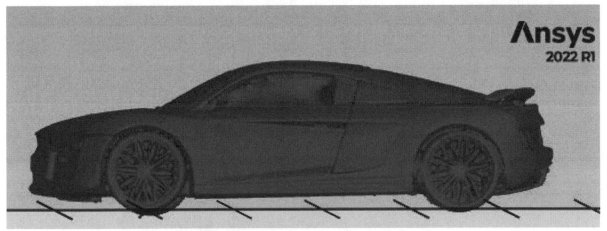

Figure 27.7d) Edge features where tires intersect the ground

8. Select Identify Regions in the Workflow. Choose the settings as shown in Figure 27.8 to Identify Regions. Select Identify Region in the lower left corner.

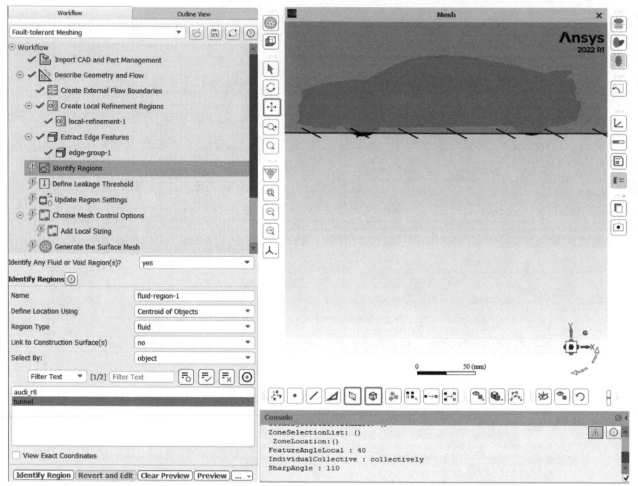

Figure 27.8 Identify regions

9. Select Define Leakage Threshold in the Workflow. Set Would you like to close any leakages for your region(s)? to Yes. Choose the settings as shown in Figure 27.9 to Define Leakage Threshold. Set Maximum Leakage Size [mm] to 4. Select Clipping Planes in the menu. Check the box for Insert Clipping Planes and then uncheck the same box. Select Define Leakage Threshold in the lower left corner.

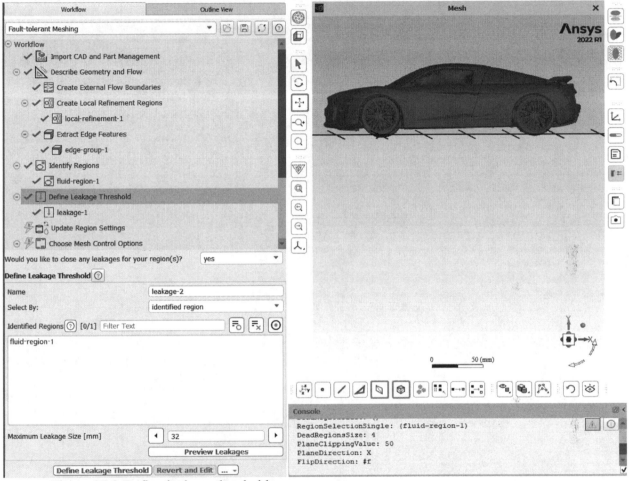

Figure 27.9 Define leakage threshold

10. Select Update Region Settings in the Workflow. Choose the settings as shown in Figure 27.10 to Update Region Settings. Select Update Regions in the lower left corner.

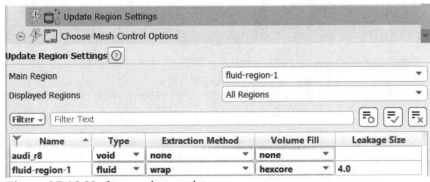

Figure 27.10 Update region settings

11. Choose the settings as shown in Figure 27.11 to Choose Mesh Control Settings. Select Choose Options in the lower left corner.

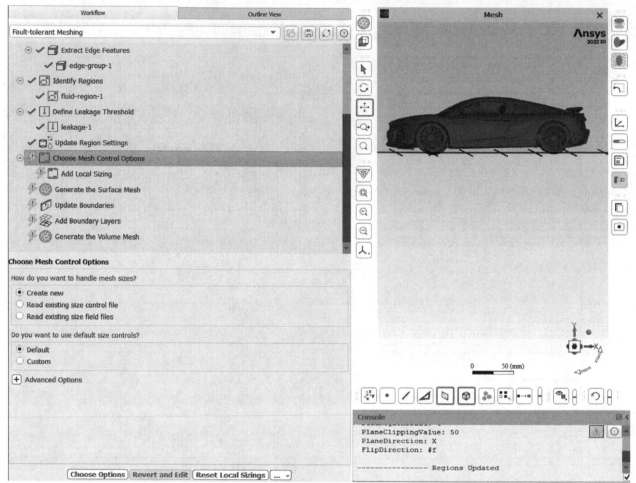

Figure 27.11 Choose mesh control options

12. Select default-curvature under Add Local Sizing in the Workflow. Select Revert and Edit. Choose the settings as shown in Figure 27.12a) to decrease the Minimum Size to 1. Select Update.

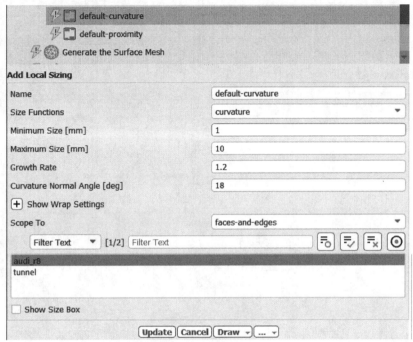

Figure 27.12a) Settings for default-curvature

Select default-proximity under Add Local Sizing in the Workflow. Choose the settings as shown in Figure 27.12b) to decrease the Minimum Size to 1. Select Update.

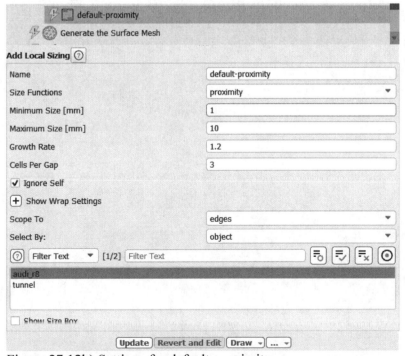

Figure 27.12b) Settings for default-proximity

Select Add Local Sizing in the Workflow. Choose the settings as shown in Figure 27.12c) to add a soft size control. Select Add Local Sizing in the lower left corner.

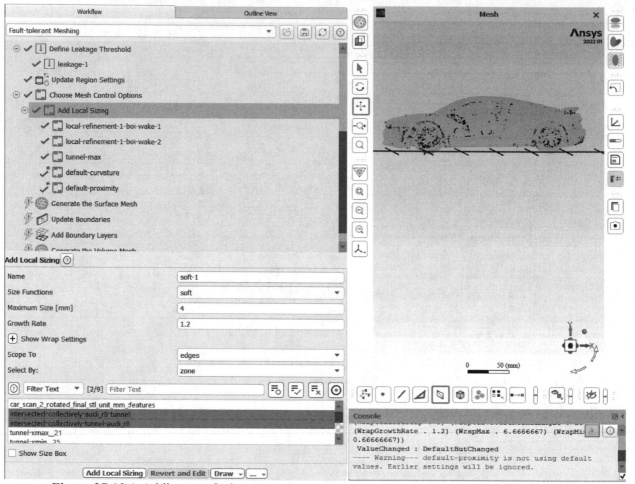

Figure 27.12c) Adding a soft size control

13. Select Generate the Surface Mesh in Workflow. Choose the settings as shown in Figure 27.13a) to Generate the Surface Mesh. Select Generate the Surface Mesh in the lower left corner.

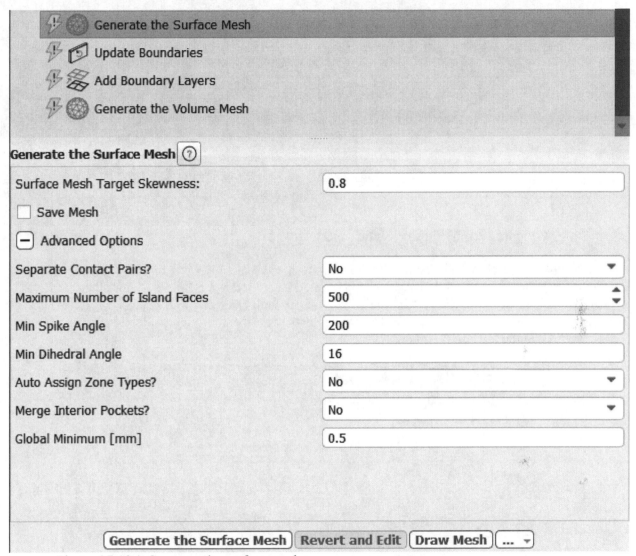

Figure 27.13a) Generate the surface mesh

Select Clipping Planes and check the box for Insert Clipping Planes. Select Limit in Y. Drag the slider to the left and right to adjust the clipping plane.

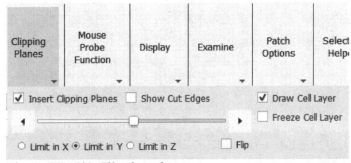

Figure 27.13b) Clipping planes

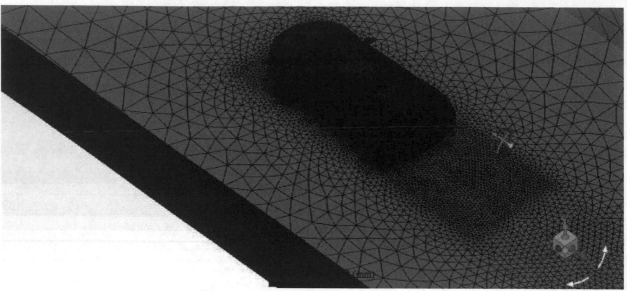

Figure 27.13b) Finished surface mesh

14. Right-click on Generate the Surface Mesh and select Insert Next Task>>Identify Deviated Faces. Select Identify Deviated Faces in the Workflow. Choose the settings as shown in Figure 27.14a) including Overlay with Geometry and select Deviation>>Compute. Select Deviation>>Draw Contour. Select Identify Deviated Faces in the lower left corner.

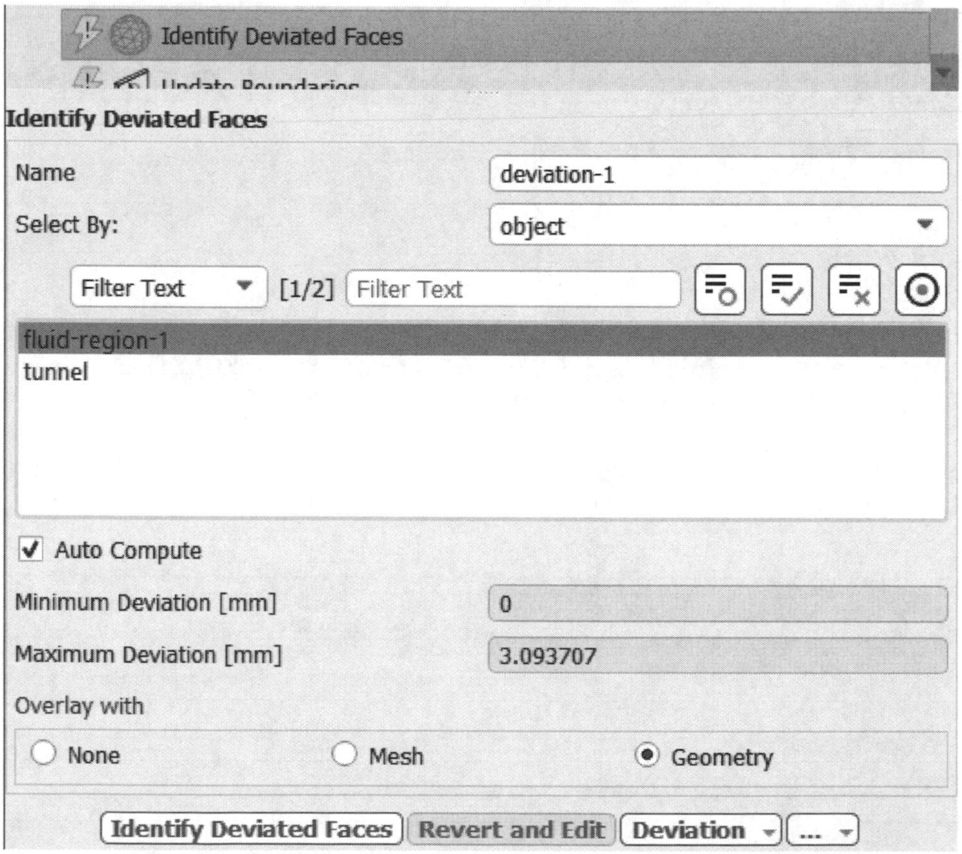

Figure 26.14a) Identify Deviated Faces

15. Select Update Boundaries in the Workflow. Choose the settings as shown in Figure 27.15. Select Update Boundaries in the lower left corner.

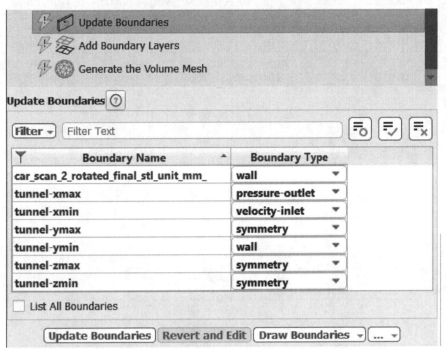

Figure 27.15 Update boundaries

16. Select Add Boundary Layers in the Workflow. Choose the settings as shown in Figure 27.16 including setting the Number of Layers to 10. Select Add Boundary Layers in the lower left corner.

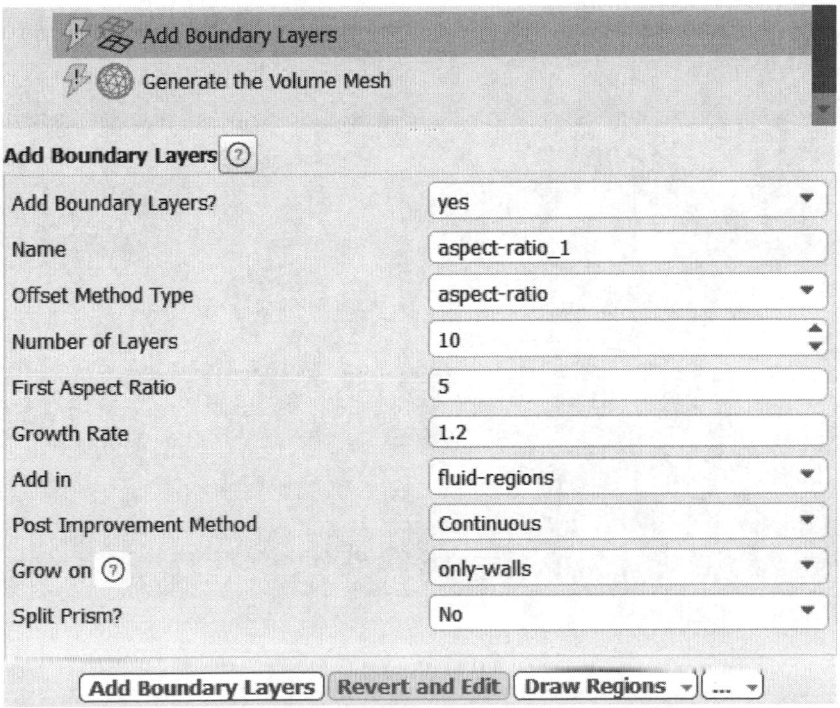

Figure 26.16 Add boundary layers

17. Select Generate the Volume Mesh in the Workflow. Choose the settings as shown in Figure 27.17 including setting the Max Cell Length to 4. This is completed by right-clicking on Max Cell Length [mm] and selecting Set Max Cell Length [mm] and entering 4 in the blank region. Select Generate Volume Mesh in the lower left corner.

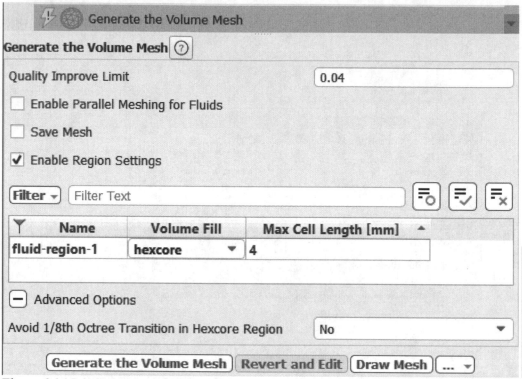

Figure 26.17a) Generate the Volume Mesh

Select Clipping Planes and check the box for Insert Clipping Planes. Select Limit in Z. Drag the slider to the left and right to adjust the clipping plane.

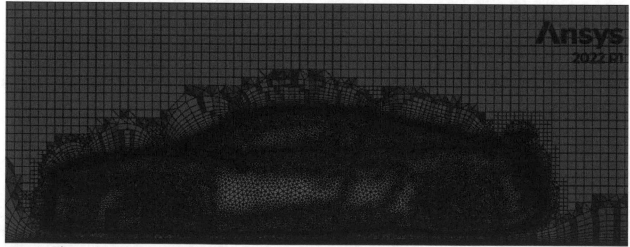

Figure 27.17b) Finished mesh

Select Mesh >> Check from the menu. Select Mesh >> Check Quality from the menu. Select File >> Write >> Mesh… from the menu. Save the mesh with the name *Audi_R8.msh.h5*. Select File >> Write >> Case… from the menu. Save the cas file with the name *Audi_R8.msh.cas.h5*. Select Solution >> Switch to Solution in the upper left corner. Answer yes to the question that you get.

```
Console

Domain extents.
  x-coordinate: min = -6.408524e+02, max = 1.077853e+03.
  y-coordinate: min = 8.288524e+00, max = 2.299090e+02.
  z-coordinate: min = 3.526365e+02, max = 9.078843e+02.
Volume statistics.
  minimum volume: 3.307817e-04.
  maximum volume: 3.923172e+04.
    total volume: 1.999019e+08.
Face area statistics.
Warning: 6 faces have warp greater than or equal to 0.5.

    minimum face area: 1.103575e-04.
    maximum face area: 1.870957e+03.
    average face area: 8.905707e+00.
Checking number of nodes per edge.
Checking number of nodes per face.
Checking number of nodes per cell.
Checking number of faces/neighbors per cell.
Checking cell faces/neighbors.
Checking isolated cells.
Checking face handedness.
Checking periodic face pairs.
Checking face children.
Checking face zone boundary conditions.
Checking for invalid node coordinates.
Checking poly cells.
Checking zones.
Checking neighborhood.
Checking interfaces.
Checking modified centroid.
Checking non-positive or too small area.
Done.
```

Figure 27.17c) Console output from checking the mesh

```
Mesh Quality:

Minimum Orthogonal Quality =  5.16854e-03
Warning: minimum Orthogonal Quality below 0.01.

Maximum Aspect Ratio =  2.98404e+02
```

Figure 27.17d) Console output from checking mesh quality

E. Ansys Fluent

18. Double click on General under Setup in the Outline View. Select Display… under Mesh on the Task Page. Select the settings as shown in Figure 27.18a) and click on Display. Close the Mesh Display window.

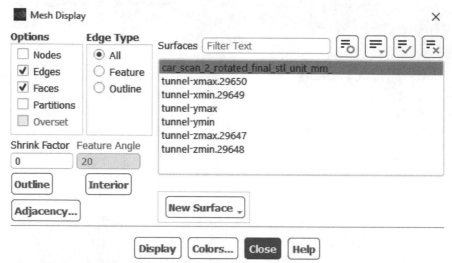

Figure 27.18a) Mesh display settings

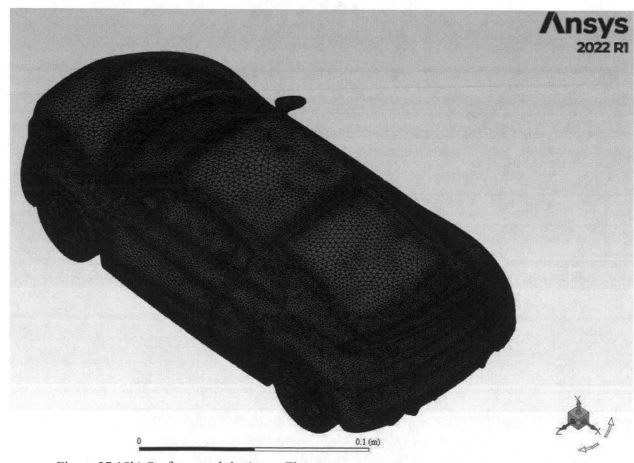

Figure 27.18b) Surface mesh in Ansys Fluent

19. Double click on Models under Setup in the Outline View. Double click on Viscous (SST k-omega) under Models. Select OK to close the window.

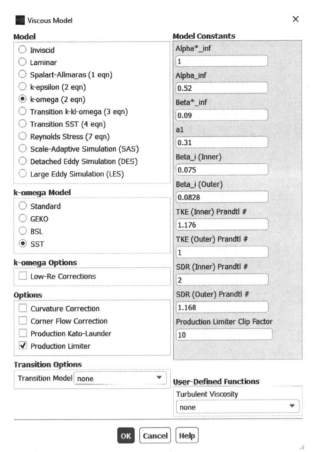

Figure 27.19 Viscous model in Ansys Fluent

20. Double click on Boundary Conditions and Inlet under Setup in the Outline View. Set the Velocity Magnitude [m/s] to 30. Select Apply and Close the Velocity Inlet window.

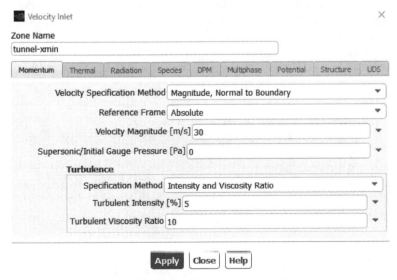

Figure 27.20 Velocity inlet settings in Ansys Fluent

21. Double click on Methods under Solution in the Outline View. Select SIMPLE as the Scheme for the Pressure-Velocity Coupling. Select the other settings as shown in Figure 27.21.

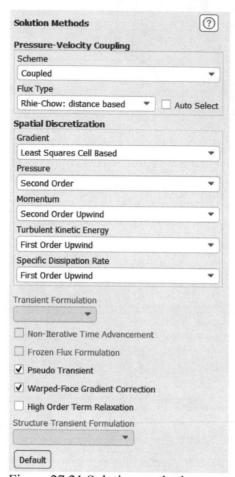

Figure 27.21 Solution methods

22. Double click on Projected Areas under Results and Reports in the Outline View. Select *car_scan_2_rotated_final_stl_unit_mm_* as the Surface and enter 0.000005 as Min Feature Size [m], select X as Projection Direction and click on Compute. The computed frontal Area [m²] is 0.00616141. Close the window.

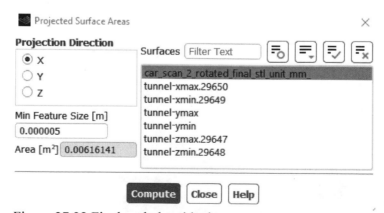

Figure 27.22 Final scaled residuals

23. Select the Physics tab from the menu and select Reference Values under Solver. Select Compute from *car_scan_2_rotated_final_stl_unit_mm_* and enter 0.00616141 as Area [m^2]. Enter 30 as Velocity [m/s]. Select fluid-region-1 as Reference Zone.

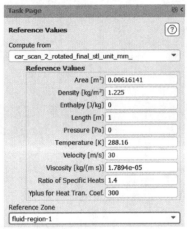

Figure 27.23 Reference values

24. Double-click on Report Definitions under Solution in the Outline View. Select New>>Force Report>>Drag... from the drop-down menu. Select *car_scan_2_rotated_final_stl_unit_mm_* under Zones. Check the box for Print to Console under Create. Select the OK button to close the Drag Report Definition window. Close the Report Definitions window.

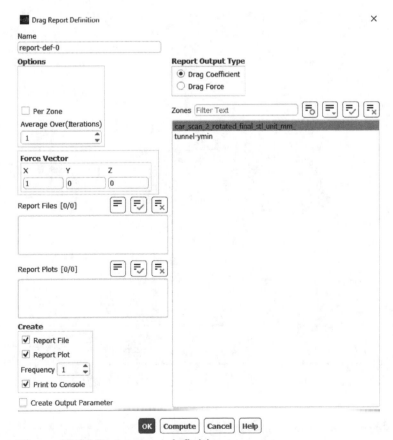

Figure 27.24 Drag report definition

25. Double click on Initialization under Solution in the Outline View. Select Hybrid Initializations and click on Initialize. Select File>>Write>>Case and enter the name *audi_r8_flow.cas.h5*

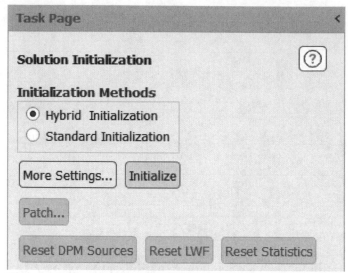

Figure 27.25 Solution initialization in Ansys Fluent

26. Double click on Run Calculation under Solution in the Outline View. Set the Number of Iterations under Parameters to 70. Select Calculate.

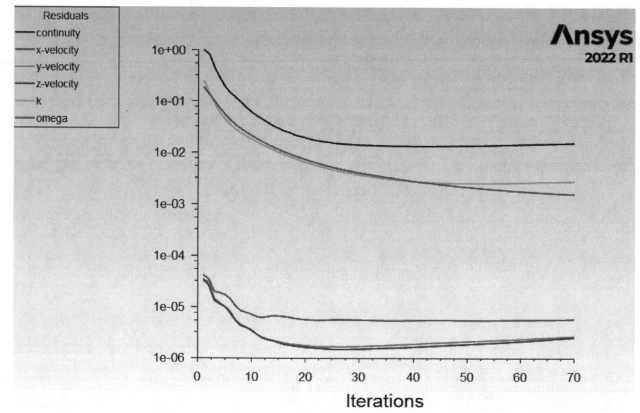

Figure 27.26 Residuals for the first 70 iterations

27. Double click on General under Setup in the Outline View. Select Transient under Time on the Task Page.

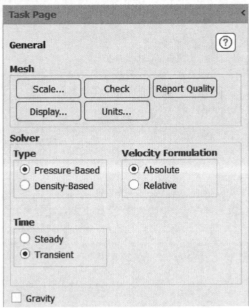

Figure 27.27 Transient simulation setting

28. Double click on Controls under Solution in the Outline View. Set the Under-Relaxation Factor for Momentum to 0.6.

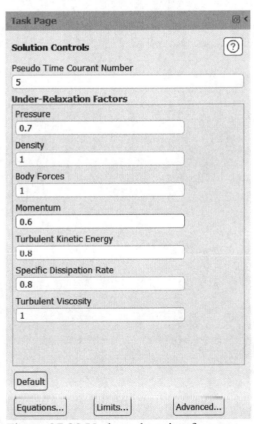

Figure 27.28 Under-relaxation factors

29. Double click on Methods under Solution in the Outline View. Set the Turbulent Kinetic Energy and Specific Dissipation Rate under Spatial Discretization to Second Order Upwind. Double click on Monitors and Residual. Set the Absolute Criteria for all Residuals to 0.0001. Select OK to close the Residual Monitor window.

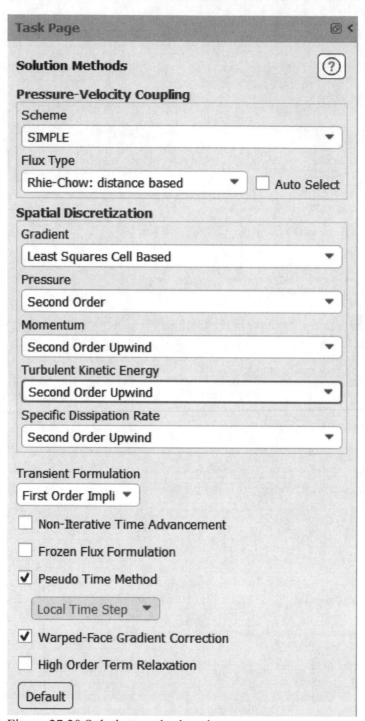

Figure 27.29 Solution method settings

30. Double click on Run Calculation. Set the Time Step Size [s] to 0.001. Set the Number of Time Steps to 200 and the Max Iterations/Time Step to 50. Select Calculate under Solution Advancement.

Figure 27.30a) Run calculation settings

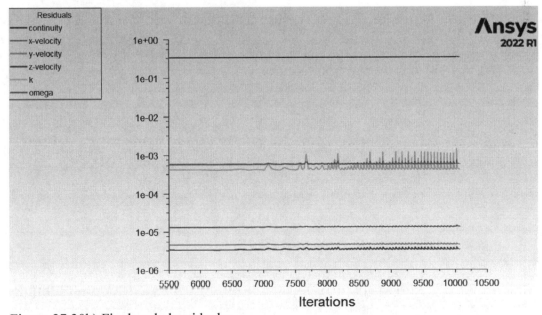

Figure 27.30b) Final scaled residuals

F. Post-Processing

31. Double click on Monitors and Report Plots under Solution in the Outline View and double click on *report-def-0-rplot* under Report Plots. Select Axes… at the bottom of the window. Select X Axis and set Precision to 3 under Number Format. Uncheck Auto Range under Options, set Minimum to 0 and Maximum to 0.2 under Range. Select Apply and select Y Axis. Set Precision for Y Axis to 2 under Number Format and select Apply. Close the Axes – Report Plots window. Select OK in the Edit Report Plot window.

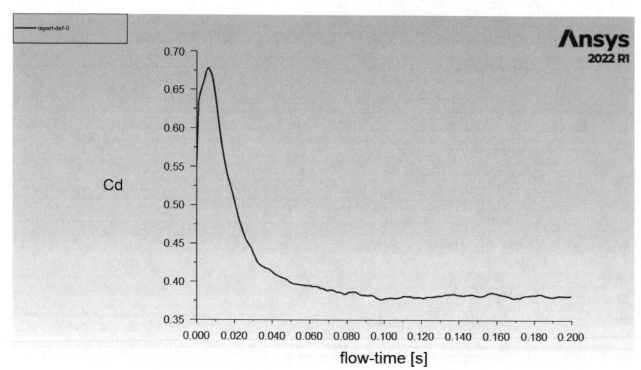

Figure 27.31a)) Drag coefficient versus flow time

Double click on Reports under Results in the Outline View and double click on Forces under Reports. Select the *car_scan_2_rotated_final_stl_unit_mm_* as Wall Zones and select the Print button. The drag coefficient is 0.382. This can be compared with the value 0.36 as listed in the technical specifications, a 6 % difference.

```
Forces - Direction Vector (1 0 0)
                              Forces [N]                                  Coefficients
Zone                          Pressure        Viscous         Total       Pressure       Viscous          Total
car_scan_2_rotated_final_stl_unit_mm  1.1747593   0.12280209  1.2975614   0.3458758      0.036155724    0.38203152
----------------------------  --------------  --------------  --------------  --------------  --------------  --------------
Net                           1.1747593       0.12280209      1.2975614   0.3458758      0.036155724    0.38203152
```

Figure 27.31b) Drag coefficient from console

32. Double click on Graphics and Contours under Results in the Outline View. Select the settings as shown in Figure 27.32a) and click on Save/Display.

Figure 27.32a) Contours window settings for static pressure

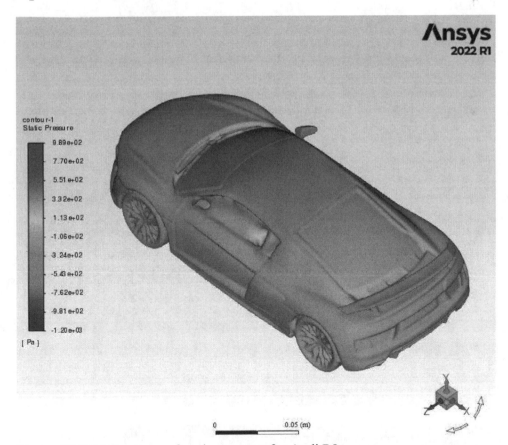

Figure 27.32b) Contours of static pressure for Audi R8.

33. Select File>>Export to CFD Post... from the menu. Select ASCII as Format and check the box for Open CFD-Post. Select fluid-region-1 under Cell Zones and highlight all Surfaces. Select Static Pressure, Pressure Coefficient, Dynamic Pressure, Absolute Pressure, Total Pressure, Velocity Magnitude, X Velocity, Y Velocity, Z Velocity, Vorticity Magnitude and Skin Friction Coefficient as Quantities.

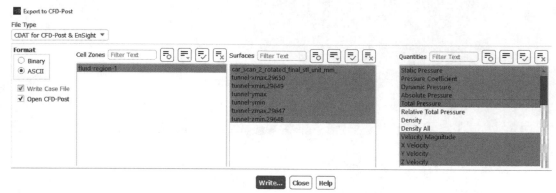

Figure 27.33 Export settings to CFD-Post

34. Check the boxes for *car_scan_2_rotated_final_stl_unit_mm_* and *tunnel ymin* under Cases, Audi_R8 and fluid region 1 in the Outline. Select Location>>Line from under the menu. Select OK in the Insert Line window. Set the locations for the Two Points in Details of Line 1:

Point 1:   0.096   0.02   0.63
Point 2:   0.096   0.075   0.63

Select Sample as Line Type and set the number of Samples to 50. Select Apply.

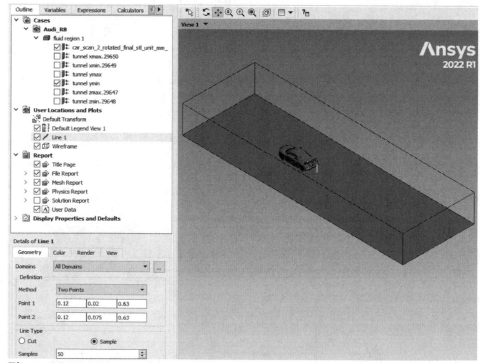

Figure 27.34a) Line settings

Select Location>>Plane from under the menu. Select OK in the Insert Plane window. Select All Domains as Domains. Select XY Plane as Method and 0.63 [m] for Z under Definition in Details of Plane 1. Select Type as None under Plane Bounds in Details of Plane 1. Select Slice as Plane Type. Select Apply.

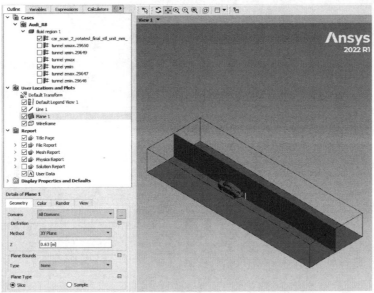

Figure 27.34b) Plane settings

Select the Streamline icon 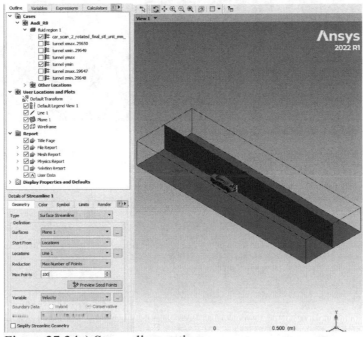 from under the menu. Select OK in the Insert Streamline window. Select Surface Streamline as Type in Details of Streamline 1. Select Plane 1 as Surfaces under Definition. Select Start From Locations, Line 1 for Locations, Max Number of Points as Reduction and set Max Points to 100. Select Velocity as Variable and Forward and Backward as Direction. Select Apply.

Figure 27.34c) Streamline settings

Select Default Legend View 1 under User Locations and Plots. Select the Definition tab in Details of Default Legend View 1. Select Horizontal. Select Location X Justification as Center and Y Justification as Top. Select the Appearance tab, Fixed and set Precision to 0 under Text Parameters. Uncheck the box for Line 1 under User Locations and Plots.

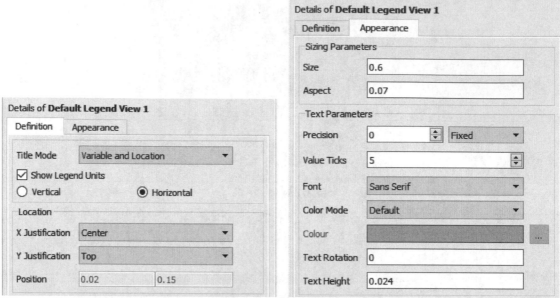

Figure 27.34d) Details of Default Legend View 1

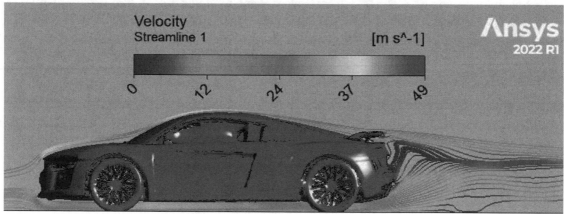

Figure 27.34e) Streamlines around Audi R8.

G. Theory

35. The drag force from aerodynamic simulations of the 2016 Audi R8 LMS has been found in a large velocity range, see [1].

$$F_d = \alpha V^2 + \beta V \qquad\qquad 0 < V < 70\ m/s \qquad\qquad (27.1)$$

, where V (m/s) is velocity and the two constants are $\alpha = 0.5338\ (\frac{Ns^2}{m^2})$ and $\beta = 0.0922\ (\frac{Ns}{m})$. We define the Reynolds number Re as

$$Re = \frac{VL\rho}{\mu} \qquad\qquad 0 < Re < 2.2 \cdot 10^7 \qquad\qquad (27.2)$$

, where $L = 4.58$ (m) is the length of the car, $\rho = 1.225\ (\frac{kg}{m^3})$ is the density of air, and $\mu = 1.79 * 10^{-5}\ (\frac{kg}{ms})$ is the dynamic viscosity of air. The drag coefficient can be determined

$$C_d = \frac{2}{A}\left(\frac{\alpha}{\rho} + \frac{\beta L}{\mu}\frac{1}{Re}\right) \qquad\qquad (27.3)$$

, where $A = 2.08$ m2 is the frontal area of the car. The drag curve is shown in Figure 27.35 and the Ansys Fluent data point has been included for reference. The drag curve approaches the value 0.419 asymptotically as the Reynolds number is increasing.

$$C_d = 2\frac{\alpha}{\rho A} = 0.419, \ Re \to \infty \qquad\qquad (27.4)$$

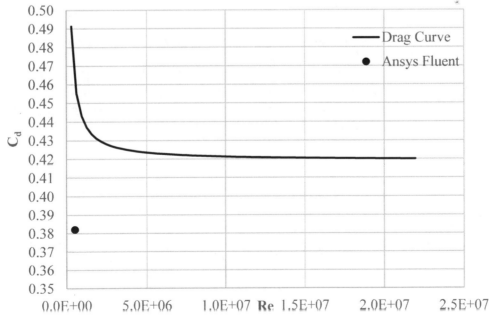

Figure 27.35 Drag curve for 2016 Audi R8 in comparison with Ansys Fluent result

CHAPTER 27. MODEL CAR: AUDI R8

H. References

1. 2016 Audi R8 LMS Aerodynamic Simulation Report, AirShaper sample report, 2019.

I. Exercise

27.1. Run Ansys Fluent calculations shown in this chapter for different velocities V (m/s) and compare your static pressures, streamlines and drag coefficient results as shown in Table 27.1. Determine the percent error between the golf ball model as shown in eq. 27.3 and Ansys Fluent results for the drag coefficient.

Student	V (m/s)	Re	Cd, eq. 27.3	Cd, Fluent	Percent Difference
A	7.5				
B	12.5				
C	17.5				
D	22.5				
E	27.5				
F	32.5				
G	37.5				
H	42.5				
I	52.5				
J	57.5				
K	62.5				
L	67.5				
M	72.5				
N	77.5				
O	82.5				
P	87.5				
Q	92.5				
R	97.5				

Table 27.1 Free stream velocity, Reynolds number, drag coefficients and percentage error for 18 different velocities

27.2. Run Ansys Fluent calculations shown in this chapter for the Audi R8 but use water as the fluid in the simulation. Compare the static pressure, streamlines and drag coefficient results with this chapter's results.

INDEX

INDEX

INDEX

INDEX

INDEX

Notes: